This is an advanced text for first-year graduate students in physics and engineering taking a standard classical mechanics course. It is the first book to describe the subject in the context of the language and methods of modern nonlinear dynamics.

The organizing principle of the text is integrability vs nonintegrability. Flows in phase space and transformations are introduced early and systematically and are applied throughout the text. The standard integrable problems of elementary physics are analyzed from the standpoint of flows, transformations, and integrability. This approach then allows the author to introduce most of the interesting ideas of modern nonlinear dynamics via the most elementary nonintegrable problems of Newtonian mechanics. The book begins with a history of mechanics from the time of Plato and Aristotle, and ends with comments on the attempt to extend the Newtonian method to fields beyond physics, including economics and social engineering.

This text will be of value to physicists and engineers taking graduate courses in classical mechanics. It will also interest specialists in nonlinear dynamics, mathematicians, engineers, and system theorists.

CLASSICAL MECHANICS
transformations, flows, integrable and chaotic dynamics

CLASSICAL MECHANICS

transformations, flows, integrable and chaotic dynamics

JOSEPH L. McCAULEY

Physics Department
University of Houston, Houston, Texas 77204-5506

CAMBRIDGE
UNIVERSITY PRESS

CAMBRIDGE UNIVERSITY PRESS
Cambridge, New York, Melbourne, Madrid, Cape Town, Singapore,
São Paulo, Delhi, Dubai, Tokyo

Cambridge University Press
The Edinburgh Building, Cambridge CB2 8RU, UK

Published in the United States of America by Cambridge University Press, New York

www.cambridge.org
Information on this title: www.cambridge.org/9780521578820

First published 1997
Reprinted (with corrections) 1998

A catalogue record for this publication is available from the British Library

Library of Congress Cataloguing in Publication data

McCauley, Joseph L.
Classical mechanics: transformations, flows, integrable, and
  chaotic dynamics / Joseph L. McCauley.
    p.   cm.
Includes bibliographical references and index.
ISBN 0 521 48132 5 (hc).   ISBN 0 521 57882 5 (pbk.)
1. Mechanics.   I. Title.
QC125.2.M39 1997
531'.01'515352–dc20   96-31574 CIP

ISBN 978-0-521-48132-8 Hardback
ISBN 978-0-521-57882-0 Paperback

Transferred to digital printing 2010

For my mother,
Jeanette Gilliam McCauley,
and my father,
Joseph Lee McCauley (1913–1985).
Ebenso
meiner Lebensgefährtin und Wanderkameradin,
Cornelia Maria Erika Küffner.

Middels klok
bør mann være,
ikke altfor klok;
fagreste livet
lever den mennen
som vet måtelig mye.

Middels klok
bør en mann være,
ikke altfor klok;
sorgløst er hjertet
sjelden i brystet
hos ham som er helt klok.

Håvamål

Contents

Foreword

Invariance principles and integrability (or lack of it) are the organizing principles of this text. Chaos, fractals, and strange attractors occur in different nonintegrable Newtonian dynamical systems. We lead the reader systematically into modern nonlinear dynamics via standard examples from mechanics. Goldstein[1] and Landau and Lifshitz[2] presume integrability implicitly without defining it. Arnol'd's[3] inspiring and informative book on classical mechanics discusses some ideas of deterministic chaos at a level aimed at advanced readers rather than at beginners, is short on driven dissipative systems, and his treatment of Hamiltonian systems relies on Cartan's formalism of exterior differential forms, requiring a degree of mathematical preparation that is neither typical nor necessary for physics and engineering graduate students.

The old Lie–Jacobi idea of complete integrability ('integrability') is the reduction of the solution of a dynamical system to a complete set of independent integrations via a coordinate transformation ('reduction to quadratures'). The related organizing principle, invariance and symmetry, is also expressed by using coordinate transformations. Coordinate transformations and geometry therefore determine the method of this text. For the mathematically inclined reader, the language of this text is not 'coordinate-free', but the method is effectively coordinate-free and can be classified as qualitative methods in classical mechanics combined with a Lie–Jacobi approach to integrability. We use coordinates explicitly in the physics tradition rather than straining to achieve the most abstract (and thereby also most unreadable and least useful) presentation possible. The discussion of integrability takes us along a smooth path from the Kepler problem, Galilean trajectories, the isotropic harmonic oscillator, and rigid bodies to attractors, chaos, strange attractors, and period doubling, and also to Einstein's geometric theory of gravity in space-time.

Because we emphasize qualitative methods and their application to the interpretation of numerical integrations, we introduce and use Poincaré's qualitative method of analysis of flows in phase space, and also the most elementary ideas from Lie's theory of continuous transformations. For the explicit evaluation of known integrals, the reader can consult Whittaker[4], who has catalogued nearly all of the exactly solvable integrable problems of both particle mechanics and rigid body theory. Other problems that are integrable in terms of known, tabulated functions are discussed in Arnol'd's *Dynamical Systems III*.[5]

The standard method of complete solution via conservation laws is taught in standard mechanics texts and implicitly presumes integrability without ever stating that. This method is no longer taught in mathematics courses that emphasize linear differential equations. The traditional teaching of mechanics still reflects (weakly, because very, very incompletely) the understanding of nonlinear differential equations before the advent of quantum theory and the consequent concentration on second order linear equations. Because the older method solves integrable *nonlinear* equations, it provides a superior starting point for modern nonlinear dynamics than does the typical course in linear differential equations or 'mathematical methods in physics'. Most mechanics texts don't make the reader aware of this because they don't offer an adequate (or any) discussion of integrability vs nonintegrability. We include dissipative as well as conservative systems, and we also discuss conservative systems that are not Hamiltonian.

As Lie and Jacobi knew, planar dynamical systems, including dissipative ones, have a conservation law. We deviate from Arnol'd by relaxing the arbitrary requirement of analyticity in defining conservation laws. Conservation laws for dissipative systems are typically multivalued functions. We also explain why phase space is a *Cartesian* space with a scalar product rule (inner product space), but without any idea of the distance between two points (no metric is assumed unless the phase space actually is Euclidean). By 'Cartesian space', therefore, we mean simply a Euclidean space with or without a metric. Finally, the development of the geometric ideas of affine connections and curvature in chapter 11 allows us to use global integrability of the simplest possible flow to define phase space as a flat manifold in chapter 13. The language of general relativity and the language of dynamical systems theory have much geometric overlap: in both cases, one eventually studies the motion of a point particle on a manifold that is generally curved.

After the quantum revolution in 1925, the physics community put classical mechanics on the back burner and turned off the gas. Following Heisenberg's initial revolutionary discovery, Born and Jordan, and also Dirac discovered the quantum commutation rules and seemingly saved physicists from having to worry further about the insurmountable difficulties presented by the unsolved Newtonian three-body problem and other nonintegrable nonlinear dynamics problems (like the double pendulum). Einstein[6] had pointed out in 1917 that the Bohr–Sommerfeld quantization rules are restricted to systems that have periodic trajectories that lie on tori in phase space, so that quantization rules other than Bohr–Sommerfeld are required for systems like the three-body problem and mixing systems that yield Gibbs' microcanonical ensemble. The interesting difficulties posed by nonintegrability (or incomplete integrability) in both classical and quantum mechanics were conveniently forgotten or completely ignored by all but a few physicists from 1925 until the mid-1970s, in spite of a 1933 paper by Koopman and von Neumann where the 'mixing' class of two degree of freedom Hamiltonian systems is shown not to admit action-angle variables because of an 'everywhere dense chaos'.

This is the machine age. One can use the computer to evaluate integrals and also to study dynamical systems. However, while it is easy to program a computer and generate wrong answers for dynamics problems (see section 6.3 for an integrable example), only a researcher with a strong *qualitative* understanding of mechanics and differential equations is likely to be prepared to compute numbers that are both correct and meaningful. An example of a correct calculation is a computation, with controlled decimal precision, of unstable periodic orbits of a chaotic system. Unstable periodic orbits are useful for understanding both classical and quantum dynamics. An example of a wrong computation is a long orbit calculation that is carried out in floating-point arithmetic. Incorrect results of this kind are so common and so widely advertised in both advanced and popular books and articles on chaos theory that the presentation of bad arithmetic as if it were science has even been criticized in the media.[7] It is a disease of the early stage of the information age that many people labor under the comfortable illusion that bad computer arithmetic is suddenly and magically justified whenever a system is chaotic.[8] That would be remarkable, if true, but of course it isn't.

The mathematics level of the text assumes that a well-prepared reader will have had a course in linear algebra, and one in differential equations that has covered power series and Picard's iterative method. The necessary linear algebra is developed in chapters 1 and 8. It is assumed that the reader knows nothing more about nonlinear systems of differential equations: the plotting of phase portraits and even the analysis of linear systems are developed systematically in chapter 3. If the reader has no familiarity with complex analysis then one idea must be accepted rather than understood: radii of convergence of series solutions of phase flows are generally due to singularities on the imaginary time axis.

Chapter 1 begins with Gibbs' notation for vectors and then introduces the notation of linear algebra near the end of the chapter. In chapter 2 and beyond we systematically formulate and study classical mechanics by using the same language and notation that are normally used in relativity theory and quantum mechanics, the language of linear algebra and tensors. This is more convenient for performing coordinate transformations. An added benefit is that the reader can more clearly distinguish superficial from nonsuperficial differences with these two generalizations of classical mechanics. In some later chapters Gibbs' notation is also used, in part, because of convenience.

Several chapter sequences are possible, as not every chapter is prerequisite for the one that follows. At the University of Houston, mechanics is taught as a one-year course but the following one-semester-course chapter sequences are also possible (where some sections of several chapters are not covered): 1, 2, 3, 4, 5, 7, 8, 9, 15 (a more traditional sequence): 1, 2, 3, 4, 6, 8, 9, 12, 13 (integrability and chaos), 1, 2, 3, 4, 6, 7, 12, 15, 16, 17 (integrable and nonintegrable canonical systems): 1, 2, 3, 4, 8, 9, 12, 13, 14 (integrability and chaos in noncanonical systems), or 1, 2, 3, 4, 5, or 6, 8, 9, 10, 11 (integrability and relativity).

The historic approach to mechanics in the first chapter was motivated by different comments and observations. The first was Herbert Wagner's informal remark that beginning German students are well prepared in linear algebra and differential equations, but often have difficulty with the idea of force and also are not familiar with the definition of a vector as it is presented in elementary physics. Margaret and Hugh Miller made me aware that there are philosophers, mathematicians, and others who continue to take Aristotle seriously. Postmodernists and deconstructionists teach that there are no universal laws, and the interpretation of a text is rather arbitrary. The marriage of cosmology and particle physics has led to the return to Platonism within that science, because of predictions that likely can never be decided empirically. Systems theorists and economists implicitly assume that commodity prices[9] and the movement of money define dynamical variables that obey mathematically precise time-evolution laws in the same sense that a collection of inanimate gravitating bodies or colliding billiard balls define dynamical variables that obey Newton's universal laws of motion. Other system theorists assert that 'laws of nature' should somehow 'emerge' from the behavior of 'complex systems'. Awareness of Plato's anti-empiricism and Aristotle's wrong ideas about motion leads to greater consciousness of the fact that the law of inertia, and consequently physics, is firmly grounded in empirically discovered universal laws of kinematics that reflect geometric invariance principles.[10] Beliefs about 'invisible hands', 'efficient markets', and 'rational expectations with infinite foresight'[11] are sometimes taught mathematically as if they might *also* represent laws of nature, in spite of the lack of any empirical evidence that the economy and other behavioral phenomena obey any known set of mathematical formulae beyond the stage of the crudest level of curve-fitting. Understanding the distinction between the weather, a complex Newtonian system subject to universal law, and mere modelling based upon the statistics of irregularities that arise from effectively (mathematically) lawless behavior like stock trading might help to free economists, and perhaps other behaviorists, from the illusion that it is possible either to predict or even to understand the future of humanity in advance from a purely mathematical standpoint, and would also teach postmodernists the difference between physics, guesswork-modelling and simulations, and pseudo-science.

I am very grateful to Harry Thomas for many stimulating, critical and therefore very helpful discussions on geometry and integrability, and for sharing with me his lecture notes on a part of the proof of Bertrand's theorem. I'm also grateful to his colleague and former student Gerhard Müller for making me aware that the double pendulum provides a relatively elementary example of a nonintegrable and chaotic Newtonian system. I am extremely grateful to Eigil Anderson for introducing me to the description of Lie's work in E. T. Bell's history of mathematics in 1992: without having read the simple sentence by Bell stating that Lie showed that every one-parameter transformation on two variables has an invariant, the main organizing principle of this book (integrability) would have remained too primitive to be effective. I am grateful to my old friend and colleague, Julian Palmore, for pointing me to his paper with Scott Burns where the conservation law for the damped linear harmonic oscillator is derived. Julian also read and criticized my heuristic definitions of solvability and global integrability and helped me to make them sharper than they were before, but they are certainly not as precise as he would have liked them to be. I have also followed his advice, given several years ago, to emphasize flows in phase space. Dyt Schiller at Siegen, Finn Ravndal in Oslo, and Herbert Wagner in München have provided me with useful general criticism of earlier versions of the manuscript. George Reiter read sections 1.1, 1.2, and 18.1 critically. I am very grateful to copy editor Jo Clegg for a careful reading of the original manuscript, to Sigmund Stintzing for proofreading chapters 9 and 15 of the typescript, and especially to Herbert Wagner and Ludwig Maxmillians Universität (München) for guestfriendship during the fall 1996 semester of my sabbatical, when the typescript was proofed and corrected. I have enjoyed a lecture by Per Bak on long-range predictability vs qualitative understanding in physics, have also benefited from conversations with Mike Fields and Marty Golubitsky on flows and conservation laws, and from a discussion of differential forms with Bob Kiehn. I am grateful to Gemunu Gunaratne for many discussions and critical questions, to J. M. Wilkes for a 1977 letter on rotating frames, to Robert Helleman for pointing me to Arnol'd's discussion of normal forms. I am especially grateful to many of the physics and engineering students at the University of Houston who have taken my classical mechanics and nonlinear dynamics courses over the last several years. Chapter 8 was read and corrected, in part, by graduate students Rebecca Ward Forrest and Qiang Gao. The computer graphics were produced as homework exercises by Ola Markussen, Rebecca Forrest, Youren Chen, and Dwight Ritums. I would also like to thank Linda Joki for very friendly help with some of the typing. Special thanks go to Simon Capelin, Publishing Director for the Physical Sciences at Cambridge University Press, for his support and encouragement.

Finally, I am extremely grateful to my stimulating and adventurous wife, Cornelia Küffner, who also longs for the Alps, the Norwegian fjords and mountains, and other more isolated places like the Sierra Madre, Iceland, and the *Færøyene* while we live, work, and write for 9 months of each year in Houston.

Joseph L. McCauley

Bibliography

1 Goldstein, H., *Classical Mechanics*, second edition, Addison-Wesley, Reading, Mass. (1980).
2 Landau, L. D., and Lifshitz, E. M., *Mechanics*, third edition, Pergamon, Oxford (1976).
3 Arnol'd, V. I., *Mathematical Methods of Classical Mechanics*, second edition, Springer-Verlag, New York (1989).
4 Whittaker, E. T., *Analytical Dynamics*, Cambridge University Press (1965).
5 Arnol'd, V. I., *Dynamical Systems III*, Springer-Verlag, Heidelberg (1993).
6 Einstein, A., *Verh. Deutsche Phys. Ges.* **19**, 82 (1917).
7 Brügge, P., Mythos aus dem Computer: über Ausbretung und Missbrauch der 'Chaostheorie', *Der Spiegel* **39**, 156–64 (1993); Der Kult um das Chaos: über Ausbretung und Missbrauch der 'Chaostheorie', *Der Spiegel* **40**, 232–52 (1993).
8 McCauley, J. L., *Chaos, Dynamics and Fractals, an algorithmic approach to deterministic chaos*, Cambridge University Press (1993).
9 Casti, J. L., and Karlqvist, A., *Beyond Belief, Randomness, Prediction, and Explanation in Science*, CRC Press, Boca Raton (1991).
10 Wigner, E. P., *Symmetries and Reflections*, Univ. Indiana Pr., Bloomington (1967).
11 Anderson, P. W., Arrow, K. J., and Pines, D., *The Economy as an Evolving Complex System*, Addison-Wesley, Redwood City, California (1988).

Acknowledgments

The writer acknowledges permission to quote passages from the following texts: Johns Hopkins Pr.: *The Cheese and the Worms* (1992) by Carlo Ginzburg; Vintage Pr.: *About Looking* (1991), *Keeping a Rendezvous* (1992), *Pig Earth* (1992), and *About our Faces* (1984) by John Berger; Wm. Morrow and Co., Inc. and the John Ware Literary Agency: *Searching for Certainty* (1990) by John Casti; Dover: *Johannes Kepler* (1993) by Max Caspar; *Symmetries and Reflections*, E. P. Wigner; Indiana Univ. Pr. (1967), permission granted by Mrs. Wigner one week after Professor Wigner's death, Addison-Wesley Publ. Co.: *Complexity: Metaphors, Models, and Reality* (1994); Santa Fe Institute, Cowan, Pines, & Meltzler, Springer-Verlag: *Mathematical Methods of Classical Mechanics* (2nd ed., 1980) by V. I. Arnol'd, and Dover Publications, Inc.: *Ordinary Differential Equations* (1956) by E. L. Ince.

Several drawings in the text are reprinted with the permission of: Addison-Wesley Publ. Co.: figures 1.5, 3.7, and 5.9a–c from *Classical Mechanics* (2nd edition, Copyright 1980, 1950), Goldstein, Springer-Verlag: figures 35, 73, & 79 from *Mathematical Methods of Classical Mechanics* (2nd edition, Copyright 1989, 1978), Arnol'd, Professors E. J. Saletan and A. H. Cromer: figures 5 & 25 from *Theoretical Mechanics* (Wiley, 1971), and McGraw-Hill: figure 4.31 from *Advanced Mathematical Methods for Scientists and Engineers* (1st edition, Copyright 1978), Bender and Orszag.

1

Universal laws of nature

In the presentation of the motions of the heavens, the ancients began with the principle that a natural retrograde motion must of necessity be a uniform circular motion. Supported in this particular by the authority of Aristotle, an axiomatic character was given to this proposition, whose content, in fact, is very easily grasped by one with a naive point of view; men deemed it necessary and ceased to consider another possibility. Without reflecting, Copernicus and Tycho Brahe still embraced this conception, and naturally the astronomers of their time did likewise.

Johannes Kepler, by Max Caspar

1.1 Mechanics in the context of history

It is very tempting to follow the mathematicians and present classical mechanics in an axiomatic or postulational way, especially in a book about theory and methods. Newton wrote in that way for reasons that are described in Westfall's biography (1980). Following the abstract Euclidean mode of presentation would divorce our subject superficially from the history of western European thought and therefore from its real foundations, which are abstractions based upon reproducible empiricism. A postulational approach, which is really a Platonic approach, would mask the way that *universal* laws of regularities of nature were discovered in the late middle ages in an atmosphere where authoritarian religious academics purported pseudo-scientifically to justify the burning of witches and other nonconformers to official dogma, and also tried to define science by the appeal to authority. The axiomatic approach would fail to bring to light the way in which *universal laws of motion* were abstracted from firmly established empirical laws and discredited the anti- and pseudo-scientific ways of medieval argumentation that were bequeathed to the west by the religious interpreters of Plato and Aristotle.

The ages of Euclid (ca. 300 BC), Diophantus (ca. 250 BC), and Archimedes (287 to 212 BC) effectively mark the beginnings of geometry, algebra, recursive reasoning, and statics. The subsequent era in western Europe from later Greek and Roman through feudal times until the thirteenth century produced little of mathematical or scientific value. The reason for this was the absorption by western Christianity of the ideas of Plato and his successor Aristotle. Plato (429 to 348 BC) developed an abstract model cosmology of the heavens that was fundamentally anti-empirical. We can label it 'Euclidean' because the cosmology was based upon Euclidean symmetry and was postulational-axiomatic, purely abstract, rather than empirical. We shall explain why Plato's cosmology and astronomy were both unscientific and anti-scientific. Aristotle (387 to 322 BC) observed nature purely qualitatively and developed a nonquantitative pseudo-physics that lumped together as 'motion' such unrelated phenomena as a rolling ball, the growth of an acorn, and the education of boys. Authoritarian impediments to science and mathematics rather than free and creative thought based upon the writings of Diophantus and Archimedes dominated academé in western

Europe until the seventeenth century. The reasons can be identified as philosophical, political, and technical.

On the political side, the Romans defeated Macedonia and sacked Hellenistic civilization (148 BC), but only fragments of the works that had been collected in the libraries of Alexandria were transmitted by the Romans into early medieval western Europe. Roman education emphasized local practicalities like rhetoric, law, and obedience to authority rather than abstractions like mathematics and natural philosophy. According to Plutarch (see Bell, 1945, or Dunham, 1991), whose story is symbolic of what was to come, Archimedes was killed by a Roman soldier whose orders he refused to obey during the invasion of Syracuse (Sicily). At the time of his death, Archimedes was drawing geometric diagrams in the dust. Arthur North Whitehead later wrote that 'No Roman lost his life because he was absorbed in the contemplation of a mathematical diagram.'

Feudal western Europe slept scientifically and mathematically while Hindi and Muslim mathematicians made all of the necessary fundamental advances in arithmetic and algebra that would eventually allow a few western European scientific thinkers to revolutionize natural philosophy in the seventeenth century. Imperial Rome bequeathed to the west no academies that encouraged the study of abstract mathematics or natural philosophy. The centers of advanced abstract thought were Hellenistic creations like Alexandria and Syracuse, cut off from Europe by Islamic conquest in the seventh century. The eventual establishment of Latin libraries north of the Alps, due primarily to Irish (St Gallen in Switzerland) and Anglo-Saxon monks, was greatly aided by the reign of *Karl der Grosse* in the eighth century. Karl der Grosse, or Charlemagne, was the first king to unite western Europe from Italy to north Germany under a single emperor crowned in Rome by the Pope. After thirty years of war Karl der Grosse decisively defeated the still-tribal Saxons and Frisians, destroyed their life-symbol *Irminsul* (Scandinavian *Yggdrasil*), and offered them either Christianity and dispersal, or death. Turning relatively free farmer-warriors into serfs with a thin veneer of Christianity made the libraries and monasteries of western Europe safer from sacking and burning by pagans, and also insured their future support on the backs of a larger peasant class. The forced conversion from tribal to more or less feudal ways did not reach the Scandinavians at that time, who continued international trading along with the occasional sacking and burning of monasteries and castles for several centuries more (Sweden fell to Christianity in the twelfth century).

Another point about the early Latin libraries is that they contained little of scientific or mathematical value. Karl der Grosse was so enthusiastic about education that he tried to compile a grammar of his own Frankish-German dialect, and collected narrative poems of earlier tribal times (see Einhard, 1984), but the *main* point is that the philosophy that was promoted by his victories was anti-scientific: St Augustine, whose *City of God* Charlemagne admired and liked to have read to him in Latin, wrote the neoPlatonic philosophy that dominated the monasteries in western Europe until the twelfth century and beyond. Stated in crude modern language, the academic forerunners of research support before the thirteenth century went overwhelmingly for religious work: assembly and translation of the Bible, organization and reproduction of liturgy, church records, and the like.

Thought was stimulated in the twelfth and thirteenth centuries by the entry of mathematics and philosophy from the south (see Danzig, 1956, and Dunham, 1991). Recursive reasoning, essential for the development and application of mathematics, had been used by Euclid to perform the division of two integers but was not introduced into feudal western Europe until the thirteenth century by Fibonacci, Leonardo of Pisa. Neither arithmetic nor recursive reasoning could have progressed very far under the unmathematical style of Roman numeration (MDCLXVI − XXIX = MDCXXXVII). Brun asserts in *Everything is Number* that one can summarize Roman mathematics as follows: Euclid was translated (in part) into Latin around 500, roughly the time of the unification and Christianization of the Frankish tribes after they conquered the Romanized Celtic province of Gaul.

The Frankish domination of Europe north of the Alps eliminated or suppressed the ancient tribal tradition of free farmers who practiced direct democracy locally. The persistence of old traditions of the *Alemannische Thing*, forerunner of the *Landsgemeinde*, among some isolated farmers (common grazing rights and local justice as opposed to Roman ideas of private property and abstract justice administered externally) eventually led through rebellion to greater freedom in three Swiss Cantons in the thirteenth century, the century of Thomas Aquinas, Fibonacci, and Snorre Sturlasson. The Scandinavian *Althing* was still the method of rule in weakly Christianized Iceland[1] in the thirteenth century, which also never had a king. Freedom/*freodom*/*frihet*/*frelsi*/*Freiheit* is a very old tribal word.

Fifteen centuries passed from the time of ancient Greek mathematics to Fibonacci, and then another four from Fibonacci through Stevin and Viéte to Galileo (1554–1642), Kepler (1571–1630), and Descartes (1596–1650). The seventeenth century mathematics/scientific revolution was carried out, after Galileo's arrest and house-imprisonment in northern Italy, primarily in the northwestern European nations and principalities where ecclesiastic authority was sufficiently fragmented because of the Protestant Reformation. History records that Protestant leaders were often less tolerant of and less interested in scientific ideas than were the Catholics (the three most politically powerful Protestant reformers were Augustinian in outlook). Newton (1642–1727), as a seventeenth century nonChristian Deist, would have been susceptible to an accusation of heresy by either the Anglican Church or the Puritans. A third of a century later, when Thomas Jefferson's Anglican education was strongly influenced by the English mathematician and natural philosopher William Small, and Jefferson (also a Deist) studied mathematics and Newtonian mechanics, the Anglican Church in the Colony of Virginia still had the legal right to burn heretics at the stake. The last-accused American witch was burned at the stake in Massachusetts in 1698 by the same Calvinistic society that introduced direct democracy at town meetings (as practiced long before in Switzerland) into New England. In the wake of Galileo's trial, the recanting of his views on Copernican astronomy, and his imprisonment, Descartes became frightened enough not to publish in France in spite of direct 'invitation' by the politically powerful Cardinal Richelieu (Descartes' works were finally listed on the Vatican Index after his death). Descartes eventually published in Holland, where Calvinists condemned his revolutionary ideas while the Prince of Orange supported him. Descartes eventually was persuaded to go to Sweden to tutor Queen Christina, where he died. The spread of scientific ideas and

---

[1] The tribal hierarchy of a small warring aristocracy (*jarl*), the large class of free farmers (*karl*), and slaves (*thrall*) is described in the *Rigsthula* and in many Icelandic sagas. In the ancient *Thing*, all free farmers carried weapons and assembled to vote on the important matters that affected their future, including the choice of leadership in war. In Appenzell, Switzerland, men still wear swords to the outdoor town meeting where they vote. The Norwegian parliament is called *Storting*, meaning the big (or main) *Thing*.

philosophy might have been effectively blocked by book-banning, heretic-burnings, and other religio-political suppression and torture, but there was an important competition between Puritanism, Scholasticism, and Cartesian–Newtonian ideas, and the reverse actually happened (see Trevor-Roper, 1968). As Sven Stolpe wrote in *Christina of Sweden,*

Then there was the new science, giving educated people an entirely new conception of the universe, overthrowing the old view of the cosmos and causing an apparently considerable portion of the Church's structure to totter. Sixteenth-century culture was mainly humanistic. In her early years, Christina had enthusiastically identified herself with it. The seventeenth century produced another kind of scholar: they were no longer literary men and moralists, but scientists. Copernicus, Tycho Brahe, Kepler, and Galileo were the great men of their time. The initially discouraging attitude of the Church could not prevent the new knowledge from spreading all over Europe north of the Alps, largely by means of the printing presses of Holland.

The chapter that quote is taken from is entitled 'French Invasion', and refers to the invasion of the ideas of free thought from France that entered Stockholm where Protestant platitudes officially held sway. It was Queen Christina who finally ended the witch trials in Sweden. To understand feudal and medieval dogma, and what men like Galileo, Descartes, and Kepler had to fear and try to overcome, one must return to St Augustine and Thomas Aquinas, the religious interpreters of Plato and Aristotle respectively.

   The Carthaginian-educated Augustine (340–420), bishop of Hippo in North Africa, was a neoPlatonist whose contribution to the west was to define and advance the philosophy of puritanism based upon predestination. Following Plotinus (see Dijksterhuis, 1986), Augustine promoted the glorification of God while subjugating the study of nature to the absolute authority of the Church's official interpreters of the Bible. We inherit, through Augustine, Plotinus's ideas that nature and the world are devilish, filled with evil, and worth little or no attention compared with the neoPlatonic idea of a perfect heavenly Christian kingdom. It was not until the twelfth century that Aristotle's more attractive pseudo-science, along with mechanistic interpretations of it (due, for example, to the Berber–Spanish philosopher 'Averröes', or Ibn-Rushd (1126–1198)), entered western Europe. It's difficult to imagine how exciting Aristotle's ideas (which led to Scholasticism) must have been to cloistered academics whose studies had been confined for centuries to Augustinian platitudes and calculating the date of Easter. Karl der Grosse's grandfather had stopped the northward advance of Islam and confined it to Spain south of the Pyrenees, but mathematics and natural philosophy eventually found paths other than over the Pyrenees by which to enter western Europe.

   There was also a technical reason for stagnation in ancient Greek mathematics: they used alphabetic symbols for integers, which hindered the abstract development of algebra. The Babylonian contribution in the direction of the decimal arithmetic was not passed on by Euclid, who completely disregarded practical empirics in favor of purely philosophic abstraction. The abstract and also practical idea of writing something like $10 + 3 + 5/10 + 9/100 = 13.59$ or $2/10 + 4/100 + 7/1000 = 0.247$ was not yet brought into consciousness. Hindu–Arabian mathematics along with older Greek scientific ideas came into northern Italy (through Venice, for example) via trade with Sicily, Spain, and North Africa. Roman numerals were abandoned in favor of the base-ten system of numeration for the purpose of calendar-making (relatively modern calendars appeared as far north as Iceland in the twelfth century as a result of the Vikings' international trading[2]). The north Italian Fibonacci was sent by his father on a trading mission to North Africa, where he learned new mathematics. He introduced both algebra and relatively modern arithmetic into Italy, although the algebra of that time was still stated entirely in words rather than symbolically: no one had yet thought of using 'x' to denote an unknown number in an equation. The introduction of Aristotle's writings into twelfth-century monasteries led part of the academic wing of feudalism to the consideration of Scholasticism, the Aristotelian alternative to Augustinian Puritanism, but it took another four centuries before axiom-saturated natural philosophers and astronomers were able to break with the rediscovered Aristotle strongly enough to discover adequate organizing principles in the form of *universal mathematical*, and therefore *truly* mechanistic, *laws of nature*. Universal mechanistic laws of nature were firmly established in the age of Kepler, Galileo, and Descartes following the Reformation and Counter Reformation (revivals of Puritanism, both) the era when Augustine's ideas of religious predestination fired the Protestant Reformation through Luther, Zwingli, and Calvin.

Plato apparently loved order and was philosophically repelled by disorder. He advanced a theoretical political model of extreme authoritarianism, and also a cosmologic model of perfect order that he regarded as a much deeper truth than the *observed* retrograde motions of heavenly bodies. His advice to his followers was to ignore all sense-impressions of reality, especially empirical measurements, because 'appearances' like noncircular planetary orbits are merely the degenerated, irregular shadow of a hidden reality of perfect forms that can only be discovered through pure thought (see Dijksterhuis, 1986, Collingwood, 1945, or Plato's *Timaeus*). The Platonic ideal of perfection resided abstractly beyond the physical world and coincided with naive Euclidean geometric symmetry, so that spheres, circles, triangles, etc. were regarded as the most perfect and therefore 'most real' forms (Plato's idea of a hidden, changeless 'reality' has been replaced by the abstract modern idea of invariance). The earth (then expected to be approximately spherical, not flat) and all earthly phenomena were regarded as imperfect due to their lack of strict Euclidean symmetry, while the universe beyond the earth's local neighborhood ('sphere') was treated philosophically as if it were finite and geometrically perfect. Augustine's idea of earthly imperfection vs a desired heavenly perfection stems from the Platonic division of the universe into two fundamentally different parts, the one inferior to the other.

Plato's main influence on astronomy is that the perfectly symmetric state of circular motion at constant speed must hold true for the 'real', hidden motions of heavenly forms. The consequence of this idea is that if empirical observations should disagree with theoretical expectations, then the results of observation must be wrong. We can label this attitude as Platonism, the mental rejection of reality in favor of a desired fantasy.

In Plato's cosmology, heavenly bodies should revolve in circles about the earth. Since this picture cannot account for the apparent retrograde motions of nearby planets, Ptolemy (100–170) added epicycles, circular 'corrections' to uniform circular motion. According to neoPlatonism this is a legitimate way of 'saving the appearance' which is 'wrong' anyway. Aristarchus (310 to 230 BC) had believed that the sun rather

[2] According to Ekeland, a thirteenth-century Franciscan monk in Norway studied pseudo-random number generation by arithmetic in the search for a theoretically fairer way to 'roll dice'. In tribal times, tossing dice was sometimes used to settle disputes (see the Saga of Olaf Haraldsson) and foretell the future.

than the earth ought to be regarded as the center of the universe, but this idea lay dormant until the Polish monk Copernicus took it up again and advanced it strongly enough in the early sixteenth century to make the Vatican's 'best seller list' (the Index).

According to Aristotle's earthly physics, every velocity continually requires a 'cause' (today, we would say a 'force') in order to maintain itself. In the case of the stars (and other planets), something had first to produce the motion and then to keep that motion circular. In neoPlatonic Christianity, the circularity of orbits was imagined to be maintained by a sequence of perfect, solid spheres: each orbit was confined by spherical shells, and motion outside the space between the spheres was regarded as impossible (this picture naturally became more complicated and unbelievable as more epicycles were added). Within the spheres, spirits or angels eternally pushed the stars in their circular tracks at constant speed, in accordance with the Aristotelian requirement that whatever moves must be moved (there must always be a mover). In other words, there was no idea of *universal* purely mechanistic laws of motion, universal laws of motion and a universal law of force that should, in principle, apply everywhere and at all times to the description of motion of all material bodies.

Augustine had taught that in any conflict between science and Christianity, Christian dogma must prevail. As Aristotle's writings began to gain in influence, it became necessary that the subversive teachings be reinterpreted and brought into agreement with official Christian dogma. The religious idea of God as the creator of the universe is based upon the model of an artist who sculpts a static form from a piece of bronze or a block of wood, with the possibility of sculptural modifications as time passes. The picture of God as interventionist is not consistent with the Aristotelian notion of the universe as a nonmathematical 'mechanism'. The forced Christianization of Aristotle was accomplished by the brilliant Italian Dominican monk *Tommaso d'Aquino* (1225–1274), or Thomas Aquinas (see Eco, 1986, for a brief and entertaining account of Tomasso). Aristotle offered academics what Augustine and Plato did not: a comprehensive world view that tried to explain rather than ignore nature (in neo-Platonism, stellar motions were regarded as 'explained' by arithmetic and geometry). Aristotle's abstract prime mover was replaced in Aquinian theology by God as the initial cause of all motion.

In Aristotelian physics, absolute weight is an inherent property of a material body: absolutely heavy bodies are supposed to fall toward the earth, like stones, while absolutely light ones supposedly rise upward, like fire. Each object has a nature (quality) that seeks its natural place in the hierarchy of things (teleology, or the seeking of a final form). The Aristotelian inference is that the earth must be the heaviest of all bodies in the universe because all projectiles eventually fall to rest on the earth. Any object that moves along the earth must be pushed so that absolute rest appears to be the natural or preferred state of motion of every body.

The Aristotelian method is to try to observe phenomena 'holistically' and then also try to reason logically backward to infer the cause that produced an observed effect. The lack of empiricism and mathematics led to long, argumentative treatises that produce wrong answers because both the starting point and the method were wrong. Thought that leads to entirely new and interesting results does not flow smoothly, pedantically, and perfectly logically, but requires seemingly discontinuous mental jumps called 'insight' (see Poincaré and Einstein in Hadamard, 1954).

An example of Aristotelian reasoning is the idea that velocity cannot exist without a force: when no force is applied then a moving cart quickly comes to a halt. The idea of force as the cause of motion is clear in this case. That projectiles move with constant velocity due to a force was also held to be true. This is approximately true for the long-time limit of vertically falling bodies where friction balances gravity, but the Aristotelians never succeeded in explaining where the driving force should come from once a projectile leaves the thrower's hand. Some of them argued that a 'motive power' must be generated within the body itself in order to maintain the motion: either a spirit or the air was somehow supposed to act to keep the object moving.

The single most important consequence of the combined Platonic–Aristotelian theory is: because the moon, sun, and other plants and stars do not fall toward the earth they must therefore be *lighter* than the earth. But since they do not rise forever upward it also follows that *the heavenly bodies must obey different laws of motion than do the earthly ones.* There was no idea at all of universal mechanical laws of motion and force that all of matter should obey independently of place and time in the universe.

The key physical idea of *inertia*, of resistance to *change of velocity*, did not occur to Galileo's and Descartes' predecessors. As Arthur Koestler has suggested, the image of a farmer pushing a cart along a road so bad that friction annihilated momentum was the model for the Platonic–Aristotelian idea of motion. Observation did play a role in the formation of philosophic ideas, but it was observation at a child-like level because no precise measurements were made to check the speculations that were advanced argumentatively to represent the truth. There was a too-naive acceptance of the qualitative appearances that were condemned as misleading by Plato. The idea that a horizontal component of *force* must be imparted to a projectile in order to make it move parallel to the earth as it falls downward seems like common sense to many people. The law of inertia and Newton's second law of motion are neither intuitive nor obvious. *Were* they intuitive, then systematic and persistent thinkers like Plato and Aristotle might have discovered correct mathematical laws of nature a long time ago.

Descartes asserted that Aristotle had asked the wrong question: instead of asking which force causes a body to move he should have asked what is the force necessary to *stop* the motion. After the twelfth century, but before Kepler and Galileo, medieval philosophers were unable to break adequately with the Platonic–Aristotelian method of searching for absolute truth by the method of pure logic and endless argumentation. The replacement of the search for absolute truth by the search for precise, empirically verifiable universal mathematical truth based upon universal laws of nature, was finally discovered in the seventeenth century and has had enormous consequences for daily life. In contrast with the idea of spirit-driven motion, it then became possible to discover and express mathematically the empirically discovered universal regularities of motion as if those aspects of nature represented the deterministic workings of an abstract machine. Before the age of belief in nature as mathematically behaving mechanism, many academics believed in the intervention of God, angels, and demons in daily life.[3] Simultaneously, weakly assimilated farmers and villagers lived under a thin veneer of Christianity and continued to believe in the pagan magic from older tribal times. The Latin words for farmer (*peasant* derives from *pagensis*) and country-side dweller (*paganus*) indicate that Christianity was primarily a religion of the cities and of the leaders of the Platonic feudal hierarchy.

The following quotes are from Carlo Ginzburg's book *The Cheese and the Worms,*

---

[3] Kepler's mother's witchcraft trial was decided by the law faculty at the University of Tübingen. His old thesis advisor at Tübingen criticized him severely for going beyond arithmetic and geometry for the explanation of heavenly motions.

and describe the inquisition[4] of a late-sixteenth century miller who could read and think just well enough to get himself into trouble. The miller lived in Friuli, a frontier region between the Dolomites and the Adriatic in northern Italy. In what follows, the Inquisitor leads the self-educated miller with questions derived from the miller's answers to earlier questions, which had led the miller to state his belief in a primordial chaos:

Inquisitor: *Was God eternal and always with the chaos?*
Menocchio: *I believe that they were always together, that they were never separated, that is, neither chaos without God, nor God without chaos.*

In the face of this muddle, the Inquisitor tried to achieve a little clarity before concluding the trial.

The aim of the questioner was to lead the accused miller into making a clear contradiction with official doctrine that could be considered proof of a pact with the devil:

Inquisitor: *The divine intellect in the beginning had knowledge of all things: where did he acquire this information, was it from his essence or by another way?*

Menocchio: *The intellect acquired knowledge from the chaos, in which all things were confused together: and then it (the chaos) gave order and comprehension to the intellect, just as we know earth, water, air, and fire and then distinguish among them.*

Finally, the question of all questions:

Inquisitor: *Is what you call God made and produced by someone else?*
Menocchio: *He is not produced by others but receives his movement within the shifting of the chaos, and proceeds from imperfect to perfect.*

Having had no luck in leading Menocchio implicitly to assign his 'god' to having been made by a higher power, namely by the devil, the Inquisitor tries again:

Inquisitor: *Who moves the chaos?*
Menocchio: *It moves by itself.*

The origin of Menocchio's ideas is unknown. Superficially, they may sound very modern but the modern mechanistic worldview began effectively with the overthrow of the physics–astronomy sector of Scholasticism in the late middle ages. Aristotle's ideas of motion provided valuable inspiration to Galileo because he set out systematically to disprove them, although he did not free himself of Platonic ideas about astronomy. For Kepler, the Aristotelian idea that motion requires a mover remained a hindrance that he never overcame as he freed himself, and us, from dependence on Platonic ideas about circular orbits, and Ptolemy's epicycles.

The Frenchman Viéte (1540–1603) took an essential mathematical step by abandoning in part the use of words for the statement of algebraic problems, but he did not write equations completely symbolically in the now-common form $ax^2 + bxy + cy^2 = 0$. That was left for his countryman Descartes (and his colleague Fermat, who did not publish), who was inspired by three dreams on consecutive winter nights in Ulm to construct analytic geometry, *the reduction of geometry to algebra*, and consequently his philosophic program calling for *the mathematization of all of science based upon the ideas of extension and motion*. Fortunately for Newton and posterity, before dreaming, Descartes had just spent nearly two years with the Hollander physicist Beeckman who

---

[4] Plato discusses both censorship and the official protection of society from dangerous ideas in *The Republic*.

understood some aspects of momentum conservation in collisions. The ancient Greeks had been blocked by the fact that line-segments of arbitrary length generally cannot be labeled by rational numbers (by integers and fractions), but arithmetic progress in the representation of irrational numbers was finally made by the Hollander Stevin (1548–1620), who really established the modern decimal system for us, and also by Viéte. Enough groundwork had been performed that Newton was able to generalize the binomial theorem and write $(1 - x)^{1/2} = 1 - x/2 - x^2/2^3 - x^3/16 - \cdots$, and therefore to deduce that $\sqrt{7} = 3(1 - 2/9)^{1/2} = 3(1 - 1/9 - 1/162 - 1/1458 - \cdots) = 2.64576\cdots$, which represents a revolutionary analytic advance compared with the state of mathematics just one century earlier. *The mathematics that Newton needed for the development of differential calculus and dynamics were developed within the half-century before his birth.* Approximate integral calculus had been already used by Archimedes and Kepler to compute the volumes of various geometric shapes. Pascal, Fermat, and Descartes were able to differentiate certain kinds of curves. Newton did not learn his geometry from Euclid. Instead, he learned analytic geometry from Descartes.

The fundamental empirical building blocks for Newton's unified theory of mechanics were the two local laws of theoretical dynamics discovered empirically by Galileo, and the global laws of two-body motion discovered empirically by Kepler. These discoveries paved the way for Newton's discovery of universally applicable laws of motion and force that realized Descartes' dream of a mathematical/mechanistic worldview, and thereby helped to destroy the credibility of Scholasticism among many educated advisors (normally including astrologers) to the princes and kings of northwestern Europe. Even Newton had initially to be dislodged from the notion that a moving body carries with it a 'force': he did not immediately accept the law of inertia while studying the works of Galileo and Descartes, and was especially hostile to Cartesian ideas of 'mechanism' and relativity of motion (see Westfall, 1980). The new mechanistic worldview was spread mainly through Descartes' philosophy and analytic geometry, and contributed to ending the persecution, legal prosecution, and torture of 'witches' and other nonconformers to official dogma, not because there was no longer an adequate supply of witches and nonconformers (the number grew rapidly as the accusations increased), but because the mysticism-based pseudo-science and non-science that had been used to justify their inquisition, imprisonment, torture, and even death as collaborators with the devil was effectively discredited by the spread of mechanistic ideas. Heretic- and witch-burnings peaked around the time of the deaths of Galileo, Kepler, and Descartes and the birth of Newton, and had nearly disappeared by the end of the seventeenth century.

The fundamental idea of *universal mechanistic laws of nature* was advocated philosophically by Descartes but was realized by Newton. The medieval belief in the supernatural was consistent with the neoPlatonic worldview of appearances vs hidden reality, including the intervention in daily life by angels and demons. Tomasso d'Aquino wrote an essay arguing that the 'unnatural' magicians must get their power directly from the devil rather than from the stars or from ancient pagan gods. His logic provided the theoretical basis for two fifteenth-century Dominican Monks who wrote *The Witches' Hammer* and also instigated the witch-hunting craze in the Alps and the Pyrenees, remote regions where assimilation by enforced Romanization was especially weak. In Augustinian and Aquinian theology, angels and demons could act as intermediaries for God and the devil. Spirits moved stars and planets from within, the angels provided personal protection. According to the sixteenth-century Augustinian

Monk Luther, the devil could most easily be met in the outhouse (see Oberman, 1992). The dominant medieval Christian ideas in the sixteenth century were the reality of the devil, the battle between the Christian God and the devil, and the approach of the Christian version of *Ragnarok*, pre-medieval ideas that are still dominant in present-day sixteenth-century religious fundamentalism that remains largely unchanged by seventeenth-century science and mathematics, even as it uses the twentieth-century technology of television, computers, and fax-machines to propagate itself.

1.2 The foundations of mechanics

The Polish monk Copernicus (1473–1531) challenged the idea of absolute religious authority by proposing a sun-centered solar system. Like all other astronomer-astrologers of his time, Brahe and Kepler included, Copernicus used epicycles to correct for the Platonic idea that the natural motion of a planet or star should be circular motion at constant speed. The arrogant, life-loving Tyge Brahe, devoted to observational astronomy while rejecting Copernicus' heliocentrism, collected extensive numerical data on planetary orbits. Brahe invented his own wrong cosmology, using epicycles with the earth at the center. In order to amass the new data, he had to construct crude instruments, some of which have survived. His observation of the appearance of a new star in 1572 showed that the heavens are not invariant, but this discovery did not rock the firmly anchored theological boat. Suppression of new ideas that seriously challenged the officially accepted explanations was strongly enforced within the religious system, but mere models like the Copernican system were tolerated as a way of representing or saving appearances so long as the interpretations were not presented as 'the absolute truth' in conflict with official Church dogma. In 1577, Brahe recorded the motions of a comet that moved in the space *between* the 'crystalline spheres', but this result was also absorbed by the great sponge of Scholasticism.

After his student days Kepler tried first to model the universe Platonically, on the basis of geometry and numerology alone. That was well before his thoughts became dominated and molded by Brahe's numbers on Mars' orbit. Without Brahe's data on the motion of Mars, whose orbit among the outer planets has the greatest eccentricity (the greatest deviation from circularity), Kepler could not have discovered his three fundamental laws of planetary motion. The story of how Kepler came to work on Mars starts with a wager on eight days of work that ran into nearly eight years of numerical hard labor, with the self-doubting Swabian Johannes Kepler in continual psychological battle with the confident and dominating personality of Tyge Brahe who rightly feared that Kepler's ideas and reputation would eventually triumph over his own. Brahe came from a long line of Danish aristocrats who regarded books and learning as beneath their dignity. He left Denmark for Prague while complaining that the king who had heaped lavish support upon him for his astronomical researches did not support him to Brahe's liking (he gave him the island Hveen where, in typical feudal fashion, Brahe imprisoned and otherwise tyrannized the farmers). Kepler, Scholastically educated at the Protestant University of Tübingen, was a pauper from a degenerating family of *Burgerlische* misfits. He fell into a terrible first marriage and had to struggle financially to survive. Kepler's Protestantism, influenced by Calvinism, caused him eventually to flee from Austria, where he was both school teacher and official astrologer in the province of Styria, to Prague in order to avoid persecution during the Counter Reformation. While in Catholic Austria he was protected by Jesuits who respected his 'mathematics': it was one of Kepler's official duties in Styria to produce astrological calendars. Kepler predicted an exceedingly cold winter and also a Turkish invasion of Austria in 1598, both of which occurred. His other predictions that year presumably were wrong, but his track record can be compared favorably with those of the professed seers of our modern age who advise heads of state on the basis of computer simulations that produce 'forecasts'[5].

In his laborious, repeated attempts at curve-fitting Kepler discovered many times that the motion of Mars lies on a closed but noncircular orbit, an ellipse with one focus located approximately at the sun. However, he rejected his formula many times because he did not realize that it represented an ellipse. Finally, he abandoned his formula altogether and tried instead the formula for an ellipse! By finally identifying his formula for an ellipse as a formula for an ellipse, Kepler discovered an approximate mathematical law of nature that divorced solar system astronomy from the realm of purely wishful thinking and meaningless mathematical modelling.

In spite of the lack of circular motion about the sun at *constant speed*, Kepler discovered that Mars traces out equal areas in equal times, which is a generalization of the simple idea of uniform circular motion. Kepler's first two laws (we state them below) were widely disregarded and widely disbelieved. Neither Brahe nor Galileo accepted Kepler's interpretation of Brahe's data. Both were too interested in promoting their own wrong ideas about astronomy to pay serious attention to Kepler, *who had unwittingly paved the way for Newton to reduce astronomy to the same laws of motion and force that are obeyed by earthly projectiles.*

Kepler did not deviate from the Aristotelian idea that a *velocity* must be supported by a *cause*. Anything that moved had to have a mover. He was inspired by Gilbert's writings on magnetism and speculated that the sun 'magnetically' sweeps the planets around in their elliptic orbits, not understanding that the planets must continually *fall toward* the sun in order to follow their elliptic paths in empty Euclidean space. His explanation of the angular speed of the planet was the rotation of the sun, which by action-at-a-distance 'magnetism' dragged the planets around it at a slower rate. According to Kepler, it was a differential rotation effect due to the sweeping action of the magnetic 'spokes' that radiated out from the sun and swept the planets along in broom-like fashion.

Kepler's success in discovering his three laws stemmed from the fact that he effectively transformed Brahe's data from an earth-centered to a sun-centered coordinate system, but without using the relatively modern idea of a coordinate system as a set of rectangular axes. Kepler, Galileo, and Descartes were contemporaries of each other and it was the Jesuit[6]-educated Descartes who first introduced (nonrectilinear) coordinate systems in his reduction of geometry to algebra.

The previous age was that of Luther, Calvin, Zwingli, and the Religious Reformation, and of the artist who so beautifully expressed the medieval belief in and fear of the devil as well as other aspects of the age, Albrecht Dürer (1471–1528). For perspective,

---

[5] Prediction and explanation do not depend merely upon 'taking enough factors into account'. A large number of variables cannot be used to predict or explain anything unless those variables obey an empirically discovered and mathematically precise universal *law*. That physics describes motion universally, and that everything is made of atoms that obey the laws of quantum mechanics, does *not* imply that, commodity prices or other behavioral phenomena also obey mathematical laws of motion (see chapter 18).

[6] The Jesuits, known outside Spain as the intellectual wing of the Counter Reformation, protected Kepler in Austria because of their respect for his 'mathematics' and Jesuit astronomers recorded evidence of sunspots at their observatory in Ingolstadt in 1609. They also did their full share in the persecutions.

the great English mathematician and natural philosopher Newton was born in 1646 while Kepler, Descartes, and Galileo died in 1630, 1666, and 1642 respectively. The fear of hell and damnation induced by the Augustinian idea of predestination is the key to understanding the mentality of the men who led the Religious Reformation. The leading Reformers were authoritarian and anti-scientific: Jean Cauvin, or John Calvin, burned a political adversary at the stake in his Geneva Theocracy. Martin Luther complained that Copernicus wanted to destroy the entire field of astronomy. Luther's relatively mild and more philosophic companion Melancthon reckoned that the state should intervene to prevent the spread of such subversive ideas. Zwingli's followers whitewashed the interiors of Swiss churches to get rid of the art (as did Luther's, eventually). Like every true puritan, Luther was driven to find evidence that he was among God's chosen few who would enter heaven rather than hell. As he lay on his deathbed, his Catholic adversaries awaited word of a sign that the devil had taken him instead! Luther did contribute indirectly to freedom of thought: by teaching that the official semioticians in Rome and in the local parish were not needed for the interpretation of the Bible, he encouraged the fragmentation of clerical and political authority, including his own.

The Italian Catholic Galileo Galilei discovered the empirical basis for the law of inertia through systematic, repeated experiments: Galileo and his followers deviated sharply from the Scholastic method by performing simple controlled experiments and recording the numbers. Abstraction from precise observations of easily reproducible experiments on very simple motions led Galileo to assert that *all* bodies should accelerate at the same constant rate near the earth's surface,

$$a_y = -g, \tag{1.0b}$$

where $g \approx 9.8 \, \mathrm{m/s^2}$. Galileo's universal 'local law of nature' violates Aristotle's notion of heaviness versus lightness as the reason why some bodies rise and others fall, but Galileo himself did not explain the reason for the difference between vertical and horizontal motions on earth. For horizontal motions, also contradicting Aristotle, Galileo discovered the fundamental idea of inertia (but without so naming it): extrapolating from experiments where the friction of an inclined plane is systematically reduced, he asserted that a body, once set into motion, should tend to continue moving at constant velocity on an ideal frictionless horizontal plane without any need to be pushed along continually:

$$v_x = \text{constant.} \tag{1.0a}$$

These ideas had occurred in more primitive forms elsewhere and earlier. It is only one short step further to assert, as did Descartes, that the natural state of motion of an undisturbed body in the universe is neither absolute rest nor uniform circular motion, but is rectilinear motion at constant speed. *This law is the foundation of physics.* Without the law of inertia, Newton's second law would not follow and there would be no physics as we know it in the sense of *universal* laws of motion and force that apply to all bodies at all times and places in the universe, independent of the choice of initial location and time.

Galileo believed rectilinear motion at constant speed to be only an initial transient. He expected that the orbit of a hypothetical projectile tossed far from the earth must eventually merge into a circle in order to be consistent with his belief in a finite and perfectly spherical universe. He was therefore able to believe that planets and stars could move in circles about the sun *as a result of inertia alone*, that no (tangential *or* centripetal) force is required in order to keep a planet on a circular track. According to Galileo, a planet experiences no net force at all along its orbit about the sun.

The idea that the natural state of undisturbed motion of a body is constant velocity motion was an abomination to organized religion because it forced the philosophic consideration of an infinite universe, and the highest religious authority in western Europe had already asserted that the universe is finite. Infinitely long straight lines are impossible in a finite universe. Following his famous dreams, Descartes, a devout Catholic libertine, made a pilgrimage and prayed to the holy Maria for guidance while abandoning Church dogma in favor of freedom of thought, especially thoughts about nature as nonspirit-driven mechanism where everything except thought is determined according to universal mathematical law by *extension and motion*. This is, in part, the origin of the modern belief that money, price levels, and other purely social phenomena should also obey objective laws of motion.

Galileo and Descartes planted the seeds that led directly to the discrediting of Aristotelian physics, but Galileo systematically ignored Kepler's discoveries in astronomy (Kepler sent his papers to Galileo and also supported him morally and openly during the Inquisition). Descartes' speculations on astronomy are hardly worth mentioning. According to Koestler (1959), Galileo was not subjected to the Inquisition because of his discoveries in physics, which were admired rather than disputed by many religious authorities including the Pope, but rather because he persisted in insinuating that the politically powerful church leaders were blockheads for rejecting Copernicus' heliocentric theory. When forced under the Inquisition to prove the truth of Copernicus' system or shut up, Galileo failed: Kepler had already shown the details of the Copernican theory to be wrong, but had no notion at all of the physical idea of 'inertia'. Kepler did not understand that a body, once set into motion, tends to *remain* in rectilinear motion at constant speed and that a force must act on that body in order to *alter* that state of motion. Galileo studied gravity from the local aspect, Kepler from a global one, and neither realized that they were both studying different aspects of the very same universal phenomenon. This lack of awareness of what they were doing is the reason for the title of Koestler's book *The Sleepwalkers*. There was, before Newton, no unification of their separate results and ideas. Even for Newton, the reasoning that led to unification was not smooth and error-free (see Westfall, 1980).

Some few of Descartes' ideas about mechanics pointed in the right direction but most of them were wrong. Descartes retained the Platonic tradition of believing strongly in the power of pure reason. He believed that the law of inertia and planetary motion could be deduced on the basis of thought alone without adequately consulting nature, hence his vortex theory of planetary motions. That is the reason for the failure of Cartesian mechanics: experiment and observation were not systematically consulted. Descartes' mathematics and philosophy were revolutionary but his approach to mechanics suggests an updated version of Scholasticism, a neoScholasticism where mathematics is now used, and where the assumption of spiritualism in matter is replaced by the completely impersonal law of inertia and collisions between bodies:

(1) without giving motion to other bodies there is no loss of motion: (2) every motion is either original or is transferred from somewhere; (3) the original quantity of motion is indestructible.

The quote is from the English translation (1942) of Ernst Mach's description of Descartes' ideas, and one can see a very crude qualitative connection with the idea of momentum conservation.

Some writers have suggested that Descartes introduced the idea of linear momentum, $p = mv$, where m is a body's inertial mass and v its speed, while others argue that he, like Galileo, had no clear idea of inertial mass. The clear introduction of the idea of inertial mass, the coefficient of resistance to change in velocity, is usually attributed to Newton. Descartes understood the action–reaction principle (Newton's third law), and also conservation of momentum in elastic collisions (Beeckman also apparently had an idea of the law of inertia). In the end, it fell to Newton to clear up the muddle of conflicting ideas and finally give us a clear, unified mathematical statement of three universal laws of motion along with the universal law of gravity.

Having formulated the basis of modern geometry, Descartes asserted *wordily* that the natural state of undisturbed motion of any body is rectilinear motion at constant speed. Newton took the next step and formulated *mathematically* the idea that a cause is required to explain why a body should deviate from motion with constant velocity. Following Newton, who had to make many discoveries in calculus in order to go further than Descartes, the law of inertia can be written as

$$\frac{d\vec{v}}{dt} = 0 \tag{1.1a}$$

where $\vec{v}$ is a body's velocity in the undisturbed state. Such a body moves at constant speed along a straight line without being pushed. If force-free motion occurs at constant velocity, then how does one describe motion that differs from this natural state? The simplest generalization of Galileo's two local laws (1.0a) and (1.0b) is

$$m\frac{d\vec{v}}{dt} = \overrightarrow{\text{cause of deviation from uniform rectilinear motion}} \tag{1.2a}$$

where m is a coefficient of resistance to any change in velocity (Newton could have introduced a tensor coefficient m_{ik} rather than a scalar one, but he implicitly assumed that Euclidean space is isotropic, an idea first emphasized explicitly by his contemporary Leibnitz (Germany, 1646–1716)). In other words, Newton finally understood that it is the *acceleration*, not the velocity, that is directly connected with the idea of force (Huygens (Holland, 1629–1695) derived the formula for centripetal acceleration and used Newton's second law to describe circular motion 13 years before Newton published anything). If a body's mass can vary, then the law of inertia must be rewritten to read

$$\frac{d\vec{p}}{dt} = 0 \tag{1.1b}$$

where $\vec{p} = m\vec{v}$ is the momentum and m is the resistance to acceleration. The law of inertia according to Newton then states that the motion of a body with mass m subject to no net force obeys the law

$$\vec{p} = m\vec{v} = \text{constant,} \tag{1.1c}$$

which is Newton's first law of motion and also the law of conservation of momentum. The consequence is that if the body should fragment into two or more parts with separate masses m_i, where $\Sigma m_i = m$, then momentum is conserved in the fragmentation:

$$\sum_{i=1}^{N} m_i \vec{v}_i = m\vec{v}. \tag{1.1d}$$

Newton's second law of motion, his generalization of Galileo's local law of gravity, then reads

$$\frac{d\vec{p}}{dt} = \vec{F} \tag{1.2b}$$

where $\vec{F}$ is a vector called the net force and stands for the cause of any deviation from uniform rectilinearity of the motion. This generalization reduces to the law of inertia under the right circumstances: $\vec{v}$ is constant whenever $\vec{F}$ vanishes if the coefficient of inertia m is constant. Whenever the net force $\vec{F}$ does not vanish then the body's motion is altered, *but not without resistance to change*. If m is constant then $\vec{F} = m\vec{a}$ where $\vec{a} = d\vec{v}/dt$ is the acceleration. The more massive the body, the greater the force required to alter the velocity of the body.

If we follow Newton then the net force that alters the motion of a body can *only* be due to the presence of other bodies whose motions are temporarily ignored in the writing of (1.2b), bodies whose own accelerations must also obey (1.2b). This idea is represented by the action–reaction principle, or Newton's third law: a body that feels a force should simultaneously exert a force on the body that causes the original force according to the rule

$$\vec{F}_{ik} = -\vec{F}_{ki}, \tag{1.3}$$

where $\vec{F}_{ik}$ is the force exerted on body i by body k. In other words, forces are not caused by properties of empty space (empty Euclidean space anyway has no features to distinguish one part of it from any other part) or by invisible animation like spirits: a force, the cause of any deviation whatsoever from constant momentum motion, should be caused by the action of another material body.

Can one discover universal, mathematically precise descriptions of forces that can be used in Newton's laws to describe any of the variety of motions that are observed in nature? The first step in this direction was also taken by Newton, who solved examples of what we would now call inverse problems: given that a body's orbit is elliptic, circular, parabolic, or hyperbolic with a motionless force-center, the sun, at one focus, what is the force law that produces the orbit? We present a modern version of Newton's approach to this problem as described in the *Principia* – see exercise 2 for Newton's first derivation of the law of gravity. Note that if we write $\vec{L} = \vec{r} \times \vec{p}$ and $\vec{N} = \vec{r} \times \vec{F}$ to define the angular momentum and torque, where $\vec{r}$ defines the position of the body then

$$\frac{d\vec{L}}{dt} = \vec{N} \tag{1.4}$$

follows directly from Newton's second law (1.2b), which means that the angular momentum vector is conserved in the absence of a net torque on the body. A central force law yields angular momentum conservation immediately: if $\vec{F} = f(r)\hat{r}$, where $\hat{r}$ is the unit vector in the r-direction, then the force on the mass m is always directed along the position vector of that body relative to the force-center, which (by the action–reaction principle) implies the presence of a second body at the force-center whose own motion, for the time being, we choose to ignore. In this case $\vec{L} = m\vec{r} \times \vec{v}$ is constant at all times and therefore defines a fixed direction $\hat{L}$ in space. Because both the velocity and linear momentum must remain forever perpendicular to the angular momentum, this conservation law confines the motion of the mass m to a fixed plane perpendicular to the angular momentum vector $\vec{L}$. Planar orbits agree with observations of planetary motions, so Newton was able to restrict his considerations to central forces.

Next, if we write Newton's law in cylindrical coordinates (r,θ) in the plane of the motion, which we take perpendicular to the z-axis, then equation (1.2b) has the form

$$m\ddot{r} - mr\dot{\theta}^2 = f(r)$$

$$\frac{\mathrm{d}L_z}{\mathrm{d}t} = 0 \qquad\qquad (1.5a)$$

where $L_z = mr^2\mathrm{d}\theta/\mathrm{d}t$ is the magnitude of the angular momentum and is constant. These equations can therefore be rewritten in the form

$$\ddot{r} - \frac{L_z^2}{m^2 r^3} = \frac{f(r)}{m}$$

$$L_z = mr^2\dot{\theta} = \text{constant}. \qquad\qquad (1.5b)$$

It is easy to show directly by differentiation and insertion into the first equation of (1.5b) that the assumption that the orbit is a conic section, that

$$\frac{1}{r} = C(1 + \varepsilon\cos\theta) \qquad\qquad (1.6)$$

where C and ε are constants (ε is called the eccentricity of the conic section), requires that $f(r)$ *must vary as the inverse square of the distance r of the body from the force-center*, which lies at one of the two focal points of the conic section (the origin of coordinates lies at one focus): $f(r) = -CL_z^2/mr^2$. If $\varepsilon < 1$ then the conic section is an ellipse (ε measures the noncircularity of the ellipse in this case), which agrees with Kepler's first law (figure 1.1), whereas $\varepsilon > 1$ yields a hyperbola. The special limiting cases $\varepsilon = 0$ and $\varepsilon = 1$ describe circular and parabolic orbits respectively. The four different types of orbits that arise as ε is varied in (1.6) are called *conic sections* because equation (1.6) can also be derived purely geometrically by asking for the shapes of the curves that define the intersections of a plane with a cone, a geometric interpretation that Newton understood. *Newton's inverse square law of gravity follows from the assumption that planetary orbits are perfect ellipses.*

Is Newton's inverse solution unique? Surprisingly enough, we shall prove in chapter 4 that there are only two possible central force laws that can yield *closed* (therefore periodic) orbits *for arbitrary initial conditions*, and that both of those closed orbits are elliptic. Namely, if

$$\vec{f} = -k\vec{r} \qquad\qquad (1.7)$$

with $k > 0$ then the orbit is elliptic with the force-center at the center of the ellipse (the proof was first worked out by Newton). This is the case of an isotropic simple harmonic oscillator: the force constant k is independent of direction and is the same in all possible directions. Any central force is a spherically symmetric force. An anisotropic harmonic oscillator would be represented by a force law $\vec{F} = (-k_1 x, -k_2 y, -k_3 z)$ where at least one of the three force constants k_i differs from the other two, representing an absence of spherical symmetry.

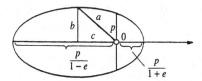

Fig. 1.1 Kepler's ellipse (fig. 35 from Arnol'd, 1989), with $p = 1/C$.

If, on the other hand,

$$\vec{F} = -\frac{k\hat{r}}{r^2} \qquad (1.8)$$

with $k > 0$ then the orbit is an ellipse with the force-center at one focus (or, as Whittaker shows, with two force-centers at both foci), which corresponds to the idealized description of a single planet moving about a motionless sun. The determination of the force laws that yield elliptic orbits for all possible initial conditions, or closed orbits of *any* kind for *arbitrary* initial conditions, is summarized by the Bertrand–Königs theorem that is proven in chapter 4 (according to Westfall, 1980, Newton knew this). Newton's solution of the Kepler problem is therefore unique. The only element missing in the analysis above is the motion of the sun itself, which we also take into account in chapter 4.

This does not really answer our original question: we have assumed *without discussion* that Newton's differential equations describing the Kepler problem, which are nonlinear, have unique solutions. Linear equations always have unique solutions but nonlinear equations do not. An example is given by $dx/dt = x^{1/2}$ with $x(0) = 0$. Three possibilities follow: $x = 0$ for all t, $x = t^2/4$ for $t \geq 0$ (with $x = 0$ for $t < 0$), or $x = 0$ for $0 \geq t < t_\circ$ with $x = (t - t_\circ)^2/4$ for $t \geq t_\circ$. The condition for uniqueness of solutions of nonlinear equations is stated in chapter 3.

Notice that spherically symmetric force laws do not imply spherically symmetric orbits: although the surfaces of constant force for $f(r) = -k/r^2$ and $f(r) = -kr$ are both spheres, the corresponding orbits are ellipses rather than circles. In general, differential equations with a given symmetry (here spherical) can give rise to solutions that exhibit less symmetry. Plato's idea of hidden forms representing ideal perfection is akin to the spherical symmetry of the gravitational force law. Two synonyms for spherically symmetric are: (i) rotationally invariant and (ii) isotropic, meaning the same in all possible directions.

According to the law of motion (1.2b) and the action–reaction principle (1.3), we may be justified in neglecting the sun's motion to within a good first approximation: for any two masses m_1 and m_2 obeying

$$m_1\vec{a}_1 = \vec{F}_{12} = -m_2\vec{a}_2, \qquad (1.9a)$$

where $\vec{F}_{ik}$ denotes the force exerted on body i by body k, then it follows that the magnitudes of the accelerations are given by the ratio

$$\frac{a_1}{a_2} = \frac{m_2}{m_1} \qquad (1.9b)$$

of the two masses. If $m_2/m_1 \ll 1$, as is true if m_1 represents the sun's mass and m_2 represents a planet's mass (or if m_1 represents the earth and m_2 represents the moon), then we can ignore the acceleration of the first body, to a first approximation, and therefore solve for the motion of the second body as if the first one forever remains unaccelerated from an initial state of rest.

We show next that Kepler's second and third laws also follow easily from Newtonian mechanics. For the proof, nothing is needed but angular momentum conservation and some analytic geometry. From conservation of angular momentum, which follows automatically for any central force regardless of the function $f(r)$, we obtain the 'areal velocity' as (figure 1.2)

$$\frac{dS}{dt} = \frac{1}{2} r^2 \dot{\theta} = \frac{L_z}{m} = \text{constant}, \tag{1.10}$$

where S is the area traced out by the body's radius vector: equal areas are therefore traced out by a planet in equal times, which is Kepler's second law. Note that the orbit under consideration need not be elliptic ($f(r)$ is arbitrary), nor need it be closed (for example, hyperbolic orbits in the Kepler problem and other open orbits in any other central force problem necessarily yield Kepler's second law).

To obtain Kepler's third law, we must integrate (1.10) over the area of a planet's elliptic orbit. Geometrically, the area of the orbit is given by $S = \pi ab$, where a and b are the semi-major and -minor axes of the ellipse and are related to each other via the orbit's eccentricity ε by the rule $b = a(1 - \varepsilon^2)^{1/2}$. Combination of these results yields that $\pi ab = L_z \tau / 2m$ where τ is the orbit's period (the orbit is closed and therefore is periodic). Kepler's third law, $\tau = 2\pi a^{3/2} m / \sqrt{CL_z} \propto a^{3/2}$, follows easily. Note that neither conservation of energy nor the numerical value of the force parameter $k = CL_z^2/m$ is needed for the proof.

According to Newton, the parameter in the inverse square law is given by $k = CL_z^2/m$, and by using Kepler's third law we can also write $k = (a^3/\tau^2)m$. Planetary data show that the ratio a^3/τ^2 is approximately constant: for the earth–sun system the numerical value is 3375 m^3/yr^2 whereas for Mars–sun and Venus–sun one finds 3357 and 3328 respectively. To within a pretty good approximation, it appears that the force parameter k is proportional to the mass m. That k/m is not precisely constant within the solar system might be due to the acceleration of the sun, which depends weakly upon the planet's mass.

Newton went much further than this. He reasoned that the trajectory of a projectile near the earth and the orbit of a planet about the sun should have exactly the same cause, and that the cause also explains the apparent lack of symmetry between the 'up–down' and horizontal directions from the standpoint of an earthbound observer. To illustrate Newton's idea, imagine standing on the Forcellina with a backpack full of apples. Imagine tossing successive apples with greater and greater initial velocity, but with negligible atmospheric friction (and imagine that there are no higher mountains along the trajectory). The first apple is tossed toward Bellinzona. The second is tossed further in the same direction toward Domodossola, and the third even further toward Herbriggen. For low enough initial speeds the trajectory is parabolic and the earth's curvature is not important, but for higher initial speeds one must take into account the curvature of the earth. Newton reasoned that if you could toss hard enough, an apple could travel around the earth, pass over Meran and come back to its starting point: given a large enough initial speed, the apple must go into an elliptic orbit with the earth

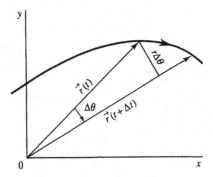

Fig. 1.2 The areal (or sectorial) velocity (fig. 3.7 from Goldstein).

at one focus[7]. Newton arrived, by similar thinking, at the speculation that the cause of Galileo's observations, the cause of the difference in acceleration along the vertical and horizontal directions near earth, may be the same as the cause of Kepler's elliptic orbits (he then checked the speculation over a long time with a lot of numerics). Therefore, in the Newtonian description a planet continually *falls* radially inward toward the sun just as a projectile continually falls vertically toward the earth, but the planet orbiting the sun has large enough azimuthal velocity that it translates just far enough transversally during each small time interval Δt to stay on the elliptic orbit as it falls.

In order that all projectiles should fall locally toward the earth with exactly the same local acceleration $g = 9.8\,\text{m/s}^2$, it is necessary that the ratio $f(r)/m$ in (1.5b) is independent of the mass m of any projectile. This can only be true if the magnitude of the force of attraction between the projectile with mass m_1 and force-center with mass m_2 has the form

$$f(r) = -\frac{G_2 m_1}{r^2},$$ (1.11a)

where G_2 depends upon the details of the second but not the first body. From the action–reaction principle we obtain that $G_2 m_1 = G_1 m_2$ where $G_1 m_2/r^2$ is the magnitude of the force exerted by body 1 on body 2. This requires that $G_k = G_{ik} m_i$ where G_{ik} may vary from one pair of bodies to the next. We know from experiment (sun–Mars, earth–projectile, earth–moon, etc.) that, to at least four decimal accuracy, $G_{ik} = G = 0.6672\cdots \times 10^{-10}\,\text{N}\,\text{m}^2/\text{kg}^2$ should be a *universal* constant, the very same constant for every pair of bodies no matter where in space the pair is located or at what time the question is asked. This reasoning yields Newton's universal law of gravity where the magnitude of the attraction between any two bodies is given by

$$f(r) = -\frac{G m_1 m_2}{r^2},$$ (1.11b)

and G is a universal constant of nature. That one never has to vary G or introduce any fudge factors in order to fit empirical orbital data illustrates the essential difference between the mere mathematical modelling of irregular phenomena and laws of nature: *research fields that are totally disconnected from physics have no universal constants.* Whether G is precisely the same number at all times and places in the universe can never be known (how can we possibly know G to 1000 or one million decimal accuracy?), but

---

[7] A parabolic orbit is therefore an approximation, over short distances, to a section of an elliptic orbit.

so far as earthbound experiments and observations over the last few centuries are concerned G is known to be constant to within reasonably high precision.

Newton proved that the neoPlatonic–Aristotelian presumption that 'heavenly physics' differs from 'earthly physics' is not only unnecessary, it is also wrong[8]: both projectile and planetary motions are accounted for via the *same* universal law of motion, equation (1.2b), and by the *same* abstract deterministic cause, the universal law of gravitational attraction

$$\vec{F}_{ik} = G \frac{m_i m_k}{r_{ik}^2} \hat{r}_{ik}, \qquad (1.11c)$$

where $\hat{r}_{ik}$ is a unit vector running from the location of the body with mass m_i to that of body with mass m_k, and $\vec{F}_{ik}$ denotes the force exerted on body i by body k.

It follows easily from (1.11c) that (with negligible atmospheric friction) the only force on a projectile with mass m is vertical, pointing toward the earth's center, and has the magnitude $F = mg$ with $g = GM/R^2$ where R is the earth's radius. The constant $g \approx 9.8 \, \text{m/s}^2$ then determines the ratio M/R^2 for the earth, given G.

Why isn't the law of gravity called Newton's fourth law[9]? In spite of his enormous success in accounting universally for trajectory and planetary motions, Newton was repelled by the idea of gravity as 'action at a distance'. Action at a distance means that two bodies like the sun and earth are supposed to attract each other instantaneously over completely empty space, but with no mechanism for *transmitting* the force from one body to the other. Both Descartes and Newton regarded the notion of action at a distance as a retreat into mysticism. Newton did not want the idea attributed to him: how could two bodies influence each other over a distance consisting only of *empty space*? Newton gave us a very useful mathematical *description* of *how* gravity effects motion but he could not explain why there is a law of gravity in the first place or how it works (how the force is actually transmitted). In order to help the reader to appreciate Newton's conceptual difficulty (and the writer's as well), and keeping in mind that, naively seen, gravity is supposed not to differ qualitatively from a tug or a push on an ox-cart, we refer the reader to exercise 1.

The British experimenter Faraday (1791–1867) also could not accept the mysterious idea of action at a distance. His reaction was to construct the mathematical idea of a field of force. If we define the vector field

$$\vec{g} = G \frac{M}{r^2} \hat{r}, \qquad (1.11d)$$

then the field lines radiate outward from the body of mass M so that we can think of the mass M as the source of this field (the field lines must end on a spherical sink at infinity). In this way of thinking, the force field of the body with mass M is present in Euclidean space (so that matter-free space is no longer a void) whether or not there is another body with mass m that is available to experience the force, which would be given by

$$\vec{F} = -m\vec{g} = -G \frac{Mm}{r^2} \hat{r}. \qquad (1.11e)$$

---

[8] No mathematical result analogous to Newton's laws has yet been discovered or established in fields like economics that are disconnected from physics.

[9] Arnol'd, in *Ordinary Differential Equations* attributes the law of gravity not to Newton, but to Hooke. We refer the reader to Westfall (1980) for an account of the controversy.

The gravitational field lines break the perfect three dimensional symmetry of empty Euclidean space and account for the difference between the vertical and the horizontal experienced by an earthbound observer. The language of fields of force provides an appealing and useful way of describing action at a distance, but it still falls short of explaining *how* the gravitational force is transmitted from one body to another through empty space.

From a purely mathematical standpoint, we can regard Newton's law of gravity as an abstract example of mechanism. One cannot see, feel, or otherwise identify any mechanical connections between two attracting masses but the bodies somehow cause each other to accelerate as if mechanical linkages were there nonetheless. The time-evolution of a Newtonian system is correspondingly deterministic: a system of masses evolving in time under Newtonian gravity evolves in a strict machine-like fashion. Newtonian gravity is an example of an abstract, universal mechanism. The machine-like nature of deterministic laws of time-evolution is explained in chapter 6: every system of differential equations digitized correctly for computation defines its own automaton that processes computable digit strings arithmetically. Arithmetic itself has a universal machine-like nature, as Turing (English, 1912–1954) showed in the 1930s. By digitizing the laws of motion for numerical computation the differential equations for gravitationally interacting bodies define automata that compute the orbit for initial conditions written as decimal expansions (digital programs), in machine-like fashion.

By showing that the forces that determine a body's motion are caused by other bodies that obey the very same laws of motion, and that the paths and speeds of both projectiles and stars must obey the same universally applicable and precise deterministic mathematical laws, Newton eliminated the opportunity for angels and demons to do any real work. As far as the motion of matter is concerned, they are at best allowed to fiddle around with initial conditions in the very distant past (Newton himself did not believe this – see chapter 17).

Newtonian mechanics stabilized and guaranteed the deterministic revolutions advocated philosophically by Descartes: for mechanical systems where the solutions of Newton's equations are finite and unique for all times t, the future and past would be completely determined by the infinitely precise specification of the positions and velocities of all bodies in the system at a single time t_o. However, precise mechanical laws of motion can tell us nothing at all about how or why one particular set of initial conditions occurred rather than other possibilities during the early stages of formation of the universe. Statistical arguments amount only to wishful thinking: there is only one universe. The attempt to explain too much about cosmology would lead us, as it led Plato and others, into errors.

A conservation law is a scalar, vector, or tensor function of the coordinates and velocities that remains constant along a trajectory. Whether or not there are any conserved quantities depends upon the mathematical form of the force law that determines the mechanical system's time-development. For one body of mass m under the influence of an external force $\vec{F}$, we have shown above that there are two possible vectorial conservation laws in Newtonian mechanics: linear momentum is conserved only if $\vec{F} = 0$, and angular momentum is conserved whenever the force produces no net torque, as is always the case for central forces.

We now show that there is (at the very least) a third possible conservation law in Newtonian mechanics, but this particular conserved quantity is a scalar, not a vector.

We restrict our considerations in what follows to bodies with constant mass m, although this restriction can easily be relaxed. The work–energy theorem follows easily from (1.2b): with

$$m \frac{d\vec{v}}{dt} \cdot \vec{v} = \frac{dm\vec{v}^2}{2dt} \qquad (1.12)$$

and

$$\vec{F} \cdot \vec{v} dt = \vec{F} \cdot d\vec{r} \qquad (1.13)$$

we see by integration that $\Delta T = W$ follows where $T = mv^2/2$ is the kinetic energy if

$$W = \int_1^2 \vec{F} \cdot d\vec{r} \qquad (1.14)$$

is the work done by the *net* force $\vec{F}$ as the body moves along its trajectory from location $\vec{r}_1$ to location $\vec{r}_2$. In general, the integral in (1.14) is not independent of the path connecting the two end points, so that the work computed for any trajectory other than the one actually followed by the body will yield a different number for W than will the natural trajectory (in the language of chapter 2, the work is a functional, not a function). The right way to say this mathematically, without further assumptions, is that the differential form $\vec{F} \cdot d\vec{r} = F_x dx + F_y dy + F_z dz$ is not an exact differential, and therefore the integral of this differential form does not yield a function that can be identified as 'the work function' $U(x)$: i.e., there is generally no function $U(x)$ satisfying the condition that $W = U(x_2) - U(x_1)$ depends only on the 'point function' U evaluated at the end points, *independently of the path chosen for the integration.*

Mathematically, the condition for the work W to be path-independent is that the integral representing the work done around any arbitrary closed path C vanishes,

$$W = \int_C \vec{F} \cdot d\vec{r} = 0. \qquad (1.15a)$$

By Stoke's theorem,

$$\int_C \vec{F} \cdot d\vec{r} = \int\int \nabla \times \vec{F} \cdot d\vec{S} \qquad (1.16)$$

where the integral on the right hand side is over the area enclosed by the loop C. For an arbitrary infinitesimal loop, it follows that the pointwise condition for the integrability of the differential form $\vec{F} \cdot d\vec{r}$ is simply that

$$\nabla \times \vec{F} = 0, \qquad (1.17a)$$

which requires that

$$\vec{F} = -\nabla U(\vec{r}) \qquad (1.17b)$$

where U is a scalar function of position called the potential energy. Such forces are called conservative forces because the work done by such a force is always path-independent,

$$W = U(\vec{r}_1) - U(\vec{r}_2). \qquad (1.15b)$$

This path-independence (integrability) leads immediately to a scalar conservation law: combining $W = \Delta T$ with $W = - \Delta U$ yields $\Delta T + \Delta U = 0$, which means that the quantity

$$E = T + U \qquad (1.18a)$$

is constant along any trajectory of a body whose law of motion is described by

$$\frac{d\vec{p}}{dt} = - \nabla U(\vec{r}). \qquad (1.19)$$

The quantity E is called the total mechanical energy, the total energy, or just the energy.

For different trajectories, representing different initial conditions, there are different constants E. In other words, E is constant along a particular trajectory but varies numerically from trajectory to trajectory. In any case, the scalar quantity $E = T + U$ is the body's 'total mechanical energy' so long as U is the potential energy of the *net* force, which is permitted to contain contributions from one or more different conservative forces. Whenever nonconservative forces also act on the body, the total energy is still given by $E = T + U$ where U is the potential energy of the net conservative force, but in this case it does not follow that E is a constant of the motion.

As an example, for a body with mass m the force of gravity acting locally $F = - mg$, can be rewritten as the gradient of a scalar potential: $F = - dU/dy$ where $U = mgy$. Gravity 'breaks' the symmetry of Euclidean space between the vertical and the horizontal by assigning larger potential energies to greater heights (translational invariance is broken along the field lines).

Newtonian gravity in the general case also provides us with an example of a conservative force: $U = - GMm/r$ yields the inverse square law force of universal gravitational attraction (1.11c) between two bodies with masses m and M. The Coulomb force is an inverse square law force and therefore also follows from a $1/r$ potential. Another example of the potential energy of a conservative force is given by Hooke's law of perfect elasticity $F = - k(x - x_o)$, where $U = k(x - x_o)^2/2$.

Note that the addition of a constant C to a potential energy U produces no observable effect, as it merely shifts the total energy E by the same constant C without changing the kinetic energy T.

We now prove another easy theorem: every one dimensional bounded motion due to a conservative force is periodic, meaning that (whatever the initial conditions) the motion repeats itself systematically after a definite time τ, and after any integral multiple of the time τ. In other words, one dimensional bounded motions always yield closed orbits. This theorem follows from energy conservation,

$$E = \frac{m}{2}v^2 + U(x) = \text{constant}, \qquad (1.18b)$$

combined with the assumption that the motion is bounded. According to figure 1.3, the motion is bounded if there are two turning points x_1 and x_2 defined by setting $T = 0$ in equation (1.18a), points where $U(x_1) = E = U(x_2)$ if $U(x) < E$ for $x_1 < x < x_2$. The motion is then confined to the interval $x_1 \leq x \leq x_2$ because motion outside that

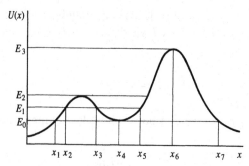

Fig. 1.3 Turning points of a conservative force for several different energies
(fig. 5 from Saletan and Cromer, 1971).

interval would require a negative kinetic energy and therefore an imaginary speed. It follows from rewriting (1.18b) in the form

$$dt = \frac{dx}{\left[\dfrac{2}{m}(E - U(x))\right]^{1/2}} \tag{1.20a}$$

and integrating that

$$\tau = 2 \int_{x_1}^{x_2} \frac{dx}{\sqrt{\left(\dfrac{2}{m}\right)(E - U(x))}} \tag{1.20b}$$

is the time required for the motion to repeat itself: $x(t + \tau) = x(t)$, $v(t + \tau) = v(t)$, and, therefore, $a(t + \tau) = a(t)$ as well. That the motion is periodic for all times after a single repetition of the initial conditions follows if solutions of the differential equation $md^2x/dt^2 = -dU(x)/dx$ are unique, a condition that can be violated easily by nonlinear equations in general but not by nonlinear equations *with bounded motion* (see the qualitative discussion of Picard's method in chapter 3). In other words, $x(t + n\tau) = x(t)$ and $v(t + n\tau) = v(t)$ follow for all integer multiples $n = 0, 1, 2, \ldots$, of the basic period τ given by (1.20b).

Energy conservation allows the solar system to run like a machine that disallows the intervention of an arbitrary or capricious god. The solar system behaves over a very long time scale like a frictionless clock and is the origin of our idea of a clock. In the quantification of the Aristotelian idea of mechanics, in contrast, there is not conservation of energy.

Aristotle's ideas of earthly physics can be summarized as follows: the earth is at rest in the universe, and all absolutely heavy bodies eventually come to rest with respect to the earth. The state of rest is the preferred state of motion because motion must be supported by force. Aristotle did not write down an equation of motion to describe his ideas, but we can take the liberty of doing that for him: the hypothetical law of motion

$$\beta \vec{v} = \vec{F}, \tag{1.21}$$

where β is a coefficient of friction representing resistance to motion and $\vec{F}$ is the direct cause of the velocity. If this idea is used to try to describe planetary motion, then one must assume a medium ('the ether') that provides a friction constant β. According to

(1.21), (i) a body remains at rest if no net force acts on it, and (ii) a net force is required in order to generate and maintain a velocity. This equation can be derived from Newton's law by assuming that the net force consists of linear friction plus a nonfrictional driving force F, by considering time intervals that are long enough that acceleration is effectively damped by friction. The Aristotelian equation of motion can be used to describe some commonplace phenomena: after a brief initial transient, a ball released from rest stops accelerating and falls downward with constant velocity through the atmosphere (Stoke's law).

According to Aristotelian critics of science, it is a negative attribute of 'Cartesian thinking' to idealize and abstract by the method of separating and isolating different causes from each other to understand how different hypothetical effects follow from different isolated causes ('reductionism'). Aristotle (whose spirit many system theorists of today implicitly follow) lumped the sprouting of acorns, the rolling of a ball, and the education of a schoolboy under the same heading of 'motion'. As Westfall (1980), writes, 'Galileo set mechanics on a new course by redefining motion to eliminate most of its Aristotelian connotations. From the discussion of motion, he cut away sprouting acorns and youths alearning.'

1.3 Conservation laws for N bodies

We now attempt to take explicitly into account $N - 1$ additional bodies that contribute, in part, to the force and torque experienced by our test body above. Writing

$$\frac{d\vec{p}_i}{dt} = \vec{F}_i = \vec{F}_i^{\text{ext}} + \sum_{j \neq i}^{N} \vec{F}_{ij} \tag{1.22}$$

for the net force on the ith body, then F_{ij} denotes the force exerted on the body i by body j within the system while F_i^{ext} denotes the net force exerted on body i by bodies that are not counted as part of the system. Here, one must assume that the latter are so massive that their reaction to the motion of the N bodies within our system can be neglected. Otherwise, we would have to take their accelerations into account and our system would consist of more than N bodies.

Beginning with Newton's second law for the dynamical system,

$$\dot{\vec{p}}_i = \vec{F}_i^{\text{ext}} + \sum_{j \neq i} \vec{F}_{ij} \tag{1.23}$$

and the third law, the action–reaction principle

$$\vec{F}_{ji} = -\vec{F}_{ij}, \tag{1.24}$$

we can sum over all bodies in the system to obtain

$$\sum_{i=1}^{N} \dot{\vec{p}}_i = \sum_{i=1}^{N} \vec{F}_i^{\text{ext}} + \sum_{i=1}^{N} \sum_{j \neq i} \vec{F}_{ij}. \tag{1.25}$$

Since

$$\sum_{i=1}^{N} \sum_{j \neq i} \vec{F}_{ij} = \sum_{i \neq j} (\vec{F}_{ij} + \vec{F}_{ji})/2 = 0, \tag{1.26}$$

this reduces to

$$\sum_{i=1}^{N} \dot{\vec{p}}_i = \sum_{i=1}^{N} \vec{F}_i^{\text{ext}} = \vec{F}^{\text{ext}}. \tag{1.27}$$

By defining the location of the center of mass of the system,

$$\vec{R} = \frac{\sum_{i=1}^{N} m_i \vec{r}_i}{M}, \tag{1.28}$$

we obtain the total momentum of the system

$$\vec{P} = M \frac{d\vec{R}}{dt}, \tag{1.29}$$

which *also* obeys Newton's second law in the form

$$\frac{d\vec{P}}{dt} = \vec{F}^{\text{ext}} \tag{1.30}$$

where $\vec{F}^{\text{ext}}$ is the net external force acting on all N bodies in the system. That there is no contribution from the internal forces guarantees that a piece of chalk at rest on a table cannot suddenly gain a net momentum due to the action of the (electrostatic) forces that hold the chalk together. Conservation of total momentum of the system then corresponds to the vanishing of the net external force.

Consider next the N-body system's total angular momentum:

$$\vec{L} = \sum_{i=1}^{N} \vec{r}_i \times \vec{p}_i. \tag{1.31}$$

The angular momentum changes in time according to

$$\frac{d\vec{L}}{dt} = \sum_i \vec{r}_i \times \dot{\vec{p}}_i = \sum_i \vec{r}_i \times \vec{F}_i. \tag{1.32}$$

Noting that

$$\sum_i \vec{r}_i \times \vec{F}_i = \sum_i \vec{r}_i \times \vec{F}_i^{\text{ext}} + \sum_{i \neq j} (\vec{r}_i \times \vec{F}_{ij} + \vec{r}_i \times \vec{F}_{ji})/2 = \sum_i \vec{r}_i \times \vec{F}_i^{\text{ext}} = \vec{N}^{\text{ext}}, \tag{1.33}$$

where $\vec{N}^{\text{ext}}$ is the net external torque, we see that the internal forces cannot contribute to any change in the N-body system's total angular momentum

$$\frac{d\vec{L}}{dt} = \vec{N}^{\text{ext}}, \tag{1.34}$$

and that conservation of total angular momentum corresponds to the vanishing of the net external torque. Conservation of angular momentum explains why, for example, if you spin on a piano stool (one with relatively good bearings) with your arms outstretched and then suddenly fold them in close to your chest, the rotation rate suddenly speeds up.

So far, we have derived two of the three geometric conservation laws for an N-body system from Newton's laws of motion. The third, conservation of energy, also follows easily. If we compute the work done by the net force along the system's trajectory during a time interval from t_1 to t_2, then

$$W_{12} = \sum_i \int_1^2 \left(\vec{F}_i^{\text{ext}} + \sum_{j \ne i} \vec{F}_{ij} \right) \cdot d\vec{r}_i, \tag{1.35}$$

which can be rewritten as

$$W_{12} = \sum_i \int_1^2 m_i \frac{d\vec{v}_i}{dt} \cdot d\vec{r}_i = \left| \sum_i m_i \frac{\vec{v}_i^2}{2} \right|_1^2 = T_2 - T_1. \tag{1.36}$$

In other words, the work done by the net forces on all of the bodies shows up as the change in total kinetic energy of the N bodies.

To go further, we find it convenient to transform to center of mass and relative coordinates. The center of mass coordinate $\vec{R}$ was defined above; the relative coordinates $\vec{r}'$ are defined as follows:

$$\vec{r}_i = \vec{R} + \vec{r}_i', \tag{1.37}$$

and the relative velocities $\vec{v}'$ are given then by

$$\vec{v}_i = \vec{V} + \vec{v}_i'. \tag{1.38}$$

This decomposition is very useful for it allows us to separate the kinetic energy into center of mass and relative contributions:

$$T = \sum_i m_i (\vec{V} + \vec{v}_i')^2 / 2 = M\vec{V}^2 / 2 + \sum_i m_i \vec{v}_i'^2 / 2, \tag{1.39}$$

which follows from the fact that

$$\sum_{i=1}^N m_i \vec{v}_i' = 0, \tag{1.40}$$

a consequence of the definition of the center of mass.

Next, we restrict to conservative external forces

$$\vec{F}_i^{\text{ext}} = - \nabla_i U_i(\vec{r}_i), \tag{1.41}$$

so that

$$\sum_i \int_1^2 \vec{F}_i^{\text{ext}} \cdot d\vec{r}_i = - \left| \sum_i U_i \right|_1^2. \tag{1.42}$$

In other words, the net work of the external forces equals a potential difference.

We also restrict to conservative internal forces

$$\vec{F}_{ij} = - \nabla_i U_{ij}, \tag{1.43}$$

where U_{ij} is the potential energy of the ith body in the force field due to the jth body. We apply the action–reaction principle in the form where the forces are presumed to act only upon the line of centers passing through the two bodies. In this case, U_{ij} can depend only upon the distance r_{ij} between the two bodies, so that with $\vec{r}_{ij} = \vec{r}_i - \vec{r}_j$,

$$\nabla_i U_{ij} \propto \vec{r}_{ij} \tag{1.44}$$

and the action–reaction principle reads

$$\vec{F}_{ij} = -\nabla_i U_{ij} = \nabla_j U_{ij} = -\vec{F}_{ji} \tag{1.45}$$

since $U_{ij} = U_{ji}$. Finally, we compute the net work done by the internal forces:

$$\sum_{i\neq j}\int_1^2 \vec{F}_{ji}\cdot d\vec{r}_i = -\sum_{i\neq j}\int_1^2 \nabla_i U_{ij}\cdot d\vec{r}_i = -\tfrac{1}{2}\sum_{i\neq j}\int_1^2 (\nabla_i U_{ij}\cdot d\vec{r}_i + \nabla_j U_{ij}\cdot d\vec{r}_j), \tag{1.46}$$

but

$$\nabla_{ij} = \nabla_i = -\nabla_j \tag{1.47}$$

and

$$d\vec{r}_{ij} = d\vec{r}_i = -d\vec{r}_j \tag{1.48}$$

so that

$$\sum_{i\neq j}\int_1^2 \vec{F}_{ji}\cdot d\vec{r}_i = -\tfrac{1}{2}\sum_{i\neq j}\int_1^2 \nabla_{ij} U_{ij}\cdot d\vec{r}_{ij} = -\tfrac{1}{2}\left|\sum_{i\neq j} U_{ij}\right|_1^2 = -\left|\sum_{i<j} U_{ij}\right|_1^2 \tag{1.49}$$

Therefore, if we define the system's total potential energy by

$$U = \sum_i U_i + \sum_{i<j} U_{ij}, \tag{1.50}$$

then the total mechanical energy

$$E = T + U \tag{1.51}$$

is conserved along every trajectory of the dynamical system. This conservation law was first derived and then used by Rudolf Clausius around 1855 as the basis for his statement of the first law of thermodynamics: the total energy of a closed, isolated system does not change with time.

Finally, we return to center of mass and relative coordinates. Beginning with the linear transformations

$$\vec{r}_i = \vec{R} + \vec{r}'_i, \tag{1.52}$$

and consequently

$$\vec{v}_i = \vec{V} + \vec{v}'_i, \tag{1.53}$$

we have earlier shown how to separate the kinetic energy into center of mass and relative contributions:

$$T = \sum_i m_i(\vec{V} + \vec{v}'_i)^2/2 = M\vec{V}^2/2 + \sum_i m_i\vec{v}'^2_i/2. \tag{1.54}$$

This decomposition is useful for application to rigid body problems and we shall use it later. For the case of two bodies, it is more convenient to rewrite the relative kinetic energy

$$T_{\text{rel}} = \sum_i m_i\vec{v}'^2_i/2 \tag{1.55}$$

in the following way. Since, by definition of total momentum, it follows that

$$m_1\vec{v}'_1 + m_2\vec{v}'_2 = 0, \tag{1.56}$$

we can add a null term to T_{rel} to get

$$T_{rel} = \sum_i m_i \vec{v}_i'^2 / 2 - (m_1 \vec{v}_1' + m_2 \vec{v}_2')^2 / M. \qquad (1.57)$$

This result can be rearranged to yield

$$T_{rel} = \mu v^2 / 2, \qquad (1.58)$$

where $\mu = m_1 m_2 / M$ is the 'reduced mass' and $\vec{v} = \vec{v}_1' - \vec{v}_2'$ is the relative velocity of the two bodies. This is the form of the relative kinetic energy that we shall use to analyze central potential problems in chapter 4. We shall also use equation (1.56) to derive another approximation in chapter 14 that allows the analytic formulation of a conceptually simple example of a deterministic mechanical system that becomes chaotic via an infinite sequence of repeated events called period doubling.

Consider the collision of two bodies moving under no net external force. Since the center of mass kinetic energy is conserved, we can concentrate on the relative motion alone. We can then write

$$T_{rel}' = \alpha^2 T_{rel} \qquad (1.59)$$

where T_{rel} and T_{rel}' denote the relative kinetic energy before and after the collision respectively, and α is Newton's 'coefficient of restitution', $0 \le \alpha \le 1$. With $\alpha = 1$ the collision is totally elastic (no kinetic energy is lost) while $\alpha = 0$ denotes a totally inelastic collision (the bodies stick together and move as a unit). By using equations (1.58) and (1.59), we conclude that the magnitude of the relative velocity obeys the relationship

$$v' = \alpha v, \qquad (1.60)$$

which says that the relative velocity is reduced in magnitude in an inelastic collision.

We turn next to a qualitative discussion of the equations of motion that emphasizes the grounding of Newton's first two laws and also the three basic conservation laws in four basic symmetries, three of which describe Euclidean geometry.

1.4 The geometric invariance principles of physics

A body always accelerates toward the earth at the same rate $g \approx 9.8 \text{ m/s}^2$, independently of *where* (near sea level) the experiment is performed and independently of *when* the experiment is performed (independently of the initial time), and independently of the body's mass, so long as air resistance is negligible. This regularity of motion made possible the discovery of universal, machine-like laws of nature. A body in free-fall (including a block sliding on a frictionless inclined plane) is also abstractly machine-like (deterministic), meaning that it behaves according to a precise and unchanging mathematical law. Determinism (combined with uniqueness of solutions) means that precisely the same initial conditions yield precisely the same trajectories. The basis for the universal law of gravity is the regularity of the planetary orbits (expressed by Kepler's first law) combined with Galileo's local laws of inertia and free-fall. The Galilean and Keplerian regularities could not hold if the geometry of space, at least locally, were not approximately Euclidean.

Galileo's law of nature $a_y = -g$ can be rephrased in terms of a conservation law: the energy of a projectile is conserved if air resistance is ignored. The law of inertia, the most fundamental regularity of motion, can also be rewritten via integration as a conservation law: momentum is conserved if no net force acts. Conservation laws and invariance principles are inseparable. A conservation law reflects the constancy in time of some dynamical quantity along a system's trajectory whereas the idea of invariance is represented by a quantity that does not change when a *transformation of coordinates* is performed on the dynamical system at a *fixed* time.

A transformation of coordinates provides an abstract way of describing an operation that may be performed on a physical system. Invariance under coordinate changes follows from symmetry. Symmetry means that an object or a system looks the same before and after some operation or sequence of operations has been performed. A cube looks the same after a ninety-degree rotation about a symmetry axis, for example, whereas a sphere looks the same after any rotation through any axis through its center. Three common examples of operations that can be performed (in reality and/or mentally and mathematically) on physical systems are: rigid rotation of a body or system of bodies about a fixed axis (change of orientation), a rigid translation along an axis (change of coordinate origin), and translation of the system at constant or nonconstant velocity (change to a moving frame of reference). The use of coordinate transformations is a way to represent mathematically the physical idea of performing an operation on a dynamical system, which may be as simple as moving a piece of laboratory equipment. The connection between invariance principles, which follow from the symmetries of a system, and the corresponding conservation laws can be understood by using coordinate transformations.

The assumption that a body can move at any speed in a straight line is equivalent to the assumption of a Cartesian axis, which requires either an infinite Euclidean space (or a Euclidean torus). Empty Euclidean space is homogeneous and isotropic because it is a featureless void. Next in symmetry would follow a curvilinear coordinate system like the polar angles (θ, ϕ) on a perfectly smooth sphere.

The use of Cartesian coordinate systems in Newtonian dynamics presumes that it is possible to extrapolate from the local perspective of our limited experience on earth and within the solar system to the abstraction of an infinitely long straight line that represents the direction along which the force-free motion of a hypothetical body would be allowed to take place in an unbounded universe. That space is at least approximately Euclidean locally, within our solar system, is indicated by the successes of Newtonian theory in explaining both projectile motions and planetary orbits from the same laws of motion and force. That this approximation should be valid throughout the entire universe beyond the neighborhood of our solar system is guaranteed by no known experiments or observations, and is still debated in cosmology circles. From a scientific standpoint, we must discover the clues to Mother Nature's geometry, like her mechanics, empirically. For the time being, we follow Galileo, Descartes, and Newton and ignore all other possibilities than Euclidean geometry.

In matter- and force-field-free space any point along a single Cartesian direction is equivalent to any other point because empty Euclidean space is uniform, homogeneous, the same everywhere along the chosen direction. A Cartesian axis in a Euclidean space defines the idea of translational invariance: nothing exists to allow one point to be distinguished from any other along that axis. Translational invariance provides the simplest illustration of a symmetry principle. The idea of symmetry is that of lack of perceptible difference once an operation has been performed. The concept of symmetry arises from Euclidean abstractions like perfect cubes, perfect spheres, and perfectly straight lines. Granted that we start our theorizing from the unrealistic abstraction of a completely empty universe, hardly any other assumption than the Euclidean one presents itself to us initially. Mentally introducing one body into empty Euclidean space, and assuming that this does not change the geometry of space and time, then that object's position along the direction of observation is labeled by a Cartesian coordinate x, starting from some arbitrarily chosen origin. In the absence of other matter and their corresponding force fields, the choice of coordinate origin cannot play any role at all because empty Euclidean space (no other bodies means no other gravitational fields) is itself completely uniform. Translational invariance of Euclidean space means that the laws of motion are independent of the choice of coordinate origin: the choice of location in empty, field-free space, where an experiment is to be performed makes no difference in the outcome of the experiment. It is assumed that the presence of a second body, the observer, does not break the symmetry of translational invariance (the hypothetical observer's gravitational field is ignored). For experiments performed on earth, translational invariance is broken vertically by the earth's gravitational field. For local experiments where the flat earth approximation $g \approx$ constant is valid we have approximate translational invariance horizontally over distances short enough that the earth's curvature is irrelevant. Translational invariance guarantees that identically prepared and identically performed mechanical experiments will always yield identical numerical results whether performed in Ventimiglia on 23 September 1889, or in Bozen on 5 November 1994, for example.

Mathematical abstractions like infinitely long and perfectly straight lines are good approximations to reality only if the universe is infinite and is assumed to be Euclidean at all distances. If, at large distances from earth, the universe should be finite and spherical, then we would be led, as was Galileo, to the notion that uniform rectilinear motion must eventually somehow give way to uniform circular motion. One can see how Galileo arrived at his idea that uniform circular motion should follow from inertia alone, without the need for a centripetal force. In a Euclidean space this is impossible. On the surface of a sphere, for example, a curved two dimensional Riemannian space, one can construct two Cartesian axes locally and approximately only in the tangent plane to any point on the sphere. By 'local', we mean a small enough but still finite region where curvature and gradients can be ignored. By 'global' we mean any finite (or possibly infinite) region over which curvature and gradients matter. Cartesian coordinates cannot be constructed globally on a sphere although they can always be constructed globally on a cylinder, which can be thought of geometrically as the Euclidean plane rolled onto itself and then pasted together. One cannot, in contrast, make a sphere out of a flat sheet of paper or rubber without cutting, tearing, and stretching the sheet before pasting it together (a piece of an exploded balloon cannot be spread flat on a table because of the intrinsic curvature of the surface).

Galileo observed that we can also perform a mechanical experiment on a ship that sails at constant velocity without changing the results. This observation is the basis for Newton's realization that the first and second laws implicitly imply and require the existence of a certain class of *preferred* Cartesian frames of reference. That Newton's second law and the law of inertia single out preferred frames of reference, called inertial frames, can easily be understood by making a coordinate transformation. Denoting p_i

and F_i as the Cartesian components of the momentum and force vectors, Newton's law of motion has the vectorial component form

$$\frac{dp_i}{dt} = F_i, \tag{1.61}$$

where x_i is the coordinate labeling the ith Cartesian axis, $p_i = mv_i$, and we take m to be constant for convenience, although this assumption is not essential. The transformation to another frame that accelerates linearly at the rate $A = dV/dt$ along axis x_i relative to the initially chosen frame is then given by

$$\left. \begin{array}{l} x_i' = x_i - \displaystyle\int V dt \\[2ex] v_i' = v_i - \displaystyle\int A dt \\[2ex] a_i' = a_i - dV/dt. \end{array} \right\} \tag{1.62}$$

In the new frame, denoted by the primed coordinates, Newton's laws are unchanged along the two Cartesian axes x_k where $k \neq i$, but along the axis x_i of the acceleration A we find that

$$\frac{dp_i'}{dt} + mA = F_i', \tag{1.63}$$

which differs from Newton's law in an inertial frame by the extra 'inertial term' mdV/dt that expresses the acceleration of the primed frame relative to the original one. Whenever $A = dV/dt = 0$, then we retrieve

$$\frac{dp_i'}{dt} = F_i' \tag{1.64}$$

in the primed frame, which tells us that we have assumed *implicitly* that Newton's law (1.61) was written down originally in an *unaccelerated* Cartesian frame. But 'unaccelerated' relative to *what*? The answer, as Newton and Leibnitz (and later Mach and Einstein) could only conclude, is relative to empty Euclidean space, also called 'absolute space'. In other words, *acceleration* relative to *completely empty Euclidean space* produces measurable physical effects in Newtonian theory. What are those effects?

First, the statement of *force-free* motion relative to an accelerated frame is that every body is observed to accelerate in that frame at the same rate $a' = - A$, independently of its mass m, where A is the acceleration of the reference frame relative to absolute space. This means that every force-free body accelerates past the 'primed' observer (an observer whose position is fixed in the primed frame) as it would be if attracted by an invisible massive body whose local gravitational acceleration is given by $g = A = dV/dt$. Second, since the primed observer is at rest, both $v' = 0$ and $a' = 0$ hold that observer in the primed system of coordinates. Therefore, also according to (1.63), a force $F' = MA$ (where M is the observer's mass) must exist in order to hold the observer in place in the accelerated frame: the observer would experience the very same effect if at rest in an unaccelerated frame on a planet whose local gravitational acceleration is given by the magnitude $g = A$. The observer's weight on that hypothetical planet would be given by $w = mA$. Since $g = Gm_{\text{planet}}/R^2$, then $A = dV/dt$ determines the ratio of mass to radius-squared for the hypothetical planet.

Summarizing, linear acceleration relative to absolute space is detectable and is physically indistinguishable from the effect of a local gravitational field. This is the famous 'principle of equivalence'. The important point for us at this stage is: mechanical measurements and predictions that are made in constantly linearly accelerated Cartesian frames are *not* equivalent to measurements and observations made in nonaccelerated Cartesian frames, but instead are equivalent to measurements made in approximate inertial frames where there is a uniform gravitational field along the direction of motion.

To emphasize that the Newtonian idea of force is not responsible for the noninertial effect described above, we show how the force F' in the accelerated frame is derived by a coordinate transformation from the force F in the original frame. If, for example, $F(x,t) = - dU(x,t)/dt$, where U is a potential energy, then it follows that $U'(x',t) = U(x,t)$ is the potential energy in the accelerated frame, where x and x' are connected by the transformation (1.62). Using the definition $F'(x',t) = - dU'(x',t)/dt$, it follows by application of the chain rule for differentiation that $F'(x',t) = - dU(x - \int V dt, t)/dt$, because $dx/dx' = 1$, so that *no extra force is introduced on the right hand side of equation (1.63) as a consequence of the coordinate transformation (1.62)*. In particular, if F is a constant force then $F' = F$ is the same constant force. In other words, the difference between equations (1.61) and (1.63) is due entirely to acceleration relative to absolute space and has nothing to do with any additional force acting on the body. *The noninertial effect is required by the law of inertia.*

To try to be internally consistent (which is possible in Cartan's reformulation of Newtonian theory – see chapter 11) we would have to follow Einstein's observation that gravity, locally, cannot be regarded as a force in the usual sense of the word: free-fall in a constant gravitational field yields all of the necessary properties of an inertial frame in spite of the acceleration toward the gravitating body! This line of thought is deferred until chapter 11.

According to equations (1.61) and (1.63), constant velocity motions (motions where $dv_i/dt = 0$) are transformed into other constant velocity motions (motions where also $dv_i'/dt = 0$) only if $A = 0$, only if the two frames in question are both 'unaccelerated relative to absolute space'. Then, and only then, is it correct to say that the trajectory of a force-free body is always rectilinear at constant speed. In other words, it is implicit in the law of inertia (Newton's first law) that this law holds, as we have stated it in section 1.1, with respect to, and only with respect to, frames that are 'unaccelerated'. Stated mathematically, $dp_i/dt = 0$ describes force-free motion in any nonaccelerated frame. In contrast, $dp_i'/dt = - mdV/dt$ is required to describe force-free motion in a linearly accelerated frame but the term on the right hand side is not a force (this can be proven experimentally by the use of scales: a force-free body in the unprimed frame accelerates in the primed frame, but is weightless there).

The abstract class of frames that are unaccelerated relative to any approximate inertial frame are preferred for the formulation of the basic principles of mechanics simply *because the law of inertia singles out rectilinear motion at constant speed as a special state of motion that is physically indistinguishable from the state of the rest.* In the end as it was in the beginning, it is the statement of the law of inertia in empty Euclidean space that leads to the prediction that acceleration relative to absolute space has precise, measurable consequences, but that constant velocity relative to absolute space has no measurable consequence at all.

Where do we find approximate inertial frames in nature? *A frame that falls freely in the gravitational field of a more massive body is approximately inertial over times that are short enough that the gradient of the gravitational field can be ignored.* Were this not true, then the law of inertia could not have been discovered by Galileo's technique. The earth spins on its axis but can be used as an approximate inertial frame for the discussion of a great many practical problems where that rotation can be ignored. However, if one wants to describe the motion of a spherical pendulum over a time interval on the order of a day or longer, then the earth can no longer be treated approximately as an inertial frame and the spinning about its own axis must be taken into account (see exercise 6 in chapter 8). The earth approximates a local inertial frame for the purpose of carrying out a great many experiments precisely *because* the earth is in free-fall about the sun, and because the effects of the moon and other planets can be treated as weak perturbations to this motion.

That equations (1.61) and (1.64) have the same vectorial form expresses the formal idea of 'covariance' of Newton's law of motion under Galilean transformations. *Galilean transformations are those that transform inertial frames into inertial frames*, and so the transformation (1.62) is an example of a Galilean transformation whenever $A = 0$, which requires that V is both time-independent and position-independent. In other words, Galilean transformations include transformations that connect inertial frames with other inertial frames via constant velocity transformations. These transformations leave the Cartesian vectorial form $dp_i/dt = F_i$ of Newton's law of motion unchanged, yielding $dp_i'/dt = F_i'$ in the primed frame, although the force component F_i' as a function of the transformed position x' is not necessarily the same as the force component F_i as a function of position x. That is, it does *not* necessarily follow that $F_i'(x') = F_i(x)$ in two or more dimensions, although this equality can easily hold in the restricted case of one dimensional motion. The correct transformation law for the force components F_i in Cartesian coordinates is discussed below in this section.

We must pay attention to initial conditions when performing coordinate transformations. The vector equations (1.61) describe the motion of a body as observed from the standpoint of an inertial observer, and equations (1.63) describe the motion of a body from the standpoint of an observer in an arbitrarily linearly accelerated frame defined by the transformations (1.62). If $A = 0$ then that frame is inertial, otherwise it is noninertial (unless $A = g!$). Here is the main point: if we should also connect the initial conditions in the two different frames by using the transformation (1.62), then the corresponding solutions of equations (1.61) and (1.64) describe the motion of one and the same body from the standpoint of two separate observers in two different inertial reference frames. However, if we should want to describe the preparation and performance of identical experiments in two different inertial frames, meaning that the primed observer and the unprimed one should use two separate but identical bodies or machines prepared under identical conditions, then we cannot transform initial conditions according to (1.62) with $A = 0$. Instead, we must set the initial conditions (x_{i_o}, v_{i_o}) and (x_{i_o}', v_{i_o}') equal to *the same numerical values*. In the latter case, one then imagines two different laboratories with identical apparati and identical preparation, but in two separate frames in relative motion with each other (for example, one laboratory is at rest on the side of the road while the second is on a train in locomotion).

What follows is a discussion of Newtonian relativity based upon Galilean transformations. We proceed via *Gedankeneksperimenten*, or 'thought experiments'. The examples discussed above are also examples of *gedanken* experiments.

Following Einstein, who taught us the importance of *gedanken* experiments for understanding relativity principles, let us assume that you are enclosed in a box so that you cannot look at the external surroundings to decide whether or not you're in motion. Suppose, in addition, that you have available all the necessary equipment to perform certain purely mechanical experiments. For example, you may have available a ball, meter sticks, and a stopwatch in order to perform measurements on the motion of the ball. The point to be understood is this: there is no experiment that you can perform to detect motion at constant velocity through 'absolute space'. In other words, the results of all possible mechanical experiments are completely independent of the choice of reference frame so long as that frame is unaccelerated relative to absolute space; is inertial.

Suppose that your box is in a uniform gravitational field and that it slides frictionlessly along a perfectly flat surface at constant velocity relative to another inertial frame, and that the surface is perpendicular to the gravitational field lines. Your box also constitutes an inertial frame of reference and the frames are connected by the transformation (1.62) with $A = 0$. The *gedanken* experiment consists of tossing a ball vertically, along the local gravitational field lines which are antiparallel to the x_k-axis, which is perpendicular to the x_i-axis. No matter whether the box is at rest relative to the field lines or moves horizontally at constant speed in a straight line, the ball first rises vertically and then falls vertically back into your hand: the results of the experiment are completely independent of the horizontal motion of the box. This is another example of the fact that one cannot devise an experiment, using the ball (or anything else that is mechanical), that will tell us whether or not the box is in rectilinear motion at constant speed relative to empty Euclidean space. Again, the results of identically prepared *purely mechanical* experiments are invariant (meaning completely unchanged) under Galilean transformations, are *identical* no matter which inertial frame we choose as our observation point. Stated mathematically, *for the same numerical choices of initial conditions, the solutions of Newton's law of motion in two different inertial frames are identical* whenever the force is invariant under the transformation (1.62) with $A = 0$, *meaning that the functional form of the force is unchanged by the transformation.*

Here's another *gedanken* experiment: suppose that the box moves vertically upward along the field lines with constant velocity relative to an approximate inertial frame that we can take to be the earth (and where we neglect air resistance). If a ball is tossed upward with velocity v'_{k_o} relative the moving origin, and from a height x'_{k_o} above that origin, then the equations of motion of the ball are $d^2 x'_k/dt^2 = -g$, exactly the same as if the moving frame were at rest relative to the parallel gravitational field lines that thread throughout space. This means that the velocity $v'_k = v'_{k_o} - g\Delta t$ and displacement $\Delta x'_k = v'_k \Delta t - g\Delta t^2/2$ are precisely the same as they would be for an experiment carried out at the same initial velocity in the frame that is fixed on the moon (we assume implicitly the vertical displacements are so small that the variation in g with height cannot be detected in the experiment).

We now introduce two more Cartesian axes labeled by the coordinates y and z (or x_2, and x_3). These axes are equivalent to the x- (or x_1-)axis because empty Euclidean space is isotropic, meaning that it is the same in all directions. In other words, empty Euclidean space is also rotationally symmetric: if you probe geometrically in one direction, then rotate through any angle and 'look' in another direction, Euclidean space appears exactly the same in every geometric sense. This way of thinking is neither automatic nor intuitive: it violates common sense because our local vertical direction, due to gravity, is not equivalent to the local horizontal directions. This way of thinking is, in fact, the product of an abstract extrapolation. Before Leibnitz, no one stated this abstraction explicitly and clearly, that the three separate directions in empty Euclidean space should, in principle, be equivalent to each other.

Isotropy of empty Euclidean space in the absence of bodies other than our 'test' body, means that if you choose any other direction for the motion of the body (x_2 or x_3 rather than x_1, for example) then the linear momentum along the new direction is still conserved. That is because empty Euclidean space is the same in all directions, and this symmetry has a mechanical consequence: the law of conservation of angular momentum. Angular momentum could not be conserved in the absence of external torque if empty space were curved or rotationally bumpy. In other words, rotational invariance is the symmetry principle upon which the conservation of angular momentum relies for its validity: *conservation of angular momentum requires the isotropy of empty Euclidean space.*

We shall prove that assertion formally in chapter 2 using a method based upon coordinate transformations. An elementary example of the connection of rotational symmetry to angular momentum is provided by the problem of the motion of two bodies of masses m_1 and m_2 in a central force. The spherical symmetry of the universal gravitational potential requires that empty Euclidean space is isotropic, is the same in all directions.

We have observed that two fundamental conservation laws of physics require underlying symmetry principles that reflect the absence of any geometric properties in empty Euclidean space. There is also a third basic conservation law, conservation of total mechanical energy. The symmetry in this case is that of 'translations in time', an idea that we can only discuss abstractly. You can translate or rotate a box in space in the laboratory, but cannot translate yourself forward or backward in time. An ideal clock or frictionless simple pendulum is, nevertheless, time-translational invariance: it ticks at exactly the same rate forever. The energy and therefore the amplitude of its oscillations is a constant of the motion, is invariant. A real pendulum is subject to friction. Its amplitude is not invariant but decreases exponentially with time.

The Aristotelian equation of motion $\beta \vec{v} = \vec{F}$ represents the idea that the natural unhindered state of a body is a state of rest. The derivation of two possible analogs of momentum conservation for this equation are assigned in the chapter 18 exercises. Depending upon the law of force, the model can be consistent with both translational and rotational invariance, but there is no energy conservation law (planets cannot fall freely about the sun and have to be pushed instead, in which case β is the friction coefficient of 'the ether').

All five of the basic geometric symmetry principles, translational invariance, rotational invariance, invariance with respect to translations among inertial frames, time translational invariance, and the principle of equivalence are all reflected in Galileo's original description of the trajectory of a body falling freely near the earth's surface: $\Delta y = v_{y_o}\Delta t - g\Delta t^2/2$, $\Delta x = v_{x_o}\Delta t$, $v_y = v_{y_o} - g\Delta t$, and $v_x = v_{x_o}$. As Wigner (1967) has pointed out, Galilean kinetics implicitly exhibit the grounding of Newtonian physics in invariance principles.

1.5 Covariance of vector equations vs invariance of solutions

We continue our discussion of Newtonian relativity by deducing the complete class of coordinate transformations called Galilean, where $dp_i/dt = F_i$ in Cartesian coordinates x_i transforms into $dp'_i/dt = F'_i$ in Cartesian coordinates x'_i. As we already have shown above, the transformation (1.62) among relatively moving frames belongs to this class if and only if the velocity V is a constant, and if and only if one of the frames is inertial, in which case both are inertial.

For single-particle motion, consider any differentiable, single-valued and invertible but otherwise arbitrary transformation of coordinates from one Cartesian frame in Euclidean space to another, $x'_i = g_i(x_1,x_2,x_3,t)$. The inverse transformation is denoted by $x_i = h_i(x'_1,x'_2,x'_3,t)$ and these transformations may be linear or nonlinear. What is the most general transformation that always reproduces the vectorial form $dp_i/dt = F_i$ of Newton's second law for the case where the force is conservative, where $\vec{F} = -\nabla U(\vec{x},t)$?

Because the potential energy is a scalar function U of the coordinates and the time, its transformation rule is simple: $U'(\vec{x}',t) = U(\vec{x},t) = U(\vec{h}(\vec{x}',t),t)$. Therefore, by the chain rule for partial derivatives, the Cartesian components F_i of the force have the transformation rule

$$F'_i = \frac{\partial U'}{\partial x'_i} = \frac{\partial U}{\partial x_k}\frac{\partial x_k}{\partial x'_i} = R_{ik}\frac{\partial U}{\partial x_k} = R_{ik}F_k \tag{1.65a}$$

where we have introduced the summation convention: sum over all repeated indices in any expression. We can summarize this by writing

$$F' = RF \tag{1.65b}$$

where R is a 3 × 3 matrix and

$$F = \begin{pmatrix} F_1 \\ F_2 \\ F_3 \end{pmatrix} \tag{1.66}$$

is a column vector, using the notation of linear algebra (equation (1.65a) uses the notation of tensor analysis).

Previously, we have used Gibbs' notation whereby vectors are denoted by arrows. In the rest of this text (except chapter 4 and several other places where Gibbs' notation is unavoidable), we will use the notation (1.65a) of tensor analysis, and (1.66) of linear algebra, to denote components F_i, of vectors, and vectors F.

This allows the development of classical mechanics in the same language and notation as are commonly used in both relativity and quantum theory.

The main point in what follows is that

$$v'_i = \frac{\partial g_i}{\partial x_k}v_k + \frac{\partial g_i}{\partial t}, \tag{1.67a}$$

or

$$v'_i = R_{ik}v_k + \frac{\partial g_i}{\partial t} \tag{1.67b}$$

does not obey the same transformation rule as $F' = RF$ unless the coordinate transformation is time-independent.

In order that Newton's second law has the same form $dp'/dt = F'$ with $p' = mv'$ in the new frame, it is necessary to restrict to the class of transformations where the acceleration, written in Cartesian coordinates, transforms in exactly the same way as the force, where

$$dv_i'/dt = R_{ik}dv_k/dt. \tag{1.68}$$

This condition determines the most general form of the transformation class that produces an equation of the form $dp'/dt = F'$ from the equation $dp/dt = F$, with $p = mv$ and $p' = mv'$, because the *vector* equation (where p is a column vector)

$$\left(\frac{dp'}{dt} - F'\right) = R\left(\frac{dp}{dt} - F\right) = 0, \tag{1.69}$$

is then guaranteed to hold independently of the coordinate system that we choose, *so long as we start with one that belongs to the class of transformations that we shall discover below.*

An example where equation (1.69) is violated is given by the transformation from an inertial to a linearly accelerated frame. In that case

$$\frac{dp_i'}{dt} + mA = F_i' \tag{1.70}$$

and the extra term on the left hand side represents the acceleration of the noninertial frame relative to the unprimed frame, which is taken to be inertial.

Since $x_i' = g_i(x_1,x_2,x_3,t)$, it follows directly by differentiation that

$$v_i' = \frac{\partial x_i'}{\partial x_k}v_k + \frac{\partial g_i}{\partial t} = v_i' = R_{ik}v_k + \frac{\partial g_i}{\partial t}. \tag{1.71}$$

Calculating the acceleration relative to the primed frame then yields

$$\dot{v}_i' = R_{ik}\dot{v}_k + \frac{\partial^2 g_i}{\partial x_l \partial x_k}v_l v_k + 2\frac{\partial^2 g_i}{\partial t \partial x_k}v_k + \frac{\partial^2 g_i}{\partial t^2}. \tag{1.72}$$

To achieve our goal, the three second derivatives on the right hand side of (1.72) must vanish in order that both the acceleration and the force transform according to the same matrix-multiplication rule. This restricts the class of transformations to transformations of the form $g_i(x_1,x_2,x_3,t) = b_i(x_1,x_2,x_3) + a_i t$ where a_i is constant and where b_i is linear in the three position variables x_i. Therefore, $R_{ik} = \partial b_i/\partial x_k$ is a constant matrix. We then obtain the requirement that

$$m\dot{v}_l' = R_{li}T_{ik}F_k', \tag{1.73}$$

where

$$T_{ik} = \frac{\partial x_i}{\partial x_k'}, \tag{1.74}$$

should be the same equation as

$$m\dot{v}_l' = F_l'. \tag{1.75}$$

Pausing, we note that only in the special case where the components $F_i'(x')$ as a function of the transformed position variables x_i' are identical with the components $F_i(x)$ as a function of x_i will the resulting second order differential equations for x_i and x_i' be identical. In general, $F_i'(x') = R_{ik}F_k(x)$ does not reduce to the simple condition that $F_i'(x_i') = F_i(x_i')$. This means that the force law generally changes its mathematical form whenever we change coordinates, so the two sets of second order differential equations for $x_i(t)$ and $x_i'(t)$ describing the motion in the two separate frames will generally differ from each other, and so therefore will their solutions.

Continuing, from the chain rule we obtain

$$R_{ij}T_{ik} = \frac{\partial x_i'}{\partial x_j} \frac{\partial x_k}{\partial x_i'} = \frac{\partial x_k}{\partial x_j} = \delta_{kj}, \tag{1.76}$$

where δ_{kj} (not Δ) is the Kronecker delta, which means that the transpose of R is the inverse of T; $\tilde{R} = T^{-1}$. Therefore, (1.75) follows automatically from (1.72). Taking into account that $\tilde{R} = T^{-1}$ then yields $\tilde{R} = R^{-1}$, which says that the matrix R is an orthogonal matrix.

Using the rule $\det(R\tilde{R}) = (\det R)^2 = \det I = 1$, where $\det R$ denotes the determinant of the matrix R, and I is the 3×3 identity matrix, we obtain that $\det R = 1$ or -1. The most general form of an orthogonal matrix is a rotation (where $\det R = 1$) or a rotation combined with an inversion (where $\det R = -1$ due, for example, to x_1 being replaced by $-x_1$). Restricting for now to motion in the Euclidean plane and to $\det R = 1$, we obtain from the orthogonality condition $\tilde{R}R = I$ that

$$\begin{pmatrix} a & b \\ c & d \end{pmatrix}\begin{pmatrix} a & c \\ b & d \end{pmatrix} = \begin{pmatrix} a^2 + b^2 & ac + bd \\ ac + bd & c^2 + d^2 \end{pmatrix} = \begin{pmatrix} 1 & 0 \\ 0 & 1 \end{pmatrix}, \tag{1.77}$$

which is solved by writing $a = \cos\theta = d, b = -\sin\theta = -c$, where θ is a rotation angle with the x_3-axis and generally can depend upon the position (x_1,x_2) in the plane. However, by considering the transformation of Newton's laws for two or more particles (see section 2.6) one finds that θ must be constant, in which case R represents a rigid rotation. In a rigid rotation, straight lines are transformed into straight lines, and circles are transformed into circles. A perceptive mathematician once wrote that 'geometry is the art of reasoning from badly drawn figures.' Every mechanics problem should begin with a badly drawn figure.

In two dimensions, the most general transformation that leaves Newton's law of motion unchanged in form is therefore the 'Galilean transformation'

$$\left.\begin{array}{l} x_1' = x_1 \cos\theta - x_2 \sin\theta + \delta x_{1_\circ} + u_1 t \\ x_2' = x_2 \sin\theta + x_2 \cos\theta + \delta x_{2_\circ} + u_2 t, \end{array}\right\} \tag{1.78a}$$

which is a rigid rotation of coordinate axes through an angle θ counterclockwise about the x_3-axis, combined with a rigid translation of the coordinate origin by an amount $(\delta x_{1_\circ}, \delta x_{2_\circ})$, and also a transformation to an inertial frame that moves with constant velocity (u_1, u_2) relative to the original inertial frame (note that (1.62) with $A = 0$ provides a special case of this transformation). In contrast, the most general linear transformation of the form

$$\left.\begin{array}{l} x_1' = ax_1 + bx_2 + \xi_1 \\ x_2' = cx_1 + dx_2 + \xi_2 \end{array}\right\} \tag{1.78b}$$

with nonvanishing determinant, $ad - bc \neq 0$, is called an affine transformation.

Having assumed that the force is conservative, let us now show that this is unnecessary. Assume that $x' = Rx + b + ut$ is the vector equation of a Galilean transformation, meaning that the matrix R and the column vectors b and u are constants. Then with $dp/dt - F(x,v) = 0$ and $dp'/dt = R dp/dt$ it follows that $0 = R(dp/dt - F) = dp'dt - RF$ so that we can *always* define $F' = RF$ as the force in the new Galilean frame. We must then substitute $x_i = R_{ki} x'_k - b_k - u_k t$ and $v_i = R_{ki} v'_k - u_k t$ into $F(x,v,t)$ in order to get F' as a function of the primed variables.

Formally stated, Newton's second law is covariant with respect to Galilean transformations: equations (1.78) define the only transformations where, starting from a Cartesian inertial frame, the simple Cartesian vector form $dp'_i/dt = F'_i$ of Newton's law is left intact by the transformation. In any other frame Newton's law does not have the form $dp'_i/dt = F'_i$: extra terms appear on the left hand side unless $dp'/dt - F' = R(dp/dt - F)$.

In curvilinear coordinates, even in an inertial frame, Newton's laws no longer have the simple component form $dp_i/dt = F_i$ where $p_i = mv_i$ because of extra acceleration terms on the left hand side of the second law. We know that we can write down $dp_i/dt = F_i$ in Cartesian coordinates in an inertial frame and then transform to noninertial and/or curvilinear coordinate systems, but is there a way *directly* to write down formal equations of motion that are correct in an *arbitrary* frame of reference, and that reduce to the simple form $dp_i/dt = F_i$ with $p_i = m_i v_i$ in a Cartesian inertial frame, but *without performing the coordinate transformation explicitly for each separate problem*? In other words, is it possible to rewrite the physical content of Newton's laws in a more abstract way so that the new equations describing classical dynamics are covariant vector equations with respect to all differentiable coordinate transformations, including transformations to curvilinear coordinate systems and/or noninertial frames? The affirmative answer to this question is provided by the work of the French mathematician Lagrange (1736–1813).

When vector equations of motion transform in covariant fashion, then the vectorial form will look formally the same in different frames. However, the second order differential equations that are obtained by writing out the components of the vectors explicitly in terms of derivatives of the position variables will generally be different in different coordinate systems. As an example, consider the covariance of Newton's law of motion under Galilean transformations: the equation $dp_i/dt = F_i$ holds in all inertial frames (this is the formal vector equation) but, because the force components $F_i(x,t)$ and $F'_i(x',t)$ are not the same functions of the x_i's and x'_i's in the two inertial coordinate frames, $F'_i(x') \neq F_i(x')$, the second order differential equations of dynamics will be *different* in different inertial frames. This leads us to the difference between *invariance* and *covariance*. To make the distinction clear we must first define what we mean by invariance of functions and equations or motion under coordinate transformations.

Whenever the coordinate transformation is Galilean then the column vector equation $dp/dt = F$ with column vector $p = mv$ transforms into the column vector equation $dp'/dt = F'$ with column vector $p' = mv'$, with the forces related by the vector transformation rule $F' = RF$. All that is required for the force to transform this way is that it transforms like the gradient of a scalar potential energy. If the potential U is merely a scalar function then the transformation rule for the potential energy is $U(x_1,x_2,t) = U'(x'_1,x'_2,t)$. A simple example is provided by the case where the potential U is not rotationally symmetric, where, for example, $U(x_1,x_2) = x_1^2 x_2$. Clearly, $U'(x'_1,x'_2,t) \neq x_1'^2 x'_2$ differs from the function U (the variables x_i and x'_i are connected by the transformation (1.78a) with both δx_{i_o} and u_i set equal to zero). Because U and U' are different functions of x and x' it follows that the second order differential equations for x_i and x'_i in the two different frames are different in spite of the *formal similarity* of the abstract vector equations $dp/dt = F(x,t)$ and $dp'/dt = F'(x',t)$. Here we have covariance without invariance. The lack of invariance follows from the fact that $F'_i(x') = R_{ik}F_k(x) = R_{ik}F_k(h(x')) \neq F_i(x')$, because $U'(x') \neq U(x')$.

The idea of *invariance* of functions, of equations of motion and their solutions, is a much stricter requirement than covariance, which is only a formal vectorial idea of similarity. In restricted cases, for example whenever the potential U is rotationally symmetric, we shall obtain not merely the formal similarity of the vector equations $dp_i/dt = F_i(x,t)$ and $dp'_i/dt = F'_i(x',t)$ but also the identity of the resulting differential equations for x_i and x'_i, and therefore the identity of their solutions as well. An example of invariance of the potential energy and correspondingly of the equations of motion (and therefore of their solutions, for the same numerical values of the initial conditions) is given by the case where $U(x_1,x_2) = x_1^2 + x_2^2$, so that $U(x_1,x_2) = x_1^2 + x_2^2 = x_1'^2 + x_2'^2 = U(x_1',x_2')$. In other words, the transformed and original potential energies U' and U are exactly the *same function* because the function U is rotationally symmetric, which simply means that the potential energy does not change its *functional form* whenever the coordinate system is rotated. Every invariant is a scalar function, but not every scalar function is invariant under some restricted class of transformations (we forewarn the reader that many writers on relativity theory have written the word 'invariant' when they should have written 'scalar'). The kinetic energy is also rotationally invariant because the scalar product of the velocity with itself is invariant under rigid rotations, so that the *resulting differential equations and their solutions are identical* for identical initial conditions in both Cartesian inertial frames.

The force-free equation of motion $dp_i/dt = 0$ is not merely covariant but is *invariant* under Galilean transformations. The invariance of the law of inertia $dp_i/dt = 0$ under Galilean transformations defines the Newtonian principle of relativity.

The rigid rotations and rigid translations included in (1.78a) are exactly the transformations that leave the basic defining figures of Euclidean geometry invariant: straight lines and circles are transformed into straight lines and circles under rigid translations and rotations, and so the shape of every solid body is also left invariant by these transformations (all scalar products of vectors, and therefore all angles and lengths, are preserved by rigid rotations and rigid translations). Rigid translations and rigid rotations are the transformations, according to F. Klein, that describe the basic symmetries of (force- and matter-free) Euclidean space because they preserve the shapes and sizes of all solid geometric figures.

The deep connection between symmetries and conservation laws was first pointed out in mechanics by G. Hamel and F. Engel based upon the methods of S. Lie, and was also emphasized by F. Klein and his school. It was Einstein who made physicists extensively aware of relativity and invariance principles.

Exercises

1. Show that if the earth were attached to the sun by a steel cable rather than attracted by gravity, then the diameter of the cable would be greater than the diameter of the earth (approximate the earth's orbit by a circle).   (Koestler, 1959)

2. Inscribe N chords of equal length within a circle of radius r. A body of mass m

travels at constant speed v along the chords and is perfectly reflected in each collision with the circle (there is no momentum change tangent to the circle).

(a) Show that the radial momentum change per collision is $\Delta p = mv \sin \theta$, where θ is the angle between tangent and chord.

(b) Show that the net centripetal force obeys $F = 2\pi mv/\tau = mv^2/r$ as $N \to \infty$.

(c) Show that if $\tau \sim r^{3/2}$ (Kepler's third law) then $F = -k/r^2$ with $k > 0$.

(Newton)

3. (a) A cylindrical station in interstellar space spins on its symmetry axis with constant angular speed ω. Explain why an astronaut of mass m standing inside the cylinder has weight $w = mr\omega^2$, where r is the cylinder's radius ('weight' is whatever a scale reads when you stand on it). If the force of static friction suddenly disappears, does the astronaut slide or continue rotating with the cylinder?

    (b) Correspondingly, calculate the percentage by which your weight mg is reduced by the earth's rotation while standing at the equator.

4. The escape velocity of a particle is the minimum initial radial velocity required at the surface of the earth in order that a body can escape from the earth's gravitational field. Neglecting the resistance of the atmosphere and the earth's rotation, use energy conservation to find the escape velocity.

5. Consider the vertical motion of a particle moving under the influence of gravity and subject to air resistance. At low enough velocities, the force of fluid friction is linear in the velocity v of the particle. Write down Newton's law of motion for the particle and show that the maximum possible velocity for a fall from rest is $v_{max} = mg/\beta$, where $f = -\beta v$ is the force due to air resistance. (Do not solve the differential equation; use basic principles of physics directly.)

6. Show, in cylindrical coordinates (r, θ), that

$$\frac{dp_r}{dt} - mr\dot{\theta}^2 = F_r,$$

where $p_r = m\dot{r}$, but that

$$\frac{dp_\theta}{dt} = F_\theta$$

where $p_\theta = mr^2\dot{\theta}$. Note in particular that $dp_r/dt \neq F_r$.

7. Prove that an orbit that is a conic section (the force-center lies at one focus),

$$\frac{1}{r} = C(1 + \varepsilon \cos \theta),$$

requires the inverse square law force

$$f(r) = -\frac{CL_z^2}{mr^2} = -\frac{k}{r^2}, \; k > 0.$$

8. Explain why several of the trajectories shown in the figure on page 513 of Koestler, *The Sleepwalkers*, are dynamically impossible. Sketch the correct trajectories, ignoring atmospheric drag.

9. Show that, for a single particle with constant mass, Newton's equation of motion implies the following differential equation for the mechanical energy $E = T + U$:

$$\frac{dE}{dt} = \vec{v} \cdot \left(\frac{d\vec{p}}{dt} - \vec{F}_c\right) = -\vec{v} \cdot \vec{f},$$

where $\vec{F}_c = -\nabla U$ and $\vec{f}$ is the net nonconservative force. Give an example of a nonconservative force that does not change the total energy E. (Hint: refer to one of the exercises of this chapter.)

10. A particle of mass m moving in one dimension is subject to the force

$$F = -kx + \frac{a}{x^3}$$

with $k > 0$ and $a > 0$. Find the potential $U(x)$ and sketch its form.
    Determine the turning points as a function of the energy E and the force constants $k > 0$ and $a > 0$.

11. Show that for an elliptic orbit with the force-center at one focus:
    (a) the eccentricity is given by

$$\varepsilon = \left(1 + \frac{2EL_z^2}{mk^2}\right)^{1/2}$$

    where $b = (a/mk)^{1/2}L_z$, and
    (b) the orbit's period is given by Kepler's third law,

$$\tau = 2\pi(m/k)^{1/2}a^{3/2}.$$

12. Show that for orbits that are conic sections (the force-center lies at one focus) the orbit is elliptic if $E < 0$, parabolic if $E = 0$, and hyperbolic if $E > 0$. What is the condition for a circular orbit?

13. Show that if the orbit is an ellipse with the force-center at the ellipse's center, then $f(r) = -kr$ with $k > 0$. (Newton)

14. Show that the orbit equations (1.5b) can be written as a single nonlinear differential equation

$$\frac{1}{f^2}\frac{d}{d\theta}\left(\frac{1}{f^2}f'(\theta)\right) - \frac{L_z^2}{f^3} = mF_r$$

for the orbit equation $r = f(\theta)$. Show that the substitution $u = 1/r$ yields

$$u''(\theta) + u(\theta) = -\frac{m}{L_z^2}\frac{F_r}{u^2}.$$

Note, therefore, that the proof of uniqueness of Newton's solution of the Kepler problem ($F_r = -ku$) is equivalent to the trivial problem of proving unique solutions for linear equations.

15. A shell of mass M is fired with an initial velocity whose horizontal and vertical components are v_{x_o} and v_{y_o}. At the highest point on the trajectory, the shell explodes

into two fragments with masses m_1 and m_2. Denoting by ΔT the kinetic energy that is gained in this 'reverse inelastic collision', show that the two fragments are separated by a horizontal distance

$$\Delta x = \frac{v_{yo}}{g}\left[2\Delta T\left(\frac{1}{m_1} + \frac{1}{m_2}\right)\right]^{\frac{1}{2}}$$

when they hit the ground.   (Saletan and Cromer, 1971)

16. A spaceship decelerates vertically at rate $a = -|a|$ on its approach to the moon. Calculate the weight of an astronaut during the deceleration.

17. Show explicitly that, under the transformation (1.78a) with $u_i = 0$, straight lines are transformed into straight lines, and circles are transformed into circles.

2

Lagrange's and Hamilton's equations

2.1 Overview

Lagrange's equations can be derived from Newton's laws. We shall show that both systems of equations follow from an extremum principle. The use of extremum principles is not confined to classical mechanics or to physics, but in physics both quantum and general relativistic field equations also follow from extremum principles. In mechanics the extremum principle of interest is called the action principle, and its integrand is a function that is called the Lagrangian.

The equations of Lagrange (French, 1736–1813) are more general than Newton's law of motion in two ways. First, the formal Euler–Lagrange equations can be derived mathematically from variational principles independently of mechanics, but in this case the integrand of the action is not the Lagrangian function of classical mechanics (the Swiss mathematician Euler lived from 1707 to 1783). Second, and within the confines of mechanics, both the Lagrangian and the action survived the quantum revolution. This was shown by Dirac, whose observation formed the basis for Feynman's path integral formulation of quantum mechanics. In the Feynman–Dirac formulation of quantum theory the action plays the central role, but a single classical mechanical trajectory dominates only in the semi-classical limit of an integrable canonical system. Whenever quantum effects are important then a complete set of nonclassical, generally nondifferentiable, trajectories must be used to describe the physics.

There is a more important sense in which the Lagrangian and Hamiltonian formulations of mechanics are *not* more general than Newton's laws: within the confines of classical mechanics the Lagrangian and Hamiltonian formulations are merely a rewriting of Newton's laws that in no way change the physical content of Newtonian theory. They allow us to free ourselves from the necessity to *start* the formulation of a problem by using Cartesian coordinates and an inertial reference frame where one can only afterward transform to other, more convenient starting points.

Newton's laws, as stated in section 1.2, are restricted to inertial frames where one typically begins with a statement of Newton's laws in Cartesian coordinates. The vector equation of motion

$$\frac{d\vec{p}}{dt} = \vec{F} \tag{2.1a}$$

presumes only an inertial frame of reference, but the component form

$$\frac{dp_i}{dt} = F_i \tag{2.1b}$$

with $p_i = mv_i$ is *only* true in a Cartesian coordinate system in an inertial frame: in other coordinate systems (elliptic, hyperbolic, etc.), there are extra terms on the left hand side. For example, for planar problems in cylindrical coordinates (2.1b) must be replaced by

$$\left.\begin{array}{c} m\left(\dfrac{d^2 r}{dt^2} - r\dot{\phi}^2\right) = F_r, \\[3ex] m\dfrac{d}{dt}(r^2\dot{\phi}) = F_\phi \end{array}\right\} \tag{2.1c}$$

where $p_r = m dr/dt$ is the radial component of linear momentum.

Because inertial frames and Cartesian coordinates are neither the most general nor the most convenient starting points for the analysis and solution of most problems in physics, we would like a more general formulation of the physical content of Newton's laws than that represented by (2.1a) and (2.1b). Some problems are best studied in spherical coordinates (or elliptic or some other system of curvilinear coordinates), and some problems are most conveniently formulated and analyzed in rotating or linearly accelerated reference frames. Lagrange's formulation of mechanics provides us with the needed generality through a rewriting of Newton's law of motion that is directly applicable in *arbitrary* coordinate systems and arbitrarily accelerated reference frames. The Lagrangian formulation is covariant with respect to arbitrary differentiable coordinate transformations in configuration space, the $3N$ dimensional Euclidean space of the positions of the N particles (see chapter 10 for the explicit proof of covariance). This is a distinct advantage over the formulation represented by (2.1a) and (2.1b). Also, Lagrange's equations are useful for expressing the deep and important connection between symmetry, invariance, and conservation laws. Noether's theorem connects conservation laws to symmetry through the invariance of the Lagrangian under a very restricted class of coordinate transformations.

At the end of this chapter, we also introduce Hamilton's canonical formulation of dynamics in phase space, the best approach to the study of conservative Newtonian systems because it is the formulation that immediately permits a direct *qualitative* analysis of the motion (chapter 3) without knowing any quantitative details of possible solutions.

Because of the deep connection between Hamilton's canonical form of the equations of motion and the representation of continuous symmetries of physical systems, an abstract generalization of canonically conjugate variables and Hamilton's equations also survived the quantum revolution, providing us directly with Heisenberg's formulation of quantum mechanics, the formulation whose roots grew directly out of the empirical attempt to analyze and understand transitions between atomic energy levels in terms of an ad hoc generalization of the old Bohr–Sommerfeld quantization rules.

2.2 Extrema of functionals

Every reader is familiar with the idea of a 'point function', or just 'function' of one or more coordinates. The usual notation is to write $y = f(x)$ in one dimension or $z = g(x,y)$ in two dimensions, e.g. for $y = f(x)$, a function f is just a rule (a 'mapping') that assigns a number y, given a number x. Of immediate interest is that there is an extension of this idea of 'mapping' to the case where one considers not merely a function evaluated at a point but a 'functional' that depends on a path. The useful aspect of ordinary functions $y = f(x)$ is, in fact, that they are completely independent of the path used to arrive at a given point x: if x is a holonomic variable and f is an ordinary function (meaning that df is an exact differential) then y is also a holonomic variable. A nonholonomic variable is one that depends on the path.

To illustrate the idea of a functional, consider a collection of curves with the same end points. An example of a functional is a rule that assigns a number to each curve, that is, to each function between the prescribed limits. An example of a functional that belongs to this category is therefore given by the integral

$$I[f] = \int_{x_1}^{x_2} f(x)dx, \tag{2.2a}$$

which is linear in the function f and has a numerical value that is independent of the details of the shape of the curve $y = f(x)$, whenever $dF(x) = f(x)dx$ is an exact differential, and depends only on the function F at the two end points. Nontrivially different functions f_1 and f_2 yield different numbers $I[f_1]$ and $I[f_2]$ only if $F_k(x_2) - F_k(x_1)$ varies with k. For a formal and more general definition of functionals based upon the idea of an abstract vector space of functions, we refer the reader to the excellent texts by Liusternik and Sobolev (1961) and by Dubrovin, Fomenko, and Novikov (1984).

In order to pave the way for Lagrangian mechanics, we must consider functionals that are nonlinear in both the integration path and its slope between two specific limits:

$$I = \int_{x_1}^{x_2} \Phi(y,y',x)dx \tag{2.2b}$$

where the curve is given by a function $y = f(x)$, and $y' = df/dx$ is the slope at the same point x. In general, the integrand $\Phi(y,y',x)dx$ is not an exact differential, and so the functional I is path-dependent. We shall generally write I in what follows to stand for the functional $I[\Phi]$. The generalization to include derivatives of higher order than the first, y'',y''', and so on, is easy once one understands the basic method of treatment.

The relation of functionals to mechanics arises through the following question: given a functional of the form (2.2a), for which paths defined by $y = f(x)$ in some general class of functions is the functional an extremum? That is, which paths yield maxima, minima, or saddle-points for the evaluation of the integral I? This question leads to an old subject called the calculus of variations, and we start on that path by reviewing briefly the related subject of locating extrema of functions of several variables.

In order to find the extrema of a function $f(x_1,\ldots,x_m)$ of m variables $(x_1,\ldots,x_m)$, one must impose the conditions for a stationary point:

$$\frac{\partial f}{\partial x_i} = 0 \tag{2.3}$$

for $i = 1$ to m. If one rewrites the integral I above as Riemann sum, then the problem of finding the extrema of that sum is like the problem of locating extrema of a function of a definite number m of variables x_i, except that the number of variables m must be permitted to approach infinity. It is the limit of infinitely many variables that leads us to the idea of a 'functional derivative', the derivative of a functional with respect to the functions that constitute its domain of definition. With this motivation in mind, we can arrive at the same idea formally and more directly without using the Riemann sum explicitly.

Consider the space of all admissible curves $y = f(x)$, all with the same end points x_1 and x_2. This requires an abstract space of functions f that can be differentiated at least once, and for which the integral I is finite. If we formally write

$$\delta I = \int_{x_1}^{x_2} \left(\frac{\partial \Phi}{\partial y} \delta y + \frac{\partial \Phi}{\partial y'} \delta y' \right) dx \qquad (2.4a)$$

then the meaning of δy is that it is the difference, at fixed x, between two different nearby *functions* $y = f_1(x)$ and $y = f_2(x)$. That is, $f_2(x) - f_1(x) = \delta f(x) = \delta y$ for each fixed value of x. The quantity δy is called the 'virtual variation' and represents the change from one function, f_1, to another nearby one f_2 at fixed x. The idea is shown graphically in figure (2.1). By $\delta y'$ we mean simply the difference in the slopes of the two nearby and nearly equal functions f_1 and f_2 at the same point x. Since the virtual variation δy is taken at fixed x, we can exchange the orders of functional variation δ and derivative d/dx:

$$\delta y' = \delta \frac{dy}{dx} = \frac{d}{dx} \delta y. \qquad (2.5)$$

In other words, the derivative with respect to x and the operation of making a virtual change in the function space commute with each other. One integration by parts then yields

$$\delta I = \left[\frac{\partial \Phi}{\partial y'} \delta y \right]_{x_1}^{x_2} + \int_{x_1}^{x_2} \left(\frac{\partial \Phi}{\partial y} - \frac{d}{dx} \frac{\partial \Phi}{\partial y'} \right) \delta y \, dx. \qquad (2.4b)$$

Because of the restriction to variations with fixed end points, we must require that $\delta y = 0$ at both x_1 and x_2 so that

$$\delta I = \int_{x_1}^{x_2} \left(\frac{\partial \Phi}{\partial y} - \frac{d}{dx} \frac{\partial \Phi}{\partial y'} \right) \delta y \, dx. \qquad (2.4c)$$

Next, we identify the 'functional derivative' of the functional I in analogy with the expression

$$\delta f = \sum_{i=1}^{m} \frac{\partial f}{\partial x_i} \delta x_i \qquad (2.6a)$$

(thinking again of the integral as a Riemann sum) as the expression

$$\delta I = \int_{x_1}^{x_2} \frac{\delta I}{\delta y(x)} \delta y(x) \, dx, \qquad (2.6b)$$

which yields the functional derivative[1] as

$$\frac{\delta I}{\delta y(x)} = \frac{\partial \Phi}{\partial y} - \frac{d}{dx} \frac{\partial \Phi}{\partial y'}. \qquad (2.7)$$

Pushing the analogy based upon the Riemann sum to its logical consequence, the condition that a function $y = f(x)$ yields an extremum for the functional I for *completely*

---

[1] It is formally possible to construct the entire functional analog of a Taylor expansion by the introduction of higher order functional derivatives.

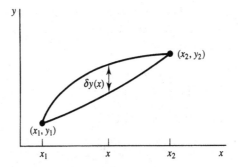

Fig. 2.1 Virtual variations at fixed x.

arbitrary and unconstrained variations $\delta y(x)$ is simply the condition that the first functional derivative of I vanishes at every point x. The consequence is that the following differential equation, called the Euler–Lagrange equation,

$$\frac{\partial \Phi}{\partial y} - \frac{d}{dx}\frac{\partial \Phi}{\partial y'} = 0, \tag{2.8}$$

must be satisfied by any function f that makes the path-dependent functional $I[f]$ an extremum. The harder question whether a particular extremum is a maximum, minimum, or saddle-point is not discussed explicitly here.

To illustrate the usefulness of variational principles in mechanics we show how to formulate the well-known Brachistochrone problem, where a bead slides frictionlessly down a wire in the presence of gravity. Consider all possible shapes of wires with the end points fixed, and find the shape that yields the least time of descent. The descent time is given by the integral

$$t_{12} = \int_1^2 \frac{ds}{v} \tag{2.9a}$$

where $ds^2 = dx^2 + dy^2 = dx^2(1 + f(x)^2)$ is the square of the arc length along the path $y = f(x)$, and $v = \sqrt{2gy}$ follows from energy conservation if we choose the zero of potential energy so that $E = mgy_o$. Therefore, $\Phi(y,y',x) = [(1 + y'^2)/2gy]^{1/2}$, and the differential equation for $y = f(x)$ follows from the Euler–Lagrange equation

$$\frac{\partial \Phi}{\partial y} - \frac{d}{dx}\frac{\partial \Phi}{\partial y'} = 0. \tag{2.9b}$$

Next, we show how to use a variational principle to formulate and find the shortest distance between two points on the surface of a sphere. The arc-length ds on a sphere of radius r is given by $ds^2 = (rd\theta)^2 + (r\sin\theta d\phi)^2 = (rd\theta)^2[1 + (f'(\theta)\sin\theta)^2]$, where we have written $\phi = f(\theta)$. With no loss of generality, we can take $r = 1$. The arc-length for an arbitrary path between points 1 and 2 on the sphere therefore has the form

$$s = \int_1^2 d\theta[1 + (f'(\theta)\sin\theta)^2]^{1/2} \tag{2.10a}$$

and depends upon the path. The Euler–Lagrange equation yields the differential equations that describe a geodesic, or great circle, on the spherical surface. Since Φ is independent of $f(\theta)$, we obtain easily that

$$\frac{f'(\theta)\sin^2\theta}{[1 + (f'(\theta)\sin\theta)^2]^{1/2}} = c, \qquad (2.10b)$$

where c is a constant. It follows easily that

$$f'(\theta) = \frac{1/\sin\theta}{\left[\dfrac{1}{c^2}\sin^2\theta - 1\right]^{1/2}}, \qquad (2.10c)$$

whose integration yields the equation of a geodesic on the unit sphere.

2.3 Newton's equations from the action principle

We begin by showing how Newton's equations of motion can be obtained from a variational principle. For discussions of the history of variational principles in mechanics and geometry we can recommend the texts by Bliss (1962), Lanczos (1949), Hamel (1967), and Sommerfeld (1947–52). We begin with a particular functional that is known in mechanics as *the action*,

$$A = \int_{t_1}^{t_2} (T - U)dt, \qquad (2.11)$$

where T and U are the kinetic and potential energies of N particles with masses m_i. For virtual variations of the particles' positions, and with both end points fixed, in an inertial frame (the spatial variations in particle positions are performed at fixed time t) we obtain

$$\delta A = -\sum_{i=1}^{N}\int_{t_1}^{t_2}(\vec{p}_i + \nabla_i U)\cdot\delta\vec{r}_i dt. \qquad (2.12)$$

The extremum condition $\delta A = 0$ is satisfied for arbitrary variations only if

$$\sum_{i=1}^{N}(\dot{\vec{p}}_i + \nabla_i U)\cdot\delta\vec{r}_i = 0, \qquad (2.13a)$$

which is a special case of a result known in the literature as D'Alembert's Principle[2], and where the second term in the equation represents the 'virtual work' done by the conservative forces. The word virtual is used because the variations in particle positions are only hypothetical displacements at fixed time rather than real motions of the mechanical system. The latter can only occur when the time t changes. In particular, the variations are fixed-time displacements that must respect any and all geometric *constraints* imposed upon the system but are otherwise arbitrary.

An example is given by the case where a marble rolls along the surface of a

---

[2] If we set the momentum derivatives equal to zero and replace gradients of potentials by arbitrary nonconstraint forces, then we have a general condition for the equilibrium of a constrained system. According to the brief sketch of history provided by Lanczos (1949), this principle, the principle of virtual work, was popular in medieval times and goes back (in disguised form) to Aristotle (see also Hamel, 1967).

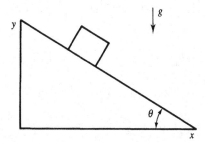

Fig. 2.2 A constrained system.

hemispherical bowl: the motion is constrained to lie on the surface of a sphere, and the equation of the constraint is just that the distance of the marble from the sphere's center is constant. The particle's trajectory is therefore not free to explore all three dimensions but is confined to a two dimensional surface.

If there are no constraints, which means that the Nd virtual displacements $\delta \vec{r}_i$ are all completely arbitrary, then Newton's law $\dot{\vec{p}}_i = -\nabla_i U$ for a conservative system follows from equation (2.13a). The reason for this conclusion is that, with completely arbitrary variations in particle position, the variational equation cannot be satisfied in any way other than term by term. Conversely, whenever there are constraints then equation (2.13a) cannot be satisfied term by term, and we shall present two methods for treating the constrained cases below.

Another elementary example of a constraint is provided by placing a block on an inclined plane: the block is prevented by the constraint force, the normal force N exerted by the plane on the block, from falling vertically downward in the presence of gravity (figure 2.2). Here we use N as the symbol for the normal force, although N is also used to denote the number of bodies in the system. In this case the constraint equation $y = x \tan \theta$ can be used systematically to eliminate one of the two Cartesian coordinates (x,y), so that we have only one independent position coordinate. Note that the force of constraint N is perpendicular to the one dimensional surface defined by the constraint equation $\phi(x,y) = y - x \tan \theta = 0$ (for a large class of constraints the force of constraint points in the direction of the normal to the constraining surface). As we have stated above, if a body is confined to move on the surface of a sphere then we have another example of a constrained system where the equation of constraint is given in spherical coordinates (with origin at the sphere's center) by the formula $r = a$, where r is the radial coordinate of the particle and a is the radius of the sphere. If we write this constraint equation in the form $\phi(r) = r - a = 0$, then again we see that the force of constraint $\vec{f}$ is perpendicular to the surface of the sphere, $\vec{f} \propto \nabla \phi$.

Next we generalize by introducing the class of constraints called holonomic, which means that the constraints are defined globally independently of any path taken by the body. This means that the constraints can be written in the form of point functions $\phi_l(\vec{r}_1, \ldots, \vec{r}_N, t) = 0$, where $l = 1, 2, \ldots, k$, and k is the number of different constraints. In this case the corresponding constraint forces f_i are always perpendicular to constraining surfaces $\phi_l(\vec{r}_1, \ldots, \vec{r}_N, t) = 0$, and therefore can be written in the form

$$\vec{f}_i = \sum_{l=1}^{k} \lambda_l \nabla_i \phi_l \qquad (2.14)$$

where the coefficients λ_l may depend upon the time t. We prove this as follows: since it follows from $\phi_l(\vec{r}_1,\ldots,\vec{r}_N,t) = 0$ that

$$d\phi_k = \sum_{i=1}^{n} \nabla_i \phi_k \cdot d\vec{r}_i + \frac{\partial \phi_k}{\partial t} dt = 0, \tag{2.15}$$

it follows from combining equations (2.14) and (2.15) that the work

$$dW_c = \sum_i \vec{f}_i \cdot d\vec{r}_i = \sum_{l,i} \lambda_l \nabla_i \phi_l \cdot d\vec{r}_i = -\sum_l \lambda_l \frac{\partial \phi_l}{\partial t} dt \tag{2.16}$$

done by the constraints for *real* displacements due to motion of the bodies is nonvanishing whenever the constraints are explicitly time-dependent. In that case, the constraint forces are *not* perpendicular to the real displacements. However, because the virtual displacements are performed at fixed time t, the virtual work vanishes,

$$\delta W_c = \sum_i \vec{f}_i \cdot \delta\vec{r}_i = \sum_{l,i} \lambda_l \nabla_i \phi_l \cdot \delta\vec{r}_i = -\sum_l \lambda_l \frac{\partial \phi_l}{\partial t} \delta t = 0, \tag{2.17}$$

simply because of the condition $\delta t = 0$. It follows that the constraint forces are perpendicular to the virtual displacements so long as the surfaces of constraint do not change with time.

Taking into account that the constraint forces have vanishing virtual work, we can then rewrite the variational equation (2.13a) in the form

$$\sum_{i=1}^{N} (\dot{\vec{p}}_i + \nabla_i U - \vec{f}_i) \cdot \delta\vec{r}_i = 0, \tag{2.13b}$$

and although the virtual variations are not all independent (because they are coupled by k constraints), *the inclusion of the constraint forces allows us to satisfy Newton's law term by term* in (2.13b), yielding Newton's second law in the form

$$\dot{\vec{p}}_i + \nabla_i U - \vec{f}_i = 0. \tag{2.18a}$$

A different way to say this is that one agrees to *choose* the k unknown coefficients λ_l in the definition (2.18a) of the constraint forces $\vec{f}_i$ precisely so that equation (2.13b) is satisfied for each separate index i. Without the explicit inclusion of the constraint forces, the variational equations (2.13b) cannot be satisfied *term by term* unless the system is completely unconstrained.

So far only conservative nonconstraint forces have been included in our discussion of the variational principle, so we must also generalize the method of derivation to permit dissipation. For systems with dissipative forces the variational principle (the action principle, also called Hamilton's principle) must be changed to read:

$$\delta \int_{t_1}^{t_2} T dt = \int_{t_1}^{t_2} \delta W dt \tag{2.19}$$

where

$$\delta W = \sum_i \vec{F}_i \cdot \delta\vec{r}_i, \tag{2.20}$$

which is a fixed-time variation, is the virtual work done by all nonconstraint forces and may include dissipative as well as conservative contributions. This yields the variational equation in the form

$$\sum_{i=1}^{N} (\dot{\vec{p}}_i - \vec{F}_i) \cdot \delta \vec{r}_i = 0. \tag{2.13c}$$

If the virtual displacements are unconstrained, then Newton's law of motion $\dot{\vec{p}}_i = \vec{F}_i$ follows where $\vec{F}_i$ is the net force on the ith body. If there are k time-independent constraints, then the virtual work done to the constraint forces vanishes (because the constraint forces are perpendicular to the virtual displacements) and the variational equation can be rewritten as

$$\sum_{i=1}^{N} (\dot{\vec{p}}_i - \vec{F}_i - \vec{f}_i) \cdot \delta \vec{r}_i = 0 \tag{2.13d}$$

to yield

$$\dot{\vec{p}}_i = \vec{F}_i + \vec{f}_i. \tag{2.18b}$$

where the right hand side represents the net external force plus the net force of constraint on the ith particle. This concludes our proof that Newton's law, in its most general form, follows from the action principle.

2.4 Arbitrary coordinate transformations and Lagrange's equations

Newton's law in the form

$$\frac{d\vec{p}}{dt} = \vec{F} \tag{2.1a}$$

requires that we use an inertial frame of reference, while the component form

$$\frac{dp_i}{dt} = F_i \tag{2.1b}$$

is correct *only* in a Cartesian inertial frame. Cartesian coordinates can be defined 'locally' (in an infinitesimal region near any point) on a nonEuclidean surface like a saddle or a sphere, but can be defined 'globally' (with the same coordinate axes extending without limit to infinity) only in a Euclidean space (meaning a 'flat' or noncurved space). The law of inertia demands a Euclidean space, but nonEuclidean spaces can only arise due to constraints that confine the motion of the system to a curved subspace that is embedded in the dN dimensional Euclidean configuration space of the positions of the N particles, where $d = 1$, 2, or 3.

Consider next a completely general transformation of coordinates from Cartesian to any other holonomic coordinate system, meaning any global coordinate system that can be constructed completely independently of any need to integrate the velocities of the bodies under consideration. Spherical, cylindrical, elliptic, or any other curvilinear set of coordinates will do, because one can define them globally and geometrically independently of the dynamical system under discussion. Denoting the new and arbitrary coordinates by the symbols q_i, we can write the transformation formally as

$$\vec{r}_i = (x_i, y_i, z_i) = \vec{R}_i(q_1, \ldots, q_f, t) = (f_i(q_1, \ldots, q_f, t), g_i(q_1, \ldots, q_f, t), h_i(q_1, \ldots, q_f, t)), \tag{2.21a}$$

when $n = dN$ is the number of Cartesian coordinates and $f = dN - k$ is the number of new coordinates q_i, k constraints having been eliminated explicitly before carrying out the transformation. Equation (2.21a), written in terms of the new set of variables $(q_1, \ldots, q_f)$, is called a 'point transformation' because points in the two different reference frames correspond to each other in one-to-one fashion (there is no path-dependence, no anholonomy, meaning that f_i, g_i, and h_i are ordinary functions rather than functionals). Configuration space may be nonEuclidean only if $k \geq 1$ but is otherwise Euclidean and is simply the space of the f variables q_i.

The coordinates q_i in (2.21a) are traditionally called 'generalized coordinates' for purely historic reasons: they are very general in the sense that any holonomic system of coordinates will do. All Cartesian and curvilinear coordinates in Euclidean space are holonomic. More generally, a holonomic coordinate is any coordinate that follows from integrating an exact differential. If, instead of starting with a 'point transformation' of the form $x_k = f_k(q_1, \ldots, q_f, t)$, we had tried to start with a transformation defined by the generalized velocities, where $\omega_\sigma = \pi_{\sigma i} dq_i / dt$, and had then tried to define a set of generalized coordinates (holonomic coordinates) q'_k as time integrals of the transformed velocities ω_σ, then the transformation matrix $\pi_{\sigma i}$ would first have to satisfy the integrability condition $\partial \pi_{\sigma i} / \partial q_k = \partial \pi_{\sigma k} / \partial q_i$. This integrability condition is non-trivial and is usually *not* satisfied by arbitrary velocity transformations, as we show in chapter 12 in the context of rigid body dynamics for the nonintegrable velocities ω_i that arise naturally in rigid body theory (chapter 9). Parameterizations of the motion that are constructed by integrating nonexact differential forms, meaning nonintegrable velocities ω_σ, are called nonholonomic coordinates, or quasicoordinates and also are discussed in chapter 12.

The main point for what follows is: because the differentiable transformation (2.21a) is completely arbitrary and may even be explicitly time-dependent, we shall arrive at a formulation of Newtonian dynamics that is valid in arbitrary coordinate systems, *even in noninertial frames.*

Because the constraints are here assumed to have been eliminated explicitly, there is no need to include the forces of constraint in our formulation and so we can start with the variational principle in the form

$$\sum_{i=1}^{N} (\dot{\vec{p}}_i - \vec{F}_i) \cdot \delta \vec{r}_i = 0, \tag{2.13c}$$

where F_i includes both conservative and dissipative effects and is the net nonconstraint force on body i. The first aim is to express both terms in the variational equation by equivalent expressions in terms of generalized coordinates. It is easy to transform the force term. Writing

$$\sum_{k=1}^{f} Q_k \delta q_k = \sum_{i=1}^{n} \vec{F}_i \cdot \delta \vec{r}_i, \tag{2.22}$$

we can then identify

$$Q_k = \sum_{i=1}^{n} \vec{F}_i \cdot \frac{\partial \vec{R}_i}{\partial q_k} \tag{2.23}$$

as the generalized force acting on the kth generalized coordinate q_k, yielding

$$\left(\sum_{i=1}^{N} m_i \frac{d^2 \vec{r}_i}{dt^2} \cdot \frac{\partial \vec{r}_i}{\partial q_k} - Q_k \right) \delta q_k = 0, \tag{2.24}$$

where from now on (unless otherwise stated) we use summation convention (sum over any and all repeated indices), meaning, for example, the sum over the repeated index k is implicit in equation (2.24). To go further, we must also transform the acceleration term in (2.24).

Noting that

$$\frac{d\vec{r}_i}{dt} = \frac{\partial \vec{R}_i}{\partial q_k}\dot{q}_k + \frac{\partial \vec{R}_i}{\partial t}, \tag{2.25a}$$

we see also that

$$\frac{d^2\vec{r}_i}{dt^2}\cdot\frac{\partial \vec{r}_i}{\partial q_j} = \frac{d}{dt}\left(\dot{\vec{r}}_i\cdot\frac{\partial \vec{r}_i}{\partial q_j}\right) - \dot{\vec{r}}_i\cdot\frac{d\partial\vec{r}_i}{dt\partial q_j}, \tag{2.26a}$$

and since

$$\frac{d\partial\vec{r}_i}{dt\partial q_j} = \frac{\partial \dot{\vec{r}}_i}{\partial q_j} \tag{2.27}$$

and

$$\dot{\vec{r}}_i = \frac{\partial \vec{r}_i}{\partial q_j}\dot{q}_j + \frac{\partial \vec{r}_i}{\partial t}, \tag{2.25b}$$

if follows that

$$\frac{\partial \dot{\vec{r}}_i}{\partial \dot{q}_k} = \frac{\partial}{\partial \dot{q}_k}\left(\frac{\partial \vec{r}_i}{\partial q_j}\dot{q}_j + \frac{\partial \vec{r}_i}{\partial t}\right) = \frac{\partial \vec{r}_i}{\partial q_k}, \tag{2.28}$$

which is a very useful result. It allows us to see, for example, that

$$m_i\frac{d^2\vec{r}_i}{dt^2}\cdot\frac{\partial \vec{r}_i}{\partial q_j} = \frac{d}{dt}\left(m_i\dot{\vec{r}}_i\cdot\frac{\partial \vec{r}_i}{\partial \dot{q}_j}\right) - m_i\dot{\vec{r}}_i\cdot\frac{d\partial\vec{r}_i}{dt\partial q_j} = \frac{1}{2}\frac{d}{dt}m_i\frac{\partial \dot{\vec{r}}_i^2}{\partial \dot{q}_j} - \frac{1}{2}m_i\frac{\partial \dot{\vec{r}}_i^2}{\partial q_j}, \tag{2.26b}$$

so that we can now combine all of our results to rewrite our variational equation, which was derived in the last section from Hamilton's principle, in the form

$$\left(\frac{d}{dt}\frac{\partial T}{\partial \dot{q}_k} - \frac{\partial T}{\partial q_k} - Q_k\right)\delta q_k = 0. \tag{2.29a}$$

Since all constraints have already been eliminated, the f variations δq_i are completely arbitrary so that each coefficient of δq_k in the variational equation must vanish term by term for $k = 1,\ldots,f$, otherwise there is no way to satisfy (2.29a) for *completely arbitrary* variations. The equations of dynamics then have the form

$$\frac{d\partial T}{dt\partial \dot{q}_k} - \frac{\partial T}{\partial q_k} = Q_k, \tag{2.29}$$

and are called Lagrange's equations of the first kind.

In Cartesian coordinates in an inertial frame, $q_i = x_i$, the kinetic energy of a single particle is $T = m\dot{x}_i\dot{x}_i/2$, and Lagrange's equations (2.29') simply reproduce Newton's second law in the usual form $md^2x_i/dt^2 = F_i$, where $Q_i = F_i$. In general, the differential equations (2.29') are second order in time for the generalized coordinates and hold for dissipative as well as for conservative systems. These equations of motion describe neither more nor less physics than do Newton's laws, because they are merely a transformation of Newton's laws to arbitrary holonomic coordinate systems. But, in contrast with our original formulation, where $dp_i/dt = F_i$, Lagrange's equations are directly applicable in the form (2.29′) in accelerated as well as in inertial frames of reference.

We now give two examples of the transformation (2.21a) to other inertial coordinate systems for a single particle. In cylindrical coordinates, where $q_1 = \rho$, $q_2 = \phi$, and $q_3 = z$, and where the hats denote unit vectors,

$$\vec{r} = \rho\hat{\rho} + z\hat{k}, \tag{2.21b}$$

yields

$$d\vec{r} = d\rho\hat{\rho} + \rho d\phi\hat{\phi} + dz\hat{k}, \tag{2.25c}$$

so that $dq_1/dt = d\rho/dt$, $dq_2/dt = \rho d\phi/dt$, and $dq_3/dt = dz/dt$. The kinetic energy is then given by

$$T = \frac{m}{2}(\dot{\rho}^2 + \rho^2\dot{\phi}^2 + \dot{z}^2) \tag{2.25c′}$$

and Lagrange's equations for a single particle then read

$$\left.\begin{array}{c} m\left(\dfrac{d^2\rho}{dt^2} - \rho\dot{\phi}^2\right) = Q_\rho, \\[2ex] m\dfrac{d}{dt}(\rho^2\dot{\phi}) = Q_\phi \\[2ex] m\dfrac{d^2z}{dt^2} = Q_z. \end{array}\right\} \tag{2.29b}$$

In spherical coordinates, where $q_1 = r$, $q_2 = \theta$, and $q_3 = \phi$,

$$\vec{r} = r\hat{r} \tag{2.21c}$$

defines the vectorial transformation function in equation (2.21a), and

$$d\vec{r} = dr\hat{r} + rd\theta\hat{\theta} + r\sin\theta d\phi\hat{\phi}. \tag{2.25d}$$

In this case, $dq_1/dt = dr/dt$, $dq_2/dt = rd\theta/dt$, and $dq_3/dt = r\sin\theta d\phi/dt$ and the kinetic energy is given as

$$T = \frac{m}{2}(\dot{r}^2 + r^2\dot{\theta}^2 + r^2\sin^2\theta\dot{\phi}^2). \tag{2.25d′}$$

Lagrange's equations then read

$$\left.\begin{array}{c} m\left(\dfrac{d^2r}{dt^2} - r\dot{\theta}^2 - r\sin\theta^2\dot{\phi}^2\right) = Q_r, \\[2ex] m\left(\dfrac{d}{dt}(r^2\dot{\theta}) - r^2\sin\theta\cos\phi\dot{\phi}^2\right) = Q_\theta, \\[2ex] m\dfrac{d}{dt}(r^2\sin^2\theta\dot{\phi}) = Q_\phi. \end{array}\right\} \tag{2.29c}$$

Any system of curvilinear coordinates can serve as 'generalized coordinates', or just 'coordinates'. For other curvilinear coordinate systems, see the very readable text by Margenau and Murphy (1956).

Returning to the physics, to the law of inertia and Newton's second law, we emphasize strongly that Lagrange's equations are merely a rewriting of Newton's second law and the law of inertia for application within a *completely general coordinate system, even a noninertial one*. No new physics is introduced by the rewriting, by the achievement of general covariance in configuration space.

For example, for a particle moving in one dimension, the transformation to a linearly accelerated frame is given by

$$x' = x - \int v(t)dt \tag{2.30}$$

and $a(t) = dv(t)/dt$ is the acceleration of the primed frame relative to the unprimed one. Since the unprimed frame is taken as inertial, then

$$T(\dot{x}) = \frac{m}{2}\dot{x}^2 = \frac{m}{2}(\dot{x}' + v(t))^2 = T'(\dot{x}',t) \tag{2.31}$$

so that

$$p = \frac{\partial T'}{\partial \dot{x}'} = m(\dot{x}' + v(t)) \tag{2.32}$$

is the linear momentum *relative to the inertial frame*, but expressed in the noninertial coordinates, and therefore

$$m(\ddot{x}' + a(t)) = F' \tag{2.33}$$

where F' is the net force on the body in the noninertial frame. The extra 'inertial term' (or, better said, *noninertial term*) $ma(t)$ represents the acceleration of the primed frame relative to absolute space (relative to the collection of all inertial frames of reference). In a linearly accelerated frame a force-free particle accelerates past any observer who is at rest in that frame, while simultaneously the observer himself experiences a force $F' = Mdv/dt$ in the direction of the acceleration that holds him in place and is qualitatively and quantitatively similar to the effect of 'weight' to a uniform gravitational field with local strength $g = dv/dt$. The physics of this phenomenon has nothing whatsoever to do with any difference in form between Newton's and Lagrange's equations, but has instead everything to do with the fact that noninertial effects occur whenever motions are observed relative to noninertial frames.

So far, we have kept the discussion general enough that dissipative forces are included. If we now restrict our considerations to conservative forces defined by a scalar potential $U(\vec{r}_1,\ldots,\vec{r}_N)$ that depends upon positions but not velocities, then with $U(\vec{r}_1,\ldots,\vec{r}_N) = \tilde{U}(q_1,\ldots,q_f)$ we have[3]

$$Q_k\delta q_k = -\frac{\partial \tilde{U}}{\partial q_k}\delta q_k. \tag{2.34}$$

Defining the integrand of the action A by the Lagrangian $L = T - U$, where L is a function of both the generalized coordinates and generalized velocities and also the time, then we can rewrite (2.29') to obtain Lagrange's equations of the second kind

---

[3] We include the tilde on U here to remind the reader that the potential energy is a scalar but is never invariant under completely arbitrary coordinate transformations. For consistency, we should also write $T(x_1,\ldots,x_f,\dot{x}_1,\ldots,\dot{x}_f,t) = \tilde{T}(q_1,\ldots,q_f,\dot{q},\ldots,\dot{q}_f,t)$. In section 2.6 we shall find it absolutely necessary to take this sort of care with our notation.

$$\frac{d\partial L}{dt\partial\dot{q}_k} - \frac{\partial L}{\partial q_k} = 0. \tag{2.35a}$$

If, rather than starting with the variational equation (2.29′), we had started at the very beginning with the action

$$A = \int_{t_1}^{t_2} L dt. \tag{2.36}$$

with the Lagrangian L expressed directly in terms of the f generalized coordinates q_i, the f generalized velocities dq_i/dt and the time t, then because the variations δq_i are completely independent of each other, and because there are no dissipative forces, the action principle $\delta A = 0$ for variations vanishing at the end points yields directly Lagrange's equations of the second kind

$$\frac{d\partial L}{dt\partial\dot{q}_k} - \frac{\partial L}{\partial q_k} = 0. \tag{2.35a}$$

As an example, in Cartesian coordinates in an inertial frame, the Lagrangian for a particle with mass m is

$$L = T - V = \tfrac{1}{2}m(\dot{x}_1^2 + \dot{x}_2^2) - U(x_1,x_2) \tag{2.35b}$$

and Lagrange's equations yield Newton's laws in their most familiar form,

$$m\frac{d^2x_i}{dt^2} = -\frac{\partial U}{\partial x_i}. \tag{2.35c}$$

In cylindrical coordinates in an inertial frame, in contrast, with

$$L = \tfrac{1}{2}m(\dot{r}^2 + r^2\dot{\theta}^2) - \tilde{U}(r,\theta) \tag{2.35d}$$

and $U(x_1,x_2) = \tilde{U}(r,\theta)$, Lagrange's equations yield directly Newton's law in their transformed form

$$mr^2\ddot{\theta} = -\frac{\partial\tilde{U}}{\partial\theta}, \quad m(\ddot{r} + r\dot{\theta}^2) = -\frac{\partial\tilde{U}}{\partial r}. \tag{2.35e}$$

Next, with a brevity that fails completely to foreshadow its importance for physics, we introduce a vector in configuration space,

$$p_i = \frac{\partial T}{\partial\dot{q}_i}, \tag{2.37a}$$

which is called the canonical momentum, the momentum conjugate to the coordinate q_i. Sometimes, this quantity is equal to the mass multiplied by the generalized velocity dq_i/dt, but p_i is generally *not* the same as mdq_i/dt. As an example, in spherical coordinates where

$$T = \frac{m}{2}(\dot{r}^2 + r^2\dot{\theta}^2 + r^2\sin^2\theta\dot{\phi}^2), \tag{2.25d′}$$

we find that

$$p_r = \frac{\partial T}{\partial\dot{r}} = m\dot{r}, \tag{2.37a}$$

is the linear momentum along the radial direction, that

$$p_\theta = \frac{\partial T}{\partial \dot\theta} = mr^2\dot\theta, \qquad (2.37c)$$

and that

$$p_\phi = \frac{\partial T}{\partial \dot\phi} = mr^2 \sin^2\theta\dot\phi \qquad (2.37d)$$

is the angular momentum about the z-axis.

With the introduction of Hamilton's equations (section 2.8 and chapter 3) and the discussion of the geometric interpretation of Lagrange's equations (chapter 10), the reader will be in a better position to appreciate the significance of the idea of the 'canonical momentum' conjugate to a generalized coordinate. Later, in section 2.6, we shall show that it is the canonical momentum p_i rather than the generalized velocity dq_i/dt that comes into play when we derive the connection between symmetry, invariance, and conservation laws by using the Lagrangian formalism.

We can also rewrite Lagrange's equations of the first kind in the form

$$\frac{dp_i}{dt} = \frac{\partial T}{\partial q_i} + Q_i, \qquad (2.38a)$$

which illustrates that the rate of change of canonical momentum usually differs from the generalized force in arbitrarily chosen coordinate systems because of position-dependent kinetic energy terms. For example, in cylindrical coordinates we obtain

$$\frac{dp_r}{dt} = r\dot\theta^2 - \frac{\partial \tilde U}{\partial r}. \qquad (2.38b)$$

Whenever there are no dissipative forces, whenever the net force on each particle is conservative, then we can introduce the Lagrangian $L = T - U$ and can rewrite Lagrange's equations of the second kind in the form

$$p_i = \frac{\partial L}{\partial \dot q_i}. \qquad (2.37e)$$

Again, we expect to find kinetic energy contributions to the right hand side of this equation.

Finally, if we apply Lagrange's equations in a linearly accelerated frame, where

$$T(\dot x) = \frac{m}{2}\dot x^2 = \frac{m}{2}(\dot x' + v(t))^2 = T'(\dot x', t), \qquad (2.31)$$

then

$$p = \dot x' + mv \qquad (2.38c)$$

is the canonical momentum, and Lagrange's equations read simply

$$\frac{dp}{dt} = m(\ddot x' + dv/dt) = F'. \qquad (2.38d)$$

In other words, the canonical momentum is the momentum in the inertial frame, but rewritten in terms of the noninertial coordinates! If we interpret p as the canonical momentum rather than as the linear momentum, then Newton's law in the simple form $dp/dt = F$ is no longer restricted to inertial frames but is covariant with respect to transformations to and among linearly accelerated Cartesian frames as well. Covariance is not a physical principle but is something that can be achieved by the reformulation of vector equations, with a careful definition of what constitutes a vector and what does not (see chapters 8 and 10).

We conclude this section with the reminder that Lagrange's equations have the same form (2.38a) in *every* coordinate system in configuration space. We discuss the geometry and transformation theory beneath this aspect of Lagrangian mechanics explicitly in chapters 8 and 10. In chapter 10 we show how general covariance in configuration space works for Lagrange's equations.

2.5 Constrained Lagrangian systems

Next we derive a method for including constraints explicitly, as an alternative to their complete elimination. The reason is that nonholonomic constraints cannot be eliminated, but holonomic ones can be treated within the same formalism. Consider a constrained Lagrangian system where $L = T - U$ in an n dimensional configuration space ($n = Nd$ when d is the dimension of physical space ($d = 1, 2,$ or 3)), and where there are k constraints of the form

$$a_{il}\delta q_i + a_{tl}\delta t = 0, \tag{2.39a}$$

for variations performed at fixed time t. If

$$a_{il} = \frac{\partial \phi_l}{\partial q_i}, \tag{2.40}$$

then the constraints are holonomic because the differential form (2.39a) is integrable (an integrating factor is allowed here). In this case (2.39a) can be integrated at fixed time to yield a point function $\phi_l(q_1, \ldots, q_n) = 0$ of the generalized coordinates. Otherwise, the differential form (2.39a) is nonintegrable (no integrating factor exists), which means that the constraint is nonholonomic. In that case, any integral of the constraint equation (2.39a) is path-dependent and so the integrations cannot be performed until the coordinates q_i are known as functions of t, that is, until after the dynamics problem has been solved. We emphasize that we include the case where the differential form (2.39a) is nonintegrable, but has an integrating factor, within the classification of holonomic constraints. Another distinct possibility for a nonholonomic constraint is one given by an inequality.

An example of a holonomic constraint is one where the radial position $r = L$ of a mass m attached to the end of a rod of length L is fixed, if we take the other end of the rod as the origin of coordinates.

An example of a nonholonomic constraint that is given by an inequality is provided by the case where a particle slides under the influence of gravity on the surface of a cylinder of radius R. In this case, the particle's radial position r, measured from the cylinder's center, obeys the constraint $r \geq R$.

An example of a nonintegrable constraint given by a differential form is provided by the motion of a disk that rolls without slipping on a plane (see figure 2.3). Here, there are two constraints,

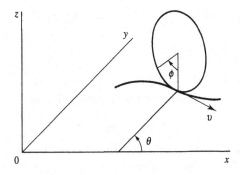

Fig. 2.3 A disc rolls without slipping (fig. 1.5 from Goldstein, 1980).

$$\left.\begin{array}{l} dx = a d\phi \cos\theta \\ dy = a d\phi \sin\theta, \end{array}\right\} \tag{2.41}$$

and we leave it as an exercise for the reader to show that this pair of differential forms is nonintegrable and has no integrating factor.

We return now to the original action principle, but for the case where the n generalized coordinates q_i cannot all be varied freely because they are subject to $k \geq 1$ constraints. As earlier in this chapter, we use the summation convention in this section (sum over all repeated indices in an expression). In the action principle, with end points fixed, one then obtains

$$\delta A = \int_{t_1}^{t_2} \left(\frac{\partial L}{\partial q_i} - \frac{d\partial L}{dt\partial \dot{q}_i} \right) \delta q_i = 0 \tag{2.42}$$

which yields the by-now familiar variational equation

$$\left(\frac{\partial L}{\partial q_i} - \frac{d\partial L}{dt\partial \dot{q}_i} \right) \delta q_i = 0. \tag{2.43}$$

However, no further simplification is possible because the k constraints couple the n virtual variations δq_i. In order to go further we must include the effects of the forces of constraint explicitly in the equations of motion. To do this in the most elegant way, we follow Lagrange and introduce his method of undetermined multipliers (the only alternative to Lagrange's method is to eliminate the constraints beforehand, and then to work with $f = Nd - k$ independent variables q_i as we did earlier).

Introducing k unknown parameters λ_k and then writing

$$\lambda_l a_{il} \delta q_i = 0 \tag{2.44}$$

for a fixed time variation, we can rewrite our variational equation (2.43) to read

$$\left(\frac{\partial L}{\partial q_i} - \frac{d\partial L}{dt\partial \dot{q}_i} + \lambda_k a_{ik} \right) \delta q_i = 0. \tag{2.45a}$$

The n variations δq_i are not independent, but the object of introducing the extra term in (2.45a) is to *agree* to choose the k undetermined multipliers (Lagrange's undetermined multipliers) λ_k so that the equation

$$\frac{d\partial L}{dt\partial \dot{q}_i} - \frac{\partial L}{\partial q_i} = \lambda_k a_{ik} \tag{2.45b}$$

is satisfied for each index $i = 1, 2, \ldots, n$. The terms on the right hand side can represent nothing other than the constraint forces $Q_{i,\text{constraint}}$. If the constraints are holonomic, then, as we proved earlier, the constraint forces are perpendicular to the surfaces defined by the constraint equations $\phi_l(q_1, \ldots, q_n) = 0$. Note, however, that we now must solve $n + k$ differential equations: the n Lagrangian differential equations given by (2.45b) along with the k (integrable or nonintegrable) first order differential equations

$$a_{ik}dq_i/dt + a_{tk} = 0, \tag{2.39b}$$

for a grand total of $n + k$ variables $(q_1, \ldots, q_n)$ and $(\lambda_1, \ldots, \lambda_k)$. Solving for the unknown multipliers λ_l is therefore an integral part of solving the dynamics problem in the Lagrange multiplier formulation of dynamics.

As an example, consider the motion of a block on a frictionless inclined plane (figure 2.2). The kinetic energy is

$$T = \frac{m}{2}(\dot{x}^2 + \dot{y}^2) \tag{2.46}$$

and the potential energy is $U = mgy$. Here, the holonomic constraint is $dy = dx \tan \theta$ so that we have $n = 2$ and $k = 1$. Lagrange's equations with one undetermined multiplier yield

$$\left.\begin{array}{l} m\ddot{x} + \lambda \tan \theta = 0 \\ m\ddot{y} - \lambda + mg = 0. \end{array}\right\} \tag{2.47}$$

We can eliminate λ to obtain

$$\ddot{x} + \ddot{y} \tan \theta + g \tan \theta = 0, \tag{2.48}$$

which can be integrated immediately. Using the constraint equation, we can then solve to find

$$\lambda = mg \cos^2 \theta. \tag{2.49a}$$

We identify $Q_x = -\lambda \tan \theta$ as the force of constraint in the x-direction and $Q_y = \lambda$ as the force of constraint in the y-direction.

Finally, let us return to the discussion of equations (2.14) to (2.17) in our earlier discussion of constraints. There we observed that the real displacements generally are *not* perpendicular to the constraint forces. We can now illustrate this effect by an example: consider a massless rod fixed at one point in space, tilted at an angle α relative to and rotating steadily about the z-axis (figure 2.4). A bead of mass m slides frictionlessly on the rod and we denote its distance from the rod's fixed point by q. The Lagrangian is

$$L = T = \tfrac{1}{2}mv^2 = \tfrac{1}{2}m\dot{r}^2 + \tfrac{1}{2}m\omega^2 r^2 \tag{2.49b}$$

where $r = q \sin \alpha$, and Lagrange's equation reads

$$\frac{d^2q}{dt^2} = q\omega^2 \sin^2 \alpha. \tag{2.49c}$$

The virtual variations are, by definition, along the direction of the rod and therefore are perpendicular to the constraint force which, because the bead slides frictionlessly, is normal to the rod. Clearly, however, the actual displacement of the bead is generally

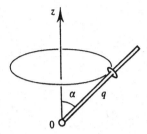

Fig. 2.4 Rotating rod (fig. 73 from Arnol'd, 1989).

not perpendicular to the constraint force and that is the main point to be made by this example.

2.6 Symmetry, invariance, and conservation laws

We return to our action principle where the action is given by

$$A = \int_{t_1}^{t_2} L \, \mathrm{d}t \tag{2.36}$$

with $L = T - U$. Instead of virtual variations and the search for an extremum, we consider only paths that are solutions of the dynamics (extremum paths), but where the variations δq_i will now represent infinitesimal transformations of coordinates[4]. The finite transformation is given by $q_i(\alpha) = F_i(q_1, \ldots, q_f, \alpha)$. Each value of α defines a different coordinate system, and all transformations in the class are generated by continuously varying the parameter α (this method is generalized to include more than one parameter in chapters 7, 8, 12 and 15). The qualitative idea is that we should imagine two different 'snapshots' of the same configuration of the particles of the system in two different coordinate systems taken at the same time t. We can take the viewpoint that the laboratory frame is transformed, while the system remains fixed in space, or one can transform the dynamical system while leaving the laboratory frame fixed in space. We denote by f the inverse transformation, $q_k = f_k(q_1(\alpha), \ldots, q_f(\alpha))$, and we take $\alpha = 0$ to yield the identity transformation $q_i(0) = F_i(q_1, \ldots, q_f; 0) = q_i$.

Up to this point, we have not been careful enough with our notation, as we have sometimes used the very same symbols L, T, and U to denote the Lagrangian, kinetic, and potential energies in any and all coordinate systems. In what follows, we must use a more refined notation, one that takes into account that whenever we transform coordinates we get a *different* Lagrangian than the one we started with: the Lagrangian is a scalar function of the generalized coordinates and velocities and therefore transforms according to the rule $L'(q', \mathrm{d}q'/\mathrm{d}t, t) = L(q, \mathrm{d}q/\mathrm{d}t, t)$ whenever we change coordinate systems. The kinetic and potential energies are themselves scalar functions: $T'(q', \mathrm{d}q'/\mathrm{d}t, t) = T(q, \mathrm{d}q/\mathrm{d}t, t)$ and $U'(q', t) = U(q, t)$. In particular, we shall denote the Lagrangian in the coordinate system $(q_1(\alpha), \ldots, q_f(\alpha))$ by $L_\alpha = L_\alpha(q(\alpha), \mathrm{d}q(\alpha)/\mathrm{d}t, t)$. Since the variations near the identity ($\alpha = 0$ defines the identity transformation) are defined to first order by

$$\delta q_i = \left[\frac{\partial q_i(\alpha)}{\partial \alpha} \right]_{\alpha = 0} \delta\alpha, \tag{2.50}$$

[4] The study of symmetry by infinitesimal coordinate transformations goes back to S. Lie.

it follows that the first variation in the action,

$$\delta A = A(\alpha) - A(0) \approx \int_{t_1}^{t_2} \frac{dL_\alpha}{d\alpha} \delta\alpha \, dt = \int_{t_1}^{t_2} \left(\frac{\partial L_\alpha}{\partial q_i(\alpha)} \frac{\partial q_i(\alpha)}{\partial \alpha} + \frac{\partial L_\alpha}{\partial \dot{q}(\alpha)_i} \frac{\partial q_i(\dot\alpha)}{\partial \alpha} \right) \delta\alpha \, dt, \quad (2.51a)$$

has the form

$$\delta A = \left[\left[\frac{\partial L_\alpha}{\partial \dot{q}_i(\alpha)} \delta q_i \right]_{t_1}^{t_2} \right]_{\alpha=0} = \left[p_i(\alpha) \left[\frac{\partial q_i(\alpha)}{\partial \alpha} \right]_{\alpha=0} \delta\alpha \right]_{t_1}^{t_2}. \quad (2.51a')$$

The time-integral is missing in (2.51a') because both the original and infinitesimally transformed paths satisfy Lagrange's equations in the two different coordinate systems: both paths are simply different coordinate descriptions of the same dynamics in configuration space. Note that the shift in the action due to a change in α at fixed time is equal to the difference in the quantity

$$G = \left[p_i(\alpha) \frac{\partial q_i(\alpha)}{\partial \alpha} \right]_{\alpha=0}, \quad (2.52a)$$

but evaluated at two different times t_1 and t_2. We are now in the right position to formulate the connection of symmetry, invariance, and conservation laws mathematically.

In general, the Lagrangian is a *scalar* function of the generalized coordinates and velocities under *arbitrary* differentiable point transformations in configuration space. This sounds impressive, but the size of this collection of transformations is much too great to be of any practical use in physics. We therefore consider a much smaller and very restricted, but much more useful, class of coordinate transformations, those that depend continuously upon one parameter and leave the Lagrangian and therefore also the action *invariant* under variations in the parameter α. The idea of a scalar function is just the idea of getting the same number: the Lagrangian at two 'image points', $q_k = f_k(q'_1, \ldots, q'_f, \alpha)$ and $q'_i = F_i(q_1, \ldots, q_f, \alpha)$, is equal to the *same number* whenever we replace the generalized coordinates and velocities by specific numbers, because $L(q, dq/dt, t) = L'(q', dq'/dt, t)$. *Invariance* is a much more restricted area that reflects, in addition, the identity of *functions* under transformation: one gets not only the same number, for a given set of values of the generalized coordinates and velocities, but the original and transformed Lagrangians are exactly the same *function*, $L(q, dq/dt, t) = L(q', dq'/dt, t)$, or just $L = L'$ (the reader can now see why sloppy notation must be eliminated). For example, in the plane, the two different ways of writing the kinetic energy, $T(\dot{x}, \dot{y}) = m(\dot{x}^2 + \dot{y}^2)/2 = m(\dot{r}^2 + r^2\dot\phi^2)/2 = T(\dot{r}, \dot\phi)$, yield the same numerical value whenever the Cartesian coordinates are connected to the cylindrical ones by the transformation $x = r\cos\phi$, $y = r\sin\phi$, but the functions T and T' are not the same function of the respective variables. If we think of a third function T'' *of the variables* $(\dot{r}, \dot{r}\phi)$, then the functions T and T'' are identical but both differ from the function T' (or, stated differently, $x^2 + y^2$ is not the same function of x and y as $x^2 + (xy)^2$).

At this stage we do not yet require that the Lagrangian is invariant, only that the action is. The invariance principle for the action under infinitesimal transformations is simply $\delta A = 0$ for arbitrary variations $\delta\alpha$, and yields

$$G = \left[p_i(\alpha) \frac{\partial q_i(\alpha)}{\partial \alpha} \right]_{\alpha=0} = \text{constant} \quad (2.52')$$

at all times, which is a conservation law! This is the content of Noether's theorem in particle mechanics, that the quantity G is *conserved* along the system's trajectories in configuration space whenever the infinitesimal transformation (2.50) leaves the action *invariant*. In other words, there is a connection between quantities that remain constant in time along dynamical trajectories and *time-independent* transformations of coordinates that leave the action invariant. There is also a missing third part to this relationship that we discover and illustrate below via two examples, namely, the idea of an underlying geometric symmetry as the reason for the invariance, and therefore also for the conservation law. Before going on to the examples that illustrate the underlying symmetries we will rewrite Noether's theorem directly as an invariance principle for the Lagrangian, rather than merely for the action.

We first approach our rewriting from a more general standpoint, one that begins without assuming that L_α is an invariant function under our coordinate transformations as α is varied. Suppose only that the kinetic energy has the quadratic form $T(q,dq/dt) = (g_{ij}dq_i/dt \, dq_j/dt^2)/2$ where $g_{ij} = g_{ji}$ generally depends upon the q_k, which is true if the original transformation from Cartesian to generalized coordinates q_k is time-independent. This result follows from substituting (2.25a) into the original kinetic energy expression in Cartesian coordinates, which then yields

$$g_{lk} = m_i \frac{\partial \vec{R}_i}{\partial q_l} \cdot \frac{\partial \vec{R}_i}{\partial q_k}. \tag{2.25a'}$$

If, in a *new* choice of generalized coordinates $(q'_1, \ldots, q'_f)$ different from $(q_1(\alpha), \ldots, q_f(\alpha))$, we should choose the parameter α as one of those coordinates, say $q'_k = \alpha$, then using the scalar property $T'(q',dq'/dt) = T_\alpha(q(\alpha),dq(\alpha)/dt)$, we obtain

$$p'_k = \frac{\partial T'}{\partial \dot{q}'_k} = \frac{\partial T_\alpha}{\partial \dot{\alpha}} = g(\alpha)_{ij}\dot{q}_i(\alpha)\frac{\partial \dot{q}_j(\alpha)}{\partial \dot{\alpha}} = g(\alpha)_{ij}\dot{q}_i(\alpha)\frac{\partial q_j(\alpha)}{\partial \alpha} = p_j(\alpha)\frac{\partial q_j(\alpha)}{\partial \alpha} = G(\alpha), \tag{2.52b}$$

where we have used the generalization of equation (2.28).

$$\frac{\partial \dot{q}_j(\alpha)}{\partial \dot{\alpha}} = \frac{\partial \dot{q}_j(\alpha)}{\partial \dot{q}'_k} = \frac{\partial q_j(\alpha)}{\partial q'_k} = \frac{\partial q_j(\alpha)}{\partial \alpha}, \tag{2.28'}$$

and where $G = G(0)$. Finally, Lagrange's equation in the primed system for the kth generalized coordinate reads

$$\frac{dG(\alpha)}{dt} = \frac{dp'_k}{dt} = \frac{\partial L'}{\partial q'_k} = \frac{\partial L'}{\partial \alpha}. \tag{2.51b}$$

So far, we have not assumed that the Lagrangian is invariant, so that $L', L_\alpha,$ and L are still permitted to be three different functions of their respective generalized coordinates and velocities. Now, we come to the main point: from (2.51b) we see that, by the right choice of coordinate system, a conserved *canonical momentum* $p'_k = G(\alpha = 0)$ can always be constructed whenever the action is *invariant* under a set of differentiable transformations. Also, invariance of the action as originally expressed in terms of the Lagrangian L_α in the $q_i(\alpha)$-coordinate system can be replaced by the statement that the Lagrangian is independent of the particular generalized coordinate $q'_k = \alpha$ in the primed system. In this case L_α is exactly the same function of the coordinates $q_i(\alpha)$ and velocities $dq(\alpha)/dt$ as is L, the Lagrangian L_α with $\alpha = 0$, of the original coordinates $q_i = q_i(0)$ and velocities $dq_i/dt = dq_i(0)/dt$. In other words, the function L_α does not change as α is varied and is *exactly the same function* as the original Lagrangian L. This is the idea of invariance under a coordinate transformation that depends continuously upon a parameter α. In outdated language q'_k was called an 'ignorable coordinate'.

We show next that invariance principles and conservation laws follow from underlying symmetries of the Lagrangian. In order to discover and illustrate this connection we turn to two of the most important cases: conservation of linear and angular momentum.

As we have seen above, whenever the Lagrangian is invariant under translations $q_k \rightarrow q_k + \delta\alpha$ along some axis q_k in configuration space, which is the same as saying that L is independent of the coordinate q_k, then

$$\frac{\partial L}{\partial q_k} = 0 \qquad (2.53a)$$

and the corresponding conservation law states simply that p_k is constant. That is, whenever the Lagrangian has translational invariance along some particular direction q_k in configuration space then the corresponding canonical momentum p_k is conserved. Conservation of linear momentum in Euclidean space (Newton's first law) is the most important special case of this result: a free particle has the Lagrangian

$$L = \frac{m}{2}\dot{x}^2 \qquad (2.53b)$$

in an inertial frame in one dimension, using Cartesian coordinates. The transformation of interest is just a translation of the coordinate origin, $x' = x - \alpha$, and the corresponding conservation law is simply that the linear momentum $p = m\mathrm{d}x/\mathrm{d}t$ is constant. The underlying reason for this result is geometric, namely, the homogeneity of Euclidean space: the x-axis looks the same everywhere (this is no longer true if you place an obstacle like a rock at some point on the axis), which is the same as saying that the x-axis is everywhere homogeneous, or uniform. This particular symmetry of Euclidean space, that one point on an axis in empty space is the same as every other point so far as the motion of a free particle is concerned, is the underlying geometric reason for the translational invariance of the free-particle Lagrangian. Translational invariance of the Lagrangian is, in turn, the reason for conservation of linear momentum and therefore for Newton's first law. Whereas symmetry is a geometric idea, invariance is algebraic, or analytic. An object has symmetry when it looks the same from two or more different perspectives, and invariance is the analytic expression of this notion: a function is invariant when it does not change under a coordinate transformation. In our example homogeneity of empty Euclidean space provides the symmetry (other particles that interact with the particle under consideration will eliminate or 'break' that symmetry), translations along a particular Cartesian axis x_i provide the invariance principle, and the corresponding conserved quantity is just the component of linear momentum mv_i along that axis. This is the first illustration of the deep connection, expressed by Noether's theorem, between symmetry, invariance, and conservation laws. One can say that symmetry implies invariance implies a conservation law. Given a conserved quantity, one can also turn the relation around and ask for the underlying symmetry principle.

Next, consider a rigid rotation of an N-body system in Euclidean space. Without loss of generality, we may use Cartesian coordinates and can take the rotation of the coordinate system (while vectors are left fixed) to be counterclockwise about the z-axis,

$$\left.\begin{array}{l} x_i(\theta) = x_i \cos \theta + y_i \sin \theta \\ y_i(\theta) = -x_i \sin \theta + y_i \cos \theta \\ z_i(\theta) = z_i \end{array}\right\} \tag{2.54a}$$

so that the infinitesimal transformations evaluated near the identity obey

$$\left.\begin{array}{l} \delta x_i = y_i \delta \theta \\ \delta y_i = -x_i \delta \theta \\ \delta z_i = 0 \end{array}\right\} \tag{2.54b}$$

for a counterclockwise rotation through an infinitesimal angle $\delta \theta$. In Cartesian coordinates,

$$T = \tfrac{1}{2} \sum_{i=1}^{N} m_i (\dot{x}_i^2 + \dot{y}_i^2 + \dot{z}_i^2), \tag{2.55}$$

and the canonical momenta conjugate to the Cartesian coordinates are just the linear momenta: $p_{xi} = m_i dx_i/dt$, $p_{yi} = m_i dy_i/dt$, and $p_{zi} = m_i dz_i/dt$. Substitution into Noether's result (2.52') yields directly that

$$\sum_{i=1}^{N} m_i (\dot{x}_i y_i - \dot{y}_i x_i) = \text{constant}. \tag{2.56}$$

The left hand side is just the component of total angular momentum L_z of the N-body system about the z-axis. In other words, if we imagine a rigid rotation of a particular configuration (a 'snapshot' at time t) of N bodies, and if the Lagrangian is invariant under rotations, then the total angular momentum L_z about the axis of rotation is a conserved quantity in the *dynamics* problem. The underlying symmetry principle in this case is the isotropy of empty Euclidean space: in the absence of other particles outside our system (particles that exert a net torque), Euclidean space is perfectly isotropic because it looks the same in all directions (in particular, all three Cartesian axes are homogeneous). In other words, the isotropy of empty Euclidean space is the geometric quality that is responsible for the conservation of angular momentum: if space is not isotropic, then angular momentum cannot be conserved.

In order to understand the above result better, we can arrive at it from a different perspective, namely, by asking, formulating, and answering the following question: what does it mean to say that the Lagrangian is rotationally invariant? If, as we have the freedom to do, we should choose the rotation angle θ as one of our generalized coordinates in configuration space, then

$$p_\theta = \frac{\partial L}{\partial \dot{\theta}} \tag{2.57}$$

and

$$\frac{dp_\theta}{dt} = \frac{\partial L}{\partial \theta}. \tag{2.58}$$

Rotational invariance means that

$$\frac{\partial L}{\partial \theta} = 0, \tag{2.59}$$

so that (by Lagrange's equation for p_θ) the corresponding canonical momentum is conserved: p_θ is constant. Now, let us evaluate p_θ in Cartesian coordinates:

$$p_\theta = \frac{\partial T}{\partial \dot\theta} = \sum_{i=1}^{N} m_i \left(\dot x_i \frac{\partial \dot x_i}{\partial \dot\theta} + \dot y_i \frac{\partial \dot y_i}{\partial \dot\theta} + \dot z_i \frac{\partial \dot z_i}{\partial \dot\theta} \right)$$

$$= \sum_{i=1}^{N} m_i \left(\dot x_i \frac{\partial x_i(\theta)}{\partial \theta} + \dot y_i \frac{\partial y_i(\theta)}{\partial \theta} + \dot z_i \frac{\partial z_i(\theta)}{\partial \theta} \right) \tag{2.60}$$

and for the infinitesimal rotation given by (2.54b), we obtain

$$p_\theta = \sum_{i=1}^{N} m_i (\dot x_i y_i - \dot y_i x_i) = L_z \tag{2.61}$$

which is an example of equation (2.52′).

Invariance under a configuration space transformation $q_i \to q_i'$ means that $L = L'$, that $L(q',dq'/dt,t) = L(q,dq/dt,t)$ is exactly the same function in two different coordinate systems. A consequence is that the Lagrangian equations of motion for $q(t)$ and $q'(t)$ are identical (which is not true if L is merely a scalar and not an invariant). The simplest example of this is provided by the description of a free particle in two different Cartesian inertial frames: in either frame, the differential equation is the same and states merely that $d(mv_i)/dt = 0$. This invariance principle is the Newtonian relativity principle and also reflects a symmetry of absolute Euclidean space, namely, its insensitivity to velocities (but not to accelerations). Taken all together, invariance under rigid translations, rigid rotations, time translations, and transformations among inertial frames is called Galilean invariance.

Lagrange's equations are vector equations in configuration space (chapter 10). A class of transformations that leaves *vector equations* unchanged in form defines *covariance* under that class of transformations. In contrast, a class of transformations that leaves certain *scalar functions* unchanged in form defines the idea of *invariance* for that class of functions. We discussed earlier that, given Newton's laws, the symmetry of Euclidean space is the underlying reason for the three geometric conservation laws, the laws of conservation of momentum, angular momentum, and total energy. We have shown that transformations that leave a certain scalar function, the Lagrangian $L = T - U$, invariant provide the ground upon which these and other conservation laws in physics may be discussed. This invariance has nothing to do with the general covariance of Lagrange's equations. Instead, the special class of transformations that leaves the Lagrangian invariant represents *symmetry operations* on the physical system defined by the Lagrangian. The Klein–Lie definition of geometry by transformation groups that leave the geometric objects of a given space invariant is very close in spirit to Noether's theorem, which connects symmetries of the Lagrangian with conservation laws. Emmy Noether, a German mathematician, lived from 1882 to 1935.

We turn now to the discussion of a nice example from Saletan and Cromer (1971), the motion of a single particle described via Cartesian coordinates in an inertial frame, where

$$L = \frac{m}{2} (\dot x_1^2 + \dot x_2^2 + \dot x_3^2) - U(x_1, x_2, x_3). \tag{2.62}$$

Under transformations in configuration space, L cannot be invariant unless T and U separately are invariant, for the former depends upon the generalized velocities while the latter does not and configuration space transformations do not mix generalized coordinates and generalized velocities. Therefore, we begin by concentrating on the kinetic energy term. The treatment of the potential energy is illustrated by the solution of exercise 13. Consider the system in two different coordinate systems $(x_1(\alpha), x_2(\alpha), x_3(\alpha))$ and (x_1, x_2, x_3) where $x_i(\alpha) = F_i(x_1, x_2, x_3, \alpha)$. For an infinitesimal transformation

$$\delta x_i(\alpha) = \frac{\partial x_i(\alpha)}{\partial \alpha} \delta \alpha, \tag{2.63}$$

the kinetic energy changes by the amount

$$\delta T = m \dot{x}_i(\alpha) \frac{\partial \dot{x}_i(\alpha)}{\partial \alpha} \delta \alpha, \tag{2.64}$$

and since our transformations are time-independent, we can use the idea of (2.25a) to obtain

$$\delta T = m \dot{x}_i(\alpha) \dot{x}_j \frac{\partial^2 F_i}{\partial x_j \partial \alpha} \delta \alpha. \tag{2.65}$$

Using the notation

$$z_i = \frac{\partial F_i}{\partial \alpha} \tag{2.66}$$

we then have

$$\delta T = m \dot{x}_i(\alpha) \dot{x}_j \frac{\partial z_i}{\partial x_j} \delta \alpha. \tag{2.67a}$$

We search now for the infinitesimal transformations that leave this expression invariant, those for which $\delta T = 0$. In particular, we restrict to the neighborhood of the identity transformation where

$$\delta T = m \dot{x}_i \dot{x}_j \frac{\partial z_i}{\partial x_j} \delta \alpha, \tag{2.67b}$$

so that the partial derivative of z_i with respect to x_j is evaluated at $\alpha = 0$. For arbitrary values of the velocities dx_i/dt, invariance requires

$$\frac{\partial z_i}{\partial x_j} = \frac{\partial z_j}{\partial x_i}, \tag{2.68}$$

and the existence of z_i as a point function of the x's then requires

$$\frac{\partial^2 z_i}{\partial x_k \partial x_j} = \frac{\partial^2 z_k}{\partial x_j \partial x_i}. \tag{2.69}$$

combining (2.68) and (2.69) yields

$$\frac{\partial^2 z_i}{\partial x_k \partial x_j} = 0, \tag{2.70}$$

which means that the z_i are linear functions of the coordinates x_j:

$$z_i = a_i + b_{ij} x_j, \tag{2.71}$$

with a_i and b_{ij} constants, but where $b_{ij} = -b_{ji}$. This antisymmetry condition on the matrix b is satisfied if we write $b_{ij} = \varepsilon_{ijk}b_k$, where ε_{ijk} is the totally antisymmetric three-index symbol: $\varepsilon_{ijk} = 1$ for all symmetric permutations of $(i,j,k) = 1,2,3$, $\varepsilon_{ijk} = -1$ for antisymmetric permutations, and $\varepsilon_{ijk} = 0$ if any two indices coincide. Clearly, the constants a_i in the infinitesimal transformation δx_i represent pure translations, and as we shall show in chapter 8, the term $b_{ij}x_j$ represents a rotation. Hence, a translation combined with a rotation is the most general time-independent transformation that leaves the kinetic energy T invariant. We will learn later that invariance principles described by transformations in *configuration space* are not the whole story.

We end with the observation that invariance of the action through invariance of the Lagrangian is *not* the most general invariance that leaves the differential equations of a dynamics problem in configuration space invariant. To see this, consider any transformation that shifts the Lagrangian by a total time derivative $L' = L + d\Lambda/dt$. It is clear from the action principle that Lagrange's equations for L' and L are identical (if we restrict to variations with fixed end points), and that the solutions of both sets of equations are also identical. What kinds of transformations in physics correspond to this kind of invariance principle? In order to answer this question, we turn now to another extremely important example.

2.7 Gauge invariance

Here we temporarily revert to Gibbs' notation for vectors. Consider a body with net charge Q in an external electromagnetic field. By 'external' we mean that we are going to ignore the reaction of our charge on the charges that create the fields under consideration. Clearly, this is only an approximation.

A charge in an electromagnetic field moves according to the Lorentz force

$$\vec{F} = Q\vec{E} + \frac{Q}{c}\vec{v} \times \vec{B} \tag{2.72}$$

where $\vec{E}$ and $\vec{B}$ denote the electric and magnetic fields created by distant charges. If we assume that our charge moves in a nonpolarizable, nonmagnetic medium, then Maxwell's equations are

$$\left.\begin{array}{c} \nabla \cdot \vec{E} = 4\pi\rho \\[2mm] \nabla \times \vec{E} = -\dfrac{1}{c}\dfrac{\partial \vec{B}}{\partial t} \\[2mm] \nabla \cdot \vec{B} = 0 \\[2mm] \nabla \times \vec{B} = \dfrac{1}{c}\dfrac{\partial \vec{E}}{\partial t} + \dfrac{4\pi}{c}\vec{j}, \end{array}\right\} \tag{2.73}$$

where ρ is the charge density and $\vec{j}$ is the current density. We can derive the fields from potentials as follows: since

$$\nabla \cdot \vec{B} = 0 \Rightarrow \vec{B} = \nabla \times \vec{A}, \tag{2.74}$$

it follows that the magnetic field can be described as the curl of a vector potential, and since

$$\nabla \times \left(\vec{E} + \frac{1}{c} \frac{\partial \vec{A}}{\partial t} \right) = 0 \Rightarrow \vec{E} = -\frac{1}{c} \frac{\partial \vec{A}}{\partial t} - \nabla \phi, \tag{2.75}$$

the electric field has contributions from both the scalar and vector potentials.

Gauge transformations are defined by

$$\left. \begin{array}{l} \vec{A}' = \vec{A} + \nabla \Lambda \\[2mm] \phi' = \phi - \dfrac{1}{c} \dfrac{\partial \Lambda}{\partial t}. \end{array} \right\} \tag{2.76}$$

Because the equations of motion depend only upon fields, and not upon the potentials, the equations of motion and therefore the trajectories as well are left invariant by a gauge transformation because fields themselves are left invariant:

$$\vec{E}' = \vec{E} \text{ and } \vec{B}' = \vec{B} \tag{2.77}$$

under any guage transformation.

By using equations (2.72), (2.74), and (2.75), the Lorentz force can be written as

$$\vec{F} = -Q\nabla \phi - \frac{Q}{c} \frac{\partial \vec{A}}{\partial t} + \frac{Q}{c} \vec{v} \times \nabla \times \vec{A}. \tag{2.78}$$

In Cartesian coordinates (x_1, x_2, x_3), we obtain

$$\vec{v} \times \nabla \times \vec{A} = \nabla \vec{v} \cdot \vec{A} - \vec{v} \cdot \nabla \vec{A} \tag{2.79}$$

so that

$$\frac{\mathrm{d}\vec{A}}{\mathrm{d}t} = \frac{\partial \vec{A}}{\partial t} + \vec{v} \cdot \nabla \vec{A} \tag{2.80}$$

and also

$$\vec{v} \times \nabla \times \vec{A} = \nabla \vec{v} \cdot \vec{A} - \frac{\mathrm{d}\vec{A}}{\mathrm{d}t} + \frac{\partial \vec{A}}{\partial t}. \tag{2.81}$$

It follows that

$$\vec{F} = -Q\nabla \phi + \frac{Q}{c} \nabla \vec{v} \cdot \vec{A} - \frac{Q\mathrm{d}\vec{A}}{c\mathrm{d}t} \tag{2.82}$$

and with

$$\frac{\mathrm{d}\partial}{\mathrm{d}t \partial v_i} \vec{A} \cdot \vec{v} = \frac{\mathrm{d}A_i}{\mathrm{d}t} \tag{2.83}$$

we obtain

$$\vec{F} = -Q\nabla \left(\phi - \frac{1}{c} \vec{v} \cdot \vec{A} \right) - \frac{Q}{c} \frac{\mathrm{d}}{\mathrm{d}t} \nabla_v \vec{v} \cdot \vec{A}. \tag{2.84}$$

Since

$$\nabla_v \phi = 0, \tag{2.85}$$

we obtain finally that

$$\vec{F} = -\nabla U + \frac{\mathrm{d}}{\mathrm{d}t}\nabla_v U \qquad (2.86)$$

where the velocity-dependent potential U is defined by

$$U = Q\phi - \frac{Q}{c}\vec{v}\cdot\vec{A}, \qquad (2.87)$$

so that if we use the Lagrangian $L = T - U$ in Cartesian coordinates, Lagrange's equations

$$\frac{\mathrm{d}\partial L}{\mathrm{d}t\partial v_i} - \frac{\partial L}{\partial x_i} = 0 \qquad (2.88)$$

yield the Newton/Lorentz equations of motion for the charge.

Consider next a gauge transformation of the electromagnetic fields. The new Lagrangian is given by

$$L' = T - U' = T - U + \frac{Q}{c}\left(\frac{\partial\Lambda}{\partial t} + \vec{v}\cdot\Lambda\right) = L + \frac{Q\mathrm{d}\Lambda}{c\mathrm{d}t}. \qquad (2.89)$$

Recall now the action principle. If we introduce any function $\Lambda(Q,t)$ of the generalized coordinates and the time, then a new Lagrangian defined by

$$L' = L + \frac{\mathrm{d}\Lambda}{\mathrm{d}t} \qquad (2.90)$$

yields an action A' that is related to the old action L by

$$A' = A + |\Lambda(q,t)|_{t_1}^{t_2}. \qquad (2.91)$$

therefore, for variations with fixed end points we obtain $\delta A' = \delta A$, so that $\delta A' = 0$ yields exactly the same equations of motion as does $\delta A = 0$. So, we see that the action itself need not be invariant in order to describe symmetries of the system, although invariance of the action is a sufficient condition for a symmetry.

An extension of Noether's theorem to include momentum and energy represented by field equations shows that gauge invariance corresponds to charge conservation (see Mercier, 1963).

2.8 Hamilton's canonical equations

Lagrange's equations are second order in the time for f generalized coordinates. Every second order system for f variables can be rewritten as a first order system for $2f$ variables. Whenever one starts with a Lagrangian system with Lagrangian $L = T - U$, then the first order system takes on the very special form of a Hamiltonian canonical system (the Irish theorist Hamilton lived from 1805 to 1865).

Beginning with the Lagrangian L written in terms of f generalized coordinates q_i, the f generalized velocities $\mathrm{d}q_i/\mathrm{d}t$, and the time t (noninertial reference frames are permitted as are certain time-dependent potentials U, e.g. one can consider a charged particle in an oscillating electric field) and the canonical momenta

$$p_i = \frac{\partial L}{\partial \dot{q}_i}, \tag{2.37e}$$

we construct the Hamiltonian H by the Legendre transform

$$H = \sum_{i=1}^{f} p_i \dot{q}_i - L. \tag{2.92}$$

The point of constructing the Legendre transform is to construct H directly as a function of the $2f$ variables $(q_1, \ldots, q_f, p_1, \ldots, p_f)$ and the time t. Therefore,

$$dH = \sum_{i=1}^{f} \left(\frac{\partial H}{\partial q_i} dq_i + \frac{\partial H}{\partial p_i} dp_i + \frac{\partial H}{\partial t} dt \right) \tag{2.93}$$

while

$$d \sum_{i=1}^{f} p_i \dot{q}_i - dL = \sum_{i=1}^{f} \left(dp_i \dot{q}_i + p_i d\dot{q}_i - \frac{\partial L}{\partial q_i} dq_i - \frac{\partial L}{\partial \dot{q}_i} d\dot{q}_i - \frac{\partial L}{\partial t} dt \right)$$

$$= \sum_{i=1}^{f} \left(dp_i \dot{q}_i - \dot{p}_i dq_i - \frac{\partial L}{\partial t} dt \right). \tag{2.94}$$

Equating coefficients in equations yields Hamilton's canonical form of the equations of motion:

$$\dot{q}_i = \frac{\partial H}{\partial p_i}, \ \dot{p}_i = -\frac{\partial H}{\partial q_i}, \text{ and } \frac{\partial H}{\partial t} = -\frac{\partial L}{\partial t}. \tag{2.95}$$

In this case, we study the motion in *phase space*, the $2f$ dimensional space of the canonically conjugate variables and momenta $(q_1, \ldots, q_f, p_1, \ldots, p_f)$.

It follows that

$$\frac{dH}{dt} = \frac{\partial H}{\partial t} = -\frac{\partial L}{\partial t}, \tag{2.96}$$

so that the Hamiltonian is a conserved quantity whenever the Lagrangian has no explicit time-dependence. That is, whenever the mechanical system has time-translational invariance. In this case, solutions depend only upon time differences $t - t_o$ and not upon the absolute value of the time t. This is the third geometric invariance principle, the one that was missing in our discussion of the connection between symmetry, invariance, and conservation laws in the last section. The reason for this is purely technical: in order to describe this symmetry from the standpoint of a variational principle, one must introduce time-dependent virtual variations (see Mercier, 1963).

For the special case where the kinetic energy is quadratic in the generalized velocities (which means that the coordinate transformations are time-independent),

$$T = \tfrac{1}{2} \sum_{i=1}^{f} g_{ij} \dot{q}_i \dot{q}_j, \tag{2.97}$$

then the coefficients g_{ij} are time-independent and so depend upon the generalized coordinates alone. With $g_{ij} = g_{ji}$, it follows that

$$p_i = \frac{\partial T}{\partial \dot{q}_i} = \sum_{j=1}^{f} g_{ij} \dot{q}_j \tag{2.98}$$

so that $L = T - U$ yields directly that

$$H = \sum_{i=1}^{f} p_i \dot{q}_i - L = 2T - T + U \qquad (2.99)$$

where

$$T = \sum_{i=1}^{f} \tfrac{1}{2} p_i \dot{q}_i. \qquad (2.100)$$

Therefore, in this special case

$$H = T + U = E \qquad (2.101)$$

is the total mechanical energy. Because g_{ij} is time-independent under our assumptions, whenever U is also time-independent, then L and H have no explicit time-dependence and the total mechanical energy E is conserved.

For a first example, we use Hamilton's equations to reproduce Newton's laws for a single particle in Cartesian coordinates. With

$$L = \frac{m}{2}(\dot{x}^2 + \dot{y}^2 + \dot{z}^2) - U(x,y,z) \qquad (2.102a)$$

the canonical momenta are given by

$$\left.\begin{aligned} p_x &= \frac{\partial L}{\partial \dot{x}} = m\dot{x} \\ p_y &= m\dot{y} \\ p_z &= m\dot{z}. \end{aligned}\right\} \qquad (2.103a)$$

The Hamiltonian is

$$H = p_x\dot{x} + p_y\dot{y} + p_z\dot{z} - L = \frac{1}{2m}(p_x^2 + p_y^2 + p_z^2) + U(x,y,z) \qquad (2.104a)$$

and Hamilton's equations are then

$$\left.\begin{aligned} \dot{x} &= \frac{p_x}{m},\ \dot{y} = \frac{p_y}{m},\ \dot{z} = \frac{p_z}{m}, \\ \dot{p}_x &= -\frac{\partial U}{\partial x},\ \dot{p}_y = -\frac{\partial U}{\partial y},\ \dot{p}_z = -\frac{\partial U}{\partial z}. \end{aligned}\right\} \qquad (2.105a)$$

As the second example, we describe the motion of the particle using spherical coordinates:

$$L = \frac{m}{2}(\dot{r}^2 + r^2\dot{\theta}^2 + r^2\sin^2\theta\,\dot{\phi}^2) - U(r,\theta,\phi) \qquad (2.102b)$$

and

$$\left.\begin{aligned} p_r &= \frac{\partial L}{\partial \dot{r}} = m\dot{r} \\ p_\theta &= mr^2\dot{\theta} \\ p_\phi &= mr^2\sin^2\theta\,\dot{\phi}, \end{aligned}\right\} \qquad (2.103b)$$

so that

$$H = p_r \dot{r} + p_\theta \dot{\theta} + p_\phi \dot{\phi} - L = \frac{1}{2m}\left(p_r^2 + \frac{p_\theta^2}{r^2} + \frac{p_\phi^2}{r^2 \sin^2\theta}\right) + U(r,\theta,\phi). \quad (2.104b)$$

Hamilton's equations of motion for the trajectory in the six dimensional $(r,\theta,\phi,p_r,p_\theta,p_\phi)$ phase space are then

$$\left.\begin{aligned}
\dot{r} &= \frac{p_r}{m}, \ \dot{\theta} = \frac{p_\theta}{mr^2}, \ \dot{\phi} = \frac{p_\phi}{mr^2\sin^2\theta}, \\
\dot{p}_r &= \frac{p_\theta^2}{mr^3} - \frac{\partial U}{\partial r}, \ \dot{p}_\theta = \frac{p_\phi^2\cos\theta}{mr^2\sin^3\theta} - \frac{\partial U}{\partial\theta}, \ \dot{p}_\phi = -\frac{\partial U}{\partial\phi}.
\end{aligned}\right\} \quad (2.105b)$$

Hamilton's canonical equations also follow directly from Hamilton's principle If we write the action as

$$A = \int_{t_1}^{t_2} (p_i\dot{q}_i - H)\mathrm{d}t \quad (2.106)$$

and agree to make variations with fixed end points in configuration space, the f dimensional subspace of phase space, then after one integration by parts we obtain

$$\delta A = \int_{t_1}^{t_2}\left[\left(\dot{q}_i - \frac{\partial H}{\partial p_i}\right)\delta q_i + \left(-\dot{p}_i - \frac{\partial H}{\partial q_i}\right)\delta p_i\right]\mathrm{d}t \quad (2.107)$$

so that independent variations (no constraints) of the coordinates and momenta yield Hamilton's equations of motion.

In order to motivate the method of the next chapter, and to show that Hamiltonian systems can arise independently of a Lagrangian, we consider briefly the flow of an ideal incompressible fluid in the Cartesian plane, a fluid with no viscosity and where a definite volume of fluid can change its shape but not its size. An incompressible fluid is one where the fluid density ρ is constant in both space and time. According to the continuity equation

$$\frac{\partial\rho}{\partial t} + \nabla\cdot\rho\vec{V} = 0,$$

incompressibility means that the velocity field $\vec{V} = (V_x,V_y)$ is divergence free, $\nabla\cdot\vec{V} = 0$, so that the velocity is given by a vector potential $\vec{V} = \nabla\times\vec{A}$. Since the flow is two dimensional, we can write $\vec{A} = \hat{n}\psi(x,y)$ where the unit vector is perpendicular to the (x,y)-plane and $\psi(x,y)$ is called the stream function of the flow. Stream functions are only possible in two dimensions. In Cartesian coordinates the flow is then described by the equations

$$V_x = \frac{\partial\psi}{\partial y}, \ V_y = -\frac{\partial\psi}{\partial x}, \quad (2.108a)$$

which bear a nonaccidental resemblance to Hamilton's canonical equations[5], except that in Hamiltonian theory the corresponding variables q and p are *not* restricted to be Cartesian. In fact, if we should consider the motion of any small particle that moves

---

[5] The motion of N point vortices in the plane is a Hamiltonian system, but there is no kinetic energy term in H. The fluid's kinetic energy *is* the potential energy of interaction of the vortices and is the Hamiltonian. Correspondingly, the canonically conjugate coordinates *and momenta* are just the Cartesian *position* coordinates of the vortices (see Lin, 1943).

along with the local flow field, then the motion of that particle is described in Cartesian coordinates by Hamilton's equations

$$\dot{x} = \frac{\partial \psi}{\partial y}, \quad \dot{y} = -\frac{\partial \psi}{\partial x}. \qquad (2.108b)$$

Streamlines, particle trajectories of (2.108b), are lines of constant ψ and the velocity field defined by (2.108a) is everywhere tangent to a streamline. The stream function ψ is time-independent and obeys

$$\frac{\partial \psi}{dt} = \frac{\partial \psi}{\partial x} V_x + \frac{\partial \psi}{\partial y} V_y = \frac{\partial \psi}{\partial x} \frac{\partial \psi}{\partial y} - \frac{\partial \psi}{\partial y} \frac{\partial \psi}{\partial x} = 0 \qquad (2.109)$$

and is therefore a conserved quantity for the flow.

We shall show in the next chapter that the theory of Hamiltonian systems is part of the more general theory of flows in phase space, where a flow can be either compressible or incompressible. There, the important idea is not that of density but of the velocity field that is tangent to the streamlines (the idea of a density can be introduced only for a very restricted class of initial conditions and is of little use in dynamical systems theory). A subclass of all flows in phase space is the class of incompressible flows generated by the 2f dimensional analog of a stream function, the Hamiltonian. There, the generalized coordinates and canonical momenta are treated formally as if they were Cartesian, even though they generally are not Cartesian in Lagrangian configuration space. Theory and applications of Hamiltonian systems are presented in parts of chapters 3, 4, 6, 7, 8, 9, 15, 16, and 17.

Exercises

1. A cord of indefinite length passes freely over pulleys at heights y_1 and y_2 above the plane surface of the earth, with a horizontal distance $x_2 - x_1$ between them. If the cord has a uniform linear mass density, then find the differential equation that describes the cord's shape, and solve it. Show also that the problem is identical with that of finding the minimum surface of revolution.

   (Goldstein, 1980)

2. Suppose that a particle falls a known distance y_o in a time $t_o = \sqrt{(2y_o/g)}$, but that the times of fall for distances other than y_o are not known. Suppose further that the Lagrangian for the problem is known, but instead of solving Lagrange's equations for y as a function of t, we guess that the solution has the form

   $$y = at + bt^2.$$

   If the constants a and b are always adjusted so that the time to fall y_o is correctly given by t_o, show directly that the integral

   $$\int_0^{t_o} L \, dt$$

   is an extremum for real values of the coefficients only when $a = 0$ and $b = g/2$. (Note: there are two separate ways to solve this problem.)   (Goldstein, 1980)

3. The equations of constraint for a rolling disk, are special cases of general linear differential equations of constraint of the form

$$\sum_{i=1}^{n} g_i(x_1, \ldots, x_n) dx_i = 0.$$

A constraint condition of this type is holonomic only if an integrating factor $M(x_1, \ldots, x_n)$ can be found that turns it into an exact differential. Clearly the function must be such that

$$\frac{\partial(M g_i)}{\partial x_j} = \frac{\partial(M g_j)}{\partial x_i}$$

for all $i \neq j$. Show that no such integrating factor can be found for either of (2.41).
(Goldstein, 1980)

4. Two wheels of radius a are mounted on the ends of a common axle of length b such that the wheels rotate independently. The whole combination rolls without slipping on a plane. Show that there are two nonholonomic equations of constraint

$$\cos\theta\, dx + \sin\theta\, dy = 0,$$

$$\sin\theta\, dx - \cos\theta\, dy = \frac{a}{2}(d\phi + d\phi'),$$

(where θ, ϕ, and ϕ' have meanings similar to those in the problem of a single vertical disc, and (x,y) are the coordinates of a point on the axle midway between the two wheels). Show also that there is one holonomic equation of constraint,

$$\theta = C - \frac{a}{b}(\phi - \phi')$$

where C is a constant. (Goldstein, 1980)

5. A particle moves in the (x,y)-plane under the constraint that its velocity vector is always directed towards a point on the x-axis whose abscissa is some given function of time $f(t)$. Show that for $f(t)$ differentiable but otherwise arbitrary, the constraint is nonholonomic. (Goldstein, 1980)

6. Two equal masses m are joined by a rigid weightless rod with length l. The center of mass of the system is constrained to move on a circle of radius a. Set up the kinetic energy in generalized coordinates. (The rod's rotations are not confined to a plane.)

7. Two mass points of mass m_1 and m_2 are connected by a string passing through a hole in a smooth table so that m_1 rests on the table surface and m_2 hangs suspended. Assuming that m_2 moves only in a vertical line but m_1 is free to move in the plane of the table, what are convenient generalized coordinates for the system? Write down the Lagrange equations for the system and then reduce the problem to a single second order differential equation. Obtain a first integral of the equation. Are there initial conditions such that m_1 does not fall through the hole? If so, how can this be possible? Next, repeat the problem for the case where m_2 is not confined to the vertical line.

8. Obtain the Lagrangian and equations of motion for the double pendulum, where the lengths of the rods are l_1 and l_2 with corresponding masses m_1 and m_2.
(Goldstein, 1980)

9. A particle is placed at the top of a vertical hoop. Calculate the reaction of the hoop on the particle by means of Lagrange's undetermined multipliers and Lagrange's equations, and find the height at which the particle falls off.   (Goldstein, 1980)

10. A uniform hoop of mass m and radius r rolls without slipping on a fixed cylinder of radius $R > r$. The only external force is due to gravity. If the hoop starts rolling from rest on top of the cylinder, use the method of Lagrange multipliers to find the point at which the hoop falls off the cylinder.

11. A charged particle moves in a constant electric field $\vec{E}$ which is parallel to a constant magnetic field $\vec{B}$. Find the equation of motion and also the general form of a trajectory. Repeat the problem for the case where $\vec{E}$ and $\vec{B}$ are perpendicular to each other.

12. A problem of constrained motion can be thought of as the limit of an unconstrained problem in which the potential energy grows very rapidly over a short distance α. Consider a particle moving in the potential

$$U(x) = k\left[\frac{1}{\alpha} - \frac{\alpha}{(x-R)^2 + \alpha^2}\right].$$

Draw a rough sketch of U for different values of α. For given energy and angular momentum obtain an expression for the maximum and minimum values of x. Find the force on the particle. Now go to the limit as $\alpha \to 0$, and show that the motion in the limit is the same as the motion for a problem with the appropriate constraint. Show also in the limit that the force on the particle has the properties we have assumed for the force of constraint.   (Saletan and Cromer, 1971)

13. From symmetry considerations alone, find a conserved quantity for the Lagrangian (in three dimensions),

$$L = \tfrac{1}{2}m\dot{\vec{x}}\dot{\vec{x}} - V_o x_1 \sin(2\pi x_3/R) - U_o x_2 \cos(2\pi x_3/R),$$

where U_o and R are constants. Does this agree with Noether's theorem? What is the symmetry of L? Find a generalized coordinate whose corresponding canonical momentum is conserved.   (Saletan and Cromer, 1971)

14. Assume that a one-particle Lagrangian has the symmetry of a uniform helix represented by the transformation $x_1 = A\cos\phi$, $x_2 = A\sin\phi$, $x_3 = c\phi$. Use Noether's theorem to show that $G = cp_3 + L_3$ is conserved.   (Arnol'd, 1989)

15. (a) In Cartesian coordinates, the divergence of a velocity is given by

$$\nabla \cdot \vec{V} = \sum_{i=1}^{n} \frac{\partial \dot{x}_i}{\partial x_i}$$

where $\vec{V} = (\dot{x}_1, \ldots, \dot{x}_n)$. Let $x_i = f_i(q_1, \ldots, q_n)$ where $q_i = h_i(x_1, \ldots, x_n)$ defines any set of curvilinear coordinates. The generalized velocity, a tangent vector in configuration space, is *defined* to be $\vec{V} = (\dot{q}_1, \ldots, \dot{q}_n)$. If $2T = g_{ij}\dot{q}_i\dot{q}_j$, then show that

$$\nabla \cdot \vec{V} = \frac{1}{\sqrt{g}} \sum_i \frac{\partial}{\partial q_i}[(\sqrt{g})\dot{q}_i]$$

where $g = \det g_{ij}$.

(b) Consider spherical coordinates $(q_1, q_2, q_3) = (r, \theta, \phi)$. In elementary vector analysis, the velocity vector is defined by

$$\vec{V}' = \frac{d\vec{r}}{dt} = \dot{r}\hat{r} + r\dot{\theta}\hat{\theta} + r\sin\theta\dot{\phi}\hat{\phi} = (\dot{r}, r\dot{\theta}, r\sin\theta\dot{\phi})$$

where $\hat{r}, \hat{\theta}$, and $\hat{\phi}$ are orthonormal vectors. Note that $\vec{V} = (\dot{r}, \dot{\theta}, \dot{\phi})$ is defined to be the tangent vector in configuration space. Show that the usual divergence formula

$$\nabla' \cdot \vec{V}' = \frac{1}{r^2}\frac{\partial}{\partial r}(r^2 V'_r) + \frac{1}{r\sin\theta}\frac{\partial}{\partial\theta}(V'_\theta \sin\theta) + \frac{1}{r\sin\theta}\frac{\partial}{\partial\phi}V'_\phi$$

agrees with the result of part (a), that $\nabla' \cdot \vec{V}' = \nabla \cdot \vec{V}$ in spherical coordinates, where $g = r^4 \sin^2\theta$ follows from the kinetic energy (see Margenau and Murphy, 1956, p. 196). As we show in chapter 10, the vectors $\vec{e}_i$ in the configuration space formula $\nabla = \Sigma \vec{e}_i \partial/\partial q_i$ are generally *not* orthonormal.

3

Flows in phase space

... the historical value of a science depends not upon the number of particular phenomena it can present but rather upon the power it has of coordinating diverse facts and subjecting them to one simple code.

E. L. Ince, in *Ordinary Differential Equations*

3.1 Solvable vs integrable

In this chapter we will consider n coupled and generally nonlinear differential equations written in the form $dx_i/dt = V_i(x_1, \ldots, x_n)$. Since Newton's formulation of mechanics via differential equations, the idea of what is meant by a solution has a short but very interesting history (see Ince's appendix (1956), and also Wintner (1941)). In the last century, the idea of solving a system of differential equations was generally the 'reduction to quadratures', meaning the solution of n differential equations by means of a complete set of n independent integrations (generally in the form of $n - 1$ conservation laws combined with a single final integration after $n - 1$ eliminating variables). Systems of differential equations that are discussed analytically in mechanics textbooks are almost exclusively restricted to those that can be solved by this method. Jacobi (German, 1804–1851) systematized the method, and it has become irreversibly mislabeled as 'integrability'. Following Jacobi and also contemporary ideas of geometry, Lie (Norwegian, 1842–1899) studied first order systems of differential equations from the standpoint of invariance and showed that there is a *universal* geometric interpretation of all solutions that fall into Jacobi's 'integrable' category. Lie's approach to the study of differential equations is based upon the idea of generating finite coordinate transformations from differential equations viewed as defining infinitesimal coordinate transformations. Lie's methodology was developed to its present advanced state by the Frenchman Cartan (1869–1951), and is used in very advanced texts like those by Fomenko (1988) and Olver (1993). Poincaré (French, 1854–1912) developed the geometric, or qualitative, method of analysis of systems of first order ordinary differential equations as flows in phase space, and also introduced iterated maps as a method for the analysis of solutions. We will refer to solution by reduction to quadratures as complete integrability (or integrability), which we define for flows in phase space along with Lie's geometric interpretation in section 3.6.

Poincaré's qualitative analysis is the best starting point and should always precede any quantitative analysis, whether the quantitative analysis is to be analytic, numeric, or mixed. The reason for this is that the only available general method for solving systems of nonlinear differential equations is the method of successive approximations, Picard's recursive method, which does *not* require Lie–Jacobi integrability and is described in many elementary advanced texts on ordinary differential equations (see Ince (1956) or Moore (1962)). Whenever solutions exist, and whenever the differential equations satisfy a Lipshitz condition, then systematic approximations to those solutions can be generated iteratively (recursively) whether a system is Hamiltonian or nonHamiltonian, whether the motion is regular or chaotic. The iterative method's strength is that it does not discriminate at all. That is simultaneously its weakness. The recursive method of solution is universal and therefore provides no geometric insight into the nature of the motion that we also want to understand qualitatively. Qualitative understanding is useful as a guide to careful numerical computation, meaning computation with error control. Most dynamics problems, whether integrable, nonintegrable, or chaotic, cannot be solved analytically and must be studied numerically, at least in part.

The main point for now is that there is geometric information about the motion that one can obtain directly from the differential equations *without any knowledge whatsoever of the quantitative form of the trajectories*. From a mathematics standpoint, there is no reason to restrict ourselves to Hamiltonian, to Newtonian, or to integrable systems, and so we begin from a much more general starting point, one that is restricted neither to conservative forces nor to problems in Newtonian particle mechanics.

3.2 Streamline motions and Liouville's theorem

We can motivate the study of flows in phase space by rewriting Newton's law of motion for a single particle of mass m in two dimensions, using Cartesian coordinates, as follows: with $(x_1, x_2, x_3, x_4) = (x, dx/dt, y, dy/dt)$ and $(F_2, F_4) = (F_x, F_y)$, we have the first order system of

$$\begin{aligned}\dot{x}_1 &= x_2 \\ \dot{x}_2 &= F_2/m \\ \dot{x}_3 &= x_4 \\ \dot{x}_4 &= F_4/m\end{aligned} \qquad (3.1a)$$

in a four dimensional space, where the forces F_i depend on (x_1, x_2, x_3, x_4). The space of the four Cartesian coordinates (x_1, x_2, x_3, x_4) is called phase space. In a generalization to N particles in d dimensions, the phase space of a Newtonian particle system will have $2Nd$ dimensions and therefore is always even (more generally, given an nth order ordinary differential equation, one can always study that equation as a dynamics problem in phase space by rewriting the equation as a first order system of n coupled differential equations). Note that equilibrium points are given by setting the right hand side of (3.1a) equal to zero. In general, there may be zero, one, or two or more equilibria in phase space depending on the force law in (3.1a) or the functions V_i in (3.2a) below.

In the Newtonian example above, the configuration subspace of phase space, the (x_1, x_3)-subspace, is Euclidean. The coordinates defining phase space in this model are all Cartesian (x_2 and x_4 also are Cartesian), but no meaning can be attached to the idea of a distance between two arbitrary points in the full four dimensional phase space because x_2 and x_4 are velocities while x_1 and x_3 are positions. In all that follows, unless a particular model (3.2a) should dictate that phase space *is* Euclidean with the metric defined by the Pythagorian theorem, then phase space is a vector space without a metric: there is generally no definition of the distance between two points in phase space, but we assume that Cartesian axes and global parallelism are possible. Phase space is an example of an inner product space. The inner product is defined below.

Nonmetric spaces are also used elsewhere in physics: the spaces of thermodynamic variables, the (p,V)-plane and the (E,S)-plane, for example, are Cartesian but not Euclidean because there is no metric: it makes no sense physically or geometrically to define a distance between two points in the (p,V)- or (E,S)-planes.

More generally, and without needing to refer to Newton or classical mechanics, and where the axes labeled by coordinates x_i are treated formally as Cartesian (although they are not necessarily Cartesian in 'physical' three dimensional Euclidean space) consider any autonomous system of n coupled ordinary differential equations

$$\dot{x}_i = V_i(x_1,\ldots,x_n) \tag{3.2a}$$

where the index runs from i to n and n can be even or odd. Zeros of all n of the V_i define equilibria of the system. Autonomous (or self-regulating) means that there is no explicit time-dependence in the functions V_i, that each V_i depends only on the point $x = (x_1,\ldots,x_n)$ and not explicitly upon t. This defines what is known in the modern literature as an 'abstract dynamical system', including Newtonian and Hamiltonian dynamics, *if* we impose the restriction that the solutions exist as *point functions of the initial conditions and the time* (the usual idea of a solution), are unique, differentiable, and remain finite for all *finite* times t. We shall discuss the condition that guarantees all of this below. First, we introduce the phase flow (streamline) picture of dynamics.

A 'dynamical system' is defined to be completely deterministic. With the conditions of uniqueness and finiteness of solutions for all times t, both past and future are completely determined by the specification of a single infinite-precision initial condition $(x_{1_o},\ldots,x_{n_o})$. There is no possibility of randomness in a deterministic system, which operates like an abstract machine. We shall see in chapters 4 and 6 that randomness may be mimicked deterministically in the form of pseudo-randomness in the simplest cases. Normally, we assume that initial conditions and parameters are 'chosen' from the mathematical continuum, which consists of 2^∞ points (think of numbers as binary strings). Clearly, this is impossible on a computer, in experiment, and in observations of nature. Whenever initial conditions and parameters are chosen to be computable, meaning that they can be generated decimal by decimal via algorithms, then the machine-like quality of the time-evolution of a dynamical system is most clearly seen in its reduction to the processing of decimal strings of numbers representing positions, $x_i(t)$, by an automaton that is defined by the differential equations. We illustrate automata for discrete iterated maps derived from differential equations in chapter 6. For example, $x(t) = x_o e^t$ is not computable unless both the initial condition x_o and the time t are chosen to be computable numbers. In all of the discussion of numerical computations of solutions that follows, we assume *implicitly* that this is the case (in floating-point arithmetic on a computer, the roundoff-truncation algorithm is unknown to the programer and was constructed by the engineer who designed the chip).

The n dimensional space is called the phase space or the state space of the dynamical system. Each point $(x_1,\ldots,x_n)$ represents a possible state (and therefore also a possible initial condition) of the dynamical system at one time t. A trajectory of the dynamical system is a curve (a one dimensional path) in the n dimensional phase space. According to (3.2a) the velocity vector $V = (V_1,\ldots,V_n)$ at any point $x = (x_1,\ldots,x_n)$ in phase space is tangent to the particular trajectory that passes through that point (the condition for uniqueness of trajectories is stated below).

In fluid flow in three dimensions the velocity field $v(r,t)$ of the flow is tangent to the streamline of a 'fluid particle' (a 'very small' mass of the fluid), and a plot of the vector

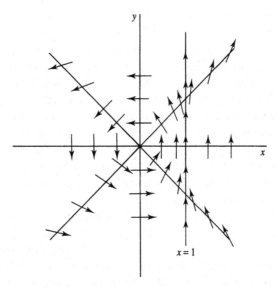

Fig. 3.1 Phase diagram for $V = (xy - y, x)$.

field $v(r,t)$ as a function of position r for one fixed time t exhibits the flow pattern at that time. The plot represents a 'snapshot' of the streamline pattern where each streamline is the trajectory followed by a hypothetical 'fluid particle'. If the velocity field is time-independent, representing a steady flow, then the flow pattern is the same for all times t. We aim to develop the following analogy: the family of trajectories in phase space is the steady streamline pattern of the time-independent velocity field V that appears in (3.2a).

Without knowing a single streamline (a phase trajectory of the dynamical system for a definite initial condition), we can get a qualitative picture of the flow pattern by plotting the *direction* of the velocity field at a lot of different points in phase space. Here, we can ignore the magnitude of the components of V except at the special points where one or more components obey $V_i = 0$.

We give as an example the flow field $V = (V_x, V_y) = (xy - y, x)$. The flow pattern is shown in figure (3.1) and was constructed from the following limited information: at $x = 0$ $V = (-y, 0)$ and at $y = 0$ $V = (0, x)$. Also, $V_x = 0$ along the vertical line $x = 1$ where $V_y = x > 0$, which therefore cannot be crossed by the flow (because there is no flow normal to this line). On the line $y = x$ we have $V = (x^2 - x, x)$ so that for $x > 0$ we get $V_x > 0$ if $x > 1$, but $V_x < 1$ if $0 < x < 1$, while $V_y > 0$. On the line $y = -x$, $V_x < 0$ if $x < 0$, but for $x > 0$ we find that $V_x < 0$ if $x > 1$ but $V_x > 0$ if $0 < x < 1$. The flow pattern in figure (3.1) shows that for $x < 1$ there is some kind of circulation about the origin (which is an equilibrium point) but for $x > 1$ trajectories appear to be attracted toward the point $(1,0)$ for $y < 0$ but then are scattered away from it for $y > 0$. We do not know if the streamlines that circulate about the origin are open or closed, but we shall see that Liouville's theorem (derived below) and some general understanding of conservative and dissipative systems (derived in sections 3.3, 3.4, and 3.5) will lead us to conclude that this is a driven flow where there may be open, spiraling orbits. Whether there may be one or more closed orbits somewhere in the phase plane requires further analysis. This illustrates the first step in qualitative analysis, the plotting of the velocity field, which is also called 'plotting the phase portrait', or 'plotting the flow diagram'.

Here are three easy examples from Newtonian mechanics. Free-particle motion $d^2x/dt^2 = 0$ is described by the velocity field $V = (x_2,0)$. The streamlines are parallel to the x_1-axis and represent a constant velocity translation along that axis. The construction of the following two flow diagrams is assigned in exercise 1. The simple harmonic oscillator $d^2x/dt^2 + x = 0$ with $m = 1$ and $\omega = 1$ has the flow field $V = (x_2, -x_1)$ and describes circulation about the origin, which is an equilibrium point (figure 3.2). Because the velocity field has equal magnitude at equal distances from the origin and is always perpendicular to the radius vector from the origin, the streamlines must be concentric circles about the origin (figure 3.2 shows the more general case of concentric ellipses). The scattering problem $d^2x/dt^2 - x = 0$ defines the flow field $V = (x_2,x_1)$, which also has an equilibrium point at the origin and yields the flow field of figure 3.3, which reminds us of hyperbolae.

It is easy to derive the equations describing the streamlines for the last two examples. For the simple harmonic oscillator, from $x_1 dx_1/dt + x_2 dx_2/dt = x_1 x_2 - x_2 x_1 = 0$ we obtain the streamline equation as $G(x,dx/dt) = x_1^2 + x_2^2 = $ constant, which describes a family of circles concentric about the origin. For the scattering problem, $x_1 dx_1/dt - x_2 dx_2/dt = x_1 x_2 - x_2 x_1 = 0$ follows and so we get the streamline equation in the form $G(x,dx/dt) = x_1^2 - x_2^2 = $ constant, describing a family of hyperbolae centered on the origin. For an arbitrary planar flow it is nontrivial to derive the streamline equation, so that plotting of the directions of the vector field V in one or more two dimensional subspaces can be an extremely valuable tool for understanding the flow pattern. For an arbitrary flow in three or more dimensions there is no guarantee that a streamline equation exists because every streamline equation is a time-independent conservation law. The existence of global[1] conservation laws is not guaranteed in three or more dimensions, as we explain below.

Summarizing, for any trajectory (any solution of the dynamical system for $t = -\infty$ to $t = \infty$), the velocity $V = (V_1,V_2,\ldots,V_n)$ is everywhere tangent to that trajectory. If we think of the collection of all trajectories as the flow of a phase fluid, then $V(x_1,\ldots,x_n)$ is just the *velocity field* of the flow, is known, and therefore can be plotted in phase space or in a two (with computer graphics, three) dimensional subspace of phase space independently of any quantitative information about solutions. One can use this technique to get limited information about what solutions do as t increases (or decreases) without knowing how a single trajectory develops quantitatively from a single precise initial condition. We now develop the language and tools of the flow picture.

Whenever the velocity field is at least once differentiable with respect to all of the n variables x_i, when all of those first partial derivatives are bounded in some finite region of phase space, then the solution of the system (3.2a) exists as a point function of the initial conditions and the time, and is also unique for a specific initial condition within the bounded region. Existence of solutions is nontrivial: an alternative to the existence of solutions as *functions* might be the idea of solutions as *path-dependent functionals*. More general than first-differentiability, if the velocity field satisfies a Lipshitz condition then both existence and uniqueness of solutions are guaranteed along with an approximate method of construction of solutions: the dynamical system is then iteratively solvable over a finite region in phase space, over a finite time interval, by Picard's method of successive approximations, regardless of what may be true of the

---

[1] A 'global' conservation law holds for all finite times. In chapter 13 we discuss 'local' conservation laws that do not.

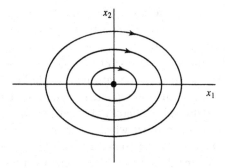

Fig. 3.2 Streamlines of a simple harmonic oscillator.

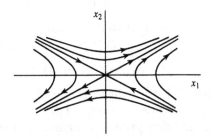

Fig. 3.3 Phase flow of a scattering problem.

motion *qualitatively*: whether the system is Hamiltonian, whether the motion is regular or chaotic, and whether or not the velocity field represents a flow in phase space are in principle irrelevant. We assume, in all that follows, that solutions, velocities, and accelerations are all bounded (but not necessarily uniformly) at all *finite* times (the solution does not blow up at any finite time, but may blow up as the time goes to infinity). *Solvable* means (i) that solutions exist as point functions of initial conditions and time (that integration of the differential equations does not lead to path-dependent answers), are unique and differentiable, and (ii) that solutions can be constructed approximately by recursion. If computable numbers are chosen as parameters and initial conditions, then solutions may be constructed to any desired degree of decimal accuracy by Picard's recursive method (in chapter 6, we point out the fundamental limitations imposed on predictability by computation time whenever a flow is chaotic).

Determinism is reflected abstractly by the fact that the solutions of (3.2a) depend only upon the initial conditions and the time difference $t - t_o$:

$$x_i(t) = \Psi_i(x_1(t_o),\dots,x_n(t_o),t - t_o), \tag{3.3a}$$

where t_o is the time at which the initial conditions are specified and where Ψ is the same function for all times and for all initial conditions. We can think of the differential equations (3.3a) as generating a one-parameter collection of coordinate transformations from holonomic coordinates $(x_1(t_o),\dots,x_n(t_o)) = (x_{1o},\dots,x_{no})$ to holonomic coordinates $(x_1(t),\dots,x_n(t))$, where the parameter in the transformation is just the time difference $t - t_o$. Each different value of the time difference $t - t_o$ defines a different coordinate transformation and path-independence of the integration of the differential equations ('existence' of solutions) guarantees that the variables $(x_1(t),\dots,x_n(t))$ are holonomic for all times t.

The idea of a flow is a fixed pattern of streamlines that threads throughout phase space and is the same at all times: the phase portrait represents the idea that the family of solutions (3.3a) of the system (3.2a) exists *as the same continuous pattern* for all times $-\infty < t < \infty$. From an analytic standpoint, this is the same as saying that the transformations (3.3a) 'exist' abstractly and can, in principle, be carried out for any and all finite times t. This is a nontrivial requirement: it requires that a flow, the coordinate transformation picture, is defined by differential equations whose solutions do not go to infinity at any *finite* time t_s. In other words, *flows are defined by equations whose solutions (3.3a) are unique and remain bounded for all finite times*. An example of a planar flow is given by the one dimensional Newtonian scattering problem

$$\left.\begin{array}{l} \dot{x} = p/m \\ \dot{p} = -U'(x) \end{array}\right\} \tag{3.1b}$$

where $U(x) = -(ax)^2$ with a real. Trajectories are given by

$$\left.\begin{array}{l} x(t) = c_1 e^{at} + c_2 e^{-at} \\ p(t) = ma(c_1 e^{at} - c_2 e^{-at}) \end{array}\right\} \tag{3.3b}$$

and are (nonuniformly) bounded at all finite times. The two constants c_i are determined by initial conditions. That trajectories are scattered to infinity at *infinite* time does not contradict the idea of a definite flow pattern at any and all finite times: the coordinate transformation (3.3a) is given by (3.3b) with $x_o = c_1 + c_2$ and $p_o = ma(c_1 - c_2)$. On the other hand, the one dimensional system

$$\dot{x} = x^2 \tag{3.2b}$$

has the solution

$$x(t) = \frac{x_o}{1 - x_o(t - t_o)} \tag{3.3c}$$

and therefore does *not* define a phase flow: the solution has singularities at finite times $t_s = t_o + 1/x_o$ so that there *is* no flow pattern that continues from $t = -\infty$ to $t = \infty$. Correspondingly, there exist no transformations of the form (3.3a) that can be carried out over time intervals that include t_s. Therefore, there is no flow diagram, no steady flow pattern that is correct independently for all times t. Finite-time singularities (also called spontaneous singularities) are a feature of nonlinear problems: the coefficients in the differential equation, which are completely analytic and even are constant in (3.2b) give no hint beforehand that singularities will appear (in retrospect, we must always ask whether quadratic terms on the right hand side produce spontaneous singularities). That linear equations cannot have spontaneous singularities is guaranteed by Fuch's theorem (see Bender and Orszag, 1978 or Ince, 1956). We prove in chapter 13 that a class of damped, driven Euler–Lagrange models that include tops (chapter 9) and the Lorenz model (introduced below) have no spontaneous singularities and therefore define flows in phase space.

If the right hand side $V(x_1,\ldots,x_n)$ is analytic and time-independent (meaning that the right hand side is a polynomial or a convergent power series in the x_i, excepting isolated singular points), then it is easy to demonstrate that Picard's recursive method merely reproduces the solution as a power series in the time interval $t - t_o$: when the right hand side of the differential equation is analytic in the coordinates x_i (and whether or not we have a flow), then the solution is also analytic in the time. *However, the radius of* convergence of the resulting power series solution is generally smaller than that of the radius of convergence of the V_i as functions of the x_i. For example $V(x) = x^2$ has infinite radius of convergence in x whereas $x(t) = x_o(1 + x_o\Delta t)^{-1} = x_o(1 - x_o\Delta t + x_o^2\Delta t^2 - ...)$ has the finite radius of convergence $\Delta t = x_o^{-1}$.

Continuing with the Lie-inspired picture of a phase flow as an infinite collection of one-parameter coordinate transformations in phase space, the inverse transformation is defined formally by

$$x_i(t_o) = \Psi_i(x_1(t), ..., x_n(t), t_o - t) \tag{3.3d}$$

and represents the idea that the differential equations can be integrated backward in time (this is possible whenever the Jacobi determinant defined below is finite). Differentiation of (3.3a) once with respect to t yields

$$\dot{x}_i(t) = d\Psi_i(x_1(t), ..., x_n(t), t - t_o)/dt. \tag{3.2c}$$

According to (3.3a) and (3.2c), we can think of both the solution $x_i(t)$ and the derivatives $dx_i(t)/dt$ as functions of the n initial conditions. This possibility is used implicitly in the construction of the Jacobi determinant below. If we set $t = t_o$ in (3.2c) then we obtain the original differential equations for *any* time $t = t_o$

$$\dot{x}_i(t_o) = d\Psi_i(x_1(t_o), ..., x_n(t_o), t - t_o)/dt)_{t=t_o} = V_i(x_{1o}, ..., x_{no}), \tag{3.2d}$$

because the zero of time is irrelevant in an autonomous system: the differential equations (3.2a) are left unchanged by (are invariant under) a shift in the time origin. The finite coordinate transformation (3.3a) is said to be 'generated infinitesimally' by the differential equations (3.2a), which can be written in the form of an infinitesimal coordinate transformation near the identity:

$$\delta x_i(t) = V_i(x_{1o}, ..., x_{no})\delta t. \tag{3.2e}$$

The finite coordinate transformation (3.3a) is called 'global' because it holds at all times $-\infty < t < \infty$ and transforms all of phase space, whereas the infinitesimal transformation (3.2e) is confined to infinitesimal displacements.

An easy example of the transformation/phase flow picture is provided by the simple harmonic oscillator where, for $t_o = 0$,

$$\left. \begin{aligned} x_1(t) &= \psi_1(x_{1o}, x_{2o}, t) = x_{1o}\cos t + x_{2o}\sin t \\ x_2(t) &= \psi_2(x_{1o}, x_{2o}, t) = -x_{1o}\sin t + x_{2o}\cos t. \end{aligned} \right\} \tag{3.3e}$$

It is left as an exercise for the reader to construct the inverse transformation that is equivalent to an integration of the harmonic oscillator differential equations backward in time. The infinitesimal transformation near the identity (near $t = 0$) is given by differentiating (3.3e),

$$\left. \begin{aligned} \delta x_1(t) &= x_{2o}\delta t \\ \delta x_2(t) &= -x_{1o}\delta t, \end{aligned} \right\} \tag{3.1c}$$

and reflects the original differential equations of the flow for times sufficiently near to $t = t_o$.

We can also write the global transformation (3.3a) symbolically as

$$x_i(t) = U(t - t_o)x_i(t_o) \tag{3.3f}$$

where $x_i(t_o)$ is the initial condition, $U(t - t_o)$ is called the time-evolution operator and symbolically represents the solution, the result of the simultaneous solution (generally via recursion) of the n differential equations that generate the transformation, leading to the functions ψ_i in (3.3a). The purpose of introducing the time-evolution operator into the theory of differential equations is to be able to represent and discuss some of the properties of the transformations (3.3a) more conveniently via purely algebraic manipulations (see chapter 7).

If the velocity field of a flow is analytic, then so is the solution and the time-evolution operator takes on a particularly simple form: we make a Taylor expansion (where the radius of convergence is necessarily determined by the nearest singularity of (3.3a) in the complex t-plane (see Ince, 1956, or Bender and Orszag, 1978)),

$$x_i(t + \Delta t) = x_i(t) + \Delta t \dot{x}_i(t) + \Delta t^2 \ddot{x}_i(t)/2 + \ldots, \tag{3.3g}$$

and also notice that we can write (using summation convention)

$$\frac{d}{dt} = \dot{x}_i \frac{\partial}{\partial x_i} = V \cdot \nabla, \tag{3.2f}$$

for the derivative with respect to t, $dF(x_1,\ldots,x_n)/dt = LF$ of any time-independent function $F(x_1,\ldots,x_n)$, where $L = V \cdot \nabla$. For example, $dx_i/dt = Lx_i = V_i$. Repeated differentiation of the differential equations (3.2a) with respect to t therefore yields the Taylor series (3.3g) in the form

$$x_i(t + \Delta t) = x_i(t) + \Delta t L x_i(t) + \Delta t^2 L^2 x_i(t)/2 + \ldots = e^{\Delta t L} x_i(t). \tag{3.3h}$$

Therefore, $U(t - t_o) = e^{\Delta t L}$ is the time-evolution operator and simply summarizes for us, in algebraic form, the infinite order differential operator that represents the idea of 'integrating', meaning solving, a system of differential equations with analytic velocity field (excepting isolated singularities). An infinite order differential operator compresses into a single formula the idea of the infinitely many coupled integrations, the abstract limit of Picard's iterative method when the number of recursive steps is allowed mentally to go to infinity[2]. We stress that the radii of convergence of formulae like (3.3g) and (3.3h) are unknown because the location of singularities of solutions in the complex time plane are a priori unknown (it is not accidental that Newton discovered the universal laws of dynamics and gravity from systems whose solutions are parabolae and ellipses).

The time-evolution operator representing the solution (3.3e) of the simple harmonic oscillator was easy to construct because the flow is linear and gives rise to solutions that are analytic for all finite real times t, so that the Taylor expansion of the solution about any point in phase space has infinite radius of convergence (Galileo's first empirical discovery of the regularities of mechanical systems came from observations of simple pendulums, whose solutions are correspondingly analytic). Except for a degenerate case, the solutions of linear equations with constant coefficients have no singularities in the complex t-plane except at infinite real or imaginary time (this is shown explicitly by the general result (3.17c) below). For the simple harmonic oscillator we can insert

$$L = x_2 \frac{\partial}{\partial x_1} - x_1 \frac{\partial}{\partial x_2} \tag{3.2g}$$

---

[2] One must define a metric in phase space in order to discuss convergence; the metric is Euclidean. Given a metric, the convergence in Picard's method corresponds to iteration toward a fixed point in a contraction mapping, whose solution is therefore path-independent.

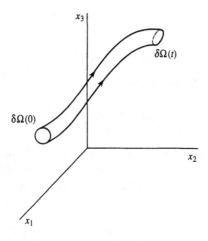

Fig. 3.4 A ball of initial conditions traces out a hypercylinder of streamlines in phase space.

in $x_i(t) = U(t) = e^{\Delta t L} x_i(0)$ to construct the Taylor expansions whose summations yield the solution of the simple harmonic oscillator problem. This is left as an easy exercise for the reader.

Although we have restricted our considerations to autonomous systems, our ideas can be extended formally to externally driven systems where the velocity field $V(x_1, \ldots, x_n, t)$ is time-dependent by writing $x_{n+1} = t$ with $V_{n+1} = 1$ and then working in an $n + 1$ dimensional phase space. An example is given below and in exercise 10.

The flow of a small but finite portion of the phase fluid is described by the (hyper)tube of streamlines traced out by the time-development starting from a finite phase space volume element $\delta\Omega(t_o) = \delta x_1(t_o) \ldots \delta x_n(t_o)$. Every point in this volume element serves as a possible initial condition for the dynamical system, so that each point in the volume element has a trajectory passing through it. Uniqueness of trajectories is guaranteed by satisfying a Lipshitz condition, which is also the condition required for the applicability of Picard's method (a Lipshitz condition is automatically satisfied if the first derivatives of the velocity field are all bounded at all finite times, which automatically requires boundedness of solutions at all finite times for polynomial velocity fields). The collection of trajectories passing through $\delta\Omega$ forms a (hyper)cylinder, a tube in phase space (figure 3.4).

In nature, we can never specify the location of a point, a state, or anything with infinite precision, so that a finite value of $\delta\Omega(t_o)$ can represent the finite ignorance that is present at the start of every experiment or sequence of observations. In numerical computation where errors are avoided (in contrast with floating-point computations), initial conditions can be specified algorithmically with arbitrarily high precision: decimal expansions of numbers can be generated to within as high a precision as computer time will permit. Even in a floating-point computation the initial conditions are determined precisely by the machine's design: the choice of a *single* floating-point initial condition on a computer does not simulate the limitation in experiment or observations of nature to finite and relatively low resolution.

For the time being, we set aside the very difficult question of how the volume element $\delta\Omega(0)$ might differ significantly in *shape* from $\delta\Omega(t)$, and concentrate instead upon the much easier question: how does the *size* of the volume element change with time? If $\delta\Omega$

is small enough, in principle infinitesimal, then thinking of equation (3.3a) as a transformation of coordinates, we can write

$$\delta\Omega(t) = J(t)\delta\Omega(t_o) \tag{3.4}$$

where we could take $t_o = 0$, and where

$$J(t) = \frac{\partial(x_1(t),\ldots,x_n(t))}{\partial(x_1(t_o),\ldots,x_n(t_o))} \tag{3.5a}$$

is the Jacobi determinant of the transformation. It would appear that we cannot use this result without first solving the differential equations, but this is not true: the Jacobi determinant also satisfies a differential equation. To see this, we need only study the form of the derivative[3]

$$\frac{dJ}{dt} = \frac{\partial(\dot{x}_1(t),x_2(t),\ldots,x_n(t))}{\delta(x_1(t_o),x_2(t_o),\ldots,x_n(t_o))} + \ldots + \frac{\partial(x_1(t),\ldots,\dot{x}_n(t))}{\partial(x_1(t_o),\ldots,x_n(t_o))}. \tag{3.6}$$

By the chain rule for Jacobi determinants, we can write

$$\frac{\partial(x_1(t),\ldots,\dot{x}_i(t),\ldots,x_n(t))}{\delta(x_1(t_o),\ldots,x_n(t_o))} = \frac{\partial(x_1(t),\ldots,\dot{x}_n(t),\ldots,x_n(t))}{\partial(x_1(t),\ldots,x_n(t)} J(t) \tag{3.7a}$$

and this reduces to

$$\frac{\partial(x_1(t),\ldots,\dot{x}_i(t),\ldots,x_n(t))}{\partial(x_1(t_o),\ldots,x_n(t_o))} = \frac{\partial\dot{x}_i(t)}{\partial x_i(t)} J(t), \tag{3.7b}$$

because the variables $(x_1(t),\ldots,x_n(t))$ are independent. The quantity

$$\nabla\cdot V = \sum_{i=1}^{n} \frac{\partial\dot{x}_i}{\partial x_i} \tag{3.8a}$$

defines the divergence of the flow field V in phase space, so we obtain the first order differential equation

$$\dot{J} = \nabla\cdot VJ \tag{3.9}$$

for the time-development of the Jacobian.

Now we come to a very important point in our formulation of the properties of phase space. By defining the divergence of the velocity field by the formula

$$\nabla\cdot V = \sum_{i=1}^{n} \frac{\partial V_i}{\partial x_i}, \tag{3.8b}$$

we have implicitly adopted the *Cartesian* definition of a scalar product, but *without* assuming that we can define the distance between any two points in phase space (the Cartesian result (3.8b) was derived from the Cartesian definition $d\Omega = dx_1\ldots dx_n$). Unless the phase space is Euclidean (which is rare), this means that phase space is an inner product space that is not a metric space. Clearly, the phase space of a canonical Hamiltonian system is not a metric space (canonically conjugate variables q and p generally do not even have the same units). The Cartesian definition $d\Omega = dx_1\ldots dx_n$ requires a space where infinite parallelism is possible, but no idea of a distance between

---

[3] This proof of Liouville's theorem was sometimes used by Lars Onsager in his statistical mechanics course at Yale.

two points is required in this definition other than that between two points that lie on the same axis x_i. Note that the Cartesian properties of phase space are preserved under affine transformations but not under general nonlinear transformations of coordinates (see exercise 5).

Conservative dynamical systems can now be *defined* by the divergence-free condition $\nabla \cdot V = 0$, which means that a conservative system is now defined to be one where the phase flow is incompressible; if we can think of the flow as that of a fluid with mass density $\rho(x_1, x_2, \ldots, x_n, t)$, then local mass conservation (or conservation of probability) corresponds to the continuity equation

$$\frac{\partial \rho}{\partial t} + \nabla \cdot \rho V = 0, \tag{3.10a}$$

where ∇ is the n dimensional gradient in phase space. If the divergence of the velocity field vanishes, then we can rewrite (3.10a) in the form

$$\dot{\rho} = \frac{\partial \rho}{\partial t} + V \cdot \nabla \rho = 0, \tag{3.10b}$$

which is the condition that $d\rho/dt = 0$, that ρ is constant in time along a streamline ($\dot{\rho} = d\rho/dt$ is the time derivative along a streamline in the flow). *A conservative flow is one where there are no sources or sinks of the flow in phase space.* This does *not* mean that every Hamiltonian system has one or more conservation laws, but it does include both energy-conserving and all other Hamiltonian systems as special cases. For example, it is easy to show that a Newtonian system where the net force is conservative (and therefore where energy is conserved) is included in our more general definition: the one dimensional Newtonian problem

$$\left. \begin{array}{l} \dot{x} = p/m \\ \dot{p} = -U'(x) \end{array} \right\} \tag{3.1b}$$

yields

$$\nabla \cdot V = \frac{\partial \dot{x}}{\partial x} + \frac{\partial \dot{p}}{\partial p} = 0. \tag{3.8c}$$

Conservative flows are also called volume-preserving flows because the size of the volume element is left invariant by the one-parameter coordinate transformations (3.3a) that define the time-development of the flow.

The canonical form of a conservative system of $n = 2f$ first order differential equations is Hamiltonian and defines a phase-volume-preserving flow where the condition of a source- and sink-free velocity field V is satisfied by the construction of the $2f$ dimensional analog of a stream function called the Hamiltonian, $H(x_1, \ldots, x_f, y_1, \ldots, y_f, t)$. The flow is then described by the canonical differential equations

$$\dot{x}_i = \frac{\partial H}{\partial y_i} \text{ and } \dot{y}_i = -\frac{\partial H}{\partial x_i} \text{ for } i = 1, \ldots, f \tag{3.11}$$

where each pair of variables (x_i, y_i) is called canonically conjugate. It follows immediately that

$$\nabla \cdot V = \sum_{i=1}^{f} \left(\frac{\partial \dot{x}_i}{\partial x_i} + \frac{\partial \dot{y}_i}{\partial y_i} \right) = \sum_{i=1}^{f} \left(\frac{\partial^2 H}{\partial y_i \partial x_i} - \frac{\partial^2 H}{\partial x_i \partial y_i} \right) = 0, \tag{3.12}$$

so that every canonical Hamiltonian flow is automatically incompressible (in chapter 15 we shall also introduce and discuss noncanonical Hamiltonian flows). In accordance with the viewpoint that the canonical equations (3.11) generate the flow infinitesimally and yield the finite transformation that takes one from the coordinate system $(x_1(t_o), \ldots, x_f(t_o), y_1(t_o), \ldots, y_f(t_o))$ to the system $(x_1(t), \ldots, x_f(t), y_1(t), \ldots, y_f(t))$, the Hamiltonian H is called the infinitesimal generator of the transformation. As an example, the Hamiltonian $H = xy$ has the equations of motion $\dot{x} = \partial H / \partial y = x$ and $\dot{y} = -\partial H / \partial x = -y$. We show in section 3.6 that every conservative planar flow is Hamiltonian. However, most Hamiltonian systems that can be constructed mathematically (like $H = xy$) cannot be Legendre-transformed into Lagrangian descriptions of Newtonian flows.

If a Hamiltonian system can be derived from a Newtonian one, then we will write $(x_1, \ldots, x_f, y_1, \ldots, y_f) = (q_1, \ldots, q_f, p_1, \ldots, p_f)$ in accordance with the notation of chapter 2, but note that we have arrived here at the idea of a Hamiltonian description completely independently of Newton's laws and Lagrangian mechanics simply by thinking of an even number of differential equations with a divergence-free velocity field as describing an incompressible, inviscid flow in a phase space of even dimension. Notice that if H depends explicitly upon the time then the dynamical system may have no conserved quantity, but the $n + 1$ dimensional flow defined by setting $x_{n+1} = t$ is still volume-preserving.

Lagrangians $L = T - U$ and the corresponding Hamiltonians in Newtonian particle mechanics represent conservative forces because these systems follow from the requirement that the generalized force is the configuration space gradient of a scalar potential (in Cartesian coordinates, the forces are curl-free). The potentials U may be time-dependent, in which case neither the Hamiltonian nor the total energy is conserved. Our general definition of a conservative system as an incompressible flow in phase space includes the case where H varies with time and is not conserved, but our definition is very general and can also be applied to flows that are not defined by Lagrangians and even to abstract dynamical systems that are represented by discrete maps with a discrete time variable and that have no velocity field at all. An example is provided by the discrete abstract dynamical system defined by the iterated map

$$\left. \begin{array}{l} x_{n+1} = f(x_n, y_n) \\ y_{n+1} = g(x_n, y_n) \end{array} \right\} \tag{3.3i}$$

where phase space is the two dimensional space of the variables (x,y). This discrete dynamical system is conservative if the map is area-preserving, i.e., if the Jacobi determinant

$$J = \begin{vmatrix} \dfrac{\partial f}{\partial x_n} & \dfrac{\partial f}{\partial y_n} \\[2mm] \dfrac{\partial g}{\partial x_n} & \dfrac{\partial g}{\partial y_n} \end{vmatrix} \tag{3.5b}$$

of the transformation (3.3i) is equal to unity. A dissipative flow is one where $J < 1$.

In order to check our definition further against the usual definition, let us ask whether the damped, driven harmonic oscillator

$$\ddot{y} + \beta\dot{y} + \omega^2 y = F(t) \tag{3.1d}$$

is conservative or dissipative. With $x_1 = y$, $x_2 = dy/dt$, and $x_3 = t$, we obtain

$$\nabla \cdot V = \frac{\partial \dot{y}}{\partial y} + \frac{\partial \ddot{y}}{\partial \dot{y}} + \frac{\partial \dot{x}_3}{\partial t} = -\beta, \tag{3.8d}$$

so that the system is nonconservative, as expected, under our broader definition. Furthermore, it follows by a direct integration of (3.9) that $J(t) = e^{-\beta t}$, so that any finite volume element obeys $\delta\Omega(t) = e^{-\beta t}\delta\Omega_0$ and vanishes as the time goes to infinity. If $F = 0$ then this is easily understood: all solutions approach the equilibrium point in phase space because friction (with $\beta > 0$) drains energy out of the system at the rate $dE/dt = -\beta v^2$ where $v = dy/dt$. This equilibrium point illustrates the idea of a zero dimensional attractor. If $F \neq 0$, if for example $F(t)$ is oscillatory, then $y(t)$ consists of two parts: a transient contribution that depends upon the initial conditions and vanishes as t goes to infinity, and a particular integral which is a solution that is dependent of the initial data. In this case, all solutions that have initial conditions in $\delta\Omega(t_o)$ approach that curve asymptotically, illustrating the idea of a one dimensional attractor in the three dimensional $(y,dy/dt,t)$-phase space, and therefore $\delta\Omega(t)$ vanishes as t approaches infinity because the attracting point set is one dimensional and therefore has zero volume.

The damped, driven harmonic oscillator illustrates something that is quite general: whenever the flow field V has a negative divergence, then every finite volume element contracts asymptotically onto an attracting point set that has a dimension less than the dimension of the phase space: the resulting condition that $\delta\Omega(t) \to 0$ as t goes to infinity is satisfied by flow into a point-sink (zero dimensional attractor), contraction onto stable limit cycle (one dimensional attractor), and by higher dimensional possibilities that are discussed in chapter 13. In a general nonconservative flow, phase space is divided into regions where driving dominates ($J > 1$) and regions where dissipation dominates ($J < 1$). Curves or closed surfaces in phase space where $J = 1$ but $J > 1$ (or $J < 1$) inside the region while $J < 1$ (or $J > 1$) outside have special significance because they lead to the idea of higher dimensional attractors (repellers). An example of an autonomous nonconservative planar flow with a one dimensional attractor, a limit cycle, is given in section 3.3.

If we set $F = 0$ and replace β by $-\beta$ with $\beta > 0$ in the damped harmonic oscillator (3.1d), then we get a driven oscillator where $\delta\Omega(t)$ increases exponentially with the time t. Instead of spiraling into the origin like the trajectories of the damped oscillator, trajectories spiral out to infinity (we prove this explicitly in the next section). In general, regions of phase space where the divergence of the velocity is negative are regions where damping dominates driving ($\delta\Omega$ decreases toward zero), while regions with positive divergence are regions where external driving dominates dissipation ($\delta\Omega$ increases with time). We shall see in sections 3.3 and 3.4 that closed orbits (periodic orbits) in phase space are generally characteristic of conservative flows, but that dissipative flows can also have closed orbits (periodic solutions called limit cycles are one example) whenever damping balances dissipation locally over some region of phase space to produce a steady state (see sections 3.3 and 3.4).

In a Newtonian or Hamiltonian description of particle mechanics in configuration space, the coordinates x_i that are treated implicitly as Cartesian in phase space may actually be cylindrical, spherical, elliptic, or some other curvilinear possibility. The configuration space of a Lagrangian system is therefore geometrically distinct from the q-subspace of a Hamiltonian system in phase space. Note that integration over the p-subspace of phase space does not leave us with the correct volume element of Lagrangian configuration space. The q-subspace of phase space is always Cartesian without a metric whereas Lagrangian configuration space has the metric tensor g_{ij} defined by $ds^2 = g_{ik}dq_idq_k = 2Tdt^2$, where ds is the arc length, whenever the transformation to generalized coordinates is time-independent. The point is that the phase space formulation of a dynamical system formally treats the canonically conjugate variables $(q_1,\ldots,p_f)$ of a canonical Hamiltonian flow as if they were Cartesian, *even if they define a curvilinear coordinate system in the configuration space of a Lagrangian system*[4]. This is illustrated in section 3.5 below where for a pendulum we plot θ vs $d\theta/dt$ as if both coordinates were Cartesian when in fact they are angular position and angular frequency.

The theoretical reason why the treatment of the q_i and p_i as Cartesian in a $2f$ dimensional phase space that is *defined* to be flat does not conflict with the fact that f dimensional Lagrangian configuration space may be curved is explained by the geometric (meaning Lie-algebraic) interpretation of the fundamental Poisson bracket relations in section 15.3. This point is not explained explicitly elsewhere in the literature, although it is implicit in every discussion of Hamiltonian flows via Poisson brackets (see chapter 15). Curvature of Lagrangian configuration spaces is discussed in chapters 10 and 11. Whereas the Cartesian properties of the phase space of non-canonical flows are preserved only under affine transformations, which are linear, the Cartesian divergence formula (3.12) and volume element $d\Omega = dq_1 \ldots dp_f$ for Hamiltonian flows are preserved under a much larger class of *nonlinear* transformations that are called canonical or symplectic (see chapters 6 and 15). For this reason, the phase space of a canonical Hamiltonian system is called a symplectic space in the modern literature (see Arnol'd, 1989).

We go on next to classification of equilibria and the stability of motions near equilibria. By stability, we mean at this stage only the idea of local stability near an equilibrium point: a trajectory that starts near enough to, but not at, equilibrium, remains near that particular equilibrium point for all times unless the equilibrium point is unstable. Below and in chapter 4 we also introduce *orbital* stability based upon the idea that a collection of *trajectories* that start near one another within an infinitesimal definite ball of initial conditions in phase space remain near to one another as the time increases without bound. Orbital stability arises from the insensitivity of solutions with respect to small changes in initial conditions.

3.3 Equilibria and linear stability theory

Consider the autonomous system

$$\dot{x}_i = V_i(x_1,x_2,\ldots,x_n). \tag{3.2a}$$

At an equilibrium point, the velocity vector $V = (V_1,\ldots,V_n)$ vanishes. Denote such a point by $\bar{X} = (\bar{x}_1,\ldots,\bar{x}_n)$. This point is stable if nearby trajectories remain near it as the time increases without bound. Otherwise it is unstable.

---

[4] Questions requiring us to define the distance between two points in phase space do not arise in the qualitative analysis of a dynamical system. One must define a metric in order to construct approximations to solutions via Picard's method, but not in order to discuss the geometric nature of the flow pattern.

In order to analyze the stability locally, we expand the right hand side of (3.2a) about the equilibrium point $\bar{X}$,

$$V_i(x_1,\ldots,x_n) = V_i(\bar{x}_1,\ldots,\bar{x}_n) + \delta x_j \frac{\partial V_i}{\partial x_j}\bigg|_{X=\bar{X}} + \ldots, \tag{3.13}$$

where $\delta x_j = x_j - \bar{x}_j$ and summation convention is used (sum over repeated indices). If we keep only the linear term in the displacement δX_i from equilibrium, we obtain the variational equation

$$\delta\dot{X} = A\delta X \tag{3.14}$$

where A is the (constant) Jacobi matrix with entries

$$A_{ij} = \frac{\partial V_i}{\partial x_j}\bigg|_{X=\bar{X}}. \tag{3.15}$$

Since A is a constant matrix

$$\delta X(t) = \begin{pmatrix} \delta x_1(t) \\ \vdots \\ \delta x_n(t) \end{pmatrix} \tag{3.16}$$

is the displacement from equilibrium, and so

$$\delta X(t) = e^{tA}\delta X(0) \tag{3.17a}$$

where $\delta X(0)$ is the initial displacement from equilibrium. We see by inspection that $U(t) = e^{tA}$ is the time-evolution operator for this example.

If we can diagonalize the matrix A, then

$$A\hat{e}_i = \lambda_i\hat{e}_i \tag{3.18a}$$

where λ_i is the eigenvalue corresponding to the eigenvector $\hat{e}_i$. When the n eigenvectors $\hat{e}_i$ are linearly independent, then from

$$\delta X(0) = \sum_{i=1}^{n} c_i\hat{e}_i \tag{3.17b}$$

and from (3.21a) and (3.21b) we obtain

$$\delta X(t) = \sum_{i=1}^{n} c_i e^{\lambda_i t}\hat{e}_i \tag{3.17c}$$

by using also the fact that $A^n\hat{e}_i = \lambda_i^n\hat{e}_i$ in the power series expansion for e^{tA}.

In the two dimensional phase plane ($n = 2$), the possible equilibria are determined by the different possible eigenvalue problems (3.18a), and there are only six that are distinct. In particular, there are only *two* topologically distinct equilibria for conservative systems: from Liouville's theorem we obtain

$$\nabla\cdot V = \sum_{i=1}^{n} \frac{\partial\delta\dot{x}_i}{\partial\delta x_i} = \sum_{i=1}^{n} A_{ii} = 0, \tag{3.8c}$$

so that traceless matrices A describe conservative systems.

Consider first the simplest case where $n = 2$. The most general traceless two by two matrix has the form

$$A = \begin{pmatrix} a & b \\ c & -a \end{pmatrix} \qquad\qquad (3.18b)$$

and the eigenvalue equation $\det | A - \lambda I | = 0$ yields $\lambda = \pm \sqrt{(bc + a^2)}$. There are only two possibilities: either $bc + a^2 > 0$ so that we have two real eigenvalues, one positive and the other negative, or else $bc + a^2 < 0$ in which case both eigenvalues are imaginary (the reader should interpret the case when $bc = -a^2$). Without loss of generality, we can set $a = 0$ here. In the former case, a direct integration of the differential equations, using $x_1 \dot{x}_1/c - x_2 \dot{x}_2/b = 0$, yields $x_1^2/c - x_2^2/b = $ constant with both b and c positive, so that the local phase portrait, the streamlines near the equilibrium point, consists of hyperbolae with asymptotes passing through the origin (see figure 3.3 for an example). In this case, the equilibrium point is called hyperbolic. In the second case, one of the pair (b,c) is positive and the other is negative, so that integration yields $x_1^2/|c| + x_2^2/|b| = $ constant and the phase portrait consists of concentric ellipses (see figure 3.2 for an example). Here, the equilibrium point is called elliptic.

The first case corresponds to the scattering by a repulsive potential $U = -x^2/2b$ of a particle with mass $m = c^{-1}$, in which case the equation of the hyperbolae represents conservation of total mechanical energy $E = mv^2/2 + U(x)$. The phase space description follows from writing $x_1 = x$ and $x_2 = v$. In the second case, the equation of the ellipses in the phase plane represents conservation of energy $E = mv^2/2 + U(x)$ for a particle with mass $m = |c|^{-1}$ that is bound to the origin by a Hook's law potential $U(x) = x^2/2|b|$.

Consider next the dissipative case, the case where the velocity field does not vanish identically so that in the linear approximation, near any equilibrium point,

$$\nabla \cdot V \approx \mathrm{Tr} A = \sum_{i=1}^{n} \lambda_i \qquad\qquad (3.18c)$$

does not vanish. If $\nabla \cdot V \neq 0$ but $\mathrm{Tr} A = 2\mathrm{Re}\lambda_i = 0$, then the prediction of the linear approximation is wrong and the problem can only be analyzed from a nonlinear standpoint. We can now catalog the different possible cases when $n = 2$. If both eigenvalues are real and have opposite sign then we have a hyperbolic point where the eigenvectors tell us the asymptotes of the hyperbolae. If both eigenvalues are real and positive (or negative) then we have flow out of a source (or into a sink). In this case, let λ_1 be larger than λ_2 in magnitude. With the exception of the case where an initial condition is chosen to lie on the eigenvector $\hat{e}_1$, the flow will leave (or enter) the equilibrium point tangent to $\hat{e}_2$, excepting the case where the flow is radial (figures 3.5 and 3.6). In the case of an elliptic point the eigenvalues and eigenvectors are complex, so that the eigenvectors are not asymptotes in the phase plane. If the eigenvalues in the dissipative case are complex, then both eigenvalues have the same real part and $\mathrm{Tr} A = 2\mathrm{Re}\lambda_1$. If $\mathrm{Re}\lambda = 0$ one retrieves an elliptic point. However, if $\mathrm{Re}\lambda > 0$ then solutions spiral out of a source at the equilibrium point whereas $\mathrm{Re}\lambda < 0$ means that the flow enters a sink (figure 3.7). In all five cases, the equilibrium point is either a zero dimensional attractor or repeller.

The case of a sink, with complex conjugate eigenvalues with negative real part, is illustrated by the damped harmonic oscillator, $\ddot{x} + \beta\dot{x} + \omega^2 x^2 = 0$, for the case of normal friction, where $\beta > 0$. Since $dE/dt = -\beta v^2$, if $\beta < 0$ then we have the case

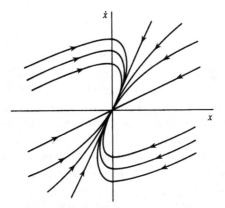

Fig. 3.5 Nonradial flow into a sink.

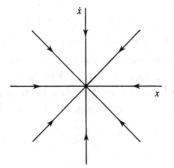

Fig. 3.6 Radial flow into a sink.

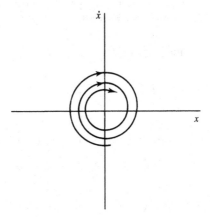

Fig. 3.7 Spiralling flow into a zero dimension attractor.

where, instead of friction, energy is dumped into the oscillator through the force $f = |\beta| v$. In this case both eigenvalues have the same positive real part and the phase flow represents the effect of a source at the origin in the (x,v)-space.

In the general case in n dimensions, if there is an equilibrium point at $\bar{X}$ *and* the lowest order approximation to the motion near $\bar{X}$ yields linear equations, then the solution of the linear equations is just

$$\delta X(t) = \sum_{i=1}^{n} c_i \exp \lambda_i t \hat{e}_i. \tag{3.17c}$$

Since bounded motions of conservative systems near equilibrium points require that those equilibria are elliptic points (if the equilibrium point is hyperbolic, then the particle is 'scattered' and eventually is repelled from equilibrium), and since elliptic points require *pairs* of imaginary eigenvalues, the dimension n of phase space must be *even* to obtain bounded motion in a conservative linearized system. In this case, we can then write $n = 2f$. In a conservative system in a phase space of old dimension bounded motion may permit the generalization of a hyperbolic point. An example of this is the linearized Lorenz model ($n = 3$) for a special choice of two of the three control parameters in that problem.

The Lorenz model is defined by the nonlinear equations

$$\left. \begin{array}{l} \dot{x} = \sigma(y - x) \\ \dot{y} = \rho x - y - xz \\ \dot{z} = -\beta z + xy \end{array} \right\} \tag{3.19}$$

in an (x,y,z)-phase space and has the three control parameters σ, β, and ρ. The Lorenz model can be derived from the Euler equations for a symmetric top by including linear dissipation and driving, as we show in chapter 13 where we also prove that the model defines a flow where the solutions are uniformly bounded.

The condition for a conservative system is satisfied by the Lorenz model whenever

$$\nabla \cdot V = \frac{\partial \dot{x}}{\partial x} + \frac{\partial \dot{y}}{\partial y} + \frac{\partial \dot{z}}{\partial z} = -\sigma - \beta - 1 = 0, \tag{3.20}$$

so that $\beta + \sigma = -1$.

There are three distinct equilibria: $\bar{x} = \bar{y} = \bar{z} = 0$ and $(\bar{x},\bar{y},\bar{z}) = (\pm [\beta(\rho - 1)]^{1/2}, \pm [\beta(\rho - 1)]^{1/2}, \rho - 1)$. When $\rho < 1$ there is only one equilibrium point, the one at the origin.

Linear stability is determined by the Jacobi matrix

$$A = \begin{pmatrix} -\sigma & \sigma & 0 \\ \rho - \bar{z} & -1 & -\bar{x} \\ \bar{y} & \bar{x} & -\beta \end{pmatrix} \tag{3.21a}$$

evaluated at an equilibrium point. With $(\bar{x},\bar{y},\bar{z}) = (0,0,0)$, we obtain

$$A = \begin{pmatrix} -\sigma & \sigma & 0 \\ \rho & -1 & 0 \\ 0 & 0 & -\beta \end{pmatrix} \tag{3.21b}$$

and the eigenvalue equation

$$(\beta + \lambda)[(\sigma + \lambda)(1 + \lambda) - \rho\sigma] = 0 \tag{3.22}$$

has (with $\rho \geq 0$) the three real roots $\lambda_1 = -\beta$ and $\lambda_{2,3} = \{-(\sigma + 1) \pm [(\sigma + 1)^2 + 4\sigma(\rho - 1)]^{1/2}\}/2$. Note that when $\sigma + \beta + 1 = 0$, the eigenvalues add up to zero, as required (the trace of a matrix equals the sum of its eigenvalues), but that either λ_1 or the pair $\lambda_{2,3}$ has a positive real part. Therefore, the equilibrium point $(0,0,0)$ is always unstable and represents a generalization to three dimensions of a hyperbolic point or a spiral. The spiral occurs whenever the argument of the square root in $\lambda_{2,3}$ is negative. If we set β equal to zero, then we retrieve the familiar case of a hyperbolic or an elliptic point.

Sources and sinks of streamlines can occur in a dissipative system (an example is the Lorenz model whenever $\sigma + \beta + 1 \neq 0$), but not in a conservative system. A sink attracts all nearby streamlines (all of them flow into it) whereas a source repels all nearby streamlines (they flow out of it). When V is divergence-free, there can be no sources or sinks of the flow. A sink can be zero dimensional (stable equilibrium point), one dimensional (stable limit cycle), or higher dimensional (attracting torus), or even chaotic ('strange' attractor). For the corresponding repellers, all of the streamlines flow out of the respective sources. This is not possible in a conservative system because it would violate Liouville's theorem: an f dimensional block of initial conditions with volume element $\delta\Omega = \delta x_1 \ldots \delta x_n$ would become equal to zero after infinite time should all of those points flow into a source ($\delta\Omega = 0$ would then follow due to the reduction in dimension). In contrast with dissipative systems conservative systems have neither attractors nor repellers. Because limit cycles are important later in the text, we now give an example of one.

Consider the damped-driven Newtonian system

$$\ddot{x} + (x^2 + \dot{x}^2 - 1)\dot{x} + x = 0, \tag{3.23a}$$

which represents a particle with unit mass attached to a spring with unit force constant, and subject to nonlinear driving and dissipative forces. The energy $E = (\dot{x}^2 + x^2)/2$ obeys the time-evolution law

$$dE/dt = -\dot{x}^2(x^2 + \dot{x}^2 - 1), \tag{3.24}$$

and from Newton's law it follows that $\dot{E} = f\dot{x}$, where the nonconservative force is identified as $f = -\dot{x}(x^2 + \dot{x}^2 - 1)$. This force represents contributions from both external driving and dissipation. With $r^2 = x^2 + y^2$, $x = r\cos\theta$, and $y = r\sin\theta$, there is a net dissipation whenever $r^2 = x^2 + \dot{x}^2 > 1$, where $\nabla \cdot V < 0$, but the system gains energy per unit time in the region where $r^2 < 1$ where $\nabla \cdot V > 0$. Therefore, the phase plane is divided into two distinct regions of energy gain and energy loss by the unit circle, because $\dot{E} = 0$ and also $\nabla \cdot V = 0$ whenever $r = 1$. On the unit circle, we have a steady state where driving is balanced by dissipation so that the energy E is maintained constant. By inspection, the unit circle $x^2 + \dot{x}^2 = 1$ is also a solution of (3.24). Next, we show that this solution is a 'stable limit cycle', a one dimensional attractor, by showing that nearby initial conditions have this solution as their asymptotic (and periodic) limit. Because the curve is closed with everywhere continuous slope, the motion on the limit cycle is periodic. Now for the 'limiting' part, the attractor aspect.

In polar coordinates it follows that

$$\left.\begin{aligned} \dot{r} &= r(1 - r^2)\sin^2\theta \\ \dot{\theta} &= (1 - r^2)\sin\theta\cos\theta - 1, \end{aligned}\right\} \tag{3.25}$$

so that the cycle at $r = 1$ attracts nearby trajectories: $\dot{r} > 0$ if $r < 1$, $\dot{r} < 0$ if $r > 1$, $\dot{\theta} \sim -1$ when $r \sim 1$, and so the direction of motion on the limit cycle is clockwise. Note that there is (by necessity) an unstable spiral at the origin. Note also that the amplitude and period of oscillation on the limit cycle are completely independent of the initial conditions, in contrast with the case of linear oscillations. Because the periodic orbit at $r = 1$ attracts nearby trajectories (figure 3.8a), the limit cycle is called stable (an

Flows in phase space

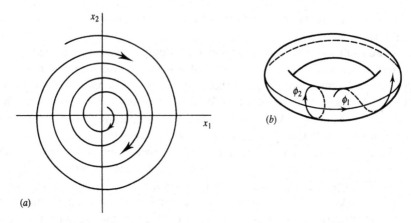

Fig. 3.8(*a*) A one dimensional attractor. (*b*) A two dimensional attractor
(from de Almeida, 1988).

unstable limit cycle repels all nearby trajectories). In other words, we have an example
of a curve that is itself a solution and also attracts other nearby solutions: a limit cycle is
a one dimensional attractor (one dimensional sink) for nearby initial conditions,
whereas an attracting equilibrium point is a zero dimensional attractor. We shall
discuss limit cycles again in chapter 13. For the time being, we need only know that a
limit cycle is a periodic solution that is also a one dimensional attractor or repeller for
nearby initial conditions of a driven dissipative system.

A three dimensional dissipative flow with an attracting 2-torus is given by

$$\left. \begin{array}{l} \dot{\phi}_1 = \omega_1 - \beta r \\ \dot{\phi}_2 = \omega_2 \\ \dot{r} = -\beta r \end{array} \right\} \qquad (3.23b)$$

where the ϕ_i are angular variables on a donut, and r is normal to the donut. As t goes to
infinity, every solution approaches the limiting solution $\phi_i = \omega_i t + \phi_{io}$, $r = 0$, which
describes stable periodic or quasiperiodic orbits that wind around the donut (figure
3.8b).

In a planar Hamiltonian system it is also possible to have periodic orbits without
elliptic points. These orbits are due to confining forces that admit a continuum of
equilibrium positions for particles at rest. The simplest example is given by a particle of
mass m confined in one dimension to a box of length L. Here, $H = E = mv^2/2$, $p = mv$
and the phase portrait consists of parallel trajectories. The equilibria are given by $p = 0$
with $0 < x < L$, representing a continuum of possibilities. The perfect reflection
represented by an infinite force at the points $x = 0$ and $x = L$ causes the discontinuity
in the phase trajectories. This discontinuity is artificial and can be eliminated by taking
into account that the deceleration from constant velocity to zero occurs in a small
region of width δ near each wall. The corrected phase trajectory would be continuous.
The main point is that there is periodicity without elliptic points: every point in the
range $\delta < x < L - \delta$ is an equilibrium point if $p = 0$. The motion is periodic with
frequency $\omega = \pi v/L$, but if we want to represent the solution by adding together simple
periodic functions, sines and cosines, then it is necessary to expand the solution in
Fourier series. In contrast with motion near an elliptic point, infinitely many simple
periodic functions with periods $\tau_n = 2\pi/n\omega$, with $n = 1,2,3,\ldots$, are required in order to

describe this sort of periodic motion mathematically. Notice also that this kind of periodic motion is stable in the same sense as the harmonic oscillator: if you change the energy slightly (by changing v slightly), or change the length L only slightly, then the result is a new orbit with nearly the same period as the old one. In other words, there is orbital stability because the geometric shapes of the trajectories and their periods are insensitive to small changes in the orbital parameters.

A particle in a one dimensional box or an oscillator with one degree of freedom both represent ideal clocks. An ideal clock is frictionless and ticks periodically forever. The Kepler problem describes the historic idea of a clock. Perfect clockwork is not the only possibility for motion in a Hamiltonian system that is completely integrable via conservation laws: stable quasiperiodic orbits, an approximate form of clockwork, occur whenever there are two or more frequencies in a mechanics problem. However, whether one considers a pair of oscillators or any larger number of oscillators the result is qualitatively the same, as we show in section 3.5.

Examples of quasiperiodic motion are easy to construct. Consider a particle with mass m moving in a rectangular box with sides of lengths a and b that are perfectly reflecting walls which is a two-frequency problem. With $H = E = m(v_1^2 + v_2^2)/2$, the two periods of the motion are $\tau_1 = 2a/v_1$ and $\tau_2 = 2b/v_2$. The trajectory of the particle is not periodic unless one period is an integral multiple of the other. If the ratio of the periods is irrational then the trajectory can never close on itself and is both quasiperiodic and orbitally stable.

3.4 One degree of freedom Hamiltonian systems

Consider the case in Lagrangian particle mechanics where $f = 1$ and where $L = mg(q)(dq/dt)^2/2 + U(q,t)$, so that $p = mg^{-1}(q)dq/dt$ and $H = p^2/mg(q) + U(q,t)$ is the total mechanical energy. So far, we have only assumed that the transformation to generalized coordinates is time-independent. If U depends explicitly upon t, then we have a problem with a three dimensional phase space. In that case deterministic chaos is possible. If, on the other hand, U is independent of t, then $E =$ constant and the problem reduces to a single integration: the differential form on the right hand side of

$$dt = \frac{dq\sqrt{g(q)}}{[2(E - U(q))]^{1/2}} \tag{3.26a}$$

is exact, and so the solution for $q(t)$ is given by inverting the result of the following integration:

$$t - t_\circ = \int_{q_0}^{q} \frac{dq\sqrt{g(q)}}{[2(E - U(q))]^{1/2}}. \tag{3.26b}$$

In other words, one conservation law determines the solution of a one degree of freedom problem completely.

Bounded motion occurs whenever there are two turning points: $E = U(q_{min}) = U(q_{max})$ holds whenever $p = 0$ so that $q_{min} < q < q_{max}$ (figure 3.9). In other words, U must decrease as q increases away from q_{min} (because $p^2 > 0$) but must decrease as p approaches zero as q approaches q_{max}. In other words, $U(q)$ must have at

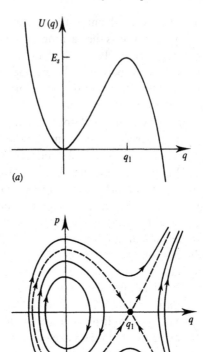

Fig. 3.9(a) A potential $U(q)$ that confines for $q < q_1$. (b) The steamlines due to $U(q)$ (from Percival and Richards, 1982).

least one minimum in the bounded region of configuration space, and the period of the trajectory is simply

$$\tau = 2 \int_{q_{min}}^{q_{max}} \frac{dq \sqrt{g(q)}}{[2(E - U(q))]^{1/2}}. \tag{3.26c}$$

That is, bounded motion for a one degree of freedom, time-independent, one-particle Hamiltonian is always periodic.

That the minimum of the potential at q_o corresponds to an elliptic point (and that a maximum defines a hyperbolic point) follows from the quadratic approximation $U(q) = U(q_o) + \delta q^2 U''(q_o)/2 + \dots$, valid near equilibrium where $U'(q_o) = 0$, follows immediately from the resulting linear dynamical equations (ignore all terms of order higher than quadratic in the Taylor expansion for $U(q)$). In this approximation we can write $\omega^2 = U''(q_o) > 0$, so that near the elliptic point, $E = p^2/2mg(q_o) + m\omega^2(q - q_o)^2/2$ and the motion is that of a simple harmonic oscillator with period $\tau = 2\pi g(q_o)/\omega$.

Except for separatrix motion where the period is infinite (see exercise 13), nothing but clockwork can occur for bounded one degree of freedom motion. In order to introduce the idea of a separatrix consider the motion of a simple pendulum that is fixed at one point in space but is free to rotate through 2π radians: $U = mgy = mgl(1 - \cos\theta) = 2mgl \sin^2(\theta/2)$ where l is the length and θ is the deviation from stable equilibrium. There is an elliptic point at $\theta = 0$ so that for low enough energies there are two turning points and the period of the motion is

$$\tau = 2 \int_{\theta_{\min}}^{\theta_{\max}} \frac{d\theta}{[2(E - U(\theta))/m]^{1/2}}, \tag{3.26d}$$

but when the energy is high enough then the pendulum makes 2π rotations. The energy that divides these qualitatively different motions defines the separatrix, a trajectory where at least one turning point is a hyperbolic point. Since it takes infinite time to arrive at a hyperbolic point, we know that $\tau = \infty$ on the separatrix (a separatrix is shown as the dashed curve in figure 3.9(b)). One can also see this directly from the integral for the period τ by using the fact that the potential is quadratic in $\delta\theta$ near a hyperbolic point.

3.5 Introduction to 'clockwork': stable periodic and stable quasiperiodic orbits

We begin by proving a general theorem, Poincaré's recurrence theorem, which states that for a conservative system with bounded motion in phase space a trajectory must return arbitrarily closely to any initial condition after a finite time, and therefore returns to visit points in a small neighborhood of its starting point arbitrarily often as time goes on.

The proof of the recurrence theorem is an easy application of Liouville's theorem: consider a finite phase volume element $\delta\Omega(t_o)$ at time t_o. Every point in the volume element is an initial condition for the phase trajectory that passes through that point. At a later time t, the size (but not the shape) of the phase volume is the same, $\delta\Omega(t) = \delta\Omega(t_o)$. If we imagine our bounded region in phase space to be partitioned mentally into N blocks, all approximately of size $\delta\Omega(t_o)$, then after a finite time t_r the phase tube that originates with the block $\delta\Omega(t_o)$ and ends with the volume elements $\delta\Omega(t)$ must intersect itself, otherwise the region of phase space under consideration would be infinite (N would be infinite). It follows that there are arbitrarily many intersection times $t_r, t_{r'}, t_{r''}, \ldots$, but there is no guarantee that these times are integral multiples of each other or even bear any simple relation to each other. Stated in other terms, every initial condition (and every other point on a bounded trajectory) must be approximately repeated infinitely often as t goes to infinity, but not necessarily periodically or even in stable fashion. Another way to say it is that the bounded orbits in phase space of a conservative system must be either periodic or quasiperiodic, but not necessarily orbitally stable. This behavior stands in contrast to that of dissipative systems where, even if $n = 3$, one may find sources and sinks that are zero, one, two, and even fractal dimensional. An example of a Newtonian system with both periodic and quasiperiodic orbits is given by a particle in a two dimensional box, which we discussed in section 3.3 above. In order to illustrate the periodicity and quasiperiodicity of bounded motions in conservative systems further we turn now to the next simplest case, the linearized description of a conservative system near an elliptic point.

Consider a conservative system in a phase space of $n = 2f$ dimensions. If we consider motion sufficiently near to an elliptic point, then the eigenvalues of the matrix A in the linear approximation are pairs of imaginary numbers, so each component of the local solution has the form

$$\delta x_i(t) = \sum_{j=1}^{n} c_{ij} \cos(\omega_j t + \theta_j). \tag{3.27a}$$

If we ask for the condition that the trajectory in phase space is periodic with period τ, for the condition that

$$\delta x_i(t + \tau) = \sum_{j=1}^{n} c_{ij}[\cos(\omega_j t + \theta_j)\cos\omega_j\tau - \sin(\omega_j t + \theta_j)\sin\omega_j\tau] \qquad (3.27b)$$

is the same as $\delta x_i(t)$, then $\omega_i\tau = 2\pi n_i$ is required for every index i where each number n_i is an *integer*. Therefore, we obtain the commensurability condition that for every pair of the $2f$ frequencies, $\omega_i/\omega_j = n_i/n_j$ is a *rational* number, a ratio of two integers. We conclude that if even one of the $2f$ frequency ratios is irrational (e, π, and $\sqrt{2}$ are common examples of irrational numbers), then the trajectory in the $2f$ dimensional phase space cannot close on itself and is therefore nonperiodic. In this case it is called quasiperiodic because for short enough times it can appear, within finite accuracy, to be periodic. The way that quasiperiodic orbits can mimic clockwork for finite times and in finite precision is analyzed in detail in the next chapter.

As a specific example, consider the case of a harmonic oscillator in the plane: the total mechanical energy $E = (dx/dt)^2/2 + \omega_1^2 x^2/2 + (dy/dt)^2/2 + \omega_2^2 y^2/2$ is constant. The phase space is the space of $(x, y, dx/dt, dy/dt)$ and the solution in configuration space, the (x, y)-subspace, is given by

$$x(t) = c_1\cos(\omega_1 t + \theta_1), \quad y(t) = c_2\cos(\omega_2 t + \theta_2). \qquad (3.27c)$$

Periodic orbits are given by $\omega_1/\omega_2 = n_1/n_2$ with n_1 and n_2 integers while quasiperiodic orbits, orbits that never close on themselves but have a two dimensional closure as t goes to infinity, are given by irrational frequency ratios. The orbits in configuration space are called Lissajou's figures.

The important idea of *orbital* stability can now be illustrated: if you shift the initial conditions only slightly, then the result is a geometrically similar orbit or exactly the same period (see figure 3.2, for example). In other words, the motion is insensitive to small changes in initial conditions. Furthermore, if you change the frequencies only slightly, then you will again get an elliptic orbit that differs but little from the original one.

We have introduced orbitally *stable* periodicity and orbitally *stable* quasiperiodicity in a restricted context: conservative dynamical systems in a $2f$ dimensional phase space in the linear approximation.

With limited experience in the theory of differential equations, one might expect that this clockwork is a special feature that disappears as soon as one introduces nonlinearity. That expectation can be true or false, depending upon the class of system one considers.

3.6 Introduction to integrable flows

... Lie showed that every infinitesimal transformation (on two variables) defines a one parameter group. He also proved that any such group has an invariant ...

E. T. Bell in The Development of Mathematics

Jacobi systematized the notion of complete integrability before Poincaré developed the geometric theory of dynamical systems theory, and there is confusion due to various different uses of the word 'integrable' in the mechanics and differential equations literature. The Lie–Jacobi idea of complete integrability in dynamical systems theory is solution via a complete set of conservation laws, also known as 'reduction to quadratures' (meaning reduction, at least in principle, of the solution to n independent integrations). A more subtle and far less restrictive idea of integrability is that of the *existence* of solutions as point functions (path-independent solutions, which are just ordinary functions), and is closer to what we mean by the word *solvable* in this text: we assume that $d\Psi_i$ in (3.2c) is an exact differential so that the variables $x_i(t)$ are holonomic, because we assume conditions under which we can construct the solution as a function of the initial data and the time recursively, via Picard's method.

By a nonintegrable flow, we therefore mean any flow whose solution is not, in principle, reducible by a single coordinate transformation, determined by conservation laws that hold for all $-\infty < t < \infty$, to n independent integrations. Every flow is solvable recursively, but a *completely integrable flow* is defined to be one where the streamlines are completely determined by $n - 1$ conservation laws that hold for all finite times t^5. Both completely integrable and 'nonintegrable' (meaning anything other than completely integrable) flows can always be solved recursively via *infinitely many coupled integrations* (by Picard's method) under the assumptions of this text: one-time differentiability of the velocity field (or, more generally, a Lipshitz condition) along with boundedness of solutions at all finite times. Convergence of any resulting series approximation is then limited by singularities in the complex time plane, because flows permit no singularities on the real-time axis.

In a 'nonintegrable' flow, the first evidence of chaotic dynamics was discovered geometrically by Poincaré around the turn of the century. The first analytic evidence of 'everywhere dense chaos' in a special class of two degree of freedom Hamiltonian systems with only one global conservation law (the Hamiltonian) was reported by Koopman and von Neumann in 1932.

An elementary and incomplete discussion of Jacobi's approach to integrability can be found in Kilmister (1964). More advanced discussions can be found in the texts by Whittaker (1965) and Duff (1962). An introduction that is similar in spirit to the one presented below can be found in the first few pages of Fomenko (1988).

In the last quarter of the last century the method of solving systems of nonlinear differential equations via conservation laws was greatly advanced by the geometer Marius Sophus Lie's systematic development of the theory of coordinate transformations that depend continuously upon one or more parameters. Lie's pioneering work was motivated by the search for general methods of solving nonlinear ordinary and partial differential equations of the first order. In that era, the main idea of a solution in mechanics was Jacobi's reduction of n coupled nonlinear differential equations to n independent integrations by the method of finding $n - 1$ constants of the motion. We present next what mathematicians would call an heuristic (as opposed to rigorous) approach to integrable flows, one that is motivated by Jacobi's method supplemented by Lie's most elementary discovery. This combination leads both to analytic and geometric definitions of integrability via conservation laws.

The simplest example of a dynamical system that is completely integrable via conservation laws is provided by any one degree of freedom Newtonian system. The most elementary example is provided by free-particle motion, where the equations of motion in phase space are

---

[5] We shall see that for a canonical Hamiltonian system, where $n = 2f$, one needs only f independent but compatible conservation laws for complete integrability, which is then called Liouville integrability.

$$\frac{dp}{dt} = 0$$

$$\frac{dx}{dt} = \frac{p}{m}. \Biggr\}$$ (3.28a)

The solution is a constant velocity translation,

$$p = \text{constant}$$

$$x = \frac{p}{m}t + c, \Biggr\}$$ (3.28b)

so that the streamlines are parallel to each other and to the x-axis. We shall eventually explain that the differential equations (3.28a) and the solution (3.28b) contain the complete *geometric* essence of the idea of integrability via a complete set of conservation laws.

A more general example of a completely integrable flow is provided by the motion of a particle of unit mass in a potential $U(x)$,

$$\ddot{x} = -U'(x).$$ (3.29a)

It follows that the energy

$$E = m\frac{\dot{x}^2}{2} + U(x)$$ (3.29b)

is left invariant by the Newtonian flow

$$\dot{x} = y \atop \dot{y} = -U'(x), \Biggr\}$$ (3.29c)

so that the solution for $x(t)$ is reduced to the definite integral

$$t - t_\circ = \int_{x_\circ}^{x} \frac{dx}{[2(E - U(x))/m]^{1/2}}.$$ (3.29d)

We can think of the complete solution as contained in the two equations

$$G(x, dx/dt) = m\frac{\dot{x}^2}{2} + U(x) = E$$

$$t + D = \int \frac{dx}{[2(E - U(x))/m]^{1/2}} = F(x, E). \Biggr\}$$ (3.29e)

The first equation in (3.29e) describes the form of the family of streamlines in phase space, while the second describes the details of the motion along any particular streamline. The same interpretation applies to equations (3.28b) above. These flows are both completely solvable for all times t by a single conservation law. After we define the idea of complete integrability to include much more general dynamical systems, we shall see that the qualitative form

$$G(x, dx/dt) = C \atop t + D = F(x, C) \Biggr\}$$ (3.30a)

of solutions (3.28b) and (3.29e) characterizes complete integrability via conservation laws for $n = 2$ and is not restricted to force-free bodies or even to Hamiltonian flows. In particular, we will discover that planar flows are completely integrable in the analytic sense of (3.29e) and in the geometric sense of (3.28b), except at isolated singular points.

Notice in particular that we can 'straighten out' the streamlines of the interacting problem (3.29c) by the coordinate transformation

$$\left. \begin{array}{l} y_1 = G(x,dx/dt) = E \\ y_2 = t + D = F(x,E) = F(x,G(x,dx/dt)) = H(x,dx/dt). \end{array} \right\} \tag{3.30b}$$

In the (y_1, y_2)-coordinate system the streamlines are all parallel to the y_2-axis, reducing the flow *abstractly and formally* to free-particle motion at constant *speed* in a coordinate system that is not necessarily Cartesian, and whose construction is equivalent to knowing the solution. We will show that this 'rectification' or 'straightening out' of the streamlines is a universal geometric feature of all completely integrable flows in regions that exclude equilibrium points and limit cycles.

By 'global' in what follows, we mean a finite but generally not small region in phase space where the velocity field varies nontrivially. By 'local', we mean a very small but still finite region of size ε, as the mathematicians would say. The neglected velocity gradients determine the size of ε in local approximations.

A one dimensional autonomous flow $dx/dt = V(x)$ is trivially integrable because $dt = dx/V(x)$ is an exact differential that can be immediately integrated (unless the integral is infinite) to yield $t + C = f(x)$, where C is a constant[6]. Following in the spirit of Jacobi and Lie, we now offer the following heuristic definition of integrable flows for $n \geq 2$: we shall call the flow defined by

$$\dot{x}_i = V_i(x_1, \ldots, x_n) \tag{3.2a}$$

completely integrable if the complete solution can be obtained by performing n independent integrations of n different *exact* differentials in *at least one* coordinate system (in the two examples above, this was exactly the procedure in the original variables). Before confidently analyzing the easy case where $n = 2$, and then advancing perilously to $n = 3$ and 4 and beyond, we begin with the basic definition of integrability from elementary calculus.

An exact differential $X_1 dx_1 + \ldots + X_n dx_n$ in n holonomic variables $(x_1, \ldots, x_n)$ is one where $\partial X_i/dx_k = \partial X_k/dx_i$ for all indices i and k where $i \neq k$. In this case, there is a 'point function', meaning a path-independent function, or just function $G(x_1, \ldots, x_n)$ such that $\partial G/\partial x_k = X_k$. Therefore, $X_1 dx_1 + \ldots + X_n dx_n = dG$ is an exact differential, and the original differential form is then called either integrable or exact. If the second cross-partials are not all equal but the differential form admits an integrating factor (or 'multiplier') $M(x_1, \ldots, x_n)$ whereby $\partial M X_i/dx_k = \partial M X_k/dx_i$, then the differential form $M(X_1 dx_1 + \ldots + X_n dx_n)$ is integrable and is equal to the exact differential dG of a function $G(x_1, \ldots, x_n)$, which means that line integrals of $M X_i dx_i$ are path-independent. If at least one pair of cross-partials is unequal *and* no integrating factor M exists, then we shall call the differential form nonintegrable, including the case where there exists an

---

[6] 'Integrable' does not mean that the resulting integrals are easy to evaluate analytically. For the analytic evaluation of the classical integrals of dynamics, we refer the reader to Whittaker (1965) and Whittaker and Watson (1963). Here, we do not evaluate those integrals explicitly and we do not care in principle whether the resulting integrals are evaluated analytically or numerically, as we concentrate upon understanding the *qualitative* properties of flows that are common to all integrable systems. It is not necessary that the integrals can be evaluated in terms of elementary functions, as physics is certainly not confined to systems described by tabulated or known functions.

integrating factor within the classification of integrable differential forms. This is unconventional terminology in the context of elementary calculus, but our adopted word-usage agrees with the meaning of the words integrable vs nonintegrable in dynamical systems theory, as we shall emphasize in section 13.2.

In the modern literature on differential forms, there is a distinction made between an exact differential form and a 'closed' one. An exact differential form gives rise to a single-valued function G, which means that all closed loop integrals of $X_i dx_i$ vanish identically. In the case of closure, $\partial X_i/dx_k = \partial X_k/dx_i$ holds almost everywhere but the function G is permitted to be multivalued so that certain closed loop integrals of the differential form $X_i dx_i$ are nonzero (see Arnol'd, 1989, Dubrovin et al., 1984 or Buck, 1956). 'Exact' is therefore reserved for differential forms where $\partial X_i/dx_k = \partial X_k/dx_i$ at all points so that there are no singularities that can make closed loop integrals of $X_i dx_i$ nonzero. In general, we need only the condition of closure, not exactness of the differential forms, in all that follows. Closed differential forms play a central role in the theory of Liouville integrability (chapter 16), and in the integrability of dissipative flows (see below).

We now arrive at the main building block in our analysis, namely, Jacobi's basic theorem: *every* two dimensional flow, $dx_1/dt = V_1(x_1,x_2)$, $dx_2/dt = V_2(x_1,x_2)$, whether dissipative or conservative, has a conservation law. This result was also discovered by Lie in the context of continuous transformation groups: every one-parameter coordinate transformation on two variables has an invariant. By rewriting the flow equations in the form

$$dt = dx_1/V_1 = dx_2/V_2 \tag{3.31a}$$

we can study the time-independent differential form

$$V_2(x_1,x_2)dx_1 - V_1(x_1,x_2)dx_2 = 0. \tag{3.31b}$$

In calculus and in thermodynamics, we learn that every differential form in two variables either is closed or else has an integrating factor $M(x_1,x_2)$ that makes it integrable (Jacobi proved this), so that we can integrate the differential form

$$M(V_2 dx_1 - V_1 dx_2) = dG \tag{3.31c}$$

to obtain $G(x_1,x_2) = C$, which is a conservation law: C is the same constant along one streamline, but varies from streamline to streamline. Even one dimensional Newtonian systems (planar Newtonian flows) where the mechanical energy E is not constant are included in this category. Synonyms for conservation law are: 'invariant (along a trajectory)', 'first integral', and 'integral of the motion'.

We now state the mathematical condition under which the required conservation law exists, which is just the condition under which Jacobi's multiplier M exists. We can rewrite the differential form $V_2 dx_1 - V_1 dx_2 = 0$ as the one dimensional differential equation

$$\frac{dx_2}{dx_1} = \frac{V_2(x_1,x_2)}{V_1(x_1,x_2)} = F(x_1,x_2) \tag{3.32a}$$

whose solution satisfies the integral equation

$$x_2 = x_{2_0} + \int_{x_{1_0}}^{x_1} F(s,x_2(s))ds. \tag{3.32b}$$

According to Picard's method, this equation can be solved by the method of successive approximations so long as F satisfies a Lipshitz condition, which for our purposes means that the ratio of the components of the velocity in phase V_2/V_1 must be at least once continuously differentiable with respect to the variable x_1. This requires that we avoid integrating over equilibria or other zeros of V_1. The resulting solution of the integral equation yields x_2 as a function of x_1. The same result follows from solving the conservation law $G(x_1,x_2) = C$ to find the solution $x_2 = f(x_1,C)$ of (3.32b), which is just the statement that the conservation law *determines* x_2 as a function of x_1 and C: the form $x_2 = f(x_1,C)$ of the conservation law represents our definition of an 'isolating integral' (the phrase 'isolating integral' is used to mean any conservation law at all in certain other texts). In other words, the required conservation law is only a rewriting of the solution of the integral equation (3.32b). In particular, differentiating the conservation law once yields

$$dG = \frac{\partial G}{\partial x_1}dx_1 + \frac{\partial G}{\partial x_2}dx_2 = 0, \tag{3.33a}$$

and comparison with the differential form $V_2 dx_1 - V_1 dx_2 = 0$ then shows that

$$-\frac{\partial G}{\partial x_1} \bigg/ \frac{\partial G}{\partial x_2} = V_2/V_1. \tag{3.33b}$$

This last equation is satisfied only if

$$V_1 = -M\frac{\partial G}{\partial x_2} \text{ and } V_2 = M\frac{\partial G}{\partial x_1}, \tag{3.33c}$$

which guarantees the existence of the integrating factor M. The existence of the conservation law follows from satisfying Picard's conditions for a unique solution of the differential equation (3.32a). This in turn guarantees, by differentiation, that the required integrating factor M exists. Excepting a few special cases there is no systematic way to construct the integrating factor $M(x_1,x_2)$ even though it is guaranteed to exist. Whether or not the integrating factor can be guessed or constructed is in principle irrelevant: we shall see that its existence has universal geometric consequences.

Notice that the condition that a two dimensional flow is phase-volume preserving,

$$\frac{\partial V_2}{\partial x_2} = -\frac{\partial V_1}{\partial x_1}, \tag{3.34a}$$

is precisely the condition that the differential form $V_2(x_1,x_2)dx_1 - V_1(x_1,x_2)dx_2 = 0$ is closed without any need for an integrating factor. This means that the velocity field of a conservative two dimensional flow can always be obtained from a stream function G:

$$V = (V_1,V_2) = \left(-\frac{\partial G}{\partial x_2}, \frac{\partial G}{\partial x_1}\right). \tag{3.35}$$

By inspection, the system is Hamiltonian with Hamiltonian G and canonically conjugate variables $(q,p) = (x_2,x_1)$, but the collapse of an integrable conservative flow into Hamiltonian form is peculiar to two dimensions and generally is not true for integrable conservative flows in higher dimensions. In particular, every odd dimensional conservative integrable flow is necessarily noncanonical.

It is easy to give examples of conservation laws for planar conservative flows because no integrating factor is necessary in order to construct the conserved quantity G. First, with $V = (x_2, -x_1)$ there is an elliptic point at the origin and $x_1 dx_1 + x_2 dx_2 = 0$ yields $G(x_1, x_2) = x_1^2 + x_2^2 = C$. The conservation law describes the family of circles that represents the simple harmonic oscillations of a particle of unit mass with energy $E = C/2$ in a Hooke's law potential $U(x) = x^2/2$. With $V = (x_2, x_1)$, representing scattering from a repulsive quadratic potential at the origin, we obtain $G(x_1, x_2) = x_2^2 - x_1^2 = C$ where $E = C/2$ is the energy of a particle with position $x = x_1$ and velocity $v = x_2$. This conservation law describes the family of hyperbolae that are the trajectories of the scattering problem in the phase plane.

If the flow is dissipative then equation (3.34a) does not hold but is replaced by the more general integrability condition

$$\frac{\partial M V_2}{\partial x_2} = -\frac{\partial M V_1}{\partial x_1}, \tag{3.34b}$$

which means that dissipative two dimensional flows are not Hamiltonian although they define a conservation law $G(x_1, x_2) = C$ where $dG = M(V_2 dx_1 - V_1 dx_2)$. This case is harder because we generally would have to guess the integrating factor in order to construct the conservation law. However, an easy example is given by $V = (x_1 - x_2, x_1 + x_2)$, representing a sink at the origin. It is then easy to check that the solution in radial coordinates is given by $dr/r = d\theta = dt$, which shows that $r = Ce^\theta$. We can rewrite this solution as a conservation law $\ln r - \theta = C$. Transforming back to Cartesian coordinates yields the conservation law for this dissipative system in the form

$$G(x_1, x_2) = [\ln(x_1^2 + x_2^2)]/2 - \tan^{-1}(x_2/x_1) = C. \tag{3.36}$$

Direct differentiation then yields the integrating factor as $M = (x_1^2 + x_2^2)^2$. We have had to put the cart before the horse, finding the integrating factor only after having found the solution. This is a possible reason why traditional texts on mechanics and differential equations that emphasize quantitative methods have ignored the existence of conservation laws for dissipative systems.

As another dissipative example, consider the damped simple harmonic oscillator

$$\ddot{x} + \beta\dot{x} + \omega^2 x = 0. \tag{3.37}$$

With $x_1 = x$ and $x_2 = dx/dt$ the corresponding differential form

$$(\beta x_2 + \omega^2 x_1)dx_1 + x_2 dx_2 = 0 \tag{3.38a}$$

is not closed but is guaranteed to have an integrating factor M that yields a time-independent conservation law $G(x, dx/dt) = \text{constant}$ via the condition

$$\frac{dG}{dt} = M[(\beta x_2 + \omega^2 x_1)\dot{x}_1 + x_2\dot{x}_2] = 0. \tag{3.38b}$$

Whether or not one is clever enough to find the integrating factor $M(x_1, x_2)$ is not the main point (both the integrating factor and the conservation law are exhibited in exercise 18): the conserved quantity G exists mathematically and *determines* the flow pattern in the phase plane. For the damped harmonic oscillator, we can plot the phase portrait to see that the streamline pattern is a family of spirals that flow into a sink at the origin. The conservation law $G(x, dx/dt) = C$ is the equation that describes that family of spirals: different constants C correspond to different spirals within the same family. By writing the solution of the damped oscillator equation only in the explicitly time-dependent form of a sum of damped sines and cosines, one never discovers the qualitative geometric information that shows how those quantitative details are the consequence of a conservation law.

That *dissipative* systems can have conservation laws is a fact that we are conditioned by education in both physics and mathematics not to expect, in spite of Ince's chapter IV (1956), which was first published in 1926. Typical modern courses on differential equations teach little or nothing about conservation laws ('first integrals'), while mechanics courses traditionally associate conservation laws primarily with systems that conserve mechanical energy, momentum, or angular momentum, as if other possible conservation laws were unimportant. In the differential equations literature, where conservation laws are mentioned at all the assumption is usually made that a function must be continuous in order to qualify as a conservation law[7]. Continuity is only an abstract mathematical idea, not a physical principle. Even in mathematics there is no reason other than convention to impose that restriction on conservation laws. We show next that conservation laws for dissipative systems can be constructed and are singular at the locations of attractors and repellers.

The flow defined by

$$\dot{x}_1 = x_1$$
$$\dot{x}_2 = x_2$$

has a source at the origin and can be rewritten in polar coordinates $x_1 = r\cos\theta$, $x_2 = r\sin\theta$, as

$$\dot{r} = r$$
$$\dot{\theta} = 0.$$

The flow is therefore radial out of the source and the conservation law (excepting the origin) is simply

$$\theta = \tan^{-1}(x_2/x_1) = \text{constant},$$

which is singular at the origin (θ is undefined at the origin). Except at one point, the origin, θ is constant along a given streamline, varies from streamline to streamline, and is a perfectly good conservation law. Note that this conservation law (and the one for the dissipative flow discussed above) is multivalued for 2π-rotations about the equilibrium point.

Very early in our discussion of phase flows we generalized the traditional definition of conservative, which in mechanics texts is restricted to the description of mechanical systems with constant mechanical energy. We generalized the definition to include all dynamical systems that generate phase-volume-preserving flows. Hamiltonian systems and systems with constant mechanical energy belong to this larger category, but the category also includes abstract dynamical systems that are not Newtonian. For example, a Hamiltonian system may have no analytic conservation law if the Hamiltonian is time-dependent and has no symmetry, yet every time-dependent Hamiltonian system leaves the phase volume element invariant. The same is true if the flow is noncanonical with no global conservation law but has vanishing divergence of the phase space velocity field. On the other hand, a flow that includes dissipative forces

---

[7] See, for example, p. 76 of *Ordinary Differential Equations* by Arnol'd.

will not leave the phase volume element invariant, but every integrable dissipative flow has conservation laws. In particular, every planar dissipative flow has one conservation law. In spite of those conservation laws, the latter flows are called nonconservative because they do not leave the phase volume element $d\Omega = dx_1 \ldots dx_n$ invariant.

The three examples discussed above explain geometrically why two dimensional dissipative flows have conservation laws: the condition $C = G(x_{1o}, x_{2o})$ fixes the constant C in terms of the two initial conditions x_{io} at time $t = t_o$ and is simply the equation of the family of streamlines of the flow, $G(x_1, x_2) = C$, in the phase plane. One value of the constant C defines one streamline and other streamlines follow from other values of C. If we can solve $G(x_1, x_2) = C$ to find the streamline equations in the form $x_2 = f_2(x_1, C)$ and/or $x_1 = f_1(x_2, C)$, then we can rewrite the differential equations in a form that is explicitly separable: $dt = dx_1/v_1(x_1, C)$ and/or $dt = dx_2/v_2(x_2, C)$. These equations are integrable (not meaning at all that the resulting integrations[8] are trivial to carry out *analytically*) and integration yields the solution in the form $t + C_i = g_i(x_i, C)$, where C_i is a constant that is fixed by t_o and x_{io}. If we can invert the solution to solve for the two variables x_i, then since the three constants C, C_1, and C_2 are fixed by the initial conditions (x_{1o}, x_{2o}) we can also write the solution of the pair of differential equations (3.2a) for $n = 2$ in the form of a one-parameter coordinate transformation on two variables, precisely the form that leads us to Lie's basic theorem on invariance under one parameter coordinate transformations: the transformation

$$\left.\begin{array}{l} x_1(t) = \psi_1(x_{1o}, x_{2o}, t - t_o) \\ x_2(t) = \psi_2(x_{1o}, x_{2o}, t - t_o). \end{array}\right\} \tag{3.3'}$$

that solves (3.3a) with $n = 2$ has a global invariant $G(x_1 x_2) = C$. Lie's theorem does not carry over to transformations on three or more variables, the case of dynamical systems with $n \geq 3$. This leaves the door open for nonintegrable flows only in three or more dimensions.

We must note that having a conservation law in the form $G(x_1, x_2) = C$ is not automatically sufficient to guarantee integrability. Complete integrability of the flow means that the conservation law, in principle, reduces the solution of a two dimensional flow to a single integration: the condition that a conservation law leads to the integrability condition $dt = dx_i/v(x_i, C)$ is that $G(x_1, x_2) = C$ must be of such a form as to determine *either* $x_1 = f_1(x_2, C)$ *or* $x_2 = f_2(x_1, C)$, in which case we call the conservation law an 'isolating integral'. Notice that isolating integrals are generally multivalued: for the harmonic oscillator, for example $x_2 = \pm (C - x_1^2)^{1/2}$. The isolating property is a nontrivial condition that is not necessarily satisfied by every conservation law, especially if $n \geq 3$. Even in elementary cases respect must be paid to the requirements of the implicit and inverse function theorems in order that the determination of one variable in terms of the other and the constant C holds at least locally (for the implicit and inverse function theorems, see Buck, 1956, or Dubrovin et al., 1984).

Next, we show that if a flow is completely integrable then the motion in phase space is *universally* determined geometrically, independently of any quantitative details that may make it hard or even impossible to find the solution in terms of the original coordinates (x_1, x_2) and to express it analytically in terms of known, elementary functions of those coordinates. The point is that there is a special coordinate system in which all integrable flows look like free-particle motion at constant speed.

---

[8] Consult Gradstein and Ryzhik (1965) for a vast compilation of known integrals in terms of known elementary functions.

The geometric interpretation that follows next was first pointed out by Lie and is based upon the construction of a special coordinate transformation in phase space (see also Olver, 1993). The result is of central importance because it provides the universal geometric basis for understanding the qualitative distinction between integrable and nonintegrable flows. Using the conservation law in the form $C = G(x_1,x_2)$, we can write $t + C_1 = g_1(x_1,C) = F(x_1,x_2)$ (or $t + C_2 = H(x_1,x_2)$). Lie's next step was to define a new pair of coordinates (y_1,y_2) by the transformations $y_1 = G(x_1,x_2)$ and $y_2 = F(x_1,x_2)$. In this new system of coordinates the original pair of differential equations is both separable and trivial to integrate: $dy_1/dt = 0$ and $dy_2/dt = 1$. The qualitative interpretation of those equations is clear. Every pair of first order differential equations (3.2a) in two variables is equivalent, by the special coordinate transformation $y_1 = G(x_1,x_2)$ and $y_2 = F(x_1,x_2)$ to a *translation* at constant speed: $y_1 = C$ and $y_2 = t + C_1$: *completely integrable flows are those where the effects of interactions, binding and scattering, can be mathematically eliminated by a special choice of coordinates that can be constructed for all times* $-\infty < t < \infty$ *from the conservation laws.* In other words, the solution of a completely integrable flow can be written in such a form to make it appear *qualitatively* as if there were no interactions at all. We provided an example earlier: the phase flow for a particle of unit mass in a potential $U(x)$ can be transformed into a family of parallel streamlines by using the variables $y_1 = G(x,y)$ and $y_2 = F(x,G(x,y)) = H(x,y)$, where the conservation law G and the function F are defined by the solution of the dynamics problem. We now discuss the qualitatively different possibilities.

Geometrically, there are only two distinct possibilities for a translation at constant speed in phase space: the solutions $y_1 = C$ and $y_2 = t + C_1$ must represent a linear velocity along an infinitely long axis if the streamlines in the original coordinates (x_1,x_2) do not close on themselves as t goes to infinity. For a conservative system, this corresponds to a vector field $V(x_1,x_2) = (V_1,V_2)$ that has either a hyperbolic point or no equilibrium point at all. For dissipative systems, this also includes the case of a streamline-source (an unstable equilibrium point with Re $\lambda_{1,2} > 0$) where the flow is unbounded as t goes to infinity, but also includes the case of a sink of streamlines (a stable equilibrium point, where Re $\lambda_{1,2} < 0$) where the flow lines are all bounded and all converge to the equilibrium point as t goes to infinity. Stable and unstable equilibria are the zero dimensional sinks and sources of a nonconserved flow. There are also the trajectories that approach (or are repelled by) one dimensional sinks (or sources) called limit cycles, and these trajectories also fall into this category because they are not closed (only the trajectory that lies directly on the limit cycle is periodic, therefore closed).

If the (x_1,x_2)-streamlines close on themselves in finite time, then the velocity field V cannot vanish along a streamline and must represent a conservative system with a periodic orbit, or else a dissipative system with a motion on a limit cycle (a trajectory on a separatrix can close on itself only in infinite time). Because of Liouville's theorem, a limit cycle can occur only in a dissipative system, and is a generalization to nonlinear oscillations of the geometry of a streamline about an elliptic point (the connection between periodic orbits in conservative systems and limit cycles in dissipative ones is explored further in section 13.4). In these two cases, the (y_1,y_2) translation is a translation about a circle with constant angular velocity, with radius fixed by $y_1 = C$. This is because the motion in phase space in the coordinates (x_1,x_2) lies on a simple closed curve that encloses an elliptic point in the conservative case (figure (3.2))[9], or is

---

[9] We can also have closed orbits without elliptic points where there is a continuum of equilibria; the particle in a box provides us with such an example.

the limit cycle that encloses a source that is called a focus in the dissipative case (a focus is an equilibrium point where Re $\lambda_i \neq 0$ and Im $\lambda_i \neq 0$ for $i = 1,2$, and the trajectories are spirals). The transformation from (x_1,x_2) to (y_1,y_2) changes all of those simple closed curves into circles. Nonclosed streamlines are not transformed into closed ones, and vice versa, by the transformations F and G, so that, geometrically seen, this is really a topologic and not a metric result: any simple closed curve is equivalent, by a continuous coordinate transformation, to a circle. In our case, a differentiable simple closed curve is equivalent to a circle by a differentiable coordinate transformation.

It is easy to understand qualitatively how a planar differentiable closed curve of arbitrary shape can be deformed smoothly into a circle by a differentiable coordinate transformation. Let t denote the arc-length along the curve, which is described by a pair of differential equations

$$\left.\begin{aligned}\frac{dx}{dt} &= f(x,y)\\[2mm]\frac{dy}{dt} &= g(x,y).\end{aligned}\right\} \tag{3.39a}$$

The solution can be written in the form

$$\left.\begin{aligned}G(x,y) &= C\\ F(x,C) &= t + D.\end{aligned}\right\} \tag{3.39b}$$

The coordinate transformation $(y_1,y_2) = (G(x,y),F(x,y))$ yields a curve $y_1 = C$, $y_2 = t + D$. We can identify y_2 according to $y_2 = R\phi$ where (R,ϕ) are polar coordinates, and where R is determined in practice by the condition $y_1 = C$. With this identification, we obtain a circle of radius R. For a closed curve, no other identification is possible.

To illustrate Lie's transformation to parallel flow we start with two well-known examples for a conservative system. First, with $V = (x_2, -x_1)$ there is an elliptic point at the origin, and $G(x_1,x_2) = x_1^2 + x_2^2 = C$ follows directly from $x_1 dx_1 + x_2 dx_2 = 0$. With $V_1 = x_2 = \pm(C - x_1^2)^{1/2}$, we obtain from $dt = \pm dx_1/(C - x_1^2)^{1/2}$ (and from a table of integrals) that $t + C_1 = \pm \sin^{-1}(x_1/C) = \pm \sin^{-1}[x_1/(x_1^2 + x_2^2)^{1/2}] = F(x_1,x_2)$, and so y_2 is in this case an angular variable. Here, $\sqrt{C}$ is the radius of the circle and the angular frequency ω on the circle is unity. In the second example, $V = (x_2,x_1)$ has a hyperbolic point at the origin and this yields $G(x_1,x_2) = x_1^2 - x_2^2 = C$. The result of integrating $dt = dx_1/V_1 = \pm dx_1/(x_1^2 - C)^{1/2}$ is that $t + C_1 = \pm \ln(x_1 + x_2) = F(x_1,x_2)$, which does not yield y_1 as an *angular* variable. Note that in both cases the transformations from (x_1,x_2) to (y_1,y_2) are singular at the equilibrium points (the effects of binding and scattering make the two transformations singular at one point). Computation of the Jacobi determinant in each case shows also that these transformations are not phase-volume-preserving $(dy_1 dy_2 \neq dx_1 dx_2)$, but the geometric idea of a Lie translation is qualitatively similar to the case studied in chapter 2 where a Lagrangian is independent of q_i and therefore p_i is conserved: in neither the Lagrangian nor the general case is it necessary that the constant speed translation represents the motion of a material body in Euclidean space. In particular, Galilean translations preserve the Euclidean metric and scalar products and therefore preserve Euclidean geometry (section 1.4), but Lie's transformation to translation variables does not preserve any scalar product or any aspect of physics other than the property of integrability: we interpret the special variables (y_1,y_2) as describing a translation in the qualitative geometric sense of topology, where no metric need be left invariant by the coordinate transformation used to define (y_1, y_2).

If the flow is dissipative, then we may or may not be able to discover the integrating factor $M(x_1, x_2)$ for the velocity field V by guesswork. In the dissipative case, there can be stable equilibria that are sinks of streamlines and unstable ones that are sources of the flow. An easy example of a source at $(0,0)$ is provided by the flow where $V = (x_1 - x_2, x_2 + x_1)$, which yields $\lambda_1 = 1 + i = \lambda_2^*$ (the source is called an unstable focus and the streamlines spiral away from it). Here, we find that $G(x_1, x_2) = \ln(x_1^2 + x_2^2)^{1/2} - \tan^{-1}(x_2/x_1) = C$, and also that $F(x_1, x_2) = \tan^{-1}(x_2/x_1)$; the proof is left to the reader as an easy exercise. The transformation functions F and G are, as expected, singular at the location of the source precisely because infinite time is needed to leave any source (or to reach any sink). The Lie transformation therefore sends sources and sinks off to infinity in (y_1, y_2)-space.

For a nonlinear flow in a three dimensional Cartesian phase space the analysis is no longer so simple. With three first order nonlinear differential equations

$$dt = \frac{dx_1}{V_1} = \frac{dx_2}{V_2} = \frac{dx_3}{V_3}, \tag{3.40a}$$

and where each component V_i of the vector field V depends upon all three variables (x_1, x_2, x_3), we are no longer guaranteed complete integrability. The search for an integrating factor follows a method due to Jacobi, but an integrating factor generally does not exist except in very special cases. Consider the case where there is a time-independent function $G_1(x_1, x_2, x_3) = C_1$ that is left invariant by the flow (3.40a), and therefore satisfies the first order linear partial differential equation

$$\frac{dG}{dt} = V \cdot \nabla G = 0. \tag{3.41a}$$

The analysis that follows excludes the Lorenz model, where the only global conservation laws that are known analytically are known only for a few special parameter values and are time-dependent. The differential equations

$$\frac{dx_1}{V_1} = \frac{dx_2}{V_2} = \frac{dx_3}{V_3} \tag{3.40b}$$

generate the characteristic curves (or just the characteristics) of the partial differential equation (3.41a) and are simply the streamlines along which G_1 is constant. That $V \cdot \nabla G_1 = 0$ means that ∇G_1 is perpendicular to V. The flow is therefore confined to a two dimensional surface that may be curved and is not necessarily orientable. The simplest examples of integrable three dimensional flows arise in rigid body theory and are confined to spheres or ellipsoids (chapters 9 and 13) or tori (chapter 13) in phase space. Spheres and tori are orientable surfaces, but nonorientable surfaces like Möbius strips and Klein bottles (or Klein tori) also may occur in nonlinear dynamics (see Fomenko, 1988).

If the conservation law is an isolating integral, e.g. if $x_3 = h_1(x_1, x_2, C_1)$, then because the resulting flow is two dimensional, a second conservation law $g_2(x_1, x_2, C_1) = G_2(x_1, x_2, x_3) = C_2$ also exists (because an integrating factor exists) and satisfies (3.41a), so that V is perpendicular to both ∇G_1 and ∇G_2:

$$MV = \nabla G_1 \times \nabla G_2 \tag{3.42}$$

where $M(x)$ is a scalar function called Jacobi's multiplier. Note that the Jacobi multiplier must satisfy the condition $\nabla \cdot MV = 0$. The most general way to satisfy this divergence-free condition would be to write $MV = \nabla \times A$ where A is a vector potential, but this case is generally nonintegrable ($MV = \nabla \times A$ does not automatically satisfy (3.42)). However, any flow field of the form $V = k\nabla G_1 \times \nabla G_2$, where G_1, G_2 and k are time-independent scalar functions, automatically satisfies (3.41a) and therefore leaves both G_1 and G_2 invariant. The components of the flow have the form $V_i = k\varepsilon_{ijk}\partial G_1/\partial x_j \partial G_2/\partial x_k$ where ε_{ijk} is the completely antisymmetric three-index symbol.

The flow on a two dimensional surface that is imbedded in three dimensional Cartesian space, defined by an isolating integral $G_1 = C_1$, is generally integrable: if $G(x_1,x_2,x_3) = C_1$ implies that $x_3 = h_3(x_1,x_2,C_2)$, where it does not matter which of the three Cartesian coordinates we label as x_3, then the flow can be reduced to a planar flow of the form $dt = dx_1/v_1(x_1,x_2,C_1) = dx_2/v_2(x_1,x_2,C_1)$, which is integrable because an integrating factor $N(x_1,x_2)$ always exists away from singular points. In other words, there are generally either two isolating integrals or none. Denote the second isolating integral by $G_2(x_1,x_2,x_3) = C_2$. One of the three differential equations, say $dt = dx_1/V_1$, can then be replaced by $dt = dx_1/v_1(x_1,C_1,C_2)$. We can integrate this to obtain $t + D = f_1(x_1,C_1,C_2) = F(x_1,x_2,x_3)$. Transforming to new variables (y_1,y_2,y_3) where $y_1 = G_1(x_1,x_2,x_3)$, $y_2 = G_2(x_1,x_2,x_3)$, and $y^3 = F(x_1,x_2,x_3)$, we obtain the solution of (3.40a) in the form of a constant speed translation along the y_3-axis: $y_1 = C_1$, $y_2 = C_2$, and $y_3 = t + D$. We defer the different possible geometric interpretations of these constant speed translations until chapters 9 and 13, where coordinates are chosen so that the flow is studied directly on the surface $G_1 = $ constant rather than as a projection onto one of the Cartesian planes.

For dissipative systems, in addition to zero dimensional attractors and repellers (equilibria) and one dimensional closed curves that attract or repel (limit cycles), there are also closed two dimensional attracting or repelling surfaces that have the geometry of spheres or tori, depending upon whether or not the velocity field has zeros (equilibria) that lie on that surface. Spheres and tori also appear as invariant sets in the conservative case, but not as attracting or repelling point sets for nearby orbits.

In the noncanonical case, we shall call an n dimensional flow generated by a vector field V completely integrable if, in at least one coordinate system, the solution of the n differential equations $dx_i/dt = V_i$ is determined by the integration of n independent closed differential forms. The underlying idea is that $n - 1$ of those differentials are represented by $n - 1$ time-independent conservation laws $dG_1 = 0, \ldots, dG_{n-1} = 0$ that are isolating integrals, so that $n - 1$ equations of the form $x_j = f_j(x_n,C_1,\ldots,C_{n-1})$ exist for all $-\infty < t < \infty$. That is, $n - 1$ of the coordinates x_i are *determined* by the nth coordinate x_n (it does not matter which coordinate we call x_n) along with the $n - 1$ constants of the motion $C_1,\ldots,C_{n-1}$. By the substitution of the $n - 1$ equations for x_i into the equation of motion for the nth variable $dx_n/dt = V_n(x_1,\ldots,x_n)$, the single remaining differential equation is reduced to the one dimensional form $dx_n/dt = v_n(x_n,C_1,\ldots,C_{n-1})$ and is an exact differential. The complete solution of the system of equations then follows (i) from inverting the result of this last integration to find $x_n(t) = \Phi_n(t,C_1,\ldots,C_{n-1})$, and (ii) by substituting this result back into the $n - 1$ equations $x_j = f_j(x_n,C_1,\ldots,C_{n-1})$ to find $x_j(t) = \Phi_j(t,C_1,\ldots,C_{n-1})$ for $j = 1,\ldots,n - 1$. Because we have solved the system by the reduction to n *independent integrations* (the first $n - 1$ integrations are represented by the $n - 1$ first integrals $G_i = C_i$), the flow is called completely integrable. Two results then follow: the motion is confined to a two-dimensional subspace of the original n dimensional phase space. By Lie's coordinate transformation the family of streamlines on that two dimensional surface can be rewritten (in any open region that excludes equilibria, attractors, and repellers) as a single translation at constant speed. Geodesic flow on a closed, bounded two dimensional surface of constant negative curvature does not admit such a coordinate transformation globally.

Nonintegrability for $n \geq 3$ is possible only for nonlinear flow fields V: we have shown above how to construct the solution of a linear equation for arbitrary n in terms of real and complex exponentials, and the same method can be used to exhibit the integrability of linear equations in n dimensions explicitly: let $(\xi_1, \ldots, \xi_n)$ denote the coordinate system in phase space where the matrix A in equation (3.14) is diagonal. It follows that $dt = d\xi_i/\lambda_i \xi_i$, so that $t + C_i = g_i(\xi_i)$, representing n independent translations with constant (unit) velocity.

We can study a nonautonomous system of the form $dx/dt = V(x,t)$ with bounded solutions as a two dimensional flow generated by $dx/dt = V(x,\phi)$, $d\phi/dt = 1$. The differential form $dx - V(x,t)dt = 0$ has an integrating factor away from singularities of the velocity field, although it is impossible in most cases of interest to construct it exactly. In other words, there is always a time-dependent conservation law of the form $G(x,t) = C$. From the perspective of Picard's method, this two dimensional flow is uniquely solvable and the solution can also be constructed to within as high a degree of accuracy as one wishes by the method of successive approximations, so long as the flow is generated by a velocity field $V(x,t)$ whose first derivatives (along with the solution) are bounded over the desired region in the (x,t)-phase space. An example of a driven one dimensional system that is not a flow is provided by the Riccati equation $dx/dt = x^2 + t$, which has an infinite number of spontaneous singularities on the positive time axis (see Bender and Orszag, 1978, for both local and global analyses of this problem). For a two dimensional flow described by $dx/dt = V(x,t)$, however, there is a direct systematic approach to the approximate construction of the conservation law $G(x,t) = C$. It goes as follows. The conservation law is time-dependent and obeys

$$\frac{dG}{dt} = V(x,t)\frac{\partial G}{\partial x} + \frac{\partial G}{\partial t} = 0. \tag{3.41b}$$

Next, assume that $G(x,0) = x$ so that $x_o = C$, and that both $G(x,t)$ and $V(x,t)$ can be expanded as power series in t,

$$\left. \begin{array}{l} G(x,t) = G_o(x) + tG_1(x) + \cdots \\ V(x,t) = V_o(x) + tV_1(x) + \cdots \end{array} \right\} \tag{3.41c}$$

where $G_o(x) = x$. Substituting (3.41c) into (3.41b) and setting the coefficient of t^n, for $n = 0, 1, \ldots$, equal to zero yields

$$\left. \begin{array}{l} G_1(x) = -V_o(x) \\ 2G_2(x) = -V_1(x) - V_o(x)G_1'(x) \\ \vdots \end{array} \right\} \tag{3.41d}$$

which can be solved recursively. This method of construction is due to Newton. The result was sent (indirectly) as a anagram in Latin to the German philosopher–mathematician Leibnitz, the father of our present-day calculus notation, and also the discoverer of binary arithmetic. Because $C = x_o$, the conservation law is trivial:

initial conditions $x_{i_o} = U(-t)x_i(t)$ satisfy the condition $dG/dt = 0$, but they do not restrict the flow to a lower dimensional manifold.

It is hard to understand how Newton accomplished as much as he did in mechanics and mathematics. He bequeathed to us the first system of universal deterministic laws of nature (the three laws of motion and the law of gravity), calculus (independently of Leibnitz), differential equations, series solutions, and applied recursive methods to the solution of both differential and algebraic equations. As Alfred North Whitehead wrote, we are still living off the intellectual capital of the seventeenth century.

We illustrate in the next chapter that any two degree of freedom canonical Hamiltonian system with only *two* (rather than three) global isolating integrals may be completely integrable. In the central force problem, both conservation of energy and angular momentum hold and the solution in that case is reduced, through the choice of coordinate system that exploits symmetry, to only *two* independent integrations via the two independent constants of the motion. In general, for the integrability of an f degree of freedom Hamiltonian flow, only f, not $2f - 1$ independent, global isolating integrals need exist. The catch to this apparent simplicity is that those f conservation laws must satisfy a nontrivial compatibility condition that is explained in detail only much later in chapter 16.

As an example, geodesic flow on a surface of constant positive curvature, the two-sphere, can be formulated as a Hamiltonian flow in a four dimensional Cartesian $(\theta, \phi, p_\theta, p_\phi)$-phase space. The Lagrangian is $L = (d\theta/dt)^2 + (\sin \theta d\phi/dt)^2$ for a particle of unit mass on a sphere of unit radius. It is easy to show that the corresponding Hamiltonian problem is integrable via two conservation laws: both H and p_ϕ yield isolating integrals. The *Lagrangian configuration space* (not the q-subspace of phase space) is the unit sphere, where the coordinates (θ, ϕ) are orthogonal but not Cartesian, and the trajectories (discussed in chapter 2 from the standpoint of a variational problem) are great circles on that sphere. We point out in chapter 13 that the analogous problem of geodesic flow on a 'compact' surface of constant negative curvature is both nonintegrable and chaotic.

If $f = 1$, the existence of a conserved Hamiltonian guarantees integrability in the two dimensional phase space, as the integrability condition coincides with the phase-volume-preserving condition $\nabla \cdot V = \partial^2 H/\partial q \partial p - \partial^2 H/\partial p \partial q = 0$. A two degree of freedom canonical Hamiltonian system is generally nonintegrable. For $f > 2$, the phase-volume-preserving condition (Liouville's theorem) does *not* guarantee integrability for conservative systems in general or for Hamiltonian systems in particular. There is a second theorem by Liouville, called Liouville's theorem on involution systems (chapter 16), that defines the integrable class of canonical Hamiltonian systems, those with f global, functionally independent, 'commuting', isolating integrals (commutation, a compatibility condition, is explained in chapters 7 and 15). There are also other integrability theorems by Lie for nonHamiltonian systems where $n > 2$, and there are examples of dissipative systems where $n \geq 3$ that are integrable for special choices of control parameters (see Tabor, 1989). We turn now to the use of conservation laws in the reduction of degrees of freedom for the study of canonical flows.

3.7 Bifurcations in Hamiltonian systems

We consider here only the special case where the Lagrangian $L = T - U$ has a kinetic energy $T = g_{ij}dq_i dq_j/2dt^2$ that is quadratic in the generalized velocities, and where the matrix g and potential U depend upon the generalized coordinates but not the time. The canonical momentum is then linear in the generalized velocities,

$$p_i = g_{ij}\dot{q}_j, \tag{3.43}$$

so that inversion yields

$$\dot{q}_k = g_{kl}^{-1}p_l \tag{3.44}$$

where g^{-1} is the inverse of the positive semi-definite matrix g that depends upon the generalized coordinates alone. The Hamiltonian then has the form

$$H = T + U = \tfrac{1}{2}g_{ij}^{-1}p_ip_j + U. \tag{3.45}$$

In addition, assume that we have found a generalized coordinate system where there are exactly c conserved canonical momenta,

$$p_i = \frac{\partial L}{\partial \dot{q}_i} = M_i = \text{constant}, \tag{3.46}$$

which means that both g_{ij} and U are independent of the c coordinates q_i that are conjugate to the conserved momenta $p_i = M_i$. In particular, g can depend only upon the remaining $f - c$ free coordinates, and so we label the remaining $f - c$ free coordinates by the indices 1 through $f - c$. Restricting further to the case where g is a diagonal in our chosen coordinate system (this is consistent with the examples of chapter 4), we can then rewrite the Hamiltonian as

$$H = \tfrac{1}{2}\Sigma' g_i^{-1}p_i^2 + \tfrac{1}{2}\Sigma'' g_k^{-1}M_k^2 + U, \tag{3.47}$$

where the primed-sum is over $f - c$ nonconserved momenta and the double-primed-sum is over the c conserved ones. Since g^{-1} and U depend only upon the remaining $(q_1,\ldots,q_{f-c})$, we have a reduced Hamiltonian for those coordinates, $H = \Sigma' g_i^{-1}p_i^2/2 + V_{\text{eff}}$, where

$$V_{\text{eff}} = \tfrac{1}{2}\Sigma'' g_k^{-1}M_k^2 + U \tag{3.48}$$

defines an effective potential for the reduced problem with $f - c$ degrees of freedom and depends only upon $f - c$ coordinates and perhaps also the time t.

Equilibria are given by

$$\dot{q}_i = \frac{\partial H}{\partial p_i} = g_{ij}^{-1}p_j = 0, \tag{3.49}$$

which requires $p_i = 0$ for all i because g^{-1} is also a positive definite matrix, and by[10]

$$\dot{p}_i = -\frac{\partial H}{\partial q_i} = -\frac{\partial g_{ij}^{-1}}{\partial q_i}p_ip_j - \frac{\partial V_{\text{eff}}}{\partial q_i} = -\frac{\partial V_{\text{eff}}}{\partial q_i} = 0. \tag{3.50}$$

Stable equilibria of this system are potential minima. Unstable equilibria are potential maxima and saddle-points.

---

[10] Setting the time derivative of the canonical momentum equal to zero guarantees equilibrium in any inertial frame, but does not guarantee that the body is at rest in the coordinate system of interest if that frame is accelerated (see chapter 8).

Consider next an effective potential V_{eff} that is time-independent but depends upon some control parameter λ. Assume also that the motion is bounded for some finite region in phase space that contains the potential minimum. It is enough here to illustrate the main idea for one degree of freedom, $f = 1$, so that there are two turning points q_{min} and q_{max} in configuration space. To go further, assume that there is a critical value λ_c of λ where the following properties hold: for $\lambda < \lambda_c$ there is a potential minimum at q_o, but $V_{eff}''(q_o,\lambda_c) = 0$ and $V_{eff}''(q_o,\lambda_c) < 0$ for $\lambda > \lambda_c$, so that the equilibrium point is unstable for $\lambda > \lambda_c$. Therefore, for $\lambda > \lambda_c$, U decreases as q decreases or increases away from q_o. However, because there are two turning points, the function $V_{eff}(q,\lambda)$ must eventually increase again as a function of q. Therefore, two new stable equilibria must come into existence on either side of q_o as λ is increased beyond λ_c.

The qualitative shape of the potential along with corresponding local phase portraits is shown as figures 3.10 and 3.11.

The phenomenon whereby equilibria become unstable as a control parameter λ is varied is called 'bifurcation' and our potential-based model yields a 'pitchfork' bifurcation: the once-stable equilibrium at q_o persists as an unstable one as λ is increased beyond λ_c, and two new equilibria at $q_1(\lambda)$ and $q_2(\lambda)$ grow out of the bifurcation point at λ_c. This is shown as the bifurcation diagram in figure (3.12), where equilibrium solutions of Hamilton's equations are plotted against the parameter λ. This mechanism for loss of stability of equilibria is quite general for Hamiltonian systems, and we shall meet it again in chapter 17 along the route from integrability to chaos in Hamiltonian systems.

Suppose that the effective potential V_{eff} is symmetric about reflection through an axis that passes vertically through the equilibrium point q_o. In this case, q_1 and q_2 must be equidistant from q_o and represent a 'broken symmetry'. The symmetry is 'broken' whenever one of these new equilibrium solutions is occupied and the other is not, for any energy $E < V_{eff}(q_o)$, as is shown schematically in the bifurcation diagram of figure 3.12. In this case, our example illustrates a 'symmetry-breaking pitchfork bifurcation'. As an example, consider the model effective potential where

$$V_{eff}(q,\lambda) = a(\lambda_c - \lambda)q^2/2 + bq^4/4, \qquad (3.51)$$

and where a and b are constants. Equilibria satisfy the equation

$$\bar{q}[a(\lambda_c - \lambda) + b\bar{q}^2] = 0 \qquad (3.52)$$

so that there is one equilibrium point at the origin. There are also two equilibria symmetrically placed about the origin which are given by

$$\bar{q} = \pm [a(\lambda - \lambda_c)/b]^{1/2}. \qquad (3.53)$$

Note that these two equilibria do not exist when $\lambda < \lambda_c$. Therefore, we have a pitchfork bifurcation at λ_c. Because the effective potential V_{eff} is symmetric about the origin, the bifurcation breaks that symmetry. The origin is a stable equilibrium point when $\lambda < \lambda_c$ but is unstable for $\lambda > \lambda_c$, as is shown in figure 3.10. However, the two equilibria given by equation (3.53) correspond to two minima of the effective potential V_{eff} and are therefore stable. Another way to say it is that an elliptic point at the origin becomes hyperbolic as λ is increased through λ_c, and two new elliptic points come into existence on either side of it as is illustrated in figure 3.11b. An example from physics that shows the same qualitative behavior as this model is given as exercise 19.

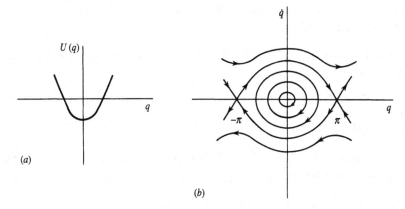

Fig. 3.10(a) Effective potential for $\lambda < \lambda_c$. (b) Steamlines for $\lambda < \lambda_c$.

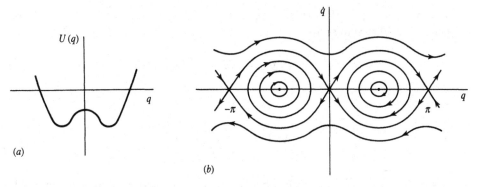

Fig. 3.11(a) Effective potential for $\lambda > \lambda_c$. (b) Steamlines for $\lambda > \lambda_c$.

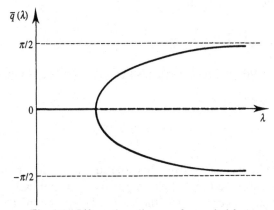

Fig. 3.12 Bifurcation diagram for a pitchfork.

Exercises

1. Draw the streamlines for a body moving at constant velocity, and a body falling freely in a uniform gravitation field.

2. Plot the vector fields $V = (x_2, - x_1)$, and (x_2, x_1), and $V = (- x - x^2, - \beta y + y^2 x)$ in the phase plane.

3. Use the time-evolution operator to derive the simple harmonic oscillator solution from the initial conditions.

4. Determine whether each of the following dynamical systems is conservative or dissipative:

   (a) $\dot{x} = 3x^2 - 2y, \ \dot{y} = 5y + x^3$
   (b) $\dot{x} = 3x^2 - 2y, \ \dot{y} = 5 - 6xy$
   (c) $p_r = 2r^2 \sin r, \ \dot{r} = p_r.$

   For the conservative systems, locate the equilibria, determine the stability, and plot the global phase portrait.

5. If $\dot{x}_1 = V_i(x_1,\ldots,x_n)$ and $y_k = f_k(x_1,\ldots,x_n)$, show that

$$\dot{y}_i = W_i(y_1,\ldots,y_n)$$

where $W_k = (\partial f_k/\partial x_i)V_i(h_1(y_1,\ldots,y_n),\ldots,h_n(y_1,\ldots,y_n))$ and $x_i = h_i(y_1,\ldots,y_n)$. With $(x_1,\ldots,x_n)$ Cartesian so that

$$\nabla \cdot V = \sum_i \frac{\partial \dot{x}_i}{\partial x_i} = \sum_i \frac{\partial V_i}{\partial x_i},$$

show that

$$\nabla \cdot V = \sum_i \frac{1}{J}\frac{\partial}{\partial y_i}(JW_i)$$

where $J = \dfrac{\partial(x_1,\ldots,x_n)}{\partial(y_1,\ldots,y_n)}$ is the Jacobian of the transformation. Show in cylindrical coordinates that $J = r$, and that

$$\nabla \cdot V = \frac{\partial \dot{x}}{\partial x} + \frac{\partial \dot{y}}{\partial y} + \frac{\partial \dot{z}}{\partial z} = \frac{\partial \dot{r}}{\partial r} + \frac{\dot{r}}{r} + \frac{\partial \dot{\phi}}{\partial \phi} + \frac{\partial \dot{z}}{\partial z}.$$

   (See Whittaker (1965), page 277; see also exercise 15 in chapter 2.)

6. Consider the model

$$\dot{x} = \sigma(y - x)$$
$$\dot{y} = \rho x - y.$$

   (a) When is the flow conservative? Find the equilibria, determine their stability, and plot the local and global phase portraits for that case.
   (b) For which parameter values is the system Hamiltonian? Derive the Hamiltonian H.

7. Construct the inverse of the transformation that describes the time-evolution of the simple harmonic oscillator.

8. A particle of mass m moves along the x-axis. Its potential energy at any point x is

$$U(x) = U_o x^2 e^{-x^2},$$

   where U_o is a constant.

(a) Find the force on the particle.

(b) Find all the points on the x-axis where the particle can be in equilibrium, that is, where the force is zero. Determine whether each equilibrium point is stable or unstable, that is, whether in its immediate neighborhood the force is toward or away from the equilibrium point. Sketch the potential and label the equilibrium points.

(c) Determine the maximum total energy E_o that the particle can have and still execute bounded motion. If $E < E_o$ is the motion necessarily bounded? Explain.

(d) Sketch the qualitatively different possible phase portraits (vary E).

9. A particle of mass m moving in one dimension is subject to the force

$$F = -kx + \frac{a}{x^3}.$$

(a) Find the potential and sketch it (assume that the constants k and a are both positive).

(b) Locate equilibria, determine their stability, and plot the phase diagram.

10. Consider the driven dissipative system $\ddot{x} + \dot{x} + x = \cos \omega t$ in the $(x, \dot{x}, \phi)$ phase space, where $\phi = \omega t$.

(a) Show that there are no equilibria (no zero dimensional attractors or repellers). By concentrating upon the 'particular integral' (the $t \to \infty$ solution) of the solution $x = Ae^{-t}\cos \omega t + 1/1 + \omega^4 \sin(\omega t + \beta)$ where $\tan \beta = [(1 - \omega^2)/2\omega]$, show that

(b) the projection of the motion into the $(x, \dot{x})$-plane lies on an ellipse,

(c) there is a one dimensional attractor that is a spiral on an elliptic cylinder that is concentric with the ϕ-axis.

11. The potential energy $U(x)$ of a conservative system is continuous, is strictly increasing for $x < -1$, zero for $|x| \leq 1$, and is strictly decreasing for $x > 1$. Locate the equilibrium points and sketch the phase diagram for the system.

(Jordan and Smith, 1977)

12. A simple pendulum strikes a wall that is inclined at an angle α to the vertical. Sketch the phase diagram, for α positive and α negative, when (i) there is no loss of energy at impact, and when (ii) the magnitude of the velocity is halved on impact.

(Jordan and Smith, 1977)

13. Sketch the potential energy and phase portrait of the dynamical system with Hamiltonian $H(q,p) = \frac{1}{2}p^2 + \frac{1}{4}q^4 - \frac{1}{2}q^2$.

Find the equation of the separatrix and give a qualitative description of the motion on the separatrix. Obtain the period of small oscillations in the neighborhood of any of the equilibria.

(Percival and Richards, *Introduction to Dynamics*, Cambridge, 1982)

14. For two independent oscillators, where

$$E = \frac{1}{2}(\dot{x}_1^2 + x_1^2) + \frac{1}{2}(\dot{x}_2^2 + \omega^2 x_2^2)$$

observe that $|x_1| \leq A$ and $|x_2| \leq B$ and

(a) find values for A and B in terms of the energies E_1 and E_2, where $E = E_1 + E_2$.

(b) Show also that the above described rectangle is inscribed within the ellipse $U \leq E$ where $U = \frac{1}{2}(x_1^2 + \omega^2 x_2^2)$.

(Arnol'd, 1989)

15. For the system defined in exercise 14, and within the inscribing rectangle, plot Lissajou's figures for $\omega = 2$, taking into account several phase differences $\phi_1 - \phi_2 = 2\pi, \pi/2, \pi$, and $3\pi/2$.

16. For the system defined in exercise 14 plot the Lissajou figures for $\omega = 1$ and $\omega = 29/31$, for both $\phi_1 - \phi_2 = 0$ and $\pi/2$.

17. Consider the flow defined by $V = (x - y, y + x)$.

(a) Show that the streamlines spiral away from a source at the origin, and that $\lambda_1 = 1 + i = \lambda_2^*$.

(b) Show that $G(x,y) = \frac{1}{2}\ln(x^2 + y^2) - \tan^{-1}y/x = C_2$ is conserved along streamlines, but with a different constant C_2 along each streamline.

(c) In Lie's transformation to the translation variables $(y_1, y_2) = (G(x,y), F(x,y))$ show that $F(x,y) = \tan^{-1}y/x = t + D$.

(Hint: use cylindrical coordinates (r, ϕ).)

18. Consider the damped harmonic oscillator $\ddot{x} + \beta\dot{x} + \omega^2 x = 0$ in the form $x_1 = x$, $x_2 = \dot{x}$ so that

$$\dot{x}_1 = x_2$$
$$\dot{x}_2 = -\beta x_2 - \omega^2 x_1.$$

Show that the equation

$$\frac{dx_1}{x_2} = \frac{dx_2}{-\beta x_2 - \omega^2 x_1}$$

becomes

$$\frac{dx}{x} + \frac{u\,du}{u^2 + \beta u + \omega^2} = 0$$

under a change of variable $x = x_1$, $u = x_2/x_1$. Setting $\alpha = \beta^2 - 4\omega^2$ and integrating, show that the conserved quantity $g(x,u) = G(x_1, x_2) = C$ is given by

$$g(x,u) = \begin{cases} \ln|x| + \frac{1}{2}\ln|u^2 + \beta u + \omega^2| - \dfrac{\beta}{\sqrt{-\alpha}}\tan^{-1}\left(\dfrac{2u + \beta}{\sqrt{-\alpha}}\right), \alpha < 0 \\[12pt] \ln|x| + \ln\left|u + \dfrac{\beta}{2}\right| + \dfrac{\beta}{2u + \beta}, \alpha = 0 \\[12pt] \ln|x| + \frac{1}{2}\ln|u^2 + \beta u + \omega^2| - \dfrac{\beta}{2\sqrt{\alpha}}\ln\left|\dfrac{2u + \beta - \sqrt{\alpha}}{2u + \beta + \sqrt{\alpha}}\right|, \alpha > 0 \end{cases}$$

corresponding to underdamped, critically damped, and overdamped oscillations respectively.

(Burns and Palmore, 1989)

19. A point mass is constrained to move on a massless hoop of radius a fixed in a vertical plane that is rotating about the vertical with constant angular speed ω.

Obtain the Lagrangian, the Hamiltonian, and construct the effective potential, taking gravity to be the only external force (other than constraint forces). Locate all equilibria and determine their stability, permitting ω to vary. Observe that the Hamiltonian has a certain reflection symmetry. Show that for $\omega < \omega_c$, where ω_c is a critical rotation speed, there is only one equilibrium point that exhibits the reflection symmetry of the Hamiltonian, but that when $\omega > \omega_c$, there are two solutions that 'break' that symmetry. Plot the phase portraits for both $\omega < \omega_c$ and $\omega > \omega_c$. Plot the equilibrium solutions as a function of ω. Sketch stable solutions as solid curves, unstable ones as dashed curves, and observe that the diagram has the form of a pitchfork. (This is an example of a symmetry-breaking pitchfork bifurcation.)

4

Motion in a central potential[1]

4.1 Overview

The great success of Newtonian mechanics was the mathematical explanation of the clockwork motion of the planets about the sun by the same universal laws of motion and force that account for projectile motions. With this success determinism and stable periodicity defined the mechanical model to be imitated. However, among arbitrary central potential problems, the periodic orbits that characterize the bounded motions of the Kepler and isotropic simple harmonic oscillator problems are a rarity. The absence of stable quasiperiodicity in both of these two degree of freedom systems is 'accidental': it is due to the analytic form of a third global conservation law, one that reflects a 'hidden' symmetry.

By a central potential we mean a sum of two or more terms that depend upon the particle separation r alone, like

$$U(r) = -\frac{k}{r} + ar^2. \tag{4.0a}$$

The rarity of perfect clockwork among the bounded, stable orbits that arise from problems with an arbitrarily chosen central potential $U(r)$ is the content of the Bertrand–Königs theorem which states that, for bounded motion, the *only* central potentials that yield periodic orbits for all values of the force constants and initial conditions are the isotropic oscillator and Kepler problems. For all other central potentials, e.g. (4.0a) above which is a linear combination of the Kepler and isotropic oscillator potentials, quasiperiodic orbits will be the outcome of an arbitrary choice of initial conditions. In the Kepler and isotropic oscillator problems, which are 'superintegrable', a third global conservation law can be used to solve for the orbit equation *purely algebraically* without evaluating a single integral. The analyticity of that conservation law forces the orbit equation $r = f(\theta, E, M)$ to be a *single-valued* function of the angular variable θ. One needs multivaluedness of the orbit function f in order to get quasiperiodicity: for a quasiperiodic orbit, $r = f(\theta, E, M)$ must be infinitely many-valued in θ.

Among the integrable systems quasiperiodic orbits are more typical than the periodic orbits that define perfect clockwork. Stable quasiperiodicity can imitate clockwork to within finite accuracy over short enough time intervals, so that periodicity is not the worst criticism of taking the Kepler problem as illustrative of classical mechanical behavior. Most systems are nonintegrable, but to go further we must first define 'orbital instability' analytically. The latter represents the geometric idea of sensitivity of long-time behavior with respect to even the smallest possible changes in initial conditions.

---

[1] In this chapter we will use Gibbs notation (arrows) for the vectors.

In harder dynamics problems than Newtonian central potential problems it is often useful to follow Poincaré and replace a system of first order differential equations by an iterated map. The properties of differential equations and their corresponding maps are in one-to-one correspondence with each other, but the maps are much easier to analyze than the entire flow. We illustrate this in part by deriving the map that universally describes motion in a central potential. An iterated map can be used to formulate analytically, and rather simply, the important idea of sensitivity with respect to small changes in initial conditions.

Deterministic chaos can be understood from the nongeometric standpoint of symbolic dynamics as deterministic disorder on a lattice, and is required for describing mechanical motions that exhibit the apparent randomness (pseudo-randomness) that gives rise to statistical behavior. In the last section of this chapter we explain how the unstable periodic and nonperiodic orbits of deterministic chaotic systems differ symbolically and qualitatively from the stable periodic and quasiperiodic orbits that characterize integrable systems.

4.2 Integration via two constants of the motion

We restrict our discussion to the motion of a body in a central force field

$$\vec{F} = -\nabla U(r) \tag{4.1}$$

where the potential $U(r)$ is rotationally symmetric and depends only upon r, the distance from the center of force to the particle whose motion we discuss. The reduced motion of two bodies of masses m_1 and m_2 gives rise to this problem, where the mass of the hypothetical particle is then $\mu = m_1 m_2/(m_1 + m_2)$.

Since

$$\frac{d\vec{M}}{dt} = \vec{r} \times \vec{F} = 0, \tag{4.2}$$

the angular momentum is conserved, which means not only that its magnitude is fixed but that its direction is fixed in space (fixed in an inertial frame). Since both the position and velocity are perpendicular to the angular momentum, this conservation law (based upon rotational symmetry) confines the particle's orbit to a plane perpendicular to the angular momentum. Therefore, we need begin only with the four dimensional phase space (r,θ,p_r,p_θ) where the configuration space (r,θ) is given by the polar coordinates of the particle in the plane. Polar coordinates are chosen with the intention of further exploiting the rotational symmetry (we have not yet used the fact that the magnitude of the angular momentum is constant in time).

The Lagrangian $L = T - U$ is

$$L = \mu\dot{r}^2/2 + \mu r^2\dot{\theta}^2/2 - U(r) \tag{4.3}$$

and the canonical momenta conjugate to (r,θ) are

$$p_r = \mu\dot{r} \tag{4.4}$$

and

$$p_\theta = \mu r^2\dot{\theta}. \tag{4.5a}$$

Because

$$\frac{\partial L}{\partial \theta} = 0, \tag{4.6}$$

p_θ is conserved (is the magnitude of the angular momentum) so that we can use

$$\dot{\theta} = p_\theta/\mu r^2 \tag{4.5b}$$

to eliminate $d\theta/dt$ from the equations of motion, but this elimination cannot be carried out in the Lagrangian. First, we must construct the Hamiltonian:

$$H = p_r\dot{r} + p_\theta\dot{\theta} - L = \mu\dot{r}^2/2 + \mu r^2\dot{\theta}^2/2 + U(r) = E \tag{4.7a}$$

where E is the total mechanical energy and is conserved. Now, we can eliminate $d\theta/dt$ to obtain

$$E = \mu\dot{r}^2/2 + V(r) \tag{4.7b}$$

where $V(r) = p_\theta^2/2\mu r^2 + U(r)$ is the effective potential for the radial motion, which can be studied in the (r,p_r)-phase-plane completely independently of the angular motion.

We have used two conservation laws in a problem with $f = 2$, and the result is a reduction of the solution of the dynamics to two integrations:

$$dt = \frac{dr}{[2(E - V(r))/\mu]^{1/2}} \tag{4.8a}$$

must be integrated and then must be inverted to find $r(t)$, and also

$$d\theta = \frac{(p_\theta^2/2\mu r^2)dr}{[2(E - V(r))/\mu]^{1/2}} \tag{4.9}$$

must be integrated to find $\theta(r)$, and then inverted to obtain $r = f(\theta)$. Whether the integrations are to be performed analytically or numerically, and whether the inversions are actually carried out is irrelevant: we started with the need for four coupled integrations, but the problem has been reduced to *two* integrations by the exploitation of *two* independent constants of the motion, and those two integrations completely determine the motion for all times t.

In what follows, we restrict to problems with bounded motion so that the effective potential $V(r)$ must have at least two turning points that confine the motion to a region of configuration space $r_{\min} < r < r_{\max}$ (figure (4.1)). *This means that the effective potential energy $V(r)$ has at least one minimum between the two turning points.* If U blows up as r goes to infinity, this works for all energies $E \geq U(r_\circ)$ where $r_\circ$ is the location of the potential minimum. If U vanishes as r blows up then $U(r_\circ) \leq E < 0$ is required (figure 4.1b). The existence of at least one minimum of $V(r)$ requires that $U(r)$ goes to minus infinity more slowly that $1/r^2$, and also that $U(r)$ vanishes negatively as r goes to infinity more slowly than $1/r^2$. The goal is now to extract as much qualitative information as possible about the nature of the motion in a central potential without evaluating either of the above integrals explicitly. Without loss of generality we now take $\mu = 1$ and also set $p_\theta = M$ in what follows. The restriction to bounded motion requires two turning points for the effective potential and so requires that $E < 0$ if $U(r)$ vanishes at infinity (or $E < E_c$ if $U(r)$ approaches any constant E_c as r goes to infinity).

First, note that there is a at least one elliptic point in the (r,p_r)-subspace of phase space

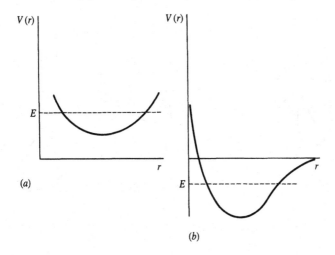

Fig. 4.1(a) A potential that binds for all energies (two turning points). (b) A potential that allows binding ($E < 0$) or scattering ($E > 0$).

at the minimum of the effective potential, at $r = r_o$ where $V'(r_o) = 0$ with $V''(r_o) > 0$. The streamlines in the phase planes are therefore closed curves with period

$$\tau_r = 2 \int_{r_1}^{r_2} \frac{dr}{[\sqrt{2(E - V(r))\mu}]^{1/2}} \tag{4.8b}$$

where $E = V(r_1) = V(r_2)$.

The motion in the (θ, p_θ)-subspace is even simpler. Since $p_\theta = M$ is constant along a streamline, the streamlines here are all parallel to the θ-axis. However, this is not a constant velocity translation: $d\theta/dt = M/\mu r^2 > 0$ is not constant except in the special case of circular motion, which can only occur at an extremum r_o of $V(r)$, where $V'(r_o) = 0$. Instead, the θ motion is periodic with nonuniform velocity: any initial condition $\theta(t_o)$ must be repeated, modulo 2π at a time $t_o + \tau_\theta$, where the period τ_θ is determined by the relationship $\theta(t_o + \tau_\theta) = \theta(t_o) + 2\pi$ (or by $\theta(t_o + n\tau_\theta) = \theta(t_o) + 2n\pi$ where n is any integer). We therefore have a two-frequency problem. It follows that the motion in the four dimensional phase space, or in the two dimensional (r,θ)-configuration-space, may be either periodic or quasiperiodic depending upon whether or not the period ratio τ_θ/τ_r is a rational number, a ratio of integers. We now reformulate this commensurability condition via a single integral.

4.3 Maps, winding numbers, and orbital stability

Since the radial motion is confined to $r_{min} < r < r_{max}$, the motion in configuration space is confined to an annulus (figure 4.2(a)), and $d\theta/dt > 0$ means that the motion within that annulus is always in the counterclockwise direction as r and θ vary with time. The integral

$$\Phi = \int_{r_{min}}^{r_{max}} \frac{dr M/r^2}{[2(E - V(r))]^{1/2}} \tag{4.10a}$$

defines the angular displacement $\Delta\theta$ between two successive turning points of the radial

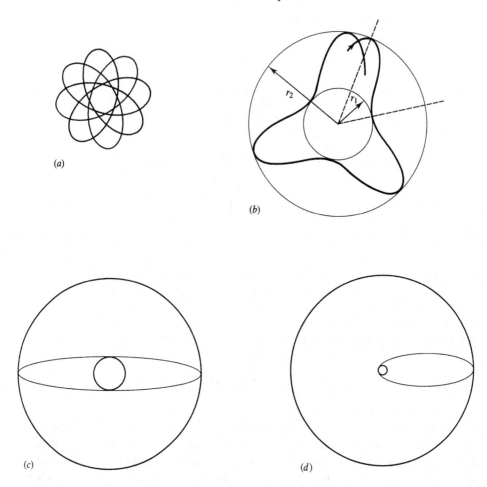

Fig. 4.2(*a*) Motion in configuration space with a winding number $W = 1/3$.
(*b*) A possible quasiperiodic orbit (fig. 3.7 from Goldstein, 1980). (*c*) Motion
with $W = 1/4$ (isotropic oscillator). (*d*) Motion with $W = 1/2$ (gravitational
free-fall).

motion (we have set $\mu = 1$ for convenience, but the same effect can be accomplished by
rescaling the variable r). By symmetry, this is the same as the angle between any two
successive turning points in figure (4.2). Now, it is possible to establish a geometric/
analytic criterion for a closed orbit in configuration space, which is to say for a periodic
orbit in the configuration plane. If, after n angular displacements of $\Delta\theta = \Phi$, the orbit
makes exactly $m2\pi$-revolutions inside the annulus and then closes on itself then the
orbit is periodic. If we write $\Phi = 2\pi W$ where, W is called the winding number of the
orbit, then the criterion for a closed orbit is just that W must be rational, $W = m/n$
where n and m are integers. It takes just a little thought to see that this is the same as
requiring the frequency ratio ω_θ/ω_r to be rational, but W and the frequency ratio are
generally not the same rational number. In contrast, if the winding number W is
irrational then the orbit can never close on itself and is quasiperiodic (perhaps as in
figure 4.2(*b*)).

Since the isotropic simple harmonic oscillator ($U(r) = kr^2/2$) has only elliptic orbits centered on $r = 0$, it follows that $W = 1/4$ (figure 4.2(c)). For $E < 0$ the Kepler problem $(U(r) = -k/r)$ has only elliptic orbits with a focus at $r = 0$, so it follows from figure 4.2(d) that $W = 1/2$.

Let us review what we mean by 'almost periodic', or quasiperiodic. There are two parts to it: if W is irrational, then (i) the orbit *seems* to have a rational winding number W_o (hence a definite period τ_o) if we only know θ to finite accuracy (which is always the case in observation), but if we wait long enough, it will become clear that W_o is *not* the right winding number. (ii) There is an idea of stability: if we make a better approximation W_1 to the irrational winding number, it will seem that the orbit is periodic with period $\tau_1 > \tau_o$, but if we wait still longer in our observation, we will see that τ_1 is not the period, but that another number τ_2 seems to be the period, and this sequence of approximations by larger and larger finite periods (that differ from each other only by last digits, and so are approximately equal to each other when decimal expansions are truncated) over longer and longer finite times can be carried out ad infinitum.

To see how quasiperiodicity can simulate periodicity with finite accuracy over short enough time intervals, consider any irrational number and construct its continued fraction expansion. With $W = \sqrt{2} - 1$, for example, this is easy: write $(1 + W)^2$, so that $1 + W = \sqrt{2}$, but $1 + 2W + W^2 = 2$ yields $W(2 + W) = 1$, which can be iterated to yield the desired continued fraction expansion:

$$W = \cfrac{1}{2 + W} = \cfrac{1}{2 + \cfrac{1}{2 + W}} = \cfrac{1}{2 + \cfrac{1}{2 + \cfrac{1}{2 + \cfrac{1}{2 + \ldots}}}}. \tag{4.11}$$

From (4.11), we get the following sequence of rational successive approximations W_n to the irrational winding number $W = \sqrt{2} - 1 = 0.4161\cdots$, which is by the very nature of the irrational numbers a nonperiodic decimal expansion (only rational numbers have periodic, including truncating decimal expansions): $W_o = 1/2 = 0.50\cdots$, $W_1 = 1/(2 + 1/2) = 2/5 = 0.400\cdots$, $W_2 = [1/(2 + 1/(2 + 1/2)] = 5/12 = 0.41666\cdots$, and so on. Since we have used base 10 arithmetic, the error in our approximation to W has the form $W - W_m \approx 10^{-N}$ so that after an exponentially large number $n \approx 2\pi 10^N$ of 2π revolutions of the orbit, the use of the rational winding number W_n will yield an error in the leading digit of $\theta(t)$. On the other hand, for discrete 'times' $n \ll 2\pi 10^N$, the orbit will appear to be periodic with rational winding number W_m, so long as the accuracy of our knowledge of $\theta(t)$ is much less than N decimal places. There is an idea of orbital stability that is implicit in this description: after several 2π revolutions, the orbit has not changed very much from its earlier shape. Next, we state the idea of orbital stability explicitly.

To quantify the idea of orbital stability, we introduce the linear circle map by writing $\theta_n = \Phi + \theta_{n-1}$ where $\theta_n = \theta(t_n)$ are the angular positions at the discrete times t_n where the orbit hits one of the turning points $r = r_{min}$ or $r = r_{max}$. Then

$$\theta_n = 2\pi W + \theta_{n-1} = 2\pi n W + \theta_o. \tag{4.12}$$

We can plot the values of θ_n on a circle of unit radius, or on any other simple closed curve for that matter. Whenever W is rational, there is only a finite number of points, which is to say that the solution θ_n of (4.12) is periodic modulo 2π, and the planar orbit of the dynamical system also closes on itself, as in figures 4.2(a), (c), and (d). If W is irrational, on the other hand, then the discrete orbit of equation (4.12) never repeats itself but instead comes arbitrarily near to every point on the circle (or closed curve) infinitely many times as n approaches infinity (this result is known as Jacobi's theorem). This is easy to understand: as one takes ever better rational approximations W_m to an irrational winding number W, then the number of points visited by the map becomes both larger and closer together on the circle, and as m goes to infinity the rational approximations to the irrational orbit become dense on the unit circle. Whether the exact solution visits points on the circle uniformly or nonuniformly depends upon the choice of winding number, but the motion on the circle is both nearly periodic (is quasiperiodic) and is ergodic: time averages on the circle may be replaced by angular averages with some (at this stage unknown) density $\rho(\theta)$, but the density $\rho(\theta)$ is not necessarily uniform. This ergodicity has nothing to do with either chaos or statistical mechanics because it is the result of *stable* quasiperiodicity. That the motion is stable is shown below. This kind of stability rules out deterministic chaos and along with it the kind of mixing in phase space that, according to Gibbs, yields statistical mechanics.

In order to formulate the idea of orbital stability explicitly, we must consider a more general circle map

$$\theta_n = F(\theta_{n-1}) \tag{4.13a}$$

where $F(\theta) = \theta + f(\theta)$ and $f(\theta + 2\pi) = f(\theta)$. The analysis that follows is not restricted to circle maps but applies equally well to arbitrary one dimensional maps (4.13a) where θ is generally not an angular variable.

Consider next the orbit $\theta_o \to \theta_1 \to \cdots \to \theta_n \to \cdots$ of the map that starts from the initial condition θ_o. Consider also another orbit starting from a nearby initial condition $\theta_o + \delta\theta_o$, whereby

$$\theta_n + \delta\theta_n = F(\theta_{n-1} + \delta\theta_{n-1}) \tag{4.13b}$$

after n iterations. If we assume that $\delta\theta_o$ is small enough, then we can track the difference between the two orbits linearly, yielding

$$\delta\theta_n \approx F'(\theta_{n-1})\delta\theta_{n-1}, \tag{4.14a}$$

which means that we expand (4.13b) about the entire original orbit, point by point. In order that the linear approximation (4.14a) is valid for large n, we must take the initial difference in initial data $\delta\theta_o$ to be sufficiently small. Equation (4.14a) can then be iterated easily to yield the solution

$$\delta\theta_n = \prod_{i=0}^{n-1} F'(\theta_i)\delta\theta_o, \tag{4.14b}$$

where the equality only denotes that we have found the exact solution of the *linear* equation. From (4.14b) it is clear that the slope F' of the map determines stability or lack of it. If the slope has a magnitude less than unity on the average, then the two orbits tend to approach each other as n increases ($\delta\theta_n$ decreases in magnitude as n increases). If, on the other hand, the average slope is greater than unity than the two orbits tend to separate as if they were repelled by each other. This is interesting, for since the slope is determined by the original orbit alone, this means that it looks as if a single orbit either attracts or repels *all* nearby orbits that have initial conditions in a 'ball' of width $\delta\theta_o$. To go further with this description, by taking the absolute value and exponentiating, we can rewrite (4.14b) as

$$|\delta\theta_n| = e^{n\lambda(\theta_o)}|\delta\theta_o|\tag{4.14c}$$

where $\lambda(\theta_o)$ is called the Liapunov exponent for the trajectory with initial condition θ_o and is given by the time average of the slope of the map along the corresponding orbit:

$$\lambda(\theta_o) = \frac{1}{n}\sum_{i=0}^{n-1}\ln|F'(\theta_i)|.\tag{4.15}$$

Whenever $\lambda < 0$, we have a very strong kind of orbital stability in the sense that all of the nearby orbits approach a common limiting orbit as n goes to infinity. This can only happen in a dissipative system with 'sinks' like attracting equilibria, stable limit cycles, and attracting tori: the orbits appear to attract each other because they are all flowing toward the same attractor in phase space. An example is provided by the tent map

$$F(x) = \begin{cases} ax, & x < 1/2 \\ a(1-x), & x > 1/2 \end{cases}\tag{4.13c}$$

with slope magnitude $a < 1$: $\lambda = \ln a$ is negative for all initial conditions whenever $a < 1$. Qualitatively speaking, such behavior represents that of an overdamped nonconservative system of differential equations, where all trajectories flow into a sink. To see the analogy, note that fixed points of the map, which are analogous to the equilibria of a set of differential equations, satisfy the condition $x = F(x)$ so that (4.13c) has two equilibria, at $x = 0$ and $x = a/(1 + a)$. Next, we need to determine the stability of these fixed points. Linearizing about an equilibrium point yields

$$\delta x_n \approx F'(x)\delta x_{n-1}\tag{4.13d}$$

with $\delta x_n = x_n - x$, which has the solution

$$\delta x_n \approx (F'(x))^n\delta x_o.\tag{4.13e}$$

Therefore, a fixed point is stable if $|F'(x)| < 1$, unstable if $|F'(x)| > 1$, and a bifurcation occurs whenever $|F'(x)| = 1$. For the tent map (4.13c), there is a bifurcation at $a = 1$ so that both equilibria are stable for $a < 1$, unstable if $a > 1$.

When $\lambda > 0$ then all nearby orbits appear to repel each other exponentially fast as n increases, so that small differences $\delta\theta_o$ are magnified into large differences $\delta\theta_n$ at later discrete times n. This is a form of orbital instability where orbits that are very near to each other initially are quantitatively and therefore qualitatively different from each other at large times. An example is provided by the tent map with slope magnitude $a > 1$, where $\lambda = \ln a > 0$. We discuss this case further in section 6.5.

When $\lambda = 0$, we have marginal stability, and this turns out to be all that can happen in a map representing an integrable canonical system. As an example, the linear circle map has slope magnitude equal to unity, so that $\lambda = 0$ for all initial conditions θ_o. Our orbits are all stable, whether periodic or quasiperiodic, as they are insensitive to small changes in initial conditions.

The goal of the next section is to illustrate that stable periodicity is rare both in nature and in dynamical systems theory.

4.4 Clockwork is rare in mechanical systems

The earth's orbit about the sun is the basis for our idea of a clock. Is it possible that other naturally occurring macroscopic systems defined by other central forces also have periodic orbits, and could just as well have served the same historic purpose? To formulate this question mathematically, we ask for the form of the potentials $U(r)$ that never yield bounded motion in the form of quasiperiodic orbits, so that *every* bounded motion of the Newtonian dynamical system is *periodic*. In other words, we look for central forces that, like the Kepler and isotropic linear oscillator problems, yield periodicity independently of initial conditions (independently of E and M) and independently of the value of the force constant k.

An arbitrary central potential of the form

$$U(r) = -\sum_i \frac{k_{-i}}{r^{\beta_i}} + k_o \ln r + \sum_i k_i r^{\alpha_j} \tag{4.0b}$$

yields bounded motion so long as $\alpha_i > 0$ and $\beta_i > 2 - \varepsilon$ where $\varepsilon > 0$. The k_i are called force constants, or coupling constants since they couple different degrees of freedom to each other. An example would be a potential energy of the form

$$U(r) = -\frac{k_{-1}}{r} + \tfrac{1}{2}k_1 r^2. \tag{4.0c}$$

Does this potential yield periodic orbits for all values of (E,M,k_{-1},k_1)? The question can be formulated as follows: is the integral

$$\Phi = \int_{r_{min}}^{r_{max}} \frac{dr M/r^2}{[2(E - V(r))]^{1/2}} \tag{4.10a}$$

equal to a rational number independent of the values of those four parameters? The answer, in general, is no: the integral defines a function $\Phi = f(E,M,k_{-1},k_1)$ that generally varies as the parameters are varied. Most integrals of the form (4.10a) with arbitrary central potentials cannot possibly be independent of the initial conditions and the force constants k_i. That Φ is completely independent of (E,M,k) in the Kepler and linear oscillator problems must therefore be rare. Are there other central potentials for which this rarity occurs?

Circular orbits are trivially periodic with a *single* constant frequency $\omega_\theta = d\theta/dt = M/\mu r_o^2$, where $V'(r_o) = 0$ and $V''(r_o) > 0$. We can look at perturbed, circular motion near a given circular orbit by writing $\delta r = r - r_o$ nd making the approximation $V(r) \approx V(r_o) + \delta r^2 V''(r_o)/2$, which yields a simple harmonic oscillator equation for the radial motion along with $d\theta/dt \approx (1 - 2\delta r/r_o)M/\mu r_o^2$. In other words, as soon as we look at anything but circular motion we have a two-frequency problem again.

Since perturbed circular orbits of potentials $U(r)$ that always yield closed orbits must also be periodic, we can begin our search for the desired class by asking an easier question: for which central potentials is the winding number W of all perturbed circular orbits a constant, independent of E, μ, M, and the coupling constants k_i in $U(r)$? Then, when is that constant a rational number?

In order to determine W for a perturbed circular orbit of a potential $U(r)$, we need not evaluate the integral (4.10) where $W = \Phi/2\pi$ is the winding number because, as the reader knows, one can read off the required frequencies that define the winding number directly from the linear oscillator equations of motion. Rather than do this directly, we transform coordinates, $x = M/r$ in the integral (4.10a), yielding $V(r) = x^2/2 + U(M/x)$ for the effective potential. The angular displacement is then given by

$$\Phi = \int_{x_{min}}^{x_{max}} \frac{dx}{[2(E - w(x))]^{1/2}} \tag{4.10b}$$

and is precisely the half-period of oscillation, $\Phi = \tau/2$, of a hypothetical particle in one dimension with the Hamiltonian

$$h = \dot{x}^2/2 + w(x) = E, \tag{4.16}$$

where $w(x) = x^2/2 + U(M/x)$ is a harmonic oscillator potential plus a generally nonharmonic contribution $U(M/x)$.

Circular orbits that correspond to $r_{min} = r_o = r_{max}$ are given by the stable equilibrium conditions $w'(x_o) = 0, w''(x_o) > 0$ and are located at $x_o^3 = MV'(r_o)$. In this case, $r_{min} = r_o = r_{max}$ and the annulus collapses onto a circle. If E is slightly greater than $V(r_o)$ then the orbits are confined to a very thin annulus and are nearly circular: for small oscillations in the potential $w(x)$, equivalent to a perturbed circular orbit (as in figures 4.2(a), (c), and (d) but with the two annuli very near to each other) in the original problem, we make the approximation $w(x) \approx w(x_o) + w''(x_o)(x - x_o)^2/2$, and the frequency of oscillations is given by $\omega = \sqrt{w''(x_o)}$, so that $\tau = 2\pi/\sqrt{w''(x_o)} = \Phi/2$.

Given that $\Phi = \pi/\omega = \pi/\sqrt{w''(x_o)}$ for perturbed circular orbits, we need only ask: when is this result independent of E, M, and the force constants? In other words, for which potentials $U(r)$ are the winding numbers W fixed independently of all parameters in the problems that are to be subject to external manipulation? Prior to the formation of the solar system, Mother Nature did not have to worry about choosing initial conditions carefully in order to produce clockwork orbits of planets, because (with $E < 0$) the gravitational potential energy $U(r) = -k/r$ takes care of that automatically for all possible initial conditions $E < 0$!

We must compute $w''(x_o)$ in order to find that condition. With

$$w''(x_o) = \frac{3Mx_o U'(r_o) + M^2 U''(r_o)}{x_o^4} \tag{4.17}$$

and using $M/x_o^3 = 1/V'(r_o)$ we obtain

$$\omega = \left[\frac{3U'(r_o) + r_o U''(r_o)}{U'(r_o)} \right]^{1/2} \tag{4.18}$$

so that

$$\Phi = \frac{\pi}{\omega} = \pi \left[\frac{U'(r_o)}{3U'(r_o) + r_o U''(r_o)} \right]^{1/2}. \tag{4.19}$$

In general, Φ depends upon M and the force constants, although not upon E, through the equilibrium position r_o. This result is independent of r_o, hence of all variables like $M, k, \ldots$, only if the expression $U'(r)/(3U'(r) + rU''(r))$ is equal to a constant independent of r for all values of r. We can easily integrate this differential equation to obtain $U(r) = kr^\alpha$ where k and α are constants. For bounded motion where $k < 0, \alpha > -2$ is required, we split the further analysis into two parts: either $U = kr^\alpha$ with $k > 0, \alpha > 0$, or else $U = -kr^{-\beta}$ with $k > 0, 0 < \beta < 2$. The case where $\alpha = 0$ is also included but corresponds to $U = k \ln r$ with $k > 0$.

Notice immediately that qualifying potentials can never be a sum of two or more terms: the potential (4.0c) will, for arbitrary initial conditions, produce a quasiperiodic orbit in spite of the fact that each contribution, taken alone, has no quasiperiodic orbits.

In all cases, for perturbed circular orbits it follows that the winding number is given by

$$W = [2(\alpha + 2)^{1/2}]^{-1} \qquad (4.20)$$

and depends only upon the exponent, not at all upon the force constant k. We get periodic perturbed orbits only if W is the ratio of two integers. This is the case only for the positive exponents $\alpha = 2, 7, 14, \ldots$, and for the negative exponents $\alpha = -1, -13/4, \ldots$. In particular, for the Kepler problem ($\alpha = -1$) we obtain $W = 1/2$, while for the isotropic simple harmonic oscillator ($\alpha = 2$) we get $W = 1/4$.

The next question is: of these rational winding numbers, which ones survive when the nonlinear terms are included, when we allow energies $E \gg w(x_0)$?

Here, we follow the 'hammer and tongs' method presented by Arnol'd (1978) (this entire section follows Arnol'd): we compute the winding number W for an energy different from the case where E is only slightly greater than $w(x_0)$, and equate the new results to the perturbation theory result $W = [2(\alpha + 2)^{1/2}]^{-1}$. The justification for this matching is that in the cases of interest, the cases where orbits are periodic for all energies E, the potential must be such that the winding number is rational and is independent of the energy E, which means that W must be the same number at very high and very low energies. This lack of dependence of W on E cannot hold for arbitrary central potentials, as we have already shown by studying the nearly circular orbits of arbitrary central potentials.

Consider first the case $U = kr^\alpha$ with $k > 0$ and $\alpha > 0$. Since W is to be independent of E, we can take and use the limit where E goes to infinity, in which case $E = w(x_{max}) \approx x_{max}^2/2$. If we set $y = x/x_{max}$, then

$$\frac{w(x)}{E} = y^2 + kM^\alpha y^{-\alpha}/x_{max}^{2+\alpha} \qquad (4.21)$$

and for large E (large x_{max}), $w(x)/E \approx y^2$. In this case, the limit where E goes to infinity, we obtain

$$W = \frac{\Phi}{2\pi} = \frac{2}{2^{3/2}} \int_0^1 \frac{dy}{(1 - y^2)^{1/2}}. \qquad (4.22)$$

This is a harmonic oscillator integral, so $W = 1/4$ follows, which means (by matching to the result (4.20)) that $\alpha = 2$ is necessary. For positive exponents $\alpha > 0$, *only* the isotropic simple harmonic oscillator has periodic orbits for all possible values of the orbit parameters. Note that $W = 1/4$ means that the orbit must close after exactly one 2π-revolution about the inner circle, with exactly 4 angular displacements of $\Phi = \pi/2$. The resulting form of the orbit therefore cannot be qualitatively different from the one shown as figure 4.2(c). Poincaré's method of qualitative analysis is very powerful: we are only lacking the detail that the closed orbit is precisely an ellipse, a fact that requires an exact quantitative analysis that is carried out in section 15.5 based upon the method of section 4.5.

We will get the same amount of information about the orbit for the Kepler problem by following the same method of qualitative analysis.

For the case $U(r) = -kr^{-\beta}$, with $k > 0$ and $0 < \beta < 2$, we can take the limit where E

goes to zero. In this case $w(x_{min}) = w(x_{max})$, so $x_{min} = 0$ and $x_{max} = (2J)^{1/(2-\beta)}$ where $J = k/M^{\beta}$. In this case,

$$\Phi = \int_{x_{min}}^{x_{max}} \frac{dx}{[2(Jx^{\beta} - x^2/2)]^{1/2}} \qquad (4.23a)$$

which can be rescaled (using $x_{max}^2/2 = Jx_{max}^{\beta}$) and transformed (using $y = x/x_{max}$) to yield

$$\Phi = \int_0^1 \frac{dy}{(y^{\beta} - y^2)^{1/2}}. \qquad (4.23b)$$

This is a known integral and has the value $W = 1/(2 - \beta)$. The only value of β for which this equals the result (4.20) for the perturbed circular orbits is $\beta = 1$, the Kepler problem. As $W = 1/2$, in two angular displacements through $\Phi = \pi$, the orbit makes exactly one 2π-loop about the inner circle and closes. The orbit therefore cannot have a form qualitatively different than that shown as figure 4.2(d). We show in section 4.5, using an algebraic method, that this closed orbit is an ellipse, a result that was discovered purely empirically by Kepler and was used by Newton to determine the $1/r^2$ dependence of the force in the law of universal gravitational attraction. What is now surprising to us is not that the inverse square law produces an ellipse, but that the inverse square law and the Hooke's law potential are the *only* central force laws that are incapable of producing quasiperiodic orbits!

Hence, of all possible central potentials and combinations of central potentials, only the Kepler problem and the isotropic simple harmonic oscillator yield strict clockwork for all possible values of the initial conditions independently of the values taken on by the energy, angular momentum, and the force constants. Had the form of the gravitational potential been any form other than $1/r$, then Mars' orbit about the sun could not be strictly elliptic and would be unlikely even to be periodic[2]. This gives us pause to think about the role played by what look like mathematical accidents in nature, in the discovery of universal laws of nature.

Although we can work out all of the consequences of the inverse square force law, we cannot explain why gravity obeys that particular law in the first place, at least not within the confines of classical mechanics. In the general case (see chapter 1), there is no requirement that a conservative force must be divergence-free in empty space. The reason for bringing up this point is that the two conservative forces that are responsible for all of the macroscopic phenomena of classical physics, Newtonian gravity and the Coulomb force, *are* divergence-free in empty space. Since the reasoning is the same for both cases, we discuss only gravity for the sake of definiteness. With $U = m\Phi$, the Newtonian gravitational potential obeys the Poisson equation

$$\nabla^2 \Phi = 4\pi\rho \qquad (4.24)$$

where ρ is the mass density of the concentration of matter that causes the potential Φ. This means that $\nabla\Phi = 0$ in empty space, outside the distribution of matter that causes the gravitational field, so that

$$\nabla \cdot \vec{F} = 0 \qquad (4.25)$$

---

[2] Newtonian gravity falls out of the Schwarzschild solution of general relativity as an approximation: the Schwarzschild solution to the field equations of general relativity yields a small $1/r^3$ correction to Newtonian gravity, and this term causes a precession of the otherwise elliptic orbit of a planet about the sun in the general relativistic equivalent of the two-body problem.

holds in empty space, in a vacuum. The divergence-free condition leads to the fact that Φ, and therefore U, are proportional to $1/r^\beta$ with $\beta = 1$ for a spherically symmetric mass distribution in three dimensions (one finds $\beta = 0$ with $\Phi \alpha \ln r$ in two dimensions, but $\beta = D - 2$ in a D dimensional Euclidean space where $D \geq 3$). Interestingly enough, considering the Bertrand–Königs theorem, we also find that U varies as r^2 inside a uniform mass distribution, $\rho = $ constant. Geometrically, the force-lines of a divergence-free field can only end on the sources of the field, which in this case are called 'gravitational mass'. If gravitational force lines could end in empty space, then potentials with an exponent β unequal to unity could also occur. As we have seen above, the importance of this fact for dynamics is that, of all possible positive exponents, $\beta = 1$ is the *only* one that yields stable *periodic* orbits for the central force problem for *all* values of the energy, angular momentum, and masses for which bounded motion occurs. If it were true, e.g. that $\beta = 1/2$, then the unperturbed motion of the earth about the sun would be likely to be quasiperiodic rather than periodic for arbitrary initial conditions.

We have left some dirt under the rug that needs to be swept up: what does it mean to get a winding number of 1/2 for an open orbit (the planar orbits are closed for $E < 0$, open for $E \geq 0$)? The interpretation must be that the orbit closes 'at infinity': this is consistent with the fact that for arbitrarily large but finite energies the orbit is closed. This interpretation is suggested by the fact that the turning point at $x_{min} = 0$ represents the point at infinity. Even here, a question arises that must be answered before the interpretation is entirely clear: we can see that for $0 < \beta < 1$ and for $\beta = 1$, there is no problem: x_{min} is not an extremum of $w(x)$. However, when $1 < \beta < 2$, then $w'(x_{min}) = 0$ and the turning point sits on a point of unstable equilibrium, although it is not the usual sort of equilibrium because $w''(x_{min}) = -\infty$. That x_{min} is a point of unstable equilibrium turns out not to present a conceptual problem, because it is easy to show that the time required to approach or leave this equilibrium point is *finite*. For motion near the turning point at $x_{min} = 0$,

$$E = 0 \approx \dot{x}^2/2 - Jx^\beta \qquad (4.26a)$$

so that

$$\dot{x} \approx \pm (2J)^{1/2} x^{\beta/2} \qquad (4.26b)$$

which yields (with $x = 0$ at $t = 0$)

$$x \propto t^{2/(2-\beta)}, \qquad (4.26c)$$

which is finite for $\beta < 2$ (we know from previous analysis of the linear problem that *infinite* time is required to reach or leave a hyperbolic point when $\beta = 2$).

This completes our discussion of the Bertrand–Königs theorem, which is just the statement that, of all possible central potentials, only the harmonic oscillator and the Kepler problem yield periodic orbits for all possible values of the parameters and initial conditions that define the system.

4.5 Hidden symmetry in the Kepler problem

My goal is to show that the heavenly machine is not a kind of divine living being but similar to a clockwork in so far as all the manifold motions are taken care of by one single absolutely simple magnetic bodily force, as in a clockwork all motion is taken care of by a simple weight. And indeed I also show how this physical representation can be presented by calculation and geometrically.

Johannes Kepler

The Bertrand–Königs theorem proves a unique kind of periodicity for the harmonic oscillator and Kepler problems, but it does not explain *why* the harmonic oscillator and the Kepler problem are singular. Why, of all possible problems with central potentials, do only these two problems yield clockwork no matter how you choose the energy, angular momentum, and the force constant? In other words, what is the *mechanism* that causes the orbits of those problems always to close on themselves, eliminating the possibility of the *quasiperiodicity* that otherwise should generally occur?

The reason, in both cases, is that there is an extra global conserved quantity, functionally independent of the Hamiltonian and the angular momentum.

Since a general four dimensional flow needs three conserved quantities to be integrable, there is no surprise in finding that there is a third conserved quantity. Normally, one does not ask questions about it because it is not needed to exhibit integrability, but the ignorance only reflects a choice motivated by convenience: in a canonical two degree of freedom system, one needs only two, not all three, compatible conservation laws in order to obtain the complete solution.

The integral of (4.9) gives us $\theta = h(r,E,M)$ which can be inverted to find the orbit equation $r = f(\theta,E,M)$ that describes the motion in the annulus. For quasiperiodic motions, the function f must be an infinitely many-valued function of θ, a function that returns arbitrarily often to the neighborhood of any point on the orbit but never exactly repeats itself: with $W = \Phi/2\pi$ irrational, a rational approximation $W \approx m/n$ means that r is almost repeated after m revolutions of θ through 2π, and there are infinitely many rational approximations to W for which this is true: the orbit passes arbitrarily closely to every possible value of r inside the annulus as the time goes to infinity.

If the orbit is closed, $W = m/n$ and r returns exactly to its initial value after m 2π-revolutions. The orbit equation must therefore be described by an m-valued function f of θ. In the case of both the Kepler and harmonic oscillator problems, $m = 1$ and the orbit equation is therefore single-valued in θ, is an analytic function of θ. The reason for the analyticity must lie in the nature of the third conserved quantity. We now show how this works for the Kepler problem (see section 15.5 for how it works in the oscillator problem). We start by guessing that the conserved quantity is a vector, although it could be a scalar or a tensor or higher rank.

We begin with Newton's second law for the linear momentum $\vec{p} = m\vec{v}$,

$$\dot{\vec{p}} = -U'(r)\hat{r} \tag{4.27}$$

for a central potential, and note that

$$\hat{r} \times \vec{M} = \mu(\vec{r}\hat{r} \cdot \vec{v} - \hat{r} \cdot \vec{r}\vec{v}). \tag{4.28}$$

Since $\vec{r} \cdot \vec{v} = r\dot{r}$, we obtain

$$\dot{\vec{p}} \times \vec{M} = -\mu U'(\vec{r}\dot{r} - r\dot{\vec{r}}) = \mu U'\left(\frac{\dot{\vec{r}}}{r} - \frac{\vec{r}\dot{r}}{r^2}\right) \tag{4.29a}$$

which has the form

$$\dot{\vec{p}} \times \vec{M} = \mu U' r^2 \frac{d\hat{r}}{dt}. \tag{4.29b}$$

Now $\dot{\vec{p}} \times \vec{M} = \dfrac{d}{dt}\vec{p} \times \vec{M}$ since angular momentum is conserved, so that for the Kepler problem, where $U = -k/r$ we obtain a new conserved quantity:

$$\frac{d}{dt}(\vec{p} \times \vec{M} - \mu k \hat{r}) = 0, \tag{4.29c}$$

which yields

$$\vec{A} = \vec{p} \times \vec{M} - \mu k \hat{r} \tag{4.30}$$

as a constant of the motion along with the energy and angular momentum. The vector $\vec{A}$ is called the Laplace–Runge–Lenz vector, or just the Runge–Lenz vector.

The orbit equation now follows from algebra: since

$$\vec{r} \cdot \vec{A} = M^2 - k\mu r, \tag{4.31}$$

then

$$r = \frac{\mu M^2/k}{\dfrac{A}{\mu k}\cos\theta + 1}, \tag{4.32a}$$

which is a *single-valued* function of θ, meaning that r is periodic rather than infinitely multivalued in θ, in accordance with the Bertrand–Königs theorem. And since $A^2 = 2\mu EM^2 + k^2\mu^2$, if we write $e = (1 + 2EM^2/k^2)^{1/2}$ along with $p' = A/\mu k$, then the orbit equation has the form

$$r = \frac{p'}{e\cos\theta + 1}. \tag{4.32b}$$

This shows that $0 < e < 1$ $(E < 0)$ yields an ellipse (Kepler's first law), $e = 1$ $(E = 0)$ a parabola, $e > 1$ $(E > 0)$ a hyperbola, while $e = 0$ yields a circle $(E = V(r_\circ)$ is a minimum).

If, in accordance with figure 1.1, we write

$$2a = p'/(1 + e) + p'/(1 - e) = 2p'/(1 - e^2) = k/|E| \tag{4.33}$$

and $e = (a^2 - b^2)^{1/2}/a$ so that $b = a(1 - e^2)^{1/2}$, then we can also derive Kepler's third law.

Approximating the section of area ΔA in figure 1.1 by the triangle shown, we obtain $\Delta A = r^2\delta\theta/2$, so that $dA/dt = M/2$ where $M = p_\theta$ is constant. Since the area of an ellipse is $A = \pi ab$, this yields

$$\tau = 2\pi ab\mu/M \tag{4.34a}$$

for the period. With $a = k/2|E|$ and $b = M/\mu(2|E|)^{1/2}$, we obtain

$$\tau = 2\pi k/(2|E|)^{3/2}. \tag{4.34b}$$

We can then eliminate E in favor of a to get Kepler's third law:

$$\tau = 2\pi a^{3/2}k^{-1/2}. \tag{4.34c}$$

By knowing that the energy and the angular momentum are conserved, one can solve the Kepler problem by two independent integrations. By knowing that the Runge–Lenz vector and the angular momentum are conserved, one can solve the Kepler problem algebraically without the need to consider the energy or to calculate an integral.

This concludes our discussion of central potentials, potentials that yield dynamics problems whose symmetry group is the group of rotations in three dimensional Euclidean space. Rotational symmetry is the obvious symmetry of these problems, but according to Noether's theorem a nontrivial third conservation law suggests that the Kepler and isotropic oscillator problems may have hidden symmetries corresponding to a symmetry group that is larger than, but contains, the rotation group. In order to discover the larger group of the Kepler problem that includes the rotation group as a subgroup and is based upon the conserved vector $\vec{A}$ (and also the larger symmetry group that yields an extra conserved *tensor* for the isotropic harmonic oscillator), we must first develop the elementary formalism of Lie algebras and Poisson brackets. We attack the theme of the hidden symmetries in chapter 15 using the mathematical tools developed in chapter 7.

Having used three rather than two nontrivial conserved quantities to solve the problem completely, the reader might now ask why we have not instead used M_x or M_y along with M_z and E to solve the problem, especially as all three components of the angular momentum are conserved. It is not possible to tackle the problem in this way, but to understand this it helps first to learn a little group theory (chapter 7). Then, in chapters 12 and 15 we explain why two or more Cartesian components of the angular momentum vector cannot be chosen simultaneously as canonical momenta. The reason is that this would require using two or more corresponding rotation–transformation angles as generalized coordinates, but we show in chapter 12 that the required rotation angles are a nonholonomic pair and therefore cannot serve simultaneously as generalized coordinates (this problem does not arise unless one tries to use more than one rotation angle as a coordinate). In other words, while all three of the Cartesian angular momentum components are compatible with the Hamiltonian, no two Cartesian components of angular momentum are compatible with each other, but compatibility is a geometric idea that is best approached from the standpoint of Lie algebras of multiparameter transformation groups.

4.6 Periodicity and nonperiodicity that are not like clockwork

Stable periodic and stable quasiperiodic orbits follow from canonical systems that are integrable, are qualitatively akin to a system of harmonic oscillators: all of the orbits in these systems are insensitive to small changes in initial conditions. Systems with f degrees of freedom are called integrable whenever the solution can be reduced to f independent integrations by finding f independent (global) conservation laws. We have examples of such systems both in mathematics and as idealizations of systems that occur in nature: the simple pendulum, the motion of the earth about the sun, and the motion of a mass attached to an elastic spring. But there is, in mathematics, a kind of unstable periodicity as well. Correspondingly, there are important systems that cannot be solved by symmetry and are therefore nonintegrable, systems where chaos is possible in portions of the phase space. The problem of three bodies interacting via Newtonian gravity provides one such example.

Consider another example, the decimal expansion of any rational number. That expansion is periodic no matter which base of arithmetic we use. Irrational numbers, on the other hand, have nonperiodic decimal expansions in every base of arithmetic. As an example, look at the decimal expansions (in base 10) for $1/7 = 10/70$ and $1/6.9 = 10/69$, which differ in magnitude by only $1/463$:

$$1/7 = 0.142859142859\cdots = 0.\overline{142859} \tag{4.35}$$

and has a basic block 142859 of length 6 that repeats infinitely often whereas

$$10/69 = 0.\overline{1449275362318840579710} \tag{4.36}$$

has a basic block of 22 digits that repeats infinitely often. In other words, by shifting the initial condition from $1/7$ only slightly to $10/69$, *the period changes from 6 to 22!* This is sensitivity with respect to small changes in initial conditions, *unstable* periodicity, and it is interesting to ask what kinds of mechanical systems have this kind of periodicity[3]. Whereas the harmonic oscillator's periodicity is like clockwork, this kind of periodicity is more like the tosses of a hypothetical 10-sided die (see the histograms in figure 4.3).

We can go further: there are also unstable nonperiodic orbits. Look at the decimal expansion for

$$\sqrt{3} = 1.73205\cdots \tag{4.37}$$

and also that for

$$\sqrt{2.99} = \sqrt{(299/100)} = 1.72916\cdots. \tag{4.38}$$

These expansions are nonperiodic. They start out with the same two first digits (the nearby initial conditions are $\sqrt{3}$ and $\sqrt{2.99}$) but diverge thereafter. By starting with $\sqrt{2.999999}$ instead of $\sqrt{2.99}$, we can mimic the decimal expansion of $\sqrt{3}$ much longer before seeing the inevitable divergence of the two results. We have here clear examples of unstable nonperiodicity.

By dividing the unit interval into 10 bins of equal width $1/10$, each decimal expansion generates a deterministic walk on the interval that is *apparently* random – only *apparently*, because there is no breakdown of cause and effect: the sequences are perfectly deterministic because they follow from an algorithm. This nonperiodicity is like the tosses of a 10-sided die. Whether or not that 'die' is fair is an unsolved problem in number theory, but the reader is invited to compute the respective histograms for the first several hundred or thousand digits in each expansion (you can do this on a computer by devising an algorithm that avoids the use of floating-point arithmetic). There is, in the absence of proof, no reason at all to believe that the histograms will be the same, even if one could take the not even computable limit of infinitely many digits[4].

---

[3] Algorithms for digit strings enter mechanics via the study of a dynamical system by the method of symbolic dynamics. Viewed from that perspective, algorithms that generate decimal expansions are the most fundamental forms of determinism, are the most basic abstract idea of mechanism. There is nothing that is more mechanical, more mechanically simple, than the generation of one digit after the other, for example, by taking the square root of an integer.

[4] One could in principle discover the limiting distributions analytically, if one were cleverer and luckier than every mathematician who's previously tried it, but the limiting distribution of digits would *not* predict correctly the histograms that one gets for *finitely* many decimals.

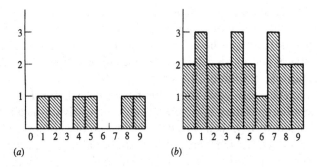

Fig. 4.3(a) Histogram for a deterministic walk on lattices of ten points,
$P/Q = 1/7$. (b) Histogram due to a slightly different initial condition
$P/Q = 10/69$.

The digit string analog of a numerically fair die is one where, in the limit of an infinitely long digit string, one digit occurs as often as any other digit $(0,1,2,\ldots,8,9$ each occur 1/10th of the time), and any block of any N digits occurs as often as any other block of N digits (for example, $00,01,02,\ldots,10,11,12,\ldots,98,99$ each occur 1/100th of the time, $000,001,\ldots,999$ each occur 1/1000th of the time, and so on). Such digit strings were called 'normal' by Lebesgue, their theory was developed by Borel (the father of modern probability theory), and we so far have only two explicit examples of normal numbers. Both were constructed by an English graduate student in mathematics, Champernowne.

It is extremely interesting to ask which conservative mechanical systems have this kind of unstable periodicity and unstable nonperiodicity, motions that are like the tosses of a 10-sided die. This is a perfectly fair question, because the tosses of dice are *not* random: there is no breakdown of cause and effect. The rolls of dice are subject to Newton's laws of motion just as are the motions of the planets about the sun. Chaotic motions arise in *nonintegrable* systems with bounded motion that have at least one positive Liapunov exponent (the Russian mathematician Liapunov lived from 1857–1918; his work on stability was continued by Poincaré). These are the dynamical systems that may mimic the instability of certain aspects of arithmetic[5], and to discover and study them one must abandon the method of solution by symmetry, the method where solutions are sought by finding f different constants of the motion. In chapters 6, 9 and 14 we show how systems of differential equations give rise to the study of iterated maps. In chapter 6, we also show by the numerical computation of a certain 'return map' how different initial conditions for the double pendulum give rise to motions that are stable periodic or quasiperiodic in some parts of the phase space, but are chaotic in other parts of the phase space. Next, we give an example of a completely chaotic map, one that generates motions that are in stark contrast with the stable periodic and stable quasiperiodic motions of the linear twist map. For this map, all of the trajectories are unstable: there are no regions in phase space where stable motions occur.

Consider the iterated map defined by $x_n = 10x_{n-1} \bmod 1$. 'Mod 1' means simply:

---

[5] The grading of arithmetic in elementary school is based upon sensitivity with respect to small changes in initial conditions: if a student makes a mistake in a single digit in the division of two numbers or in the extraction of a square root, the rest of the (computed) decimal expansion is completely wrong and the teacher typically awards the student with a zero for that problem. This was true until the granting of permission to use floating-point arithmetic (hand calculators and desktop computers) in the classroom.

whenever $x_n > 1$, then subtract off the integer part, so that $0 < x_n < 1$ always holds. The phase space is then the unit interval $[0,1]$, but with mod 1 this is the same as identifying the points 0 and 1 as the same point, so that a circle of unit circumference is really the phase space. The map is called the decimal (or base 10) Bernoulli shift. We can write down the map's exact solution: it is simply $x_n = 10^n x_0 \bmod 1$, where x_0 is the initial condition. Next, consider any initial condition that is rational, $x_0 = P/Q$ where P and Q are integers. The solution is then $x_n = 10^n P/Q \bmod 1$. If we divide the unit interval into Q different bins of length $1/Q$, then each iterate falls into one of those Q bins, for each iterate has the form P'/Q, and there are only Q possible values of P', which are 1 or 2 or ... or $Q - 1$. Therefore, the orbit is periodic with an integer period that is less than or equal to Q. For example, for $x_0 = 1/7$, the orbit is the 6-cycle $1/7, 3/7, 2/7, 6/7, 4/7, 5/7$, where $x_7 = x_1 = 1/7$. It follows that irrational initial conditions yield nonperiodic orbits. All of these orbits are unstable because the Liapunov exponent for each is $\lambda = \ln 10 > 0$: a small change δx_0 in an initial condition is magnified exponentially after a discrete time $n \approx |\delta x_0|/\ln 10$, yielding a qualitatively different orbit. For example, for $x_0 = (\sqrt{3}) - 1$, the orbit after only three iterations is different than the orbit of $x_0 = (\sqrt{2.99}) - 1$. The dynamics are very simple: one iteration means in the decimal expansion of the initial condition, shift the decimal point one digit to the right. In other words, if $x_0 = 0.\varepsilon_1\varepsilon_2\cdots\varepsilon_N\cdots$ where $\varepsilon_i = 0,1,2,\ldots,9$, then $x_1 = 0.\varepsilon_2\varepsilon_3\cdots\varepsilon_N\cdots$, and $x_n = 0.\varepsilon_n\varepsilon_{n+1}\cdots$. In other words, $\lambda = \ln 10$ means here that the nth digit of the initial condition becomes the first digit of the trajectory after n iterations! This instability shows up for periodic orbits in the following way: if we shift the initial condition for P/Q to P'/Q', where $Q' > Q$ but the difference $\delta x_0 = P/Q - P'/Q'$ is small compared with $x_0 = P/Q$, then the result is typically an orbit of much longer period. An example is given by the two initial conditions $x_0 = 1/7$ and $x'_0 = x_0 + \delta x_0 = 10/69$, whose decimal expansions (see equations (4.35) and (4.36) above) are periodic, and we get a 6-cycle and a 22-cycle respectively. The unit interval is densely filled with rational numbers, so that the unstable periodic orbits densely permeate the phase space. And, there is a continuum of unstable nonperiodic orbits because the irrational numbers yield the continuum.

Clearly, we are not used to chaotic behavior from our study of integrable systems. Nor has the freshman chemistry instructor's advice about 'least significant digits' prepared us for the computation of chaotic motions: truncating a decimal expansion for an initial condition means throwing away the chance to predict even the first digit of the system's trajectory at some later time. We see why the name 'Bernoulli' applies: J. Bernoulli was one of the great contributors to probability theory, and the Bernoulli shift with $\lambda = \ln 10$ mimics the tosses of a 10-sided die. Whether or not the die is fair is the question whether or not the initial condition is effectively normal to base 10. A number that is effectively normal to a given base yields a roughly even distribution of blocks of digits, for finitely many digits in the number string, but need not be normal when infinitely many digits are taken into account.

Exercises

1. Consider the Newtonian two-body problem with $M_z = 0$. Is gravitational collapse between two point masses possible in finite time? Analyze the integral f in $t + D = f(r,E)$ to estimate the time variation of r for small initial separations r_0 near zero. Is this a phase flow?

2. A particle of mass m is constrained to move under gravity without friction on the inside of a paraboloid of revolution whose axis is vertical. Derive the one dimensional problem that describes the motion. What condition must be imposed on the particle's initial velocity to produce circular motion? Find the periods τ_r and τ_θ for small oscillations about this circular motion. When is the perturbed circular orbit periodic?

3. A particle with mass m moves in a constant gravitational field $\vec{g} = (0, -g)$ along the curve $y = ax^4$. Find the equation that describes small oscillations about equilibrium and find also the period. Is the equilibrium an elliptic point? If not, plot the phase portrait near equilibrium for the linearized problem.

4. A particle moves in a central force field given by the potential

$$U = -k\frac{e^{-ar}}{r},$$

where k and a are positive constants. Using the method of the effective one dimensional potential, discuss the nature of the motion, stating the ranges of M and E appropriate to each type of motion. (a) When are circular orbits possible? (b) Find the periods of radial and angular oscillations for motions near circular motion. (c) Find at least one periodic orbit. (d) For $a = 0$, show that part (c) yields $W = 1/2$. Sketch the orbit.

5. (a) Show that if a particle describes a circular orbit under the influence of an attractive central force directed toward a point on the circle, then the force varies as the inverse fifth power of the distance.
   (b) Show that for the orbit described the total energy of the particle is zero.
   (c) Find the period of that motion.
   (d) Find $\dot{x}, \dot{y}$, and v as a function of angle around the circle and show that all three quantities are infinite as the particle goes through the center of force.

   (Goldstein, 1980)

6. For the central potential problem, sketch qualitatively the configuration space orbits (r,θ), for winding numbers

   (a) $W = \frac{1}{6}$,
   (b) $W = \frac{1}{3}$, and
   (c) $W = \frac{1}{5}$.

7. For a central potential, the configuration space orbit for $W = 1/\sqrt{2}$, does not close on itself. Sketch several of the orbits corresponding to rational winding numbers given by truncating the continued fraction

$$W^{-1} = \cfrac{1}{2 + \cfrac{1}{2 + \cfrac{1}{2 + \ldots}}}.$$

(Note: Writing $\sqrt{2} = 1 + x$ and squaring yields $x = \dfrac{1}{2+x}$. Iteration then yields

$$x = \cfrac{1}{2 + \cfrac{1}{2 + \dots}}.$$

8. A uniform distribution of dust in the solar system adds to the gravitational attraction of the sun on a planet an additional force

$$\vec{F} = -mC\vec{r}.$$

where m is the mass of the planet, C is a constant proportional to the gravitational constant and the density of the dust, and $\vec{r}$ is the radius vector from the sun to the planet, both considered as points. (The additional force is very small when compared with the direction sun–planet gravitational force.)

(a) Calculate the period of a circular orbit of radius r_o of the planet in this combined field.
(b) Calculate the period of radial oscillations for slight disturbances from this circular orbit.
(c) Show that nearly circular orbits can be approximated by a precessing ellipse and find the precession frequency. Is the precession in the same or opposite direction to the orbital angular velocity?
(d) Is the winding number for this problem rational or irrational?

(Goldstein, 1980)

9. Show that the motion of a particle in the potential field

$$V(r) = -\frac{k}{r} + \frac{h}{r^2},$$

where h and k are positive constants, is the same as that of the motion under the Kepler potential alone when expressed in terms of a coordinate system rotating or precessing around the center of force.

For negative total energy show that if the additional potential term is very small compared to the Kepler potential, then the angular speed of precession of the elliptical orbit is

$$\Omega = \frac{2\pi m h}{M^2 \tau}.$$

The perihelion of Mercury is observed to precess (after correction for known planetary perturbations) at the rate of about 40" of arc per century. Show that this precession could be accounted for classically if the dimensionless quantity

$$\eta = \frac{h}{ka}$$

(which is a measure of the perturbing inverse square potential relative to the gravitational potential) were as small as 7×10^{-8}. (The recent eccentricity of Mercury's orbit is 0.206, and its period is 0.24 year.)   (Goldstein, 1980)

10. The additional term in the potential behaving as r^{-2} in the previous problem looks very much like the centrifugal barrier term in the equivalent one dimensional potential. Why is it that the additional force term causes a precession of the orbit while an addition to the barrier through a change in angular momentum does not?

(Goldstein, 1980)

5

Small oscillations about equilibria

5.1 Introduction

Every linear system of ordinary differential equations can be solved by the method of series expansion (see Fuch's theorem in Ince, 1956, or Bender and Orszag, 1978). Moreover, every linear system with constant coefficients can be solved in closed form in terms of real and complex exponentials (see chapter 3), whether conservative or dissipative. If a conservative Hamiltonian system with f degrees of freedom is linear, then it is trivially integrable (only nonlinear systems can be nonintegrable) and therefore can be solved by constructing f independent constants of the motion. For a linear system, the Hamiltonian is quadratic and the f conserved quantities are obtained easily by a certain transformation of coordinates: in the transformed system, the Hamiltonian decouples into f independent quadratic Hamiltonians. If the motion is bounded, then one has the normal mode decomposition into f independent simple harmonic oscillators. Although a finite number of simple harmonic oscillators cannot equilibrate in energy and therefore is not a statistical mechanical system, this result provides the *purely formal* basis for the famous equipartition theorem in classical statistical mechanics[1]. The main point in what follows is to construct the transformation to the normal mode system, which we approach from the standpoint of Lagrangian mechanics.

5.2 The normal mode transformation

Beginning with Hamilton's equations

$$\dot{q}_i = \frac{\partial H}{\partial p_i}, \ \dot{p}_i = -\frac{\partial H}{\partial q_i}, \tag{5.1a}$$

and the Hamiltonian

$$H = \sum_{i=1}^{f} p_i \dot{q}_i - L \tag{5.1b}$$

as constructed from a Lagrangian

$$L = T - U, \tag{5.2}$$

we temporarily restrict our discussion to a kinetic energy that is quadratic in the generalized velocities:

---

[1] In that case, it is necessary to assume that the oscillators share energy with each other by the mechanism of a weak interaction that makes the system completely nonintegrable. The cat map in chapter 6 provides a discrete model of a completely nonintegrable system: phase space is permeated densely by a set of unstable periodic orbits, orbits with positive Liapunov exponents.

$$T = \tfrac{1}{2}\sum_{i,j} g_{ij}(q_1,\ldots,q_f)\dot{q}_i\dot{q}_j. \qquad (5.3)$$

We also restrict our considerations in this chapter to interaction energies $U(q_1,\ldots,q_f)$ that depend upon the generalized coordinates alone. The kinetic energy is given by a q-dependent symmetric matrix g, $g_{ij} = g_{ji}$, that is positive definite because $T \geq 0$. The canonical momenta follow from

$$p_i = \frac{\partial L}{\partial \dot{q}_i} = \tfrac{1}{2}\sum_j g_{ij}\dot{q}_j + \tfrac{1}{2}\sum_j g_{ji}\dot{q}_j \qquad (5.4a)$$

so that we obtain

$$\sum_{i=1}^{f} p_i\dot{q}_i = \sum_{i,j} g_{ij}\dot{q}_i\dot{q}_j = 2T \qquad (5.5)$$

and therefore

$$H = T + U = E. \qquad (5.6)$$

In addition, the matrix g is symmetric and positive definite and so is its inverse g^{-1}, where

$$\dot{q}_i = g_{ik}^{-1}p_k. \qquad (5.4b)$$

In other words, under these circumstances, the Hamiltonian is the total mechanical energy of the dynamical system.

Equilibria of the system follow from

$$\left.\begin{aligned} \dot{q}_i &= \frac{\partial T}{\partial p_i} = g_{ik}^{-1}p_k = 0 \\ \dot{p}_i &= \tilde{p}\,\frac{\partial g^{-1}}{\partial q_i}p - \frac{\partial U}{\partial q_i} = 0, \end{aligned}\right\} \qquad (5.7a)$$

and as we know from chapter 3, hyperbolic and elliptic points are the only possible equilibria. With a positive definite matrix $g^{-1}dq_i/dt = 0$ requires that $p_k = 0$ for all indices k, so that the equilibrium conditions can be stated as

$$\left.\begin{aligned} p_k &= 0 \\ \frac{\partial U}{\partial q_i} &= 0. \end{aligned}\right\} \qquad (5.7b)$$

Hyperbolic points in phase space correspond to either saddles or maxima of the potential U, whereas elliptic points correspond to potential minima.

We note that if one or more canonical momenta are conserved, then these must be replaced in the Hamiltonian by constants and those kinetic energy terms must then be included along with U to define an effective potential V. In this case, it is V rather than U that appears in equations (5.7a) and (5.7b). We can write $f = dN - n$ as the effective number of degrees of freedom if n is the number of conserved canonical momenta.

Small oscillations are possible motions near an elliptic point, that is, near a point of stable equilibrium $(q_{1o},\ldots,q_{fo})$ but we also permit unstable equilibria in what follows. To analyze the stability of an equilibrium point q_o we write $\delta q_i = q_i - q_{io}$ and expand the potential to second order in the small displacements

$$U(q_1,\ldots,q_f) \approx U(q_{1_o},\ldots,q_{f_o}) + \tfrac{1}{2} \sum_{i,j=1}^{f} \delta q_i \delta q_j \frac{\partial^2 U}{\partial q_i \partial q_j} + \ldots \qquad (5.8)$$

This approximation yields linear equations that describe the motion sufficiently near to the equilibrium point.

Replacing $H - U(q_{1_o},\ldots,q_{f_o})$ by H, we obtain the energy as

$$H \approx T(p_1,\ldots,p_f) + \tfrac{1}{2} \sum_{i,j=1}^{f} B_{ij} \delta q_i \delta q_j \qquad (5.9)$$

where $T(p)$ is obtained from (5.3) by replacing q by the constant q_o in g. In this approximation, both T and U are quadratic in the small displacements p and δq from equilibrium.

The matrix B is symmetric, $B_{ij} = B_{ji}$, and is also positive semi-definite whenever we expand about an elliptic point. Since $dq_i/dt = d\delta q_i/dt$, we can introduce the column vector

$$\eta = \begin{pmatrix} \delta q_1 \\ \vdots \\ \delta q_f \end{pmatrix} \qquad (5.10)$$

and write $L = T - U$, where

$$T = \tfrac{1}{2} \dot{\bar{\eta}} g \dot{\eta} \text{ and } U = \tfrac{1}{2} \bar{\eta} B \eta. \qquad (5.11)$$

The Lagrangian equations of motion are therefore

$$g\ddot{\eta} = -B\eta. \qquad (5.12)$$

To search for solutions of the form

$$\eta = \text{Re} \sum_{k=1}^{f} c_k e^{i\omega_k t}, \qquad (5.13)$$

we write

$$\eta \approx e^{i\omega t} \qquad (5.14)$$

to obtain the eigenvalue equation

$$\omega^2 g \eta = B\eta, \qquad (5.15)$$

which we can rewrite, with $\lambda = \omega^2$, as

$$B\eta = \lambda g \eta. \qquad (5.16)$$

Because B and g are real and symmetric, each can be diagonalized separately by different transformations, but here we need to know whether they can be *simultaneously* diagonalized (matrices that do not commute with each other generally are not diagonalized by the same transformation). With the unitary scalar product between any two complex vectors ζ and η defined by

$$\zeta^+ \eta = \sum_{i=1}^{f} \zeta_i^* \eta_i, \qquad (5.17)$$

we obtain

$$\eta_j^+ B\eta_i = \lambda_i \eta_j^+ g\eta_i \text{ and } \eta_j^+ B\eta_i = \lambda_j^* \eta_j^+ g\eta_i \tag{5.18}$$

where η_i is the eigenvector corresponding to the eigenvalue λ_i. It follows that

$$(\lambda_i - \lambda_j^*)\eta_j^+ g\eta_i = 0, \tag{5.19}$$

from which two conclusions also follow: (i) if we set $i = j$, then $\eta_i^+ g\eta_i \geq 0$ because g is positive semi-definite, so that $\lambda_i = \lambda_i^*$ (the eigenvalues are all real). (ii) If $\lambda_i \neq \lambda_j$, then η_j is perpendicular to the vector $g\eta_i$. If there is degeneracy, if $\lambda_i = \lambda_j$, then η_j and $g\eta_i$ are still linearly independent and so we can use them to construct two new orthogonal eigenvectors. Henceforth, we shall assume that any necessary orthogonalization procedures have been performed, and that η_j and $g\eta_i$ are perpendicular if $i \neq j$. We are free to choose the normalization, so we agree here to set $\eta_i^+ g\eta_i = 1$. Only if $g = I$ is the identity (which is always true in a Cartesian coordinate system) does this reduce to the usual choice of normalization whereby $\eta_i^+ \eta_i = 1$ (eigenvectors of unit length).

To find the f eigenvalues $\lambda_i = \omega_i^2$, we must solve for the f roots of the polynomial given by $\det[B - \omega^2 g] = 0$.

Next, we construct the normal mode transformation (the 'principal axis transformation'). After solving for the f eigenvectors, we use them to construct the matrix

$$N = \begin{pmatrix} \eta_{11} & \eta_{21} & \cdots & \eta_{f1} \\ \vdots & & & \\ \eta_{1f} & \eta_{2f} & \cdots & \eta_{ff} \end{pmatrix}, \tag{5.20}$$

where the entries of the kth eigenvector η_k form the kth column of the matrix. By construction, N satisfies the matrix equation

$$BN = gN\Lambda \tag{5.21}$$

where Λ is the diagonal matrix whose entries are the eigenvalues: $\Lambda_{ij} = \delta_{ij}\lambda_k$.

The reason for constructing N is to investigate the linear transformation $\eta = N\zeta$. First, note that $U = \eta^+ B\eta/2 = \zeta^+ N^+ BN\zeta/2$ where $N_{ij}^+ = \eta_{ji}^*$ (N^+ is the Hermitean adjoint of N). Since $N^+ BN = N^+ gN\Lambda$, and $\eta_l^+ B\eta_k = \lambda_l \eta_l^+ g\eta_k = \lambda_l \delta_{lk}$, we obtain the result that $N^+ BN = I\Lambda = \Lambda$ is a diagonal matrix. Therefore,

$$U = \zeta^+ B\zeta/2 = \sum_{k=1}^{f} \omega_k^2 \zeta_k^* \zeta_k. \tag{5.22}$$

In other words, the transformation $\zeta = N^{-1}\eta$ diagonalizes not only the kinetic energy matrix g, which is the metric tensor in configuration space (see chapter 10), but also, simultaneously, the potential energy matrix B. With

$$T = \dot{\zeta}^+ \dot{\zeta}/2, \tag{5.23}$$

we obtain the energy as a sum over f *independent* linear oscillators:

$$E = T + U = \sum_{k=1}^{f} (\dot{\zeta}_k^* \dot{\zeta}_k + \omega_k^2 \zeta_k^* \zeta_k)/2. \tag{5.24}$$

The transformation $\zeta = N^{-1}\eta$ is called the normal mode transformation and the f independent coordinates $\zeta_k(t) = c_k e^{i\omega_k t}$ (we need only the real parts) are called the normal modes, and the frequencies ω_k are called the natural frequencies of oscillation of the system.

Each normal mode defines a two dimensional subspace $(d\zeta_k/dt, \zeta_k)$ of the $2f$ dimensional phase space, and each two dimensional subspace so-defined has only the trajectories of an elliptic point. If we study the motion in the f dimensional configuration space $(\zeta_1, \ldots, \zeta_f)$, then we will also obtain results that are recognizable from the analyses of chapters 3 and 4. As an example, consider the projection of the configuration space trajectory onto any two dimensional subspace (ζ_b, ζ_k):

$$(\zeta_l(t), \zeta_k(t)) = [c_l \cos(\omega_l t + \theta_l), c_k \cos(\omega_k t + \theta_k)]. \tag{5.25}$$

If we ask whether the projected trajectory is periodic or not, the condition $(\zeta_l(t + \tau), \zeta_k(t + \tau)) = (\zeta_l(t), \zeta_k(t))$ yields, after an easy calculation, the familiar condition $\omega_1/\omega_2 = n_1/n_2$ where n_1 and n_2 are both integers. Otherwise, the projected trajectory is quasiperiodic and never closes. The reader should determine the relationship of the frequency ratio to the winding number W.

5.3 Coupled pendulums

Next, we study an example in detail in order to illustrate the application of the method described above: consider two simple pendulums of equal mass m and length l coupled by a massless spring whose elastic constant is k (figure 5.1). The kinetic energy of the two degree of freedom system is

$$T = \tfrac{1}{2}ml^2(\dot{\theta}_1^2 + \dot{\theta}_2^2), \tag{5.26}$$

and the potential energy, which provides the coupling, is given as

$$U = mgl(1 - \cos\theta_1 + 1 - \cos\theta_2) + \tfrac{1}{2}kl^2(\theta_1 - \theta_2)^2. \tag{5.27}$$

The small angle approximation then yields the Lagrangian $L = T - U$ as

$$L = \tfrac{1}{2}ml^2(\dot{\theta}_1^2 + \dot{\theta}_2^2) - \frac{mgl}{2}(\theta_1^2 + \theta_2^2) - \frac{kl^2}{2}(\theta_1 - \theta_2)^2. \tag{5.28}$$

Before proceeding to analyze this or any problem, the first step is to transform to dimensionless variables. that way, one can identify the strong couplings and also the weaker ones that may be treated as perturbations under the right circumstances. With $L' = L/mgl$, $t'^2 = t^2 g/l$, and $\alpha = kl/gm$, L', t', and α are dimensionless and we obtain

$$L' = \tfrac{1}{2}\left[\left(\frac{d\theta_1}{dt'}\right)^2 + \left(\frac{d\theta_2}{dt'}\right)^2\right] - U' \tag{5.29}$$

where

$$U' = \tfrac{1}{2}(\theta_1^2 + \theta_2^2) + \frac{\alpha}{2}(\theta_1 - \theta_2)^2. \tag{5.30}$$

With

$$\eta = \begin{pmatrix} \theta_1 \\ \theta_2 \end{pmatrix}, \tag{5.31}$$

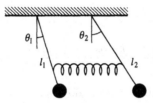

Fig. 5.1 Coupled pendulums (fig. 79 from Arnol'd, 1989).

we have $T' = \eta^+ g\eta/2$ but with the simplification that $g = I$. Likewise, $U' = \eta^+ B\eta/2$ with

$$B = \begin{pmatrix} 1 + \alpha & -\alpha \\ -\alpha & 1 + \alpha \end{pmatrix}.$$ (5.32)

The eigenvalue problem to be solved then yields the determinant

$$\begin{vmatrix} 1 + \alpha - \lambda & -\alpha \\ -\alpha & 1 + \alpha - \lambda \end{vmatrix} = 0,$$ (5.33)

from which we obtain the eigenvalues $\lambda_1 = 1 + 2\alpha$, $\lambda_2 = 1$. The natural frequencies are given by $\omega_1 = 1$ and $\omega_2 = (1 + 2\alpha)^{1/2}$.

By solving for the two eigenvectors given by $B\eta_k = \lambda_k \eta_k$, we can construct the principal axes:

$$\eta_1 = \frac{1}{\sqrt{2}} \begin{pmatrix} 1 \\ -1 \end{pmatrix} \text{ and } \eta_2 = \frac{1}{\sqrt{2}} \begin{pmatrix} 1 \\ 1 \end{pmatrix}.$$ (5.34)

The normal mode transformation is therefore given by

$$\zeta_1 = \frac{\theta_1 + \theta_2}{\sqrt{2}}, \zeta_2 = \frac{\theta_1 - \theta_2}{\sqrt{2}},$$ (5.35)

and the inverse transformation is given by

$$\theta_1 = \frac{\zeta_1 + \zeta_2}{\sqrt{2}}, \theta_2 = \frac{\zeta_1 - \zeta_2}{\sqrt{2}}.$$ (5.36)

Finally, we obtain the Lagrangian in the form of two decoupled simple harmonic oscillators

$$L' = \tfrac{1}{2}(\dot{\zeta}_1^2 + \dot{\zeta}_2^2) - \tfrac{1}{2}(\omega_1^2 \zeta_1^2 + \omega_2^2 \zeta_2^2),$$ (5.37)

whose solution we discussed qualitatively in chapter 3, albeit in the much more general context of linear oscillations of general conservative dynamical systems.

In order to exhibit the phenomenon of beats, which is interference of two signals in time at a fixed position in space, we write the solution in the form

$$\left.\begin{array}{l} \zeta_1 = a_1 \sin t + b_1 \cos t, \\ \zeta_2 = a_2 \sin \omega t + b_2 \cos \omega t \end{array}\right\}$$ (5.38)

where $\omega = (1 + 2\alpha)^{1/2}$, and a_i and b_i are constants.

The inverse transformation then yields

$$\left.\begin{aligned}\theta_1 &= \frac{a_1 \sin t + b_1 \cos t + a_2 \sin \omega t + b_2 \cos \omega t}{\sqrt{2}}, \\[2mm] \theta_2 &= \frac{a_1 \sin t + b_1 \cos t - a_2 \sin \omega t - b_2 \cos \omega t}{\sqrt{2}}.\end{aligned}\right\} \tag{5.39}$$

With the initial conditions $\theta_1(0) = \theta_2(0) = d\theta_2(0)/dt = 0$ and $d\theta_1(0)/dt = c$, we obtain

$$\left.\begin{aligned}\theta_1 &= \frac{c}{2}\left(\sin t + \frac{1}{\omega}\sin \omega t\right), \\[2mm] \theta_2 &= \frac{c}{2}\left(\sin t - \frac{1}{\omega}\sin \omega t\right).\end{aligned}\right\} \tag{5.40}$$

Beats can occur whenever there is a small frequency difference, which means that α must be small. In this limit, we have

$$\left.\begin{aligned}\frac{\omega_1 - \omega_2}{2} &\approx -\frac{\alpha}{2}, \\[2mm] \frac{\omega_1 + \omega_2}{2} &\approx 1,\end{aligned}\right\} \tag{5.41}$$

and obtain

$$\left.\begin{aligned}\theta_1 &\approx 2\sin t \cos \frac{\alpha t}{2} \\[2mm] \theta_2 &\approx -2\sin \frac{\alpha t}{2}\cos t.\end{aligned}\right\} \tag{5.42}$$

We can also write $\theta_1(t) \approx A_1(t)\sin t$, $\theta_2(t) \approx A_2(t)\cos t$, where the amplitudes A_i are oscillatory with a period $\tau \approx 4\pi/\alpha$ that is very large compared with the period 2π of $\sin t$ and $\cos t$, whose oscillations are said to be amplitude-modulated by the low-frequency factors A_i.

The generalization of the normal mode transformation to dissipative flows is described in chapter 5 of Arnol'd (1983).

Exercises

1. A mass particle moves in a constant vertical gravitational field along the curve defined by $y = ax^4$, where y is the vertical direction. Find the equation of motion for small oscillations about the position of equilibrium. Write down the integral that describes the period of both small and large oscillations.

2. A pendulum is attached to the ceiling of a boxcar which accelerates at a constant rate a. Find the equilibrium angle of the pendulum and also the frequency of small oscillations.

3. Obtain the normal modes of vibration for the double pendulum shown in figure 5.1, assuming equal lengths, but not equal masses. Show that when the lower mass is small compared to the upper one the two resonant frequencies are almost equal. If the pendulum is set into motion by pulling the upper mass slightly away from the vertical and then releasing it, show that subsequent motion is such that at regular

intervals one pendulum is at rest while the other has its maximum amplitude. This is the familiar phenomenon of 'beats'.

4. Consider a double pendulum with rods of equal length but with unequal masses m_1 and m_2 attached at the end points. The pendulum is attached at one point in space but both rods are free to make 2π-rotations.
   (a) Locate all equilibria.
   (b) Use the normal mode equations to compute the normal mode frequencies and thereby determine the stability of the four different equilibria.
   (c) Sketch, in every case, the pendulum configurations that correspond to *stable* oscillations (include *all* stable modes even when there is a second unstable mode).

5. Two particles, each of mass m, move in one dimension at the junction of three elastic springs confined to a straight line. The springs all have unstretched lengths equal to a, and the force constants are k, $3k$, and k.
   Find both the eigenfrequencies and eigenvectors for the normal modes of oscillation of the system. (Goldstein, 1980)

6. Consider two pendulums (rod lengths l_1 and l_2, masses m_1 and m_2) coupled by an elastic spring with potential energy $\frac{1}{2}\alpha(\phi_1 - \phi_2)^2$. Find the normal mode equations and solve them approximately for the cases of weak ($\alpha \to 0$) and strong ($\alpha \to \infty$) coupling. (Arnol'd, 1989)

6

Integrable and chaotic oscillations[1]

Hamiltonian dynamics is geometry in phase space.

V.I. Arnol'd

6.1 Qualitative theory of integrable canonical systems

An integrable flow[2] has a special coordinate system where the motion is a translation at constant speed. For any integrable four dimensional noncanonical flow

$$\frac{dx_i}{dt} = V_i(x_1,\ldots,x_4) \qquad (6.1)$$

with three time-independent conservation laws $G_i(x_1,\ldots,x_4) = C_i$, two of the conservation laws confine the motion to a two dimensional surface and the third describes the streamlines of the flow on that surface. Eliminating three of the four coordinates x_i by using the three conservation laws, the solution for the fourth coordinate has the form of a constant speed translation,

$$h(x_4,C_1,C_2,C_3) = \int \frac{dx_n}{v_4(x_4,C_1,C_2,C_3)} = t + D. \qquad (6.2)$$

By again using the three conservation laws to eliminate the three constants C_i in favor of the three variables (x_1,x_2,x_3), we can also write $h(x_4,C_1,C_2,C_3) = F(x_1,x_2,x_3,x_4) = t + D$, so that Lie's transformation to the explicit translation variables $(u_1,\ldots,u_n)$ has the form $u_i = G(x_1,\ldots,x_n) = C_i$ for $i = 1, 2,$ and 3, and $u_4 = F(x_1,\ldots,x_4) = t + D$.

When the four dimensional system is canonical with a time-independent Hamiltonian H, then only one global constant besides H is needed in order to exhibit integrability. This was illustrated by the central potential problems in chapter 4 where, with two degrees of freedom, we used the energy E and angular momentum M to express the complete solution in terms of two independent integrations.

For a canonical system, two independent translation coordinates are consistent with integrability based upon the use of only two independent constants of the motion: denote the special coordinate system by the variables (Q_1,Q_2,P_1,P_2) and the Hamiltonian in these variables by K. Then with

$$\dot{Q}_i = \frac{\partial K}{\partial P_i} = \text{constant} \qquad (6.3a)$$

---

[1] This chapter provides some numerical and some analytic realization of the ideas discussed qualitatively in sections 3.5 and 4.6.

[2] Until chapter 13 we will write 'integrable' when we mean 'completely integrable'. 'Nonintegrable' means not completely integrable, and is clarified in section 13.2.

for $i = 1,2$, it follows directly that K cannot depend upon Q_1 or Q_2 and can only depend upon P_1 and P_2. Therefore,

$$\dot{P}_i = -\frac{\partial K}{\partial Q_i} = 0 \tag{6.3b}$$

so that there are two conserved canonical momenta, P_1 and P_2. Clearly, if one can construct the transformation to such a coordinate system then one needs only two[3] conserved quantities in order to exhibit integrability and determine the complete solution.

We now give two simple examples where not only Q_i but also P_i and K can be identified explicitly without the need for the advanced transformation group theory of chapters 7, 15, and 16: the simple harmonic oscillator, and a particle in a box. In the first case $H = E = (p^2 + \omega^2 q^2)/2$, and E and ω are independent of each other. Therefore,

$$\omega = \frac{\partial K}{\partial P} \tag{6.4}$$

means that $K = P\omega = E$, or $P = E/\omega$. Here, the new generalized coordinate $Q = \omega t + Q_o$ is an angle in the (q,p)-plane and is not defined by the coordinate q alone: one needs the entire phase space in order to picture the translation coordinate Q geometrically. Presumably, this is the origin of the term 'phase space' because Q is here the phase of a trigonometric function.

In the second example, consider a particle of mass m in a box of length L, so that the particle's position is given by $0 \leq x \leq L$. By inspection, the motion is periodic with frequency $\omega = \pi v/L$ where $H = E = mv^2/2$. In this case, there is no elliptic point and the generalized coordinate $Q = \omega t + Q_o$ is the argument of the trigonometric functions in a Fourier series representation of the sawtooth curve that represents the motion,

$$x(t) = \sum_{n=-\infty}^{\infty} x_n e^{inQ}. \tag{6.5}$$

The time derivative of this Fourier series,

$$\frac{p(t)}{m} = \frac{\omega}{m} \sum_{n=-\infty}^{\infty} x_n in e^{inQ}, \tag{6.6}$$

must then represent the momentum. In central potential problems, the identification of the two coordinates Q_i is easy but the construction of the two canonical momenta is nontrivial (chapter 16).

In contrast with the noncanonical case, every integrable four dimensional canonical flow has the same universal *geometric* features independent of all other details of the Hamiltonian. This universality is made explicit in the special (Q_1,Q_2,P_1,P_2)-coordinate system where the Q_i are translations (there is no corresponding geometric universality for noncanonical flows). So long as one restricts to easy exactly solvable examples like the Kepler problem, the isotropic harmonic oscillator, or Lagrange's top (chapter 9), one has little or no motivation to understand the geometry of integrable flows in phase space. Most mechanical systems are nonintegrable and must

---

[3] By writing $G_4(x_1,\ldots,x_4,t) = F - t = D$, we can always construct a trivial fourth time-dependent conservation law. In a canonical system, both of the 'missing' constants of the motion may be trivial integration constants, although the third constant in the Kepler problem is nontrivial.

be approached numerically with little or no help from an exact quantative analysis. Lacking qualitative understanding, one has no idea what to expect from numerical integrations performed on a computer.

We first explain the connection of conservation laws to orbital stability. An integrable canonical flow with two degrees of freedom has a coordinate system in phase space where the coordinates Q_i are two constant velocity translations. Translation coordinates are possible only because there are two frequencies in the problem that describe *stable* periodic or *stable* quasiperiodic motion. One way to describe the orbital stability is as follows: both frequencies ω_i depend smoothly upon two conserved quantities, for example on E and M in the central potential problems, and therefore cannot change rapidly from one orbit to a nearby one. There is no way to get a doubling or halving of a period merely by changing a least significant digit in one of the conserved quantities. If we recall the idea of orbital instability discussed in the last section of chapter 4 illustrated by decimal expansions for numbers, then we see that that kind of wildly variable behavior[4] is ruled out whenever we can characterize two degree of freedom motion by two frequencies that depend smoothly upon two conserved quantities. Still another way to say this is that our examples above are all described by linear circle maps with vanishing Liapunov exponents. That there are no positive Liapunov exponents is intimately connected with the fact that there are two frequencies that vary smoothly with two constants of the motion: a linear circle map directly reflects the effect of two frequencies that vary smoothly (or remain constant) with small changes in two conserved quantities.

What is the qualitative nature of the subspace defined by the translation variables Q_1 and Q_2? Compatibility of the two conservation laws is also necessary in order that we can construct a coordinate system where there are two conserved canonical momenta P_k (we explain in chapter 15 why two orthogonal components of the angular momentum cannot serve simultaneously as canonical momenta). In this coordinate system, the Hamiltonian K can depend only upon the momenta P_k and not upon the generalized coordinates Q_k,

$$\dot{P}_k = -\frac{\partial K}{\partial Q_k} = 0 \tag{6.7a}$$

so that

$$\dot{Q}_k = \frac{\partial K}{\partial Q_k} = \omega_k = \text{constant.} \tag{6.7b}$$

The f constant generalized velocities ω_k can therefore depend only upon the f conserved quantities P_i. As in the noncanonical case, *an integrable canonical Hamiltonian flow is one where the interactions can be transformed away*: there is a special system of generalized coordinates and canonical momenta where the motion consists of f independent global translations $Q_k = \omega_k t + Q_{k_o}$ along f axes, with both the ω_k and P_k constant.

Irrespective of the details of the interactions, all integrable canonical flows are geometrically equivalent by a change of coordinates to f hypothetical free particles moving one dimensionally at constant velocity in the $(Q_1, \ldots, P_f)$ coordinate system. This coordinate system does not exist unless the Hamiltonian flow is integrable,

---

[4] Behavior whereby a change in a 'least significant digit' in an initial condition produces another discrete periodic orbit, but one with a wildly different period than the first orbit.

because the construction of the transformation to this coordinate system for an f degree of freedom canonical system relies explicitly upon having f independent, compatible (meaning 'commuting') constants of the motion (chapter 16). Deferring the explanation of compatibility, we ask what is the geometric nature of the f dimensional subspace in phase space where these f independent translations occur? In the same way as the original canonical variables $(q_1,\dots,p_f)$, the canonically conjugate variables $(Q_1,\dots,P_f)$ define $2f$ Cartesian axes in phase space, and the motion is restricted to the f-axes labeled by Q_i. It is the fact that the axes are Cartesian that determines the geometry of the surface that these constant velocity translations are confined to.

There are only two distinct geometric possibilities for any one of the f independent translations: the constant ω_k may represent a Cartesian velocity corresponding to unbounded motion in a scattering problem, in which case there is a linear velocity ω_k. If the motion is bounded, then there must be a period whereby $Q_i(t + \tau_i) = Q_i(t)$ modulo 2π where $\tau_i = 2\pi/\omega_i$. If the motion in the original $(q_1,\dots,p_f)$ coordinate system is bounded in all $2f$ directions, then so must be the motion in the $(Q_1,\dots,P_f)$ coordinate system, so that there are f angular frequencies ω_k. It follows that the geometry of the f-surface to which those f 'translations' are confined can only have the geometry of an f dimensional torus in phase space: the Q_k are the translation variables on the f dimensional torus, and the P_k-axes are all orthogonal to the torus. We prove this quite generally in chapter 16 but will now explain it directly for the cases $f = 2$ and $f = 1$.

We consider first the case where $f = 2$. The two orthogonal translations $Q_i = \omega_i t + Q_{io}$, define two Cartesian axes because these translations are defined for all times t from minus to plus infinity, with no exceptions. If the motion is unbounded in all directions in the original configuration space of the variables (q_1,q_2), then we need go no further: we have two translations at constant velocities v_1 and v_2 that are determined by the conserved canonical momenta P_1 and P_2. The motion is described as two independent free-particle motions in our (Q_1,P_1,Q_2,P_2)-system, but in the original (q_1,p_1,q_2,p_2)-system we will generally have a scattering problem (one example is provided by the Kepler problem with energy $E \geq 0$): all integrable scattering problems can be transformed to a coordinate system where there is no scattering at all.

Next, suppose that the motion is unbounded in one direction in the (q_1,q_2)-configuration space but is bounded in the other direction. This corresponds to the combination of a periodic motion in the variable Q_1 on a circle, and an unbounded motion in the Cartesian coordinate Q_2 (or the reverse). To understand the geometry of the two dimensional surface in phase space where this motion occurs, join the two opposite sides of a piece of graph paper with perpendicular axes Q_1 and Q_2 together so that the points $Q_1 = 0$ and $Q_1 = 2\pi$ of the Q_1-axis coincide and are exactly the same point. The resulting two dimensional surface is a cylinder. The motion on that cylinder is simply the combination of a translation at constant velocity along the Q_2-axis combined with a circular motion at constant frequency ω_1 along the Cartesian Q_1-axis. The possible motions are just spirals on the cylinder. Of course, we have a perfectly circular cylinder in terms of the coordinates (Q_1,P_1,Q_2,P_2), but when we transform back to the original (q_1,p_1,q_2,p_2)-system the cylinder is continuously deformed, may even become stretched, compressed, twisted, and smoothly bent in different sections, and so is not circular in the original four coordinates although the motion remains *topologically* cylindrical.

Next, suppose that the motion in the original (q_1,p_1,q_2,p_2)-phase space is bounded in all four directions. In this case, both Q_1 and Q_2 can only be Cartesian coordinates on a *completely closed* two dimensional surface. What is the geometry of that surface? We must think of the transformation of any two dimensional surface described by the variables (Q_1,P_1,Q_2,P_2) back to one described by the coordinates (q_1,p_1,q_2,p_2) qualitatively in the following way: in these coordinate transformations in phase space, the two dimensional surfaces are continuously and even differentiably deformed into each other without cutting or tearing (cutting and tearing would violate Liouville's theorem, but stretching, twisting, and smoothly bending are permitted). Therefore, we are free to imagine that our sketch of the perpendicular Q_1- and Q_2-axes is not made on paper, as above in the case of the cylinder, but instead is made on a sheet of rubber. Since the points $Q_1 = 0$ and 2π are identical, then we must join and glue two opposite sides of the rubber sheet just as we imagined doing above, where paper was sufficient. In addition, we must now join and glue the remaining two opposite sides of the rubber sheet together so that the two points $Q_2 = 0$ and 2π also coincide with each other. The result is a closed two dimensional surface like a donut, a two dimensional torus, and this is the *only* closed two dimensional surface that can be constructed from a piece of rubber without cutting or tearing (keep strongly in mind that our transformations in phase space do not tolerate cutting or tearing). In the very special (Q_1,P_1,Q_2,P_2)-system, the donut is perfectly circular. We can draw a 2-torus in a three dimensional phase space (figure 3.8b), but in the case where $f = 2$ we cannot draw a picture of the whole donut all at once because it is a two dimensional donut in a *four* dimensional space (note that the same is true of the cylinder described above). Pictorially, the best that one can do is to look for evidence of the donut in the form of projections of orbits onto configuration space (chapters 4 and 9), or the way that a cross-section of the donut shows up whenever the torus intersects the (q_1,p_1)-plane or the (q_2,p_2)-plane (see figure 6.1). Quite generally, whenever one finds two frequencies that depend only upon two constants of the motion in a two degree of freedom problem, then that is analytic evidence for integrability and for tori in phase space.

In the original (q_1,p_1,q_2,p_2)-coordinate system, the 2-torus is no longer perfectly circular but is deformed (different cross-sections of it can be stretched, bent, and compressed in different ways). What is important is that the donut feature, a closed surface with a hole, cannot be changed by a transformation of coordinates: holes are examples of 'topologic invariants'. Any surface that has a hole in it in one coordinate system will have a hole in it in any other coordinate system that is reached from the original system by a continuous transformation. A sphere cannot be continuously transformed into a donut and vice versa, but a perfect donut can be continuously transformed into a highly deformed one. Our transformations in phase space are not merely continuous: they are always differentiable as well.

There are two frequencies ω_1 and ω_2 for the two translations on the 2-torus. If ω_1/ω_2 is rational, then the motion is periodic on the torus (the orbit on the torus closes on itself) and is periodic in the original variables as well: the solution $(q_1(t),p_1(t),q_2(t),p_2(t))$ has the general form

$$q_i(t) = \tilde{\phi}_i(\omega_1 t + Q_{1_0},P_1,\omega_2 t + Q_{2_0},P_2)\}$$
$$p_i(t) = \tilde{\psi}_i(\omega_1 t + Q_{1_0},P_1,\omega_2 t + Q_{2_0},P_2)\} \qquad (6.8a)$$

written as the transformation from one coordinate system to the other and depends upon two independent frequencies ω_1 and ω_2, but is *also* a periodic function of the time with a period τ:

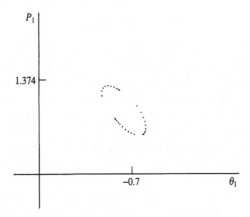

Fig. 6.1 Poincaré section of a stable orbit on a torus, initial conditions:
$\theta_1 = -1.7$, $\theta_2 = 1.4$, $p_1 = 1.374$, and $p_2 = 0$.

$$\left.\begin{array}{c} q_i(t+\tau) = q_i(t) \\ p_i(t+\tau) = p_i(t). \end{array}\right\} \tag{6.8b}$$

If, on the other hand, the frequency ratio ω_1/ω_2 is irrational, then the orbit on the torus never closes and is quasiperiodic. As we saw in chapter 4, central potential problems with bounded motion fall into the general category described here. Every integrable two degree of freedom flow, no matter how nonlinear Hamilton's equations may be in the original coordinate system, can be transformed (can be understood *qualitatively*) into a problem of two *independent* oscillators.

For $f = 1$, the geometry in phase space is easy: if the motion is bounded in the original (q,p)-system, then the single translation $Q = \omega t + Q_o$ with $P = $ constant can be a rotation about a circle with radius fixed by P (the circle is a 1-torus). This behavior reflects the presence of an elliptic point in the original (q,p)-coordinate system or, if there is no elliptic point, then there are confining forces, like those represented by perfectly reflecting walls. In the latter case, as in the former, $\tau = 2\pi/\omega$ is the time for one complete cycle of the motion.

Generalizing to f degrees of freedom, we assert without proof that the only possible *closed* f dimensional surface in phase space where f independent, Cartesian translations can occur has the topology of an f-torus. This is discussed in chapter 16 based upon a geometric result from chapter 11. On the f-torus, the motion is either periodic (all frequency ratios ω_l/ω_k are rational) or quasiperiodic (at least one frequency ratio is irrational). This result is universal in the qualitative sense that it holds for all integrable Hamiltonian flows independently of the details of the Hamiltonian, and shows directly the connection of bounded motion in integrable flows with a collection of f independent harmonic oscillators: there are f independent *frequencies* ω_k whose ratios determine the nature of the motion. This is at once very general and very limiting: it means that if you choose the right coordinate system, then the Kepler problem, certain rigid body problems, and a lot of other problems look qualitatively the same as each other. There is another way to say this: so long as we restrict our studies to integrable flows, then we are not looking at motions that are dynamically more *complex* than a collection of independent translations in phase space (free particles) or a collection of simple harmonic oscillators.

6.2 Return maps

The bounded integrable motion of a two degree of freedom autonomous Hamiltonian flow is confined to a two dimensional torus in the four dimensional phase space. Consequently, there are two frequencies ω_1 and ω_2 in the problem (these are generally not known beforehand), and the motion on the torus is periodic if the frequency ratio ω_1/ω_2 is rational, quasiperiodic if it is irrational. If the motion is nonintegrable then there may be parts of phase space where this description still holds, but the motion can be considerably more complicated in other parts of phase space. In this section, we introduce return maps as tools for analyzing and characterizing mechanical systems where nonintegrability occurs.

   In order to study two degree of freedom problems in a way that allows us to see the evidence for tori in a two dimensional subspace numerically, and also to provide a more convenient starting point for studying nonintegrable flows, we follow Poincaré and introduce the idea of a return map for our two degree of freedom system. Our particular discrete map is to be constructed by the following rule, which is not unique: every time that q_1 passes through 0 with $p_1 > 0$ we plot a point in the (q_2,p_2)-plane. Based upon the idea that a trajectory intersects the (q_2,p_2)-plane at discrete intervals, we shall obtain a discrete set of points by using this method. How those points organize themselves, whether onto a simple closed curve or in a more complicated (including pseudo-random) manner, is the main subject of this chapter.

   We denote the intersection times of the trajectory with the (q_2,p_2) plane by $t_1,t_2,\ldots,t_n,\ldots$, and we denote the corresponding intersections $(q_2(t_n), p_2(t_n))$ by (q_n,p_n). Unless there is periodicity, the time t_n is not a simple multiple of the time t_{n-1}. Especially for chaotic orbits, there is no simple, known analytic relationship between the different intersection times t_n. In fact, we generally cannot know these times without first solving the dynamics problem. In spite of that ignorance, the determinism of Hamilton's equations can be used to show that there should be a functional relationship of the form

$$\left.\begin{array}{l} q_n = G_1(q_{n-1},p_{n-1}) \\ p_n = G_2(q_{n-1},p_{n-1}), \end{array}\right\} \tag{6.9}$$

which is a two dimensional iterated map.

   We can argue as follows that there should be such a map. Denote by t_{n-1} a time at which $q_1 = 0$ and by t_n the succeeding time when q_1 vanishes again, with $p_1 > 0$ at both times (if q_1 does not vanish on the trajectory in question, then any other suitable numeric value c of q_1 may be chosen in place of zero). Next, write the formal solution at time t_n in terms of its previous value (initial condition) at time t_{n-1}:

$$\left.\begin{array}{l} q_\sigma(t_n) = \phi_\sigma(0,p_1(t_{n-1}),q_2(t_{n-1}),p_2(t_{n-1}),t_n - t_{n-1}) \\ p_\sigma(t_n) = \psi_\sigma(0,p_1(t_{n-1}),q_2(t_{n-1}),p_2(t_{n-1}),t_n - t_{n-1}), \end{array}\right\} \tag{6.10}$$

where $\sigma = 1,\ldots,f$ and we have used $q_1(t_{-1}) = 0$. Setting $q_1(t_n) = 0$ in the transformation (6.10) then yields a functional relation of the form[5]

$$t_n - t_{n-1} = f(p_1(t_{n-1}),q_2(t_{n-1}),p_2(t_{n-1})). \tag{6.11}$$

---

[5] One must respect the requirements of the implicit function theorem in order to be able to do this.

In order to go further, the conservation law $H(0,p_1(t_{n-1}),q_2(t_{n-1}),p_2(t_{n-1})) = E = $ constant implies a relation of the form

$$p_1(t_{n-1}) = g(q_2(t_{n-1}),p_2(t_{n-1}),E), \tag{6.12}$$

called an 'isolating integral', and where the condition $p_1(t_{n-1}) > 0$ must be employed at some stage to get a sign right in (6.12). Substitution of (6.11) and (6.12) into the transformations (6.10) for $q_2(t_n)$ and $p_2(t_n)$ then yields the two dimensional (energy-dependent) iterated map

$$\left. \begin{array}{l} q_n = G_1(q_{n-1},p_{n-1}) \\ p_n = G_1(q_{n-1},p_{n-1}), \end{array} \right\} \tag{6.13}$$

where $q_n = q_2(t_n)$ and $p_n = p_2(t_n)$. This formal argument suggests that a two degree of freedom Hamiltonian system can be studied in part via a two dimensional iterated map. The two dimensional map does not contain complete information about the dynamical system, but it contains enough information about the dynamics for us to be able to use it in what follows to characterize both integrability and nonintegrability in mechanical systems. The formal argument can also be extended to suggest that there should be higher dimensional maps for higher degree of freedom systems.

If there is an additional conserved quantity $G(q_1,q_2,p_1,p_2) = C_1$, then we can rewrite it in the form $G(0,q_n,g(q_n,p_n,E),p_n) = C_1$, and therefore as $G_1(q_n,p_n,E) = C_1$. If the conservation law is an isolating integral that holds for all finite times, then we can solve it for either $q_n = f(p_n,E,C_1)$, or $p_n = h(q_n,E,C_1)$. In this case, the points in the (q_n,p_n)-plane described by the map (6.13) must fall on a *single curve* for fixed values of E and C_1. By the reasoning of the last section, we expect this curve to be a *simple closed curve*. We have called conservation laws that permit the determination $q_n = f(p_n,E,C_1)$ or $p_n = h(q_n,E,C_1)$ of one variable in terms of the other 'isolating integrals' because they are first integrals of the motion that isolate one degree of freedom in a way that permits the elimination of that variable, hence a reduction in dimensionality of the dynamics problem (we assume that the Hamiltonian is an isolating integral). If the implicit function theorem is not satisfied so that the second conservation law cannot be used to isolate one degree of freedom for all finite times, then the trajectory of the map (6.13) for a fixed value of E will generally not be confined to a simple closed curve and may appear to wander over a finite region of the (q_n,p_n)-plane, at least for some initial conditions. One purpose of this chapter is to describe in part the nature of the wandering that may occur whenever a discrete trajectory of the map is not constrained to a one dimensional curve by a second isolating integral.

6.3 A chaotic nonintegrable system

The idea of randomness may be a historic but misleading concept in physics.

M. J. Feigenbaum

Every linear system of equations, regardless of the number of degrees of freedom, is trivially integrable and is trivially solvable in terms of real and complex exponentials (see chapters 3 and 5). But there are not only linear, but also nonlinear systems with a large number of degrees of freedom that can never approach statistical equilibrium (equipartition of energy is impossible in an integrable system with finitely many degrees of freedom). This was discovered numerically for a certain set of coupled nonlinear

Integrable and chaotic oscillations

oscillators by Fermi, Pasta, and Ulam. As Poincaré discovered geometrically around 1900, and Koopman and von Neumann discovered analytically in 1932, deterministic chaos does not require a large number of degrees of freedom: in a canonical Hamiltonian system $f = 2$ is adequate. We give numerical evidence of this below. The main point is that there are *low* dimensional deterministic dynamical systems that are nonintegrable and can generate statistical behavior while there are high degree of freedom systems that are integrable and therefore cannot: the consequence of integrability is qualitative equivalence to a translation at constant speed, which is the very opposite of chaotic dynamical behavior.

We now illustrate that nonintegrability and chaos are common in mechanics by studying a very simple conservative system that is very incompletely discussed in many standard texts: the double pendulum. The reason that this problem is often neglected is that other texts do not use the qualitative approach to the theory of differential equations along with numeric computation.

Consider first the Lagrangian of two massless rods of length l with two equal masses m attached at each end, and with the two rods attached to each other and to a pivot to form a double pendulum. Each rod makes an angle θ_i with the vertical, so that

$$L = \tfrac{1}{2}ml^2[2\dot{\theta}_1^2 + \dot{\theta}_2^2 + 2\dot{\theta}_1\dot{\theta}_2\cos(\theta_1 - \theta_2)] - mgl(3 - 2\cos\theta_1 - \cos\theta_2). \quad (6.14)$$

We have chosen as generalized coordinates the angles $q_1 = \theta_1$ and $q_2 = \theta_2$, and the corresponding canonical momenta are given by

$$\left.\begin{array}{l} p_1 = ml^2[2\dot{\theta}_1 + \dot{\theta}_2\cos(\theta_1 - \theta_2)] \\ p_2 = ml^2[\dot{\theta}_2 + \dot{\theta}_1\cos(\theta_1 - \theta_2)]. \end{array}\right\} \quad (6.15)$$

The Hamiltonian then follows as

$$H = \frac{1}{2ml^2}\frac{p_1^2 + 2p_2^2 - 2p_1p_2\cos(\theta_1 - \theta_2)}{1 + \sin^2(\theta_1 - \theta_2)} + mgl(3 - 2\cos\theta_1 - \cos\theta_2) \quad (6.16)$$

and is conserved. If the system has one additional global isolating integral that is compatible with H (describes a symmetry that commutes with the Hamiltonian, in the language of chapters 7, 15, and 16), then the motion is integrable. That is, the trajectories are confined to a 2-torus in the four dimensional phase space and are either stable periodic or stable quasiperiodic on that torus. On the other hand, if there is no second isolating integral then the trajectories are merely confined to the three dimensional energy surface $H(\theta_1,\theta_2,p_1,p_2) = E = $ constant.

Next, we ask: what is the nature of the double pendulum's motion in phase space? Because we combine qualitative analysis with a numerical integration to investigate the return map, we can in part answer this question. First, some qualitative analysis.

For any initial condition for which there is a 2-torus (tori occur at low energies in the small oscillations about stable equilibria, for example), imagine the intersection of the torus with the (q_2,p_2)-plane. In this case, we would expect the following behavior for our return map (6.13): a periodic orbit would show up as a discrete set of points on a simple closed curve in that plane whereas a quasiperiodic orbit would more and more fill in the curve and would yield a dense set of points on that curve as n goes to infinity. In both cases, the points lie on a *simple closed curve* that is the intersection of the donut with the (q_2,p_2)-plane. Only finite numbers of points can be obtained in computation, and only periodic sets of points can be generated by performing machine calculations while using floating-point arithmetic in fixed precision. Before getting into a more serious discussion of what can happen in computation, we first ask and answer the main qualitative question: why should we expect the points in a Poincaré section to organize themselves onto a *simple closed curve*?

In the special coordinate system where the f canonical momenta P_k are constants and the f generalized coordinates are independent translations, $Q_k = \omega_k t + Q_{ko}$, the radii of the torus are constant, and so the cross-section of the torus is circular. In this coordinate system, the return map (6.13) has a very simple form: since the points $Q_1 = 0$ and 2π coincide modulo 2π, we simply plot $Q_n = Q_2(t'_n)$ and $P_n = P_2(t'_n)$ whenever $t'_n = n\tau_1$ where $\tau_1 = 2\pi/\omega_1$ is the period of Q_1. The return map therefore has the form of a linear twist map

$$\left.\begin{aligned} Q_n &= Q_{n-1} + 2\pi W \\ P_n &= P_{n-1} \end{aligned}\right\} \tag{6.17a}$$

with winding number $W = \omega_2/\omega_1$, and can also be written in terms of its solution

$$\left.\begin{aligned} Q_n &= Q_o + 2\pi n W \\ P_n &= P_{n-1}. \end{aligned}\right\} \tag{6.17b}$$

For $f = 2$, this *circular* torus is imbedded in the *four* dimensional (Q_1,Q_2,P_1,P_2)-phase space, but cross-sections of it do not look perfectly circular when transformed back into the original (q_1,q_2,p_1,p_2)-coordinate system: the coordinate transformation

$$\left.\begin{aligned} q_i(t) &= \phi_i(\omega_1 t + Q_{1o}, P_1, \omega_2 t + Q_{2o}, P_2) \\ p_i(t) &= \psi_i(\omega_1 t + Q_{1o}, P_1, \omega_2 t + Q_{2o}, P_2), \end{aligned}\right\} \tag{6.18}$$

which is just the solution, is a continuous, differentiable deformation that neither cuts, tears, nor repastes sections of the torus (the transformation itself must respect Liouville's theorem, as we show implicitly in chapter 15). The intersection of the torus with the (q_2,p_2)-plane is therefore a continuous deformation of the circle and can in principle be any simple, differentiable closed curve.

An example of a Poincaré section is shown as figure 6.1. The figure agrees with the expected intersections of deformed circular 2-tori with the (q_2,p_2)-plane, as if there were an additional isolating integral that confines the motion to that torus. The Poincaré section in figure 6.1 is in qualitative agreement with what we would obtain for an integrable two degree of freedom problem. However, this is not the whole story as the rest of the available phase space has not been probed.

That the truth is considerably more complicated is shown by the Poincaré sections (which in reality are due to *pseudo-orbits* for a reason that is explained below) in figure 6.2(a). There, we see a scatter of points that reminds us superficially of something like coin-tossing or dart-throwing and suggests that if there is a second conserved quantity $G_1(q_n,p_n,E) = C_1$ then it does not act as an isolating integral for the initial conditions considered here.

Given the discussions of Liapunov exponents and unstable periodic and nonperiodic orbits in chapter 4, we should now look for evidence of exponential instability of orbits with nearby initial conditions. Evidence in that direction is provided by the two time series for $\theta_1(t)$ shown as figure 6.3 for two slightly different initial conditions (one could as well compare two analogous time series for $p_2(t)$, $\theta_2(t)$, or $p_1(t)$). The two time series track each other very well for any finite time $t \ll \tau$, but are very sensitive to their

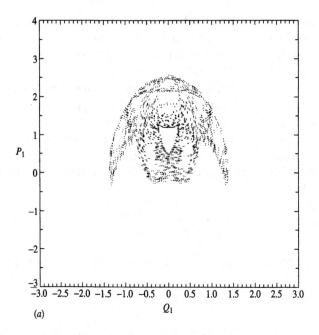

(a)

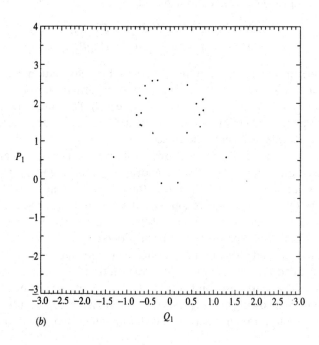

(b)

Fig. 6.2(a) Poincaré section of a chaotic orbit, initial conditions: $\theta_1 = 30°$, $\theta_2 = 50°$, $p_1 = 1$, and $p_2 = 1.5$. (b) As (a) but restricted to a forward integration time short enough that backward integration recovers the first digit of the initial conditions.

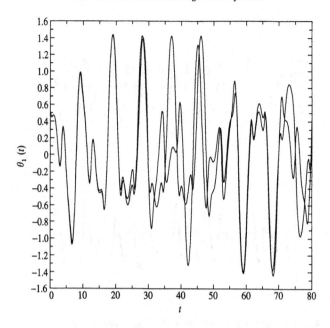

Fig. 6.3 Time series for a chaotic orbit, initial conditions $\theta_1 = 30°$, $\theta_2 = 50°$, $p_1 = 1$, $p_2 = 1.5$, and also for a nearby chaotic orbit where $\Delta\theta_1 = 0.5°$.

difference in starting points and diverge completely from each other for times $t \gg \tau$, where we define τ as the time when $|\delta\theta_1(\tau)| \approx 1$. That is, τ is defined as the time when the two time series $\theta_1(t)$ and $\theta_1(t) + \delta\theta_1(t)$ starting from different nearby initial conditions do not even agree to within first digit accuracy. This instability of long-time behavior as a result of small changes in initial conditions is to be compared with the two periodic and insensitive time series in figure 6.4(a)&(b) that occur for two nearby initial conditions that lie on a torus. In the case of initial data on a torus, a numerical integration that runs long enough to generate significant errors will show up as a 'smearing out' of the simple closed curve into a thin annulus, as is shown in figure 6.5. Figure 6.2(b) represents the part of figure 6.2(a) that is obtained if one restricts the forward integration time so that backward integration recovers the initial data to within at least one digit accuracy.

The behavior whereby all initially nearby time series diverge from each other so rapidly after a short time τ that even their *leading* digits are different, we can colloquially call chaotic or disorderly, because there is no way in a corresponding laboratory experiment or observation of nature to obtain enough information about initial conditions to be able to predict any such trajectory at very long times, even though the system is perfectly deterministic. That two nearby orbits diverge exponentially fast initially certainly means that a *single* orbit cannot be predicted with any accuracy at all whenever $t \gg \tau$ if we only know the orbit's initial condition to within some finite precision, which is always the case in observations of nature.

In computation one can write down initial conditions precisely, but floating-point computation on a machine (or truncations and roundoffs made by hand) cannot produce a correct orbit at times $t \gg \tau$ because the machine's roundoff/truncation mechanism (for binary arithmetic on digital computers, decimal on hand calculators) introduces errors into the calculation at every iteration of the numerical integration scheme. At times $t \gg \tau$

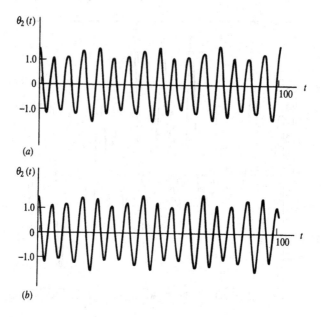

Fig. 6.4(a) Time series for a stable orbit, initial conditions: $\theta_1 = -0.65$, $\theta_2 = 1.4$, $p_1 = 1.419$, $p_2 = 0$, and $\Delta t = 0.2$. (b) Time series for a nearby stable orbit.

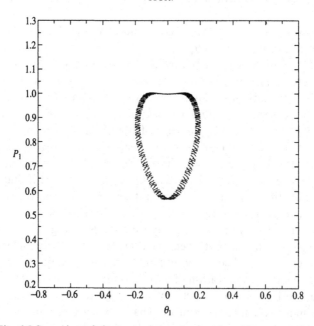

Fig. 6.5 Smearing of the torus due to accumulated innaccuracies of floating-point arithmetic, initial conditions: $\theta_1 = 0.072 = \theta_2$, $p_1 = 1.5$, and $p_2 = 1$.

the machine-produced orbit is dominated by roundoff/truncation error and is called a pseudo-orbit: at times $t \gg \tau$, even the first digit in the pseudo-orbit is wrong (see exercises 6 and 7). On the other hand, the roundoff/truncation algorithm of a computer is deterministic (integer-based algorithms are the simplest form of determinism) and cannot introduce 'randomness' into the dynamics. That the roundoff/truncation algorithm of a computer is in no way the source of the sensitivity of the trajectories of deterministic chaotic dynamical systems is shown by the *exact* solutions, calculated precisely in digital arithmetic, of simple models like the Bernoulli shift (chapter 4), the tent map (see below), and Arnol'd's cat map (see also below).

The analysis of section 6.4 leads us to understand why we should expect that the trajectory represented by the time series in figure 6.3 has a positive Liapunov exponent $\lambda \approx \tau^{-1}$. In a two degree of freedom integrable Hamiltonian system with bounded motion, there are two frequencies ω_1 and ω_2 that suffer only small changes whenever initial conditions are shifted slightly. In a two degree of freedom Hamiltonian system with chaotic oscillations, there must be (for each chaotic trajectory) two finite Liapunov exponents, one positive and one negative, as we show in our discussion of iterated maps in the next section. There, the periods of the unstable periodic orbits can change drastically whenever initial conditions are shifted only slightly, as we illustrated symbolically in section 4.6 and show explicitly for iterated maps in sections 6.4 and 6.5 (see also exercises 11 and 12).

Since the Jacobi determinant is unity, our conservative differential equations can be integrated backward in time. This provides the computer programer with a method for checking whether or not the numbers make sense: beginning with any point on a computed trajectory, you can use that point as the initial condition and then integrate numerically backward in time to try to retrieve the original initial condition. If the original initial condition at time t_o is recovered to within several digit accuracy, then that tells you the accuracy of your machine calculation forward in time. When there is a positive Liapunov exponent λ, then for times $t \gg \lambda^{-1}$ you will not even be able to recover the first digits of the initial conditions in a backward integration in a machine calculation that simulates mathematics by introducing errors, as does 'floating-point arithmetic' or fixed precision 'real arithmetic'.

Energy and other constants of the motion will be conserved to within good numeric accuracy in any calculation where one computes accurately enough to be able to retrieve the initial data to within several digits by integrating backward in time on the machine. *Merely* using energy conservation as a check on accuracy in a numerical calculation is inadequate and can be highly misleading: energy is often conserved to good accuracy in pseudo-orbit calculations that extend to times $t \gg \lambda^{-1}$ where the computed pseudo-orbit does not approximate the correct orbit at all. Even in a nonchaotic system, energy may be approximately conserved for times that are much longer than the time over which initial conditions can be recovered to one digit accuracy by a backward integration. The point is: the mathematical dynamical system retraces its path perfectly when one performs a computation that integrates the equations backward in time without introducing errors. The method for iterating a chaotic map forward in time without the introduction of errors is illustrated in the last section of this chapter by using a simple chaotic map.

The chaotic behavior indicated in figure 6.2(*b*) is nonintegrable: it is not the projection onto the phase plane of motion on a 2-torus. Hence, we expect that the double pendulum does not have a second global isolating integral that is compatible with the Hamiltonian, although there may be a local compatible isolating integral or even a globally incompatible one. The double pendulum is an example of something that is rather common and basic, namely, that low dimensional deterministic chaos is generated by relatively simple mechanical systems.

It is also necessary to understand that merely having one or even several nearby orbits that repel each other is not sufficient for deterministic chaos: an unstable limit cycle repels all nearby orbits but those orbits are generally not chaotic. For deterministic chaos, the orbits must all be uniformly bounded in some region of phase space and there must be no stable orbits available to attract the orbits that are repelled by the unstable ones. We can define (orbital) stability by Liapunov exponents. A stable orbit attracts nearby orbits with a negative Liapunov exponent, an unstable one repels nearby orbits with a positive Liapunov exponent, and a vanishing Liapunov exponent represents borderline or marginal stability. A sure recipe for chaos is uniformly bounded motion with all orbits inside the bounded region unstable. The simplest case of this, algebraically, is provided by the Bernoulli shift introduced in chapter 4. The double pendulum is considerably more complicated, as there are regions in phase space where the orbits are chaotic along with others where there is no chaos at all.

6.4 Area-preserving maps

In the beginning of chapter 3, we asserted that any area-preserving map in two dimensions is a discrete model of a conservative dynamical system. If we should try to model the double pendulum's Poincaré section analytically, then we would be led directly to the study of area-preserving maps of the form (6.13). That the map (6.13) *is* area-preserving follows as a special case of Poincaré's theorem on integral invariants: integrals of the form

$$I_2 = \sum_{i=1}^{f} \int \int dq_i dp_i \tag{6.19a}$$

are time-invariant and are, more generally, invariant under general canonical transformations. The proof, requiring calculus and also the recognition that certain quantities are invariant under canonical transformations, is formulated below.

First, let us see why the time-invariance of the integral (6.19a),

$$\sum_{i=1}^{f} \int \int dq_i(t_{n-1}) dp_i(t_{n-1}) = \sum_{i=1}^{f} \int \int dq_i(t_n) dp_i(t_n) \tag{6.19b}$$

for any two times t_{n-1} and t_n, implies that the map (6.13) is area-preserving. Since points in the surface of integration S_{n-1} at time t_{n-1} transform in one-to-one fashion by the transformation

$$\left. \begin{array}{l} q_i(t_n) = \phi_i(q_1(t_{n-1}), p_1(t_{n-1}), q_2(t_{n-1}), p_2(t_{n-1}), t_n - t_{n-1}) \\ p_i(t_n) = \psi_i(q_1(t_{n-1}), p_1(t_{n-1}), q_2(t_{n-1}), p_2(t_{n-1}), t_n - t_{n-1}), \end{array} \right\} \tag{6.20}$$

(the solution of Hamilton's equations) into points of the integration surface S_n at time t_n, we need only prove that the differential form $\Sigma dq_i dp_i$ is time-invariant, that

$$\sum_{i=1}^{f} dq_i(t_{n-1}) dp_i(t_{n-1}) = \sum_{i=1}^{f} dq_i(t_n) dp_i(t_n). \tag{6.19c}$$

If we then restrict our discussion to the surface of section defined by equation (6.13), where t_{n-1} and t_n are chosen so that all of the points of the return map lie in the (q_2, p_2)-subspace, then it follows from (6.19c) that $dq_{n-1} dp_{n-1} = dq_n dp_n$, which is the area-preserving condition for the map (6.13) for a two degree of freedom system.

We next show how to reduce the proof of (6.19b) to the proof of a certain general result in section 15.2. Denoting by (u,v) the two parameters needed to integrate over the two dimensional surface S_{n-1} and over the transformed surface S_n, (6.19c) follows if

$$\sum_{i=1}^{f} \frac{\partial(q_i(t_{n-1}),p_i(t_{n-1}))}{\partial(u,v)} = \sum_{i=1}^{f} \frac{\partial(q_i(t_n),p_i(t_n))}{\partial(u,v)}, \tag{6.19d}$$

where we use the notation of chapter 3 for the Jacobi determinant of a transformation. Expanding the right hand side of equation (6.19d) yields

$$J = \sum_{i=1}^{f} \frac{\partial(q_i(t_n),p_i(t_n))}{\partial(u,v)} = \sum_{i,j,k} \left[\left(\frac{\partial\phi_i}{\partial q_j}\frac{\partial q_j}{\partial u} + \frac{\partial\phi_i}{\partial p_j}\frac{\partial p_j}{\partial u} \right) \left(\frac{\partial\psi_i}{\partial q_k}\frac{\partial q_k}{\partial v} + \frac{\partial\psi_i}{\partial p_k}\frac{\partial p_k}{\partial v} \right) \right.$$
$$\left. - \left(\frac{\partial\phi_i}{\partial q_j}\frac{\partial q_j}{\partial v} + \frac{\partial\phi_i}{\partial p_j}\frac{\partial p_j}{\partial v} \right) \left(\frac{\partial\psi_i}{\partial q_k}\frac{\partial q_k}{\partial u} + \frac{\partial\psi_i}{\partial p_k}\frac{\partial p_k}{\partial u} \right) \right], \tag{6.21a}$$

where we have used the transformation functions in (6.20) to denote the q's and p's at time t_n, and also have denoted the q's and p's at time t_{n-1} simply as (q_1,p_1,q_2,p_2). Equation (6.21a) may be expanded and recombined in another way to yield

$$J = \sum_{i,j,k} \left[\frac{\partial q_j}{\partial u}\frac{\partial q_k}{\partial v} \left(\frac{\partial\phi_i}{\partial q_j}\frac{\partial\psi_i}{\partial q_k} - \frac{\partial\phi_i}{\partial q_k}\frac{\partial\psi_i}{\partial q_j} \right) + \frac{\partial p_j}{\partial u}\frac{\partial p_k}{\partial v} \left(\frac{\partial\phi_i}{\partial p_j}\frac{\partial\psi_i}{\partial p_k} - \frac{\partial\phi_i}{\partial p_k}\frac{\partial\psi_i}{\partial p_j} \right) \right.$$
$$\left. + \left(\frac{\partial q_j}{\partial u}\frac{\partial p_k}{\partial v} - \frac{\partial q_j}{\partial v}\frac{\partial p_k}{\partial u} \right) \left(\frac{\partial\phi_i}{\partial q_j}\frac{\partial\psi_i}{\partial p_k} - \frac{\partial\phi_i}{\partial p_k}\frac{\partial\psi_i}{\partial q_j} \right) \right]. \tag{6.21b}$$

Defining the Lagrange bracket of two functions (or two variables) F and G with respect to the canonical variables $(q_1(t_n),\ldots,p_f(t_n))$ by

$$[F,G] = \sum_{i=1}^{f} \frac{\partial(q_i(t_n),p_i(t_n))}{\partial(F,G)}, \tag{6.22}$$

we see that we can then write

$$J = \sum_{j,k=1}^{f} \left(\frac{\partial q_j \partial q_k}{\partial u \partial v}[q_j,q_k] + \frac{\partial p_j \partial p_k}{\partial u \partial v}[p_j,p_k] + \left(\frac{\partial q_j \partial p_k}{\partial u \partial v} - \frac{\partial q_j \partial p_k}{\partial v \partial u} \right)[q_j,p_k] \right). \tag{6.23}$$

Now, it is trivially true that

$$[q_j(t_n),q_k(t_n)] = 0, \ [p_j(t_n),p_k(t_n)] = 0, \ [q_j(t_n),p_k(t_n)] = \delta_{jk}, \tag{6.24a}$$

and these are called the fundamental Lagrange bracket relations. If the fundamental Lagrange brackets were time-invariant, it would also follow that

$$[q_j(t_{n-1}),q_k(t_{n-1})] = 0, \ [p_j(t_{n-1}),p_k(t_{n-1})] = 0, \ [q_j(t_{n-1}),p_k(t_{n-1})] = \delta_{jk}, \tag{6.24b}$$

which is a nontrivial condition on the transformation (6.20), as indeed it should be. The insertion of (6.24b) into (6.23) would then yield

$$J = \sum_{k=1}^{f} \left(\frac{\partial q_k(t_{n-1})}{\partial u}\frac{\partial p_k(t_{n-1})}{\partial v} - \frac{\partial q_k(t_{n-1})}{\partial v}\frac{\partial p_k(t_{n-1})}{\partial u} \right), \tag{6.25}$$

which yields in turn the area-preserving property (6.19d), hence (6.19c), that we want to prove.

We show in section 15.2 that the fundamental Lagrange bracket relations are not merely time-invariant, they are also invariant under the entire class of transformations in phase space that is called canonical. As we show in that chapter, the time-evolution transformation (6.20), the solution of Hamilton's equations, provides one example of a canonical transformation.

Therefore, in accordance with what is required for understanding Hamiltonian systems with bounded motion in phases space, let us restrict our further considerations in this section to area-preserving maps with bounded discrete trajectories.

Poincaré's recurrence theorem (see chapter 3) applies to every conservative system, to area-preserving maps as well as to continuous flows. Therefore, any block of initial conditions of an area-preserving map must intersect itself after some finite time, and thereafter infinitely often as the discrete time n goes to infinity: aside from any possible separatrix motions, all of the map's orbits are either periodic or quasiperiodic. However, any map that models the double pendulum must generate a fair degree of variety: in parts of phase space, the map must produce only stable periodic and quasiperiodic orbits (orbits with two frequencies, where both Liapunov exponents vanish) corresponding to the behavior shown as figure 6.1. But that is not all: in other parts of the phase space, corresponding for example to figure 6.2, the map must generate unstable orbits with one positive and one negative Liapunov exponent. As preparation for studying maps where the Liapunov exponents vary strongly over phase space, let us formulate the way to extract the two Liapunov exponents for an orbit of a two dimensional map.

For an arbitrary two dimensional discrete map of the plane,

$$\left.\begin{array}{l} x_{n+1} = f(x_n, y_n) \\ y_{n+1} = g(x_n, y_n), \end{array}\right\} \tag{6.26}$$

we begin by linearizing about any exact orbit $(x_o, y_o) \rightarrow (x_1, y_1) \rightarrow \ldots \rightarrow (x_n, y_n) \rightarrow \ldots$. The linear equation

$$\delta X_{n+1} = A_n \delta X_n \tag{6.27}$$

describes the time-evolution of the difference for small enough times, where the vector δX_n is given by

$$\delta X_n = \begin{pmatrix} \delta x_n \\ \delta y_n \end{pmatrix} \tag{6.28}$$

and the Jacobi matrix

$$A_n = \begin{pmatrix} \dfrac{\partial f}{\partial x_n} & \dfrac{\partial f}{\partial y_n} \\ \dfrac{\partial g}{\partial x_n} & \dfrac{\partial g}{\partial y_n} \end{pmatrix}, \tag{6.29}$$

which follows from linearizing the equation of motion for the perturbed initial condition along the original trajectory, is evaluated for each iteration at the point (x_n, y_n) *on* the exact trajectory, which itself starts from the initial condition (x_o, y_o).

Consider next the solution by iteration of the linearized equations,

$$\delta X_n = A_{n-1}A_{n-2}\ldots A_o \delta X_o = J_n \delta X_o \qquad (6.30)$$

in terms of the matrix $J_n = A_{n-1}A_{n-2}\ldots A_o$.

The eigenvalues $\mu_i(n)$ of J_n vary with n, as do the eigenvectors $e_i(n)$, and both are obtained by solving the eigenvalue problem given by

$$\left.\begin{array}{l} \det|J_n - \mu(n)I| = 0, \\ J_n e_i(n) = \mu_i(n)e_i(n). \end{array}\right\} \qquad (6.31)$$

Note that the directions of the eigenvectors are generally not constant along the trajectory original $(x_o,y_o) \to \ldots \to (x_n,y_n) \to \ldots$. If the eigenvectors $e_1(n)$ and $e_2(n)$ (which we assume to be normalized, so that $\| e_i(n) \| = 1$) are linearly independent, then we can expand the difference in the two initial conditions in the form

$$\delta X_o = \sum_{i=1}^{2} c_i e_i(n) \qquad (6.32)$$

and then obtain an equation for the difference δX_n between the nth iterates,

$$\delta X_n = J_n \delta X_o = \sum_i c_i \mu_i(n)e_i(n). \qquad (6.33)$$

The Liapunov exponents λ_1 and λ_2 are then defined by

$$\mu_i(n) \sim e^{n\lambda_i} \qquad (6.34)$$

for large n, so that

$$\lambda_i \sim \frac{1}{n}\ln \mu_i(n), \; n \gg 1. \qquad (6.35)$$

The exponents λ_i characterize the trajectory passing through (x_o,y_o) and (except for maps with constant derivatives) will vary when the initial conditions (x_o,y_o) are changed. Geometrically, a positive exponent $\lambda_i > 0$ corresponds to a local expansion of small areas along the direction of $e_i(n)$ while $\lambda_k < 0$ yields a local contraction of small areas along the direction $e_k(n)$. That is because areas $\Delta a = \Delta x \Delta y$ transform according to the rule

$$\Delta a_n = |J_n| \Delta a_o, \qquad (6.36a)$$

where $\Delta a_o = \Delta x_o \Delta y_o$ an initial element of area in phase space, and $|J_n| = \det J_n$ is the Jacobian determinant of the transformation from (x_o,y_o) to (x_n,y_n). It follows that $\det J_n \equiv \mu_1(n)\mu_2(n)$, so that for large enough n we can rewrite (6.36a) as

$$\Delta a_n \cong e^{n(\lambda_1 + \lambda_2)}\Delta a_o. \qquad (6.36b)$$

Hence, $\lambda_1 + \lambda_2 = 0$ corresponds to an area-preserving map whereas $\lambda_1 + \lambda_2 < 0$ describes a dissipative map. The condition for deterministic chaos is sensitivity of trajectories with respect to small changes in initial conditions and requires the action of at least one positive Liapunov exponent. The condition $|\mu_1| = |\mu_2|$ is therefore both necessary and sufficient for orbitally stable motion in an area-preserving map, requiring $\lambda_1 = \lambda_2 = 0$. For dissipative systems, nonchaotic motion requires $\lambda_i < 0$ for $i = 1,2$, which corresponds to a zero dimensional attractor in the two dimensional phase space. For a chaotic area-preserving map we need $\lambda_1 > 0$, and then $\lambda_2 = -\lambda_1 < 0$ follows.

In general, it is difficult to study area-preserving maps that model systems like the double pendulum where phase space is divided in a complicated way into regions of orbital stability and regions where orbits have nonvanishing Liapunov exponents, $\lambda_1 > 0$ and $\lambda_2 = -\lambda_1 < 0$, and so we defer that subject until chapter 17. It is much easier to model systems that are 'completely chaotic' in the sense that they exhibit pseudo-random behavior over all of their available phase space. In a completely chaotic system, there are no regions in phase space with stable orbits: all of the system's orbits are unstable. In one dimension, the Bernoulli shift is an example of a completely chaotic system. A two dimensional analog of the Bernoulli shift that is area-preserving is Arnol'd's cat map.

The cat map is defined by the transformation

$$\left. \begin{array}{l} x_{n+1} = 2x_n + y_n \bmod 1 \\ y_{n+1} = x_n + y_n \bmod 1 \end{array} \right\} \tag{6.37a}$$

and is area-preserving (the Jacobi determinant of the transformation from (x_n,y_n) to (x_{n+1},y_{n+1}) is unity). 'Mod 1' in equation (6.37a) simply means that one must always subtract off the integer parts, so that x_n and y_n always lie between 0 and 1, in which case the phase space is the unit square. The part of (6.37a) that is responsible for the chaos is simply the one dimensional map

$$x_{n+1} = 2x_n \bmod 1, \tag{6.37b}$$

the binary Bernoulli shift. Now, let us look at some details of the cat map.

By an application of the method of proof of section 4.6, where we proved that rational numbers used as initial conditions for the decimal Bernoulli shift yield periodic orbits (or the proof of the next section where we show that rational initial conditions yield periodic orbits for the binary tent map), if follows that rational initial conditions $(x_o,y_o) = (P/Q,P'/Q')$ for the cat map yield periodic orbits. It then follows that any irrational initial condition, e.g. $(x_o,y_o) = (\sqrt{2} - 1,\sqrt{3} - 2)$, yields a nonperiodic orbit, a discrete orbit that is confined to the unit square but that never closes on itself. All of these orbits, both periodic and quasiperiodic, are unstable because all of them have one positive (and therefore one negative) Liapunov exponent. It is easy to compute the two Liapunov exponents:

$$\lambda_{1,2} = \ln\left(\frac{3 \pm \sqrt{5}}{2}\right) = \pm \ln\left(\frac{3 + \sqrt{5}}{2}\right). \tag{6.38}$$

Since every trajectory has the same exponents, and since one exponent is positive, we have a model of an area-preserving Poincaré map where the phase space, the unit square, is 'filled up' by unstable periodic and unstable nonperiodic orbits. Because the system has no stable orbits, arbitrary shifts in the last digits of any initial condition will generally produce a completely different orbit with a completely different period or quasiperiod than the one we started with (see examples in the next two sections).

6.5 Computation of exact chaotic orbits

Arnol'd's cat map is the area-preserving transformation

$$\left.\begin{array}{l} x_{n+1} = 2x_n + y_n \bmod 1 \\ y_{n+1} = x_n + y_n \bmod 1 \end{array}\right\} \tag{6.37a}$$

of the unit square.

Because of the modulo 1 condition on both x and y, the motion is confined to the surface of a two dimensional torus: the points $x = 0$ and 1 are identical modulo an integer, as are the points $y = 0$ and 1. We shall use this map to illustrate two points: first, closed form algebraic solutions are not ruled out just because a system is chaotic (closed form algebraic solutions are not necessarily possible just because a system is integrable). More important is that exact chaotic orbits can be computed as precisely as one wishes, within the limits set by computer time. Information need not 'get lost' in a chaotic system as one computes (as it did in the floating-point computation of the double pendulum problem above) so long as one takes precautions and carries out the computation correctly. As we showed above, however, the length of time that one can see into the future when a trajectory has a positive Liapunov exponent is very severely limited by computer time.

Recall that the map has two nonzero Liapunov exponents, $\lambda_1 = \ln\left[(3 + \sqrt{5})/2\right]$ $= -\lambda_2$. This follows from writing

$$\begin{pmatrix} x_{n+1} \\ y_{n+1} \end{pmatrix} = \begin{pmatrix} 2 & 1 \\ 1 & 1 \end{pmatrix} \begin{pmatrix} x_n \\ y_n \end{pmatrix} = A \begin{pmatrix} x_n \\ y_n \end{pmatrix} \bmod 1 \tag{6.39}$$

where $(3 \pm \sqrt{5})/2$ are the eigenvalues of the matrix A. Hence, the map is uniformly chaotic, and the unit square is densely filled with unstable periodic orbits. Both periodic and nonperiodic orbits of the map can be computed exactly, as we shall show.

Note that

$$\begin{pmatrix} x_n \\ y_n \end{pmatrix} = A^n \begin{pmatrix} x_o \\ y_o \end{pmatrix} \bmod 1 \tag{6.40}$$

for any initial condition (x_o, y_o), where

$$A^n = \begin{pmatrix} F_{2n} & F_{2n-1} \\ F_{2n-1} & F_{2n-2} \end{pmatrix} \tag{6.41}$$

and F_n is a Fibonacci number: $F_{n+1} = F_n + F_{n-1}$ with $F_o = F_1 = 1$. Equation (6.41) yields the exact solution, in closed algebraic form, of the cat map.

Consider first the periodic orbits. All rational initial conditions can be written in the form

$$(x_o, y_o) = (P/N, Q/N) \tag{6.42}$$

where the unit square is partitioned into N^2 equal squares, each of side $1/N$, and P and Q are integers: $0 \le P,Q \le N$. Because there are N^2 cells, and A^n is an integer matrix (all entries are integers), the cell labeled by P and Q must be revisited after not more than N^2 iterations of the map, and by uniqueness, the motion must then repeat. Hence, all rational initial conditions yield periodic orbits, and all these orbits are unstable, are chaotic, because $\lambda_1 > 0$: if you shift an initial condition slightly from one rational pair of numbers to another pair with larger denominators, then the resulting orbit usually has a much longer period than that of the original orbit. As a typical example, the initial condition $(1/7, 1/7)$ lies on an 8-cycle (eight iterations of the map are necessary before the initial condition is repeated) whereas $(1/9, 1/9)$ lies on an 11-cycle, and a much longer cycle will be found if you have the patience to compute the orbit for the initial condition $(10/91, 1/9)$, for example. Such drastic changes in period cannot occur for small changes in initial conditions of a completely stable, hence nonchaotic, mechanical system.

All cycles start from rational initial conditions and are unstable, while irrational starting points yield unstable nonperiodic chaotic orbits: there are no regions on the unit torus where stable orbits occur. Like the Bernoulli shifts and tent maps, this dynamical system is 'uniformly hyperbolic' (has a dense set of unstable periodic orbits with a constant positive Liapunov exponent). Arnol'd (1983) has shown that the map's invariant density is uniform, meaning that a 'measure one' set of irrational initial conditions yields orbits whose distribution of iterates covers the torus uniformly as $n \to \infty$. However, other irrational initial conditions produce nonuniform (even multifractal) distributions of the map's iterates over the torus. An arbitrary choice of initial conditions in computation or experiment differs from the idealized and unattainable notion of a random draw of initial conditions from the mathematical continuum.

The main idea in what follows is that the digit strings for the initial conditions must be computed along with those for the iterates of the map. Because the map is piecewise linear and $\lambda_1 > 0$, the information in the Nth bit of (x_o, y_o) is transformed at a finite rate until in relatively short time it becomes information in the first bits of x_n and y_n. Therefore, to know (x_n, y_n) to N-bit accuracy, the initial condition (x_o, y_o) must be computed to within roughly $n + N \ln 2 / \ln \lambda_+$ bit accuracy, and this is easy to do for the cat map. We need only use an algorithm for an irrational number (or a pair of algorithms), such as the algorithm for $\sqrt{(2)} - 1$ to write x_o and y_o in (6.37a) as N-bit binary strings. The exact solution is then given by

$$\left.\begin{array}{l} x_n = F_{2n} x_o + F_{2n-1} y_o \bmod 1 \\ y_n = F_{2n-1} x_o + F_{2n-2} y_o \bmod 1 \end{array}\right\} \tag{6.43}$$

where the coefficients F_m are Fibonacci numbers.

All of the required arithmetic can be done as binary-point shifts on x_o and y_o combined with addition of finite-length binary strings. However, we must keep exactly as many bits in x_o and y_o as we need in order to compute both x_n and y_n accurately to N-bits, where N is our desired accuracy. That $F_m x_o$ consists only of binary-point shifts plus addition on $x_o \pmod 1$ is easy to see: with

$$F_m = \sum_{i=0}^{\mu} \varepsilon_i 2^i = \varepsilon_\mu \varepsilon_{\mu-1} \ldots \varepsilon_o . 0, \tag{6.44}$$

where $\varepsilon_i = 0$ or 1, we have $F_3 = 11 = 10 + 1$, $F_4 = 101 = 100 + 1$, $F_5 = 1000$, $F_6 = 1101 = 1000 + 100 + 1$, and so on. For example, given $x_o = \sqrt{(2)} - 1 \simeq 0.0110$, then $F_4 x_o = 0.10 + 0.01 = 0.11$ will be exact to 2-bits if there are no 'carries' in the neglected terms. If we want to know $F_3 x_o$ to 3-bits we have to use $\sqrt{(2)} - 1 = 0.01101$, so that $F_3 x_o = 0.101 + 0.011 = 0.000$. For example, given $x_o = \sqrt{(2)} - 1 = 0.0110101000001001 \cdots$ to 16 bits, we can compute $F_4 x_o = 10 x_o + x_o \bmod 1 = 0.00010010001011 \cdots$ to 14-bit accuracy. The reader should note that it is necessary to have a method for handling 'carries'. For example, if we use $x_o = 0.0110\cdots$ to 4-bits to compute $F_4 x_o = 0.10 + 0.01 = 0.11\cdots$ to 2-bits, this is wrong, but $x_o = 0.01101\cdots$ to 5-bits yields the correct 3-bit result $F_3 x_o = 0.000$. So, (x_n, y_n) can be computed from (6.43) to any prescribed precision N by using an algorithm that generates the digits of the initial condition (x_o, y_o) in binary arithmetic. The resulting computed orbit is a *uniform approximation* to an exact nonperiodic chaotic orbit so long as one uses an algorithm to generate the number of bits of an irrational initial condition that the action of the positive Liapunov exponent demands.

Clearly, to start with a point (x_n, y_n) and iterate the cat map backward to recover the first N bits of an initial condition (x_o, y_o), we must first have computed the nth iterate to about $n + N \ln 2 / \ln \lambda$ bits. Information cannot get lost if one computes a chaotic orbit correctly. On the contrary, information about the dynamics is gained: the information encoded as the high order bits in the initial condition defines a binary program that determines the long-time behavior of the dynamical system seen as a simple automation. Throwing away those bits by using floating-point arithmetic is akin to truncating a computer program and throwing it into a loop that prevents you from obtaining the information that you aimed to get by writing the program in the first place. Deterministic chaos means deterministic unpredictable motion in nature, but this does not imply motion that is unpredictable on a computer in a correctly formulated and correctly carried out calculation. The idea that deterministic chaotic motion is not computable is a misconception that has been widely and carelessly propagated throughout much of the literature. Deterministic chaotic orbits can be computed to within as many decimals as one wants *within the limits imposed by the time needed for the computation*. The inability to compute chaotic orbits at long times is not that the decimals cannot be generated and computed *in principle* (all of the digits are mathematically *determined*), rather, it is the amount of computation time that is needed to generate and compute them correctly that limits our ability to see where a chaotic orbit is going. A numerical computation is not the same thing as a real experiment, especially when a system is chaotic. In the former case initial conditions are specified precisely and the system is closed whereas in the latter case initial conditions are imprecisely known and the system is open. Simulations of nature are never the same as nature and the systematic errors introduced into computation by the use of floating-point arithmetic do not account for the difference.

6.6 On the nature of deterministic chaos

Because the properties that we identify as chaotic in the cat map can be understood by first understanding the Bernoulli shift, and because those properties are characteristic of an entire *class* of one dimensional maps, we now turn to noninvertible maps of the interval in order to clarify what we mean by deterministic chaos. The differential equations defining a flow have unique solutions and are time-reversible. The maps that they give rise to are uniquely invertible. Chaotic behavior also occurs in noninvertible one dimensional maps, but noninvertible maps cannot be derived from flows except perhaps by making drastic approximations to higher dimensional maps in dissipative systems.

In order to understand how the exponentially rapid separation of trajectories works numerically, we study a one dimensional map

$$x_{n+1} = f(x_n), \tag{6.45}$$

where the iterations x_n are uniformly bounded and therefore lie in some definite interval for all values of the discrete time n. By rescaling the iterates x_n we can always replace this interval by the unit interval $[0,1]$. The condition that the motion on the unit interval (which is our phase space) is bounded is satisfied by maps $f(x)$ that peak once or more in the interval $[0,1]$ but satisfy $0 \leq f(x) \leq 1$, that is, $f_{max} \leq 1$. It then follows directly that $0 \leq x_{n+1} = f(x_n) \leq 1$ if $0 \leq x_n \leq 1$. Therefore, we have self-confined motion with at least a two-valued inverse f^{-1}.

The Liapunov exponent was defined in chapter 4 by starting with two exact orbits $x_o \rightarrow x_1 \rightarrow x_2 \rightarrow \ldots \rightarrow x_n \rightarrow \ldots$ and $x_o + \delta x_o \rightarrow x_1 + \delta x_1 \rightarrow x_2 + \delta x_2 \rightarrow \ldots \rightarrow x_n + \delta x_n \rightarrow \ldots$ and then choosing $|\delta x_o| \ll 1$ in order to linearize the map along the first orbit. If we satisfy the conditions $|\delta x_o| \ll |\delta x_n| \ll 1$, then we can study the linearized equation

$$\delta x_n \approx f'(x_{n-1})\delta x_{n-1} \tag{6.46}$$

whose solution

$$|\delta x_n| \approx |\delta x_o| e^{n\lambda} \tag{6.47}$$

defines the Liapunov exponent

$$\lambda(x_o,n) \approx \sum_{i=0}^{n-1} \ln|f'(x_i)| = \frac{1}{n}\ln|f^{(n-1)\prime}(x_o)|, \text{ for } n \gg 1, \tag{6.48}$$

where $f^{(n)}(x_o) = x_n$ is the nth iterate of f starting with x. Except for certain very simple maps, the Liapunov exponent λ varies with the choice of initial condition x_o (see chapter 14 for simple examples).

It is typical that one does not obtain *pointwise* convergence to a definite number $\lambda(x_o)$ that is completely independent of n for large but finite n in computation whenever the map's slope f' varies along a trajectory (deterministic noise). The main point is that the map's average slope determines everything: the Liapunov exponent is just the time average of the logarithm of the slope along a trajectory, and whenever that average is positive (whenever the magnitude of f' is larger than one, on the average), then we get exponential magnification of small differences in initial conditions.

A positive Liapunov exponent describes the exponentially rapid separation of two initially nearby trajectories and is the mechanism for the instability of long-time behavior of two different orbits that begin very near to each other: if one plots the discrete time series x_n vs n for the two nearby initial conditions, then the iterates will track each other for times $n \ll -\ln|\delta x_o|/\lambda$ but will diverge from each other when $n \gg \lambda^{-1}$ (where equations (6.46 and 6.47) no longer hold).

If one knows the initial state of a dynamical system to within finite resolutions, which is at best the case in any observation of nature, then one can only predict the time-evolution of the system with accuracy over a time interval n that is given by $e^{n\lambda}\delta x_o \ll 1$, where the uncertainty δx_o represents the resolution of the experiment.

In computation, in contrast with observations and experiments on physical systems, $\delta x_o = 0$ because one starts with a single precisely defined initial condition given in the form of a finite length digit string $x_o = 0.\varepsilon_1(0)\varepsilon_2(0)\cdots\varepsilon_N(0)$ in some integer base μ of arithmetic. By using an algorithm (say, the rule for computing the digits one by one of $\sqrt{2}$) rather than floating-point arithmetic on a computer, one can then generate as many digits $\varepsilon_i(0)$ as one pleases. In any case, the digits $\varepsilon_i(0)$ take on any one of μ possible values $1,2,3,\ldots,\mu-1$, and we can choose freely to work in any base $\mu=2,3,4,\ldots$. This freedom of choice of base of arithmetic is somewhat analogous to the freedom to choose to study a partial differential equation in rectangular, spherical, elliptic, or any other set of curvilinear coordinates. Just as there may be a coordinate system where the solution of a particular partial differential equation is the simplest, there may be a base of arithmetic where the behavior of a particular map is most transparent.

In digital arithmetic, which von Neumann called 'an application of information theory', to what extent can we (or can't we) compute the first N digits of the nth iterate

$$x_n = 0.\varepsilon_1(n)\varepsilon_2(n)\cdots\varepsilon_N(n)\cdots \tag{6.49}$$

of a chaotic map without making a single error? This is an interesting question because the system is deterministic, yet it is supposed to be in some sense unpredictable. If there is a limit to our ability to see into the future, what determines that limit whenever the system has bounded trajectories with positive Liapunov exponents? It is quite easy to see that, for arbitrary maps that yield bounded motion, the only limitation is that of computation time.

In order to understand that limitation, we must also discretize the information propagation equation (6.47), where $\lambda(x_o,n)$ is the Liapunov exponent for the particular trajectory under consideration after n iterations. By δx_o we mean μ^{-N}, *the last computed digit* $\varepsilon_N(0)$ in our initial condition $x_o = 0.\varepsilon_1(0)\varepsilon_2(0)\cdots\varepsilon_N(0)\cdots$. Since the maximum error occurs when the first digit of x_n is on the order of unity, when $\delta x_n \approx 1/\mu$ in magnitude, then we obtain from (6.47) the condition

$$1 \approx e^{n\lambda}\mu^{-N+1}, \tag{6.50a}$$

which means that (for N large)

$$n \approx N\ln\mu/\lambda \tag{6.50b}$$

is roughly the number of iterations that one can perform without committing an error in the first digit of x_n, *given only the first N digits of* x_o. This is an information flow property, one that follows whenever the Liapunov exponent is positive. When the exponent is negative, then one gains accuracy in computation as n is increased. In the latter sort of system, $\varepsilon_N(0)$ is the least significant digit. In the chaotic system, one can with justification say that $\varepsilon_1(0)$ is the least significant digit!

The information propagation condition (6.50b) tells us that the most fundamental limitation on the ability to compute and therefore to predict the future from the past in any deterministic system is the limitation of computation time. This limitation is nontrivial because it means that it is necessary to increase the precision of the computation (as we demonstrated with the cat map), by systematically computing more and more digits in the initial condition x_o as one iterates the map, and also by holding more and more digits of every other computed iterate as well (because the 'present condition' x_n is the initial condition for future states x_{n+m}, and so on). If truncation and roundoff are used (e.g. floating-point arithmetic), then uncontrollable errors will be introduced into the calculation. As we stated in section 4.6, chaotic dynamical systems are precisely those systems that have orbits that reflect the complexity of the digit strings of both rational and irrational numbers. Contrary to popular expectations, this conclusion is not at all changed by considering nonlinear maps where the slope is not piecewise constant (see chapter 9 in McCauley, 1993).

The time series in figure 6.2 for the map of the double pendulum at long times do not represent true orbits but rather pseudo-orbits. Pseudo-orbits arise any time that roundoff/truncation errors are introduced into a computation, because those errors are then magnified exponentially by the map's positive Liapunov exponents. A machine program that uses floating-point or fixed precision arithmetic only produces an inexact 'simulation' of the desired mathematics and will easily produce a pseudo-orbit whenever the system is chaotic. Floating-point arithmetic is only a weak side-stepping of the difficult problem of how to handle digit strings of numbers. This erroneous method may well be abandoned in favor of accuracy as computer technology evolves further.

Another way to look at this is that every pseudo-orbit is an *exact* orbit of some other map that the machine's roundoff/truncation algorithm has invented to try to simulate your computation. Is chaos an artifact of bad machine arithmetic? Does the 'chaos' go away if the computation is performed correctly? That the answer to this question is no can be understood most easily by computing the solutions of simple maps like the Bernoulli shift, the tent map, and the cat map correctly by hand, or by restricting to integer operations when programming a computer. These piecewise linear maps can be solved exactly and easily by using decimal expansions in a definite base of arithmetic.

To see how this works, consider the binary tent map

$$x_{n+1} = 2x_n \text{ if } x_n < 1/2, \; x_{n+1} = 2(1 - x_n) \text{ if } x_n > 1/2 \qquad (6.51)$$

where the Liapunov exponent is $\lambda = \ln 2$ for every trajectory. If we choose a rational initial condition, $x_o = P/Q$ where P and Q are integers, then either $x_1 = 2P/Q$ or $2(Q - P)/Q$, each of which is one of the Q fractions $0, 1/Q, 2/Q, \ldots, (Q - 1)/Q$, as is every higher iterate. Therefore, after at most Q iterations x_n repeats one of the earlier iterates and the orbit is periodic. Conversely, irrational initial conditions yield nonperiodic orbits. All orbits of the binary tent map are unstable because $\lambda = \ln 2 > 0$, and we want to understand better how this instability works.

Consider as an example the orbit starting from the point $x_o = 1/5$: it is easy to show that this is a 3-cycle with points $1/5, 2/5$, and $4/5$. Next, try to track this orbit by starting from a nearby initial condition, say $x_o + \delta x_o = 10/51$. The result is another periodic orbit, but one with a cycle length much greater than 3! Matters get even worse if you take the initial condition of be $100/501$, which is even nearer to $1/5$. This is in stark contrast to stable periodic orbits of integrable systems where a small change in initial conditions produces (through a small shift in the constant canonical momenta P_k) only a small change in frequencies and periods. It means that you can never have enough information about initial conditions in experiment or observation to know where a chaotic system is going in the long run. This is very bad news for people who want to believe that the future is predictable. One can understand both the short- and long-time behavior of systems with chaotic behavior, and one can even compute chaotic trajectories precisely, but one cannot in any practical way predict *which* trajectory will occur whenever we encounter one of these systems in nature.

In observations and experiments, statistics also come into play because one cannot control initial conditions with absolute precision: in a *gedanken* experiment, the repetition of N identically prepared experiments means that there are N different trajectories that start near to each other and emanate from a tiny 'ball' of initial conditions in phase space (from a small interval, in the case of one dimensional maps). If we divide the phase space into N different bins of finite width (not all of which need have the same size and shape), then we normally would plot a histogram to reflect how often a given bin in phase space is visited by the different time series that originate from each initial condition in the interval. Each time series represents the repetition of an identically prepared experiment, but the experiments are identical only to within the finite precision variation in the initial conditions (the size of the ball of initial conditions is determined by the precision within which the 'identical repeated experiments' are prepared). This is stated in the spirit of the collection of experimental or observational data. However, there is a much deeper reason why there are statistics in the first place: each *exact* orbit of a chaotic dynamical system generates its own histogram! This was discussed briefly in section 4.6, but we can add some details here. The different classes of statistics that can be generated for different classes of initial conditions is one way to characterize these systems.

For any map (6.26) we can write the initial condition and all subsequent iterates in some definite base μ of arithmetic $(\mu = 2,3,4,\ldots),x_n = 0.\varepsilon_1(n)\varepsilon_2(n)\cdots\varepsilon_N(n)\cdots$, where each digit $\varepsilon_i(n)$ takes on one of the μ values $0,1,2,\ldots,\mu-1$. In this case, when a *single* orbit $\{x_n\}$ has a positive Liapunov exponent, then

$$n \approx N \ln \mu/\lambda \tag{6.50b}$$

is the number of iterations that one can perform before the information stored as the Nth digit (more generally, stored as several of the first N digits) of the initial condition x_o will determine the first digit of the nth iterate x_n. Since the decimal Bernoulli shift above (where $\lambda = \ln 10$) consists only of a digit shift per iteration when the iterates are written in base 10, $\varepsilon_i(n) = \varepsilon_{i+1}(n+1)$, it is precisely the nth digit $\varepsilon_n(0)$ in the initial condition x_o that, after n iterations, becomes the first digit $\varepsilon_1(n)$ of the nth iterate x_n,

$$x_n = 0.\varepsilon_n(0)\varepsilon_{n+1}(0)\cdots. \tag{6.52}$$

The statistics of the orbit follow easily, but we describe the general method of histogram construction below in the context of the binary tent map.

The binary tent map (6.51) is completely transparent in binary arithmetic (base $\mu = 2$) where $\varepsilon_i(n) = 0$ or 1. In this case, the map can be replaced by a simple auto-mation, $\varepsilon_i(n) = \varepsilon_{i+1}(n-1)$ if $\varepsilon_1(n-1) = 0$, but $\varepsilon_i(n) = 1 - \varepsilon_{i+1}(n-1)$ if $\varepsilon_1(n-1) = 1$, which is only slightly more complicated than reading the digits of an initial condition one by one, as we have done for the decimal Bernoulli shift above. The tent map's automation follows easily by noting that, in binary arithmetic,

$$1.00000\cdots0\cdots = 0.111111\cdots1\cdots, \tag{6.53a}$$

which is just the statement that

$$1 = 1/2 + 1/4 + 1/8 + \cdots, \tag{6.53b}$$

and therefore that

$$1 - x_n = 0.(1 - \varepsilon_1(n))(1 - \varepsilon_2(n))\cdots(1 - \varepsilon_N(n))\cdots. \tag{6.53c}$$

For the binary tent map, therefore, the first bit $\varepsilon_1(n)$ of the nth iterate is determined by the *pair* of bits $\varepsilon_{n-1}(0)\varepsilon_n(0)$ in the initial condition: $\varepsilon_1(n) = 0$ if $\varepsilon_{n-1}(0)\varepsilon_n(0) = 00$ or 11 but $\varepsilon_1(n) = 1$ if $\varepsilon_{n-1}(0)\varepsilon_n(0) = 01$ or 10. Here, the information flows at the rate of a bit per iteration, corresponding to the fact that the Liapunov exponent is $\lambda = \ln 2$. Informa-tion flow at the average rate of a digit per iteration in an integral base μ of arithmetic that is roughly e^λ is the very essence of deterministic chaos.

We can now systematically derive the statistics generated by a *single* orbit that starts from a *definite* initial condition x_o. Divide the unit interval into 2^N bins of equal width 2^{-N} and label the bins from left to right by the 2^N N-bit binary words $00\cdots0$, $00\cdots01$, $00\cdots10,\cdots,11\cdots1$. Whenever the first N bits $\varepsilon_1(n)\cdots\varepsilon_N(n)$ of x_n coincide with one of these words, then the bin labeled by that word is visited by the map. If we denote by $P(\varepsilon_1\cdots\varepsilon_N)$ the frequency with which the bin labeled by the N bit word $\varepsilon_1\cdots\varepsilon_N$ is visited, then the required 2^N probabilities $\{P(\varepsilon_1\cdots\varepsilon_N)\}$ are given by the histogram that follows from iterating the map from a definite initial condition. For $N = 1$, we get 2 bins and the coarsest level of description of the statistics generated by the dynamics, but finer coarse-grained descriptions are provided by increasing the precision N. In experiment and in computation, both the precision N and the time n are always finite: the limits where n and N are infinite are misleading abstractions that have little or nothing to do with nature, and certainly have nothing to do with computation. The bin probabilities that we encounter in any case are *always* empirical frequencies (they follow from histograms) and cannot arbitrarily be replaced by infinite time, infinite precision limits, even if those limits are known mathematically. Finally, the bin probabilities $\{P(\varepsilon_1\varepsilon_2\cdots\varepsilon_N)\}$ are not independent of the choice of initial data, but are different for different classes of initial data.

Normal numbers were defined in section 4.6. The property of a normal number is as follows: the number is expanded in some base of arithmetic, base $\mu = 2,3,4$, or any higher integer, and the number then has the form $x = 0.\delta_1\delta_2\cdots\delta_N\cdots$ where the digits δ_i take on the discrete set of values $0,1,2,\ldots,\mu - 1$. Consider next a string of digits of length M, $\varepsilon_1\varepsilon_2\cdots\varepsilon_M$. As the ε_i can take on μ different values, there are μ^M different possible strings of length M. Let $P(\varepsilon_1\varepsilon_2\cdots\varepsilon_M)$ denote the frequency with which a particular block $\varepsilon_1\varepsilon_2\cdots\varepsilon_M$ occurs as the string is read left to right by sliding an M-digit window one digit at a time (see McCauley, 1993, chapter 9). For a normal number, $P(\varepsilon_1\varepsilon_2\cdots\varepsilon_M) = \mu^{-M}$ for all M for the infinite length string. It is not known whether the number π is normal to any base, but it is known that π is effectively normal to base 10 for $M \le 5$. By effectively normal, we mean that $P(\varepsilon_1\varepsilon_2\cdots\varepsilon_M) \approx 10^{-M}$ for a reading of finitely many digits (clearly, it is effective normality and not the abstract limiting case that is of interest for science). 'Effective normality' for π occurs for a string of length $N > 500$ digits. Therefore, the histograms obtained from iterating the decimal tent map, starting from the initial condition $x_o = \pi$, are approximately even for a division of the unit interval into 10^M bins, with $M = 1, 2, 3, 4$, or 5, the same distribution that one would 'expect' to obtain for a large number of tosses of a fair 10-sided die.

One does not know whether either $\sqrt{2}$ or $\sqrt{3}$ is normal to any base of arithmetic, but it is easy to verify by a correctly written computer program that both numbers are effectively normal to base 2 for up to at least 5000 bits. For an arbitrary map, the bit distribution of the initial condition is not the same as the histograms that one obtains from iterating a map although one can prove (see chapter 9 in McCauley, 1993) that, for the binary tent map, a normal initial condition also yields even bin statistics when the map is iterated. Therefore, iteration of the binary tent map, starting from either $x_o = \sqrt{(2)} - 1$ or $\sqrt{(3)} - 1$ will yield histograms that reflect the statistics that one would 'expect', on the basis of the law of large numbers (see Gnedenko and Khinchin, 1962), to find for a large number of tosses of a fair coin. However, we also point out that it is extremely easy to construct initial conditions that yield very uneven histograms and sometimes histograms where certain bins are never visited: $x_o = 0.101001000100001000001\cdots$ is one example; $x_o = 0.101100111000111110000\cdots$ is another.

Finally, you cannot eliminate deterministic chaos by a coordinate transformation: periodic orbits transform into periodic orbits, and nonperiodic orbits transform into nonperiodic ones as well (the periods only involve counting and so are topologic invariants). A generalization to other maps of the digital arithmetic performed above for the tent map leads to the study of the 'symbolic dynamics' of a whole class of topologically related maps (maps that are related by continuous but generally highly nondifferentiable coordinate transformations). Symbolic dynamics is a method that concentrates upon the topologically invariant properties of chaotic systems. This is important, for one reason, because it allows us to use the simplest map in a class to understand the periodic orbits, the possible statistics and other properties of every system in the same class.

Liapunov exponents (and fractal dimensions) are not topologic invariants: they are not invariant under arbitrary continuous but nondifferentiable coordinate transformations. However, you cannot transform a dynamical system with finite Liapunov exponents into a one with vanishing exponents by a nonsingular transformation. The proof is as follows: consider a noninvertible map

$$x_{n+1} = f(x_n) \tag{6.54a}$$

and also a map

$$z_{n+1} = g(z_n) \tag{6.54b}$$

which is obtained from the original map (6.54a) by a coordinate transformation,

$$z_n = h(x_n), \tag{6.55}$$

where $h' \neq 0$. That is, the transformation h is invertible. We denote by λ and σ the Liapunov exponents of the maps (6.54a) and (6.54b) respectively, i.e.,

$$\sigma \cong \frac{1}{n} \sum_{i=0}^{n-1} \ln |g'(y_i)|. \tag{6.56}$$

The question is: how is σ related to λ, where λ is the exponent for the corresponding trajectory of map f (trajectories map one to one)? From (6.54a) we obtain

$$\frac{dx_{n+1}}{dx_n} = f'(x_n) \tag{6.57a}$$

and likewise,

$$\frac{dz_{n+1}}{dz_n} = g'(z_n) \tag{6.57b}$$

follows from (6.54b), so that from $dz_n = g'(z_n)dz_n$ we obtain

$$\frac{dz_{n+1}}{dz_n} = \frac{h'(x_{n+1})dx_{n+1}}{h'(x_n)dx_n} \tag{6.58}$$

which becomes

$$g'(z_n) = \frac{h'(x_{n+1})}{h'(x_n)} f'(x_n) \tag{6.59}$$

when the above results are combined. Inserting the above results yields

$$\sigma \approx \frac{1}{n} \sum_{i=0}^{n-1} \left(\ln |f'(x_i)| + \ln \left| \frac{h'(x_{i+1})}{h'(x_i)} \right| \right) \tag{6.60}$$

which can be rewritten in the form

$$\sigma = \lambda + \ln \left| \frac{h'(x_n)}{h'(x_1)} \right| /n. \tag{6.61}$$

It follows that $\sigma \to \lambda$ for large n. For times $n \gg 1$, the Liapunov exponent, for trajectories that map onto each other, is invariant under *differentiable* coordinate transformations which means that you cannot transform chaos away. In chapter 13 we show that integrability is also invariant under differentiable coordinate transformations.

Is chaos computable; e.g. is $x_n = 2^n x_o \bmod 1$ computable? Is the solution $x(t) = x_o e^t$ of $dx/dt = x$ computable? Both are either computable or noncomputable together: if x_o is generated decimal by decimal (or faster) by an algorithm (like the various rules for computing $x_o = \sqrt{7}$, of which Newton's method is optimal), then both x_n and $x(t)$ are computable (granted, in the second case t must also be chosen to be a computable number). If x_o is assumed to be noncomputable (most numbers defined to belong to the continuum belong to this category), then neither $x_n = 2x_{n-1} \bmod 1$ nor $dx/dt = x$ have computable solutions. Computable numbers form a countable subset of the continuum, and they are all that we have available for the numerical study of either periodic or nonperiodic orbits (see Minsky, 1967, chapter 9 or McCauley, 1993, chapter 11). In fact, both problems are equivalent: if we study the solutions of $dx/dt = x$ on the unit circle, so that $x(t) = e^t x_o \bmod 1$, and in addition strobe the solution at times $t_n = n \ln 2$, then $x(t_n) = e^{n \ln 2} x_o \bmod 1 = 2^n x_o \bmod 1 = x_n$.

Exercises

1. For a particle of mass m in a box with perfectly reflecting walls, $0 \le x_1 \le a_1$, $0 \le x_2 \le a_2$, show that the canonical momenta P_i conjugate to the generalized coordinates $Q_i = \omega_i t + Q_{io}$ are given by $P_i = mav_i/\pi$, and therefore that $K = (\pi^2/2m)[(P_1^2/a_1^2) + (P_2^2/a_2^2)]$. (Hint: use $\omega_i = \pi v_i/a_i$ along with $H = m(v_1^2 + v_2^2)/2$.)

2. For the Kepler problem, compute a Poincaré section numerically and using Cartesian coordinates. Choose initial conditions corresponding to the order of magnitude of the earth–sun system, but use dimensionless variables in order to integrate numerically.

3. For the double pendulum, with $l_1 = l_2 = 1$, $g = 1$, and $m_1 = m_2 = 1$, compute numerically and also plot the Poincaré sections for the following initial conditions:
    (a) $\theta_1(t_o) = -0.65, -0.7$, and -0.85 with $\theta_2(t_o) = 1.4, p_2(t_o) = 0$, and $E = 2.24483$
    (b) same as (a) except that $\theta_1(t_o) = -0.822, \theta_2(t_o) = 1.4335$
    (c) $\theta_1(t_o) = -0.5, \theta_2(t_o) = 1.4$, otherwise same as (a)
    (d) $\theta_1(t_o) = 0.5 p_2(t_o) = p_1(t_o) = 0, \theta_2(t_o) = \pi$.
    For cases (a), (b), and (c), plot the Poincaré sections only for times that are small enough that you can recover most of the digits of your initial condition by in-

tegrating backward in time. For case (d), ignore that constraint on accuracy until exercise 6.

4. Plot and compare the time series $\theta_2(t)$ vs t for the following cases:
    (a) The initial data in exercise 3(b) above and also for the case where $\theta_2(t_o) = 1.4400$.
    (b) The initial data in exercise 3(d) above and also for the case where $\theta_2(t_o) = 3.1416$.

5. (a) For exercise 3(a), integrate far enough forward in time that energy E is not conserved even to one digit accuracy. Is the torus in the Poincaré section 'smeared out'?
    (b) Compare the integration time scale for one-digit energy conservation with the time scale for one-digit initial condition recovery in backward integration. What does this mean?

6. Denote by $\theta_2(t)$ the time series in exercise 3(d) starting from $\theta_2(t_o) = \pi$ and by $\theta_2(t) + \delta\theta_2(t)$ the time series that starts from $\theta_2(t_o) + \delta\theta_2(t) = 3.1416$. Let τ denote the time t for which $|\delta\theta_2(t)| \gtrsim 1$. Treating the point $\theta_2(t)$ for *any* time $t \geq \tau$ as the initial condition, integrate the differential equations backward in time and attempt to recover the original initial condition $\theta_2(t_o) = \pi$ to within several digits (check your result for three different times $\tau < t_1 < t_2 < t_3$).
    (a) Show that when $t \geq \tau$ even the leading digit of the 'recovered initial condition' is wrong.
    (b) Show that if $t \ll \tau$, then as many digits as you like of $\theta_2(t_o)$ can be recovered. *Note*: Be aware that, in single precision on a binary computer, π is represented by about 25 binary digits and the rest of the (infinitely long) binary string is truncated and is approximated by a 'floating exponent' M in 2^M. Note also that 25 binary digits are equivalent to about 7 base 10 digits: $2^{25} = 10^N$ means that $N = 25\ln 2/\ln 10 = 7.52$.

7. If
    $$I_n = \int_0^1 x^n e^x dx$$
    where $l_o = e - 1$ and $l_1 = 1$, then show that $l_n = e - nl_{n-1}$ if $n \geq 2$ is an integer.
    (a) Starting with e computed from a PC or hand calculator, compute l_{13} by recursion from l_o. Compare with the exact result known to several decimals in a table of tabulated functions.
    (b) Show that this map effectively has a Liapunov exponent $\lambda \sim \ln(n)$ for forward iterations. How, then, can one compute l_{5000} recursively?
    (c) As a first approximation, set $l_{5000} \approx 0$ and compute l_{4995} by backward iteration of the map. To how many decimals is your result accurate?
    (Julian Palmore)

8. Compute the Liapunov exponents for the bakers' transformation
    $$x_{n+1} = 2x_n \bmod 1$$
    $$y_{n+1} = \begin{cases} ay_n, & 0 \leq x_n < \frac{1}{2} \\ ay_n + \frac{1}{2}, & \frac{1}{2} < x_n \leq 1. \end{cases}$$
    Is the system conservative or dissipative?

9. For the bakers' transformation with $a = \frac{1}{2}$, draw a sequence of diagrams in phase space that shows how the block of initial conditions $0 \leq x \leq 1$, $0 \leq y \leq \frac{1}{2}$ is transformed by the map (for $n = 1,2,3,\ldots$). Identify graphically the effects of the two Liapunov exponents in the resulting figures.

10. Using the idea of a return map and the example of section 7.5, explain how forward and backward integrations in time can be carried out so that one does not 'lose' the initial condition $\theta_2(t_o)$ in the double pendulum problem. Show in detail how this can be done for the cat map for three forward iterations to 6-bit accuracy (where the first 6 bits of (x_3, y_3) are to be computed exactly for all iterates).

11. Find and compare the periods of the cat map's orbits when $(x_o, y_o) = (\frac{1}{5}, \frac{1}{5})$ and $(\frac{1}{7}, \frac{1}{7})$.

12. We can construct a three dimensional histogram for the cat map in the following way. Writing binary expansions

$$x_n = 0.\varepsilon_1(n)\cdots$$
$$y_n = 0.\delta_1(n)\cdots$$

for the map's iterates, subdivide the unit square into four bins and label each bin by the symbols $\varepsilon_1(n)\delta_1(n) = 00,10,01,11$. Then, compute and plot the histograms for the initial conditions $(x_o, y_o) =$

(a) $(\frac{5}{12}, \frac{5}{12})$

(b) $(\frac{12}{29}, \frac{12}{29})$

(c) $(\frac{29}{70}, \frac{29}{70})$

Use an algorithm to generate $\sqrt{2}$ in binary, and compute the histograms for 500 iterates of the initial condition $(\sqrt{2} - 1, \sqrt{2} - 1)$.

7

Parameter-dependent transformations

Kunsten alene kan skape verker, som aldrig forældes.

F. Engel (1922)

7.1 Introduction

Marius Sophus Lie was born the son of a Lutheran priest in Nordfjordeid, a small village near the Sunmøre Alps. He grew up near Oslo and spent his vacations in the still-typical Norwegian way: ski tours in the winter and foot tours in the summer. After his student days at the University of Oslo he was influenced by the writings of the German geometer Plücker, whose student Felix Klein (1849–1925) he came to know well in Berlin. In Paris, where Lie and Klein roomed together, Lie met the outstanding French mathematicians Darboux and Jordan, and began his work on contact transformations while there. He set out on foot alone to make a *Wanderung* to Italy but was arrested outside Paris and jailed for a month due to the onset of the Franco–Prussian war: he carried with him several letters on mathematics from Klein that caused French police to suspect him of being a German agent with encoded information. Back in Oslo, Lie finished his doctor's degree and then taught and wrote for several years at the University. After sending Engel to work with him, Klein left Leipzig for Göttingen and arranged that Lie replace him in Leipzig. It was at Leipzig that Lie and his German collaborators Engel and Scheffers wrote several voluminous works on transformation theory and differential equations. Lie's program was taken up and greatly advanced by the Frenchman E. Cartan. Quantum physicists use Cartan's contributions to Lie algebras but most physicists remain unaware that Lie groups and Lie algebras originated in the search for solutions of nonlinear first order differential equations. Poisson (French mathematician, 1781–1840) and Jacobi contributed forerunners of the Lie algebra of operators defined by vector fields.

Galois (1811–1832) had developed group theory in order to put the search for solutions of algebraic equations on a systematic footing. Lie set for himself the same task for differential equations. The result of that effort is the theory of Lie groups. The study of a Lie group in the tangent space near the identity transformation is called a Lie algebra and leads to a system of differential equations. Lie classified systems of differential equations according to their invariants. An invariant is a conservation law.

7.2 Phase flows as transformation groups

A vector field V defines a system of differential equations as a flow in phase space. The solutions

$$x_i(t) = \Psi_i(x_1(t_\circ), \ldots, x_n(t_\circ), t - t_\circ) \tag{7.1a}$$

of the first order system

$$\dot{x}_i = V_i(x_1,\ldots,x_n) \tag{7.2}$$

also describe a transformation from a coordinate system $(x_1(t_o),x_2(t_o),\ldots,x_n(t_o))$ defined by the parameter value t_o to another coordinate system $(x_1(t),x_2(t),\ldots,x_n(t))$ defined by the parameter t. Since all of these coordinate transformations are defined by specifying different values of the time difference $t - t_o$, we have a one-parameter collection of coordinate transformations. The condition for the existence of the streamline picture is that these transformations can be performed for all finite values of the parameter, which is in this case the time t. In the language of differential equations, for any specified initial condition, the solutions (7.1a) of (7.2) must yield a transformation from the initial condition to any point on the trajectory that passes through it, either forward or backward in time, for all finite times. Whenever this is true, as we show next, then the transformations form a group.

This collection of transformations has certain properties: first, there is an identity transformation

$$x_i(t_o) = \Psi_i(x_1(t_o),\ldots,x_n(t_o),0). \tag{7.1b}$$

Second, we can go backward in time (the differential equations (7.2) can be integrated backward in time whenever the Jacobi determinant does not vanish), yielding

$$x_i(t_o) = \Psi_i(x_1(t),\ldots,x_n(t),t_o - t) \tag{7.3}$$

as the inverse of the transformation (7.1a). Third, we can reach the final destination in more than one way: we can get there directly, as in equation (7.1a), but we can also transform indirectly by using an intermediate step,

$$x_i(t_1) = \Psi_i(x_1(t_o),\ldots,x_n(t_o),t_1 - t_o), \tag{7.4}$$

$$x_i(t) = \Psi_i(x_1(t_1),\ldots,x_n(t_1),t - t_1), \tag{7.5}$$

so that the transformation functions must also satisfy the closure condition

$$\Psi_i(x_i(t_o),\ldots,x_n(t_o),t - t_o)$$
$$= \Psi_i(\Psi_1(x_1(t_1),\ldots,x_n(t_1),t - t_1),\ldots,\Psi_n(x_1(t_1),\ldots,x_n(t_1),t - t_1),t_1 - t_o). \tag{7.6}$$

Fourth, and finally, the order of combination of three or more transformations made in sequence does not matter in the following sense:

$$\Psi(\Psi(\Psi(x(t_o),t_1 - t_o),t_2 - t_1),t - t_2) = \Psi(\Psi(x(t_o),t_1 - t_o),t - t_1), \tag{7.7}$$

which is an associativity rule (we have introduced an obvious shorthand notation in equation (7.7)).

The four rules stated above are equivalent to saying that our collection of transformations forms a group. *The condition for the validity of the streamline picture is exactly the condition that the solutions, a collection of coordinate transformations, form a one-parameter group where the time difference $t - t_o$ is the group parameter.*

In order to state the group properties algebraically, we can represent the solutions of equations (7.2) by using the time-evolution operator $U(t - t_o)$:

$$x_i(t) = U(t - t_o)x(t_o), \tag{7.8}$$

where the operator U stands for the transformation Ψ. Denoting time differences $t - t_o$ by τ, for a one-parameter collection of transformations $U(\tau)$ where τ varies continuously the following rules must be obeyed by the time-evolution operator in order that the transformations form a one-parameter group.

(1) There is an identity element $U(0) = I$: $IU(\tau) = U(\tau)I = U(\tau)$.
(2) Each element $U(\tau)$ of the group has a unique inverse $U(\tau)^{-1} = U(-\tau)$, so that $U(\tau)U(-\tau) = U(-\tau)U(\tau) = I$.
(3) There is a group combination law so that every combination ('product') of two transformations in the group is again a member of the group (this is called the closure rule for the group): $U(\tau_1)U(\tau_2) = U(\tau_1 + \tau_2)$.
(4) The group combination law is associative: $U(\tau_1)(U(\tau_2)U(\tau_3)) = (U(\tau_1)U(\tau_2))U(\tau_3)$.

As we saw in chapter 3, for the case of linear differential equations

$$\frac{dX(t)}{dt} = AX(t) \tag{7.9}$$

with constant matrix A, then $X(t) = e^{tA}X(t_\circ)$ so that the time-evolution operator is given explicitly by

$$U(t) = e^{tA} = I + tA + t^2 A^2/2! + \ldots + t^n A^n/n! + \ldots, \tag{7.10}$$

where I is the identity matrix.

Whenever the time-evolution operator of (7.10) represents bounded solutions for all finite times (when the solutions of the system of differential equations do not blow up at any finite time), then the motion is geometrically equivalent to a flow in phase space, and the phase flow picture takes us quite naturally to the study of parameter-dependent transformation groups.

The condition for the existence of the time-evolution operator and the phase flow picture is that the solution is not interrupted by a singularity at any *finite* time. Singularities at infinite time are consistent with the phase flow picture. All of this was explained in chapter 3 and is discussed further in section 7.6. We assume that the velocity field satisfies a Lipshitz condition so that solutions exist, are unique, and $U(\tau)$ can be constructed approximately to any desired degree of accuracy over a finite region in phase space (whose size is unknown, a priori) by Picard's method.

The study of a flow leads to a one-parameter group, but every one-parameter group of differentiable transformations also defines a flow in phase space. From the finite transformations, a set of differential equations can be derived by differentiation, treating the group parameter formally or abstractly as if it were the time. A group is Abelian[1] if all of the group elements commute with each other. Every continuous one-parameter group is Abelian and is also a Lie group.

7.3 One-parameter groups of transformations

Consider the continuous transformations

$$x(a) = \psi(x(0),a) \tag{7.11}$$

where a is a parameter, x and ψ stand for the coordinates $(x_1,\ldots,x_n)$ and the

---

[1] A soaring statue of Niels Henrik Abel, who died at age 27 in 1829, stands just inside Slottsparken (the palace grounds) along Bygdøy Alle in Oslo. A modest statue of Sophus Lie, who died at 57 in 1899, stands outside the Matematisk Institutt at the University of Oslo, and another in Nordfjordeid near the Sunmøre Alps.

transformation functions $(\psi_1, \ldots, \psi_n)$. We show below how a set of differential equations generates the flow infinitesimally.

The identity transformation is given by setting $a = 0$, so that $x(0) = \Psi(x(0),0)$ (the choice $a = 0$ is convenient but is not necessary), the inverse is given by

$$x(0) = \psi(x(a),d), \tag{7.12}$$

and every combination of transformations is again a transformation:

$$x(c) = \psi(x(0),c), \tag{7.13}$$

$$x(c) = \psi(x(a),b), \tag{7.14}$$

so that the group closure rule is

$$\psi(\psi(x(0),b),a) = \psi(x(0),c). \tag{7.15}$$

Therefore, there must be a definite combination law for the group parameter:

$$c = \phi(a,b), \tag{7.16}$$

where $b = 0$ yields the identity: $a = \phi(a,0)$. An infinitesimal transformation $x + dx = \psi(x,\delta a)$ follows from setting $b = \delta a$, whereby $c = a + da$ and $x(c) = x(a) + dx$, so that

$$dx_i = V_i(x)\delta a \tag{7.17}$$

and $V_i(x) = \partial\psi_i(x,b)/\partial b$ is to be evaluated at $b = 0$. From $a + da = \phi(a,\delta a)$ we find

$$da = \delta a v(a) \tag{7.18}$$

where $v(a) = \partial\phi(a,b)/\partial b$ is to be evaluated at $b = 0$. A *canonical* parameter is *additive*, like the time. In general, however, ϕ is not linear in a but a canonical parameter θ can be constructed by integrating $d\theta = da/v(a)$.

An infinitesimal transformation from x to $x + dx$ induces in any scalar phase space function the transformation $F \rightarrow F + dF$ where

$$dF/d\theta = LF \tag{7.19}$$

and the linear operator L, the 'infinitesimal generator' or 'infinitesimal operator' of the group, is *determined by the velocity field* $V(x)$:

$$L = \sum_1^n V_i(x)\frac{\partial}{\partial x_i} = V \cdot \nabla. \tag{7.20}$$

The differential equations that describe the flow locally are given by applying (7.16) and (7.17), with a canonical parameter θ, to the n coordinates x_i,

$$dx_i/d\theta = V_i(x), \tag{7.21}$$

so that we do not deviate qualitatively from the study of dynamical systems in phase space: every flow in phase space describes a hypothetical dynamical system, and every flow can be described as a one-parameter group of transformations.

Time-independent conserved quantities $G(x_1, \ldots, x_n)$ are functions of the phase space coordinates that are left invariant by the transformations and therefore obey the first order partial differential equation

$$LG = \sum_1^n V_i(x)\frac{\partial G}{\partial x_i} = 0. \tag{7.22a}$$

Geometrically, this requires that the velocity vector V is tangent to the surface G = constant, because this equation can be understood as

$$V \cdot \nabla G = 0, \tag{7.22b}$$

and because the gradient of G is normal to the surface G = constant.

A simple example of a two dimensional phase flow is given by rotations in the Euclidean plane. With Cartesian coordinates x and y, an infinitesimal rotation is described by $dx(\theta) = y(\theta)d\theta$, $dy(\theta) = -x(\theta)d\theta$. In this case, the rotation angle θ is the canonical parameter and the vector field that defines the flow locally is given by $V = (y, -x)$. The infinitesimal operator is $L = y\partial/\partial x - x\partial/\partial y$. Integration of (7.22a) yields $F(x(\theta),y(\theta)) = e^{\theta L}F(x,y)$, and direct application of the exponential of the operator to $x(\theta) = e^{\theta L}x$ and $y(\theta) = e^{\theta L}y$ yields, by way of the power series definition of the exponential of an operator that,

$$x(\theta) = \psi_1(x,y,\theta) = e^{\theta L}x = x\cos\theta + y\sin\theta \tag{7.23a}$$

and

$$y(\theta) = \psi_2(x,y,\theta) = e^{\theta L}y = -x\sin\theta + y\cos\theta, \tag{7.23b}$$

which hold for all values of θ (the radii of convergence of the power series expansions of $\sin\theta$ and $\cos\theta$ are infinite).

All functions that are rotationally invariant satisfy the first order partial differential equation

$$LG = x\frac{\partial G}{\partial y} - y\frac{\partial G}{\partial x} = 0, \tag{7.24}$$

and the method of characteristics (see Duff, 1962, or Sneddon, 1957) yields

$$d\theta = dy/x = -dx/y, \tag{7.25}$$

which can be integrated to yield $x^2 + y^2$ = constant. This is no surprise: all rotationally invariant functions of x and y alone must have the form $G(x,y) = f(x^2 + y^2)$.

Consider the simplest case where $G = (x^2(\theta) + y^2(\theta))/2$. The quantity G is invariant under the infinitesimal rotation, $dx/d\theta = y$ and $dy/d\theta = -x$, and we can use the differential equations to rewrite $G = (xdy/d\theta - ydx/d\theta)/2$, which is proportional to the angular momentum about the z-axis, M_z, if θ is proportional to the time t. Here, we have used the fact that since both θ and t are canonical parameters (are both additive), then $\theta = \omega t + \theta_o$ must hold, where ω and θ_o are constants. This is easy to prove: if $\theta = f(t)$, and $f(t_1) + f(t_2) = f(t_1 + t_2)$ holds, it follows by differentiation that $f'(t)$ must be a constant. Hence $f(t) = \omega t + \theta_o$ with ω and θ_o constants.

Notice also that we can use these very same equations to describe the one dimensional simple harmonic oscillator in the phase plane: with variables (Q,P) rescaled so that the ellipse in the phase plane becomes a circle, $H = P^2/2 + Q^2/2$ and $V = (P, -Q)$, so that the time-evolution operator of the simple harmonic oscillator is simply the rotation operator in the (Q,P)-plane. The infinitesimal generator of rotations is the angular momentum operator in phase space,

$$M = V \cdot \nabla = P\frac{\partial}{\partial Q} - Q\frac{\partial}{\partial P}. \tag{7.26}$$

Because P and Q are always treated in phase space as if they were Cartesian coordinates, P in the oscillator problem is analogous to x in the angular momentum, and Q is analogous to y. We could go further with the analogy and write $H = (PdQ/dt - QdP/dt)/2$, the form of the generator of rotations in the (Q,P)-plane. Since it is also true that $G = (x^2 + y^2)/2$, we can use Hamilton's equations to describe infinitesimal rotations,

$$\frac{dx}{d\theta} = \frac{\partial G}{\partial y}, \frac{dy}{d\theta} = -\frac{\partial G}{\partial x}. \tag{7.27}$$

Here, we have an example of a Hamiltonian system that has no Lagrangian. It is indicated by this example, and is demonstrated further in chapters 15 and 16, that Hamiltonian systems have a validity as stream functions (or infinitesimal generating functions) that extends well beyond their derivation in chapter 2 from a Lagrangian formulation of Newtonian dynamics. The description of rotations as a Hamiltonian flow in phase space is not accidental: rotations are area-preserving flows ($dxdy$ is the 'volume-element' in the phase plane), and any area-preserving flow has a divergence-free velocity field V which automatically yields the differential equations of the flow in canonical form, as we saw in section 3.6.

In what follows, we always assume that the canonical parameter has been found, because the differential equations of the flow take on their simplest form in that case.

7.4 Two definitions of integrability

There are at least three separate meanings of the idea of global integrability in the theory of differential equations. The most general idea defines global coordinate systems from the standpoint of a set of differential forms and is guaranteed by existence theorems. The second is the idea of complete integrability discussed in section 3.6, meaning the reduction of the solution of n differential equations to n independent integrations, for all finite times $-\infty < t < \infty$, via a single set of $n-1$ isolating integrals. Complete integrability of a flow implicitly presumes integrability of the existence sort, but the reverse is not true. By 'isolating integrals', we mean specifically $n-1$ conservation laws $G_i(x_1,\ldots,x_n) = C_i$ that determine $n-1$ variables in the form $x_j = f_j(x_n,C_1,\ldots,C_{n-1})$, $j = 1,\ldots,n-1$.

A flow in phase space is generated locally by a vector field. The streamlines are everywhere tangent to the vector field of the flow. The solution of the differential equations defines a one-parameter coordinate transformation from one set of holonomic coordinates (the initial conditions) to another (the positions at time t). Flows with $n > 3$ are generally not completely integrable but all flows considered here are *solvable*, meaning that $U(t)$ *exists* and can be *approximated* iteratively by polynomials in t. To make matters even more complicated, there is a third idea of integrability that we shall discuss in chapter 13.

The global coordinate transformations (7.1a) generated infinitesimally by a flow represent the most general idea of integrability: the transformations exist as *functions* of the initial conditions and the time, meaning that the $x_i(t)$ are holonomic coordinates for all t. Existence theorems guarantee that the variables $x_i(t)$ that arise from integration of the differential equations are path-independent. By a solvable flow we mean in this text that:

(1) Solutions exist as differentiable functions of initial data and the time t over a finite domain.
(2) Solutions are unique (this could be relaxed, but not for a flow).
(3) There is a systematic method for constructing the transformations approximately (Picard's iterative method).
(4) There are no singularities at real finite times t.

The first three points are guaranteed over finite regions in phase space whenever the velocity field satisfies a Lipshitz condition over a generally larger finite region. Consider an alternative to point (1), where solutions of differential equations do not exist as global coordinate transformations: given a set of holonomic coordinates $(x_1,\ldots,x_n)$, if the differential form $\omega_i dt = B(x)_{ik}dx_k$ is nonintegrable (is not an exact differential form and no integrating factor exists) then there is no global coordinate system $(\alpha_1,\ldots,\alpha_n)$, where $d\alpha_i = \omega_i dt$, because the line integrals of the right hand side of the differential form are path-dependent. Since integration does not yield the functions required to define a coordinate transformation, the α_i are called nonholonomic variables. Nonintegrable velocities of the form $\omega_i = d\alpha_i/dt$ are used as coordinates to study rigid body motions as integrable and nonintegrable flows in chapter 9, and are discussed theoretically in chapters 11, 12, and 13.

The Lie–Jacobi idea of complete integrability *presumes* the existence of *holonomic* coordinate systems $(x_{1_o},\ldots,x_{n_o})$ and $(x_1(t),\ldots,x_n(t))$ for all values of t and t_o and therefore takes the more general kind of integrability for granted. For $n \geq 3$ the transformations (7.1a) describe a *completely integrable flow* in the sense of Lie and Jacobi if the time-evolution operator can be transformed into a *translation* operator describing a single constant speed translation: a *completely integrable flow* is one where there are $n - 1$ time-independent invariants $G_i(x_1,\ldots,x_n) = C_i$ that isolate $n - 1$ of the variables in the form $x_i = f_i(x_n,C_1,\ldots,C_{n-1})$, for all finite times, $i = 1,\ldots,n - 1$, so that the differential equation for the variable x_n is an exact differential and can be integrated to yield the solution in the form

$$\left.\begin{array}{l} G_i(x_1,\ldots,x_n) = C_i, i = 1,\ldots,n - 1 \\ F(x_1,\ldots,x_n) = t + D. \end{array}\right\} \tag{7.28a}$$

In this case, a single transformation

$$\left.\begin{array}{l} y_i = G_i(x_1,\ldots,x_n), i = 1,\ldots,n - 1 \\ y_n = F(x_1,\ldots,x_n) \end{array}\right\} \tag{7.28b}$$

transforms the flow to a coordinate system where the operator of the group $U(t) = e^{td/dy_n}$ describes a constant speed translation, for all finite times t, over a global region in phase space[2], meaning a finite region that is much greater in size than 'ε' (or ε^n), a region large enough that the gradients of the velocity field matter. A completely integrable flow generates a global coordinate transformation to a special coordinate system where the motion is a constant speed translation along one axis of a set of orthogonal coordinates $(y_1,\ldots,y_n)$. For a specific set of values of the constants $(C_1,\ldots,C_{n-1})$, the flow is confined to one two dimensional surface imbedded in the n dimensional phase space. A nonintegrable flow defines a one-parameter group that is *not* globally equivalent to a single

---

[2] Eisenhart (1961, p. 34) states that every one-parameter group is a translation group, but this claim is demonstrably false.

translation group: the translation variables y_i do not exist as holonomic coordinates for all $-\infty < t < \infty$ in that case.

Our choice of language leaves open the possibility that 'nonintegrable' may include singular conservation laws that permit separate transformations of the form (7.28b) piecewise for finite times $\ldots - t_{i-1} < t < t_i, t_i < t < t_{i+1}, t_{i+1} < t < t_{i+2} \ldots$.

7.5 Invariance under transformations of coordinates

We distinguish between scalar functions and a function that is, in addition, *invariant* under some restricted class of transformations. Scalar products of vectors are examples of scalars that are invariant only under certain restricted transformations.

Consider a transformation of coordinates $x' = Tx$ where T is an operator. In the language of chapter 2, $x' = Tx$ can be shorthand notation for a linear or nonlinear transformation $q'_i = F_i(q_1,\ldots,q_n)$, or, in the language of this chapter, it can stand for $x_i(a) = \psi_i(x_1,\ldots,x_n,a) = U(a)x_i$. Next, consider any scalar function $g(x)$ of the coordinates x_i (a scalar function evaluated at a point x is just a number). Since $g(Tx)$ is in general a different function of x, is not the same function as g, we formally introduce the operator O_T that transforms the old function g into the new function g':

$$g'(x') = O_T g(Tx) = g(x). \tag{7.29}$$

We can rewrite this as

$$g(x') = O_T^{-1} g(x). \tag{7.20b}$$

An example of this relationship is given by the equation of section 7.3 $F(x(\theta),y(\theta)) = e^{\theta L} F(x,y)$ where $L = V \cdot \nabla$. Therefore, $O_T = e^{-\theta L}$ in this case.

The idea of a function that is invariant under a coordinate transformation is expressed by $O_T g(x) = g(x)$, so that $g(T^{-1}x) = g(x)$, or

$$g(x) = g(Tx) = g(x'), \tag{7.30}$$

which expresses the invariance of the function g explicitly in the language of chapter 2. Every scalar product of two vectors is invariant under the group of transformations that defines the geometry of a space and therefore is used to define those objects as vectors[3]. As an example, $g(x,y) = x^2 + y^2$, the Euclidean scalar product of a position vector with itself, is invariant under rotations about the z-axis, as is $h(x,y,dx/dt,dy/dt) = x\,dy/dt - y\,dx/dt$, but $g(x,y) = x^3y$ is not rotationally invariant although x^3y is a perfectly good scalar function. The function x^3y is, however, invariant under the one-parameter rescaling transformation $x' = \lambda x$ and $y' = \lambda^{-3}y$.

Consider next the action of a coordinate-dependent operator $L(x)$ on a scalar function, $L(x)g(x) = h(x)$. Then

$$O_T L(x)g(x) = O_T h(x) = h(T^{-1}x) = L(T^{-1}x)g(T^{-1}x). \tag{7.31}$$

Therefore,

$$O_T L(x)O_T^{-1}O_T g(x) = L(T^{-1}x)O_T g(x) = L'(x)O_T g(x), \tag{7.32}$$

---

[3] In chapters 8 and 10, objects are defined as vectors, tensors, scalars, etc. according to their behavior under a specific group of coordinate transformations. In this refined picture, merely having n components is not enough to qualify an object to be counted as a vector. The generalized velocity does not transform as a vector under time-dependent coordinate transformations in configuration space, but the canonical momentum does.

where O_T^{-1} is the inverse of O_T, so that the transformation rule for operators is given by

$$L'(x) = L(T^{-1}x) = O_T L(x) O_T^{-1}. \tag{7.33}$$

Therefore, an operator is invariant under a collection of coordinate transformations whenever

$$O_T L(x) O_T^{-1} = L(x), \tag{7.34}$$

that is, when it has exactly the same form in both coordinate systems. An example is provided by the Laplace operator

$$\Delta = \frac{\partial}{\partial x^2} + \frac{\partial}{\partial y^2} \tag{7.35}$$

under rotations of Cartesian coordinates about the z-axis, as is easy to verify. We will use the idea of invariance of operators under a group of transformations in the next section.

7.6 Power series solutions of systems of differential equations

We have started formally with the idea of the global flow, the collection of solutions of the differential equations, and have differentiated to obtain the differential equations that generate the flow locally. The differential equations of the flow are just the infinitesimal transformations of the group expressed in terms of a canonical group parameter. If one could know all possible flows in advance, then there would not be any reason to begin an analysis of a dynamics problem with differential equations. Except in the simplest of cases, we cannot do that because we need only the vector field in order to describe a flow locally, whereas the entire future of a global flow is encoded digitally in the initial conditions whenever a nonintegrable flow is chaotic: there, knowing an infinite time orbit would require knowledge of infinitely many digits in all n of the initial conditions $x_i(t_o)$, an amount of knowledge that is unnecessary for solving the Kepler problem.

Globally, the flow is characterized by all possible different solutions and these follow from the application of the group operator to different initial conditions:

$$x_i(\theta) = \psi_i(x(0),\theta) = U(\theta)x_i(0). \tag{7.36}$$

The group operator U, in turn, is generated by the differential operator L, that is, by the vector field V. If $F(x)$ can be expanded in a Taylor series

$$F(x(\theta)) = F(x) + \theta dF(x)/d\theta + \theta^2 d^2 F(x)/d\theta^2/2 + \ldots \tag{7.37}$$

where $x = x(0)$ means set $\theta = 0$ after differentiating $F(x(\theta))$ with respect to θ, then by successive differentiations of $dF/d\theta = LF$ with respect to θ, we can substitute L for $d/d\theta$, re-sum the power series, and obtain $F(x(\theta)) = F(U(\theta)x) = U(\theta)F(x)$, where $U(\theta) = e^{\theta L}$. This also requires that the velocity field is analytic in the x_i, since we need to differentiate $V_i(x)$ infinitely many times in order to expand the exponential in a power series. We have also used the fact that the infinitesimal generator L is invariant under the group that it generates:

$$L = \sum_1^n V_i(x(a)) \frac{\partial}{\partial x_i(a)} = \sum_1^n V_i(x(0)) \frac{\partial}{\partial x_i(0)}, \tag{7.38}$$

*and have also assumed that the power series (7.37) converges for all values of θ and for all
initial conditions.*

Most of the time, the power series has a finite radius of convergence or diverges, but
when the group is compact then the exponentiation of L to obtain U is possible for all
values of θ (the infinite series converges).

In general, *if* there were no convergence problem, *then* every nonlinear system of
self-regulating differential equations defining a flow via an analytic velocity field could
be solved from the initial data $(x_1,\ldots,x_n)$ in the form $x_i(t) = U(t)x_i$, which is defined by
the power series,

$$x_i(t) = \exp\left(t\sum_1^n V_i \partial/\partial x_i\right)x_i = x_i + tV_i(x) + t^2\sum_1^n V_j(x)\frac{\partial V_i}{\partial x_j}/2 + \ldots. \qquad (7.39)$$

This is precisely the approach that solved the problem of integrating the equations of
infinitesimal rotations to construct the transformation equations that describe finite
rotations in section 7.2. Why cannot the same method be used to solve every problem of
differential equations with an analytic right hand side? The problem with this approach
is that *power series typically do not converge for all times for arbitrary solutions of
systems of differential equations*: in fact, radii of convergence are generally limited by
singularities in the complex time plane because (for a flow) there are no singularities on
the real-time axis. If the velocity field components are all analytic at some point in
phase space (can be expanded in Taylor series about that point in phase space
coordinates), then the system of differential equations also has an analytic solution in
the neighborhood of that point but there is no way except by explicit construction of
the power series to find out how large is its radius of convergence. In other words, the
radius of convergence of the velocity field as a function of the x_i does not tell us the
radius of convergence of the solution as a power series in $t - t_o$.

Even if the series-solution's radius of convergence is finite, the motion at large times
usually cannot be discovered from a power series expansion about a point at $t = 0$.
Whenever a power series solution has a finite radius of convergence then one can
proceed by analytic continuation to later times, but even if the power series should
converge for all times, more and more terms in the series would be required in order to
discover the long-time behavior of the solution. All of this is true with a vengence
whenever the motion is chaotic (whenever the system has at least one positive Liapunov
exponent). To summarize the main point, unless the Taylor expansion of the solution in
the time has an infinite radius of convergence, then the time-evolution operator written
in the simple form $U(t) = e^{tL}$ for all values of the time and for all values of the initial
conditions requires analytic continuation.

Finally, whenever the velocity field satisfies a Lipshitz condition (as do the Lorenz
model and the double pendulum, two dynamical systems that are chaotic in parts of
their phase space), then the solution can be constructed in the form of a sequence of
successive approximations, a form that is particularly suitable for computation so long
as one uses a method that controls the errors. This result can be used to prove formally
that the solution is analytic in some neighborhood of a point whenever the velocity field
is analytic at that point, but it provides us with no way of actually computing the radius
of convergence. When the system is autonomous and the velocity field is analytic then
the power series of the solution in $t - t_o$ is generated by Picard's method. When, in
addition, the system defines a flow, then there are no spontaneous singularities, but we

still cannot say what the radius of convergence is because it is determined by singularities in the complex t-plane.

The reason that deterministic chaotic solutions are extremely hard to compute correctly even at moderate times is that they present a problem that is related to the computation of digit strings of irrational numbers: if you make a mistake in the computation of even one digit of $\sqrt{2}$, for example, then the rest of the computed expansion will be completely wrong. The same is true of deterministic chaotic orbits.

Coordinate transformations can be generalized to include multiparameter groups. That generalization was constructed by Lie, leads to his principal theorems, and includes dissipative as well as conservative dynamical systems.

7.7 Lie algebras of vector fields

We now generalize to the case of two or more flows. The reason for this generalization is easy to motivate. Consider three rotations, three different transformations in three different Cartesian planes. Each rotation is itself a flow, and the group parameter, the analog of the time, is just the rotation angle. Now, instead of three separate rotations taken alone, consider (as indeed we shall in chapters 8 and 9) a transformation that is itself some *combination* of the three distinct rotations. What are the properties of *that* collection of transformations, given that they depend upon all *three* parameters simultaneously? To answer this question, we may as well consider the problem of an arbitrary multiparameter collection of transformations that go beyond rotations in chapters 15 and 16. Now, we shall find out what is meant by 'noncommuting flows', or 'incompatible flows'.

Consider r different flows $U_k(\theta_k)$, each with its own canonical parameter θ_k and with vector field $V_k(x) = (V_k^1,\ldots,V_k^m)$, $k = 1,\ldots,r$:

$$V_k^i(x) = \frac{\partial U_k}{\partial \theta_k} x_i \text{ at } \theta_k = 0. \tag{7.40}$$

The corresponding infinitesimal operator for the kth flow is given by

$$L_k = \sum_{i=1}^{n} V_k^i(x)\frac{\partial}{\partial x_i}, \tag{7.41}$$

and the differential equations that describe the flow locally are $dx_i/d\theta_k = V_k^i(x)$. Generally, *two distinct flows are not compatible in the sense that they do not commute*: $U_1(\theta_1)U_2(\theta_2)$ usually differs from $U_2(\theta_2)U_1(\theta_1)$. This is the global picture. In the local (meaning infinitesimal) picture the degree of incompatibility is measured by the non-commutivity of the infinitesimal operators L_1 and L_2: for a function $F(x) = F(x_1,\ldots,x_n)$, we find

$$\frac{\partial^2}{\partial\theta_1\partial\theta_2}\{F(U_1U_2x) - F(U_2U_1x)\} = (L_1,L_2)F(x) \text{ when } \theta_1 = \theta_2 = 0, \tag{7.42}$$

where $(L_2,L_1) = L_2L_1 - L_1L_2$ is called the commutator[4] of L_2 and L_1. The commutator is not, as it might appear, a second order differential operator. Rather,

$$(L_2,L_1) = \sum_{i,k=1}^{n}\left(V_2^i\frac{\partial V_1^k}{\partial x_i} - V_1^i\frac{\partial V_2^k}{\partial x_i}\right)\frac{\partial}{\partial x_k} \tag{7.43}$$

---

[4] The use of the Klammersymbol $(L_i,L_k) = L_iL_k - L_kL_i$ for two differential operators L_i and L_k in mechanics and differential equations goes back to Jacobi.

is also a first order differential operator. So, $L_3 = (L_2, L_1)$ is the infinitesimal operator of a *third* flow whose vector field V_3 is given by the *Poisson bracket* $V_3 = [V_1, V_2]$ of the vector fields V_1 and V_2 of the first two flows:

$$[V_1, V_2]_i = V_3^i = \sum_{k=1}^{n} \left(V_2^k \frac{\partial V_1^i}{\partial x_k} - V_1^k \frac{\partial V_2^i}{\partial x_k} \right). \qquad (7.44a)$$

That is, the infinitesimal generator of the new flow is given by

$$L_3 = \sum_{k=1}^{n} [V_1, V_2]_k \frac{\partial}{\partial x_k} = \sum_{k=1}^{n} V_{3k} \frac{\partial}{\partial x_k}. \qquad (7.45a)$$

Notice that with the Cartesian inner product of phase space we can write these two equations as vector equations in the form

$$[V_1, V_2] = V_2 \cdot \nabla V_1 - V_1 \cdot \nabla V_2 = V_3 \qquad (7.44b)$$

and

$$L_i = V_i \cdot \nabla. \qquad (7.45b)$$

Because the Jacobi identity

$$[V_1, [V_2, V_3]] + [V_2, [V_3, V_1]] + [V_3, [V_1, V_2]] = 0 \qquad (7.46a)$$

is satisfied by the Poisson brackets of the vector fields, as the reader can easily check, it follows that the commutators of the infinitesimal generators also satisfy the Jacobi identity:

$$(L_1, (L_2, L_3)) + (L_2, (L_3, L_1)) + (L_3, (L_1, L_2)) = 0. \qquad (7.46b)$$

Our antisymmetric vector algebra under the Poisson bracket operation (or, equivalently, operator algebra under the commutator operation) generates a linear algebra that is called a Lie algebra. In other words, the vector fields of the flows generate the *same* Lie algebra under the Poisson bracket operation as do the operators L_i under commutation. Whether one discusses Lie algebras in terms of Poisson brackets of vector fields or in terms of commutators of operators is only a matter of choice. As we shall see, the idea of a Lie algebra is one of the most powerful unifying ideas in the theory of dynamical systems.

In general, the very general Lie algebra defined above does not close with any finite number of infinitesimal generators: for two arbitrarily chosen flows V_1 and V_2 the generator L_3 of the third flow is usually not a linear combination of L_1 and L_2. As an example of a Lie algebra that closes with a finite number of commutation rules, consider real rotations in three mutually perpendicular Cartesian planes. The vector fields that define the three flows locally are $V_1 = (z, -y)$, $V_2 = (-x, z)$, and $V_3 = (y, -x)$. It follows from direct calculation that the corresponding infinitesimal generators obey the 'angular momentum' commutation rules

$$(L_i, L_j) = \varepsilon_{ijk} L_k, \qquad (7.47)$$

where we have used summation convention (sum over the repeated index k) and ε_{ijk} is the completely antisymmetric three-index symbol (or third rank tensor): $\varepsilon_{ijj} = 0$ if any two indices coincide, $\varepsilon_{ijk} = 1$ for symmetric permutations of $(i,j,k) = (1,2,3)$, and

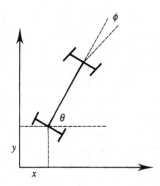

Fig. 7.1 Wheel–axle system (from Burke, 1985).

$\varepsilon_{ijk} = -1$ for antisymmetric permutations of (1,2,3). An arbitrary flow generated by (7.47) has the form

$$U(\eta_1,\eta_2,\eta_3) = U_1(\theta_1)U_2(\theta_2)U_3(\theta_3) = \exp\left(\sum_{i=1}^{3} \eta_i L_i\right), \tag{7.48}$$

and this means that any combination of three generally noncommuting rotations $U_1(\theta_1)U_2(\theta_2)U_3(\theta_3)$ is equivalent to a *single* rotation through a *single* angle η about a *single* axis $\hat{\eta}$. Euler's theorem on rigid body motion follows from this result (sections 8.4 and 9.1).

We started with the assumption that each global flow U_k defines a one dimensional rotation group (rotations in a single Cartesian plane). According to equation (7.48) an arbitrary product of two or more noncommuting rotation operators for different Cartesian planes is still a rotation operator. In other words, the three separate one dimensional flows combine to form a larger group, the group of rotations in three dimensions which is itself a very small subset of the set of all possible global flows in phase space. The fact that we have a three dimensional subgroup is reflected locally by the fact that the three infinitesimal operators L_i form a *complete set* under commutation: the commutator of any two of the L_i and L_j is given by the third operator L_k.

To go further, consider two translations U_1 and U_2 in the Euclidean plane with infinitesimal generators d/dx and d/dy: $U_1(\Delta x) = e^{\Delta x d/dx}$ and $U_2(\Delta y) = e^{\Delta y d/dy}$. These translations commute, so that the sequence $U_2^{-1}U_1^{-1}U_2U_1$ traces out a rectangle, a closed curve in the (x,y)-plane. We think of a sequence of rotations as a path traced out on a sphere: imagine rotating a single vector that intersects the sphere's surface; the path is just the curve traced by the point of intersection. However, the sequence $U_2^{-1}U_1^{-1}U_2U_1$, with $U_i = e^{\theta_i L_i}$, does not yield a closed path because L_1 and L_2 do not commute. In fact, $U_2^{-1}(\delta\theta_2)U_1^{-1}(\delta\theta_1)U_2(\delta\theta_2)U_1(\delta\theta_1) \approx 1 + \delta\theta_1\delta\theta_2(L_2,L_1)$ for a sequence of four infinitesimal transformations, which explains why the path cannot close on itself. We can also look at noncommuting transformations in the Euclidean plane. They can be very useful: you cannot parallel park a car without them (the following example is taken from Burke, 1985), although with a car where the rear wheels and front wheels both rotate together through 90°, parallel parking would be a trivial problem equivalent to only two (commuting) perpendicular translations $e^{\Delta x d/dx}$ and $e^{\Delta y d/dy}$.

Imagine instead a typical car with a rigid rear axle and with front-wheel steering (see figure 7.1). A straight line translation of the car is generated by

$L_T = \cos\theta\partial/\partial x + \sin\theta\partial/\partial y$ where the relevant group parameter is r, the position vector of any point P in the car. The steering is represented by the operator $L_{St} = \partial/\partial\phi$. A general motion of the car requires both steering and forward-driving and is given by a combination of a translation and a rotation, $L_D = L_T + (\tan\phi/L)\partial/\partial\theta$. This corresponds to figure 7.1, where L is the distance from rear to front axle. Note that $(L_{St},L_D) = (L\cos^2\phi)^{-1}\partial/\partial\theta = L_R$ is a rotation operator. The car cannot rotate through a large angle θ in a small space in a single motion (in a single finite transformation) but can accomplish the same result by means of the sequence of finite transformations $\dots U_{St}^{-1}U_D^{-1}U_{St}U_D$, even though (one might also say because) L_{St} and L_D do not commute with each other. Note next that $(L_D,L_R) = (L\cos^2\theta)^{-1}(\sin\theta\partial/\partial x - \cos\theta\partial/\partial y) = L_{Sl}$ and corresponds to the operation of a slide perpendicular to the car itself because the unit vector $(\cos\theta,\sin\theta)$ that defines L_T is perpendicular to the one $(\sin\theta, -\cos\theta)$ that defines L_{Sl}. Although a slide is not directly possible without breaking traction, a result that is geometrically equivalent to a slide can be effected by the repeated performance of the sequence of transformations $U_R^{-1}U_D^{-1}U_RU_D$, for example $\dots U_R^{-1}U_D^{-1}U_RU_DU_R^{-1}U_D^{-1}U_RU_DU_R^{-1}\ U_D^{-1}U_RU_D$, which corresponds to a possible parallel parking sequence. Parallel parking is possible but requires intricate manipulations because of the noncommuting nature of steer and drive. To complete the story, $(L_R,L_{Sl}) = (L^2\cos^3\phi)^{-1}L_T$, where L_T is a linear combination of L_D and L_R, and $(L_{St},L_R) = 2\sin\phi L_R$, so that our six operators close under commutation to generate a Lie algebra (an algebra of driving).

We ask next: what does the Lie algebra look like in the general case of an r parameter transformation group? Here, every product $U_k(\theta_k)U_l(\theta_l)$ of two or more (generally noncommuting) flows is again a flow that belongs to the group. In particular, we want to state the condition under which the infinitesimal transformations that define the Lie algebra can be integrated independently of path to generate a global Lie group (this is *not* integrability in the sense of a flow, but is only integrability in the sense that the transformation functions should exist independently of path). Each group element is an element of a one-parameter subgroup, but these r subgroups can combine to form a larger group that is a multiparameter Lie group. The conditions for a set of transformations

$$x^i(\theta) = \psi^i(x^1,\dots,x^n;\theta^1,\dots,\theta^r), \tag{7.49}$$

depending continuously and differentiably upon r parameters $\theta = (\theta^1,\dots,\theta^r)$ to form a Lie group are: (i) that the group combination law $\psi(\psi(x,\theta),\theta') = \psi(x,\theta'')$ is possible because there is a combination law for the parameters

$$\theta''^\sigma = \phi^\sigma(\theta^1,\dots,\theta^r;\theta'^1,\dots,\theta'^r), \tag{7.50}$$

(ii) there is an identity, $x = \psi(x,\theta_o)$, where, without loss of generality, we can take $\theta_o = 0$; and (iii) each transformation has a unique inverse $x(\theta) = \psi(x(\theta),\phi)$.

Lie's idea was to study the flows locally in the tangent space near the identity, which is determined by the n vector fields with components V_σ^i:

$$dx^i = V_\sigma^i\delta\theta^\sigma. \tag{7.51a}$$

Although in phase space it makes no difference, if we wish to consider transformations in Lagrangian configuration space (as we do in chapters 10 and 12) then we must pay attention to whether a given vector is covariant (lower indices) or contravariant (upper indices). For this purpose a certain tensor g_{kl} must be introduced. The method of construction of the matrix g along with the method of construction of covariant and contravariant components of vectors is displayed in section 10.3. In configuration space, if the frame is obtained from an inertial frame by a time-independent transformation, then g is the metric tensor that appears in the kinetic energy: $ds^2 = 2T (dt)^2 = g_{ij}dq^i dq^j$ is the infinitesimal distance squared between two points in configuration space. If configuration space is Euclidean and Cartesian coordinates are used then $g_{ij} = \delta_{ij}$ and (as is always the case in phase space) there is no distinction between covariant and contravariant components of vectors.

The group parameter combination law yields

$$d\theta^\sigma = v_\rho^\sigma \delta\theta^\rho. \tag{7.52}$$

Consider the matrix v with entries v_ρ^σ, and let λ denote its inverse; both v and λ depend upon $\theta = (\theta^1, \dots, \theta^r)$. Then

$$\delta\theta^\rho = \lambda_\sigma^\rho d\theta^\sigma, \tag{7.53}$$

and so it follows that

$$dx^i = V_\sigma^i \lambda_\tau^\sigma(\theta) d\theta^\tau. \tag{7.54}$$

The transformation that is induced in any scalar function $F(x) = F(x_1, \dots, x_n)$ is therefore given by writing

$$dF = \delta\theta^\sigma X_\sigma F, \tag{7.55}$$

where

$$X_\sigma = V_\sigma^i(x) \frac{\partial}{\partial x^i} \tag{7.56}$$

is the infinitesimal generator that is determined by the velocity field whose components are given by $V_\sigma^i(x)$. We are now prepared to turn to Lie's integrability condition for the existence of the global transformations (7.49).

The condition that the transformations exist is a path-independence condition that is expressed by the integrability conditions

$$\frac{\partial^2 \psi^i}{\partial\theta^s \partial\theta^t} = \frac{\partial^2 \psi^i}{\partial\theta^t \partial\theta^s}, \tag{7.57}$$

which must be satisfied for all i and for all s,t. These conditions guarantee that the transformations (7.49) exist as point functions and yield, after a little manipulation, that

$$[V_\kappa^i(x), V_\gamma^i(x)] = \left(\frac{\partial\lambda_\rho^\tau}{\partial\theta^\sigma} - \frac{\partial\lambda_\sigma^\tau}{\partial\theta^\rho}\right) v_\kappa^\rho v_\gamma^\sigma V_\tau^i(x), \tag{7.58}$$

which is the closure relation for the Lie algebra expressed in terms of Poisson brackets of the vector fields. If we write

$$c_{\kappa\gamma}^\tau = \left(\frac{\partial\lambda_\rho^\tau}{\partial\theta^\sigma} - \frac{\partial\lambda_\sigma^\tau}{\partial\theta^\rho}\right) v_\kappa^\rho v_\gamma^\sigma, \tag{7.59}$$

then the quantities $c_{\kappa\gamma}^\tau$ are independent of θ, because there is no explicit θ-dependence on the left hand side of the Poisson bracket relation above (note that the ϕ-dependent coefficients in the driving-algebra above do not satisfy the global integrability condition). The constants $c_{\kappa\gamma}^\tau$ are called the structure constants of the Lie algebra. The closure relation

$$[V_\kappa^i(x), V_\gamma^i(x)] = c_{\kappa\gamma}^\tau V_\tau^i(x), \qquad (7.60)$$

for the Lie algebra means simply that the Poisson bracket of any two vectors in the tangent space is itself a linear combination of the r linearly independent vector fields that define the Lie algebra. The corresponding infinitesimal generators satisfy the commutation rules $(X_\sigma, X_\gamma) = c_{\sigma\gamma}^\tau X_\tau$. The vector fields of the Lie algebra provide r linearly independent (but not necessarily orthogonal) basis vectors in the tangent space near the group identity, in terms of which any other vector in the tangent space can be expanded.

If the $c_{\sigma\gamma}^\tau$ are constants, then the partial differential equations

$$\frac{\partial \psi^i}{\partial \theta^\rho} = V_\sigma^i \lambda_\rho^\sigma \qquad (7.51b)$$

are integrable in the sense that they define *global* coordinate transformations (integrability of the first kind). Otherwise, one has a Lie algebra that does not generate a global coordinate transformation.

Poisson brackets will be used in chapter 15 to formulate the connection between symmetry, invariance, and conservation laws in Hamiltonian mechanics. We know that the Hamiltonian is the stream function of a vector field that generates the flow that corresponds to some mechanical system. We shall discover that the conserved quantities in a Hamiltonian system are stream functions that define other flows (representing other possible mechanical systems), and that these flows have vanishing Poisson brackets with each other and with the Hamiltonian (section 15.2). Rotations on a sphere do not commute and will be used in chapter 9 to study rigid body motions. We show in chapter 16 that operators representing a complete set of commuting translations in phase space can be used to construct the solution of any integrable canonical Hamiltonian system. There, the compatibility condition is put to work.

Exercises

1. Show by direct calculation that the simple harmonic oscillator solution $x_i = \psi_i(x_{1o}, x_{2o}, t - t_o)$ satisfies the closure and associativity conditions (7.6) and (7.7).

2. Observe that the velocity u in the Galilean transformation $x' = x - ut$ is a canonical parameter (why?). Show that $x' = U(u)x$ where

    $$U(u) = e^{-uL} \text{ with } L = -t\partial/\partial x.$$

3. For a Newtonian two-body problem, construct the infinitesimal generator of Galilean transformations and use it to construct the functional form of invariants. Identify the basic physically significant invariant.

4. Show directly by substitution that $f(x,y) = x^2y + y^3$ is not a rotationally invariant scalar function.

5. For the Lorentz transformations

    $$x' = \gamma(x - ut)$$
    $$t' = \gamma(t - ux/c^2),$$

    with $\gamma = (1 - u^2/c^2)^{-\frac{1}{2}}$,

(a) Derive the group parameter combination law (with $\beta = u/c$)

$$\beta_{13} = \frac{\beta_{12} + \beta_{23}}{1 + \beta_{12}\beta_{23}}.$$

(b) Note that β is not a canonical parameter. Show that $\theta = \frac{1}{2}\ln[(1 + \beta)/(1 - \beta)]$ is canonical, and show that this is a 'rotation angle' in the pseudo-Cartesian (x,ct)-plane. (Hint: write $\cos\phi = \gamma/(1 - \beta^2)^{\frac{1}{2}}$ and show that $\theta = i\phi$.)

(c) Show that $L = -(ct\partial/\partial x) - (x\partial/\partial ct)$ follows from the transformations. Show that the transformations therefore can be rewritten as

$$x' = U(\phi)x, \quad ct' = U(\phi)ct,$$

with $U(\phi) = e^{i\phi L}$.

(d) Show that invariant scalars, $LF(x,ct) = 0$, are of the form $F(x^2 - c^2t^2)$.

(e) Show that the wave operator

$$0 = \frac{1}{c^2}\frac{\partial^2}{\partial t^2} - \frac{\partial^2}{\partial x^2}$$

is Lorentz invariant.

6. Derive equations (7.42), (7.43), and (7.44a).

7. Prove mathematically that parallel parking is possible for a car that tows a boat-trailer. (This is easier mathematically than in reality!)

8. Show that the operators L_1, L_2, and L_3, where $L_k = V_k \cdot \nabla$ is defined by the velocity field $V_k = (V_k^1(x_1,\dots,x_n),\dots,V_k^n(x_1,\dots,x_n))$, satisfy the Jacobi identity (7.46b). Show also that the velocity fields satisfy (7.46a).

9. Show that the integrability condition (7.57) yields the closure relation (7.60) for a Lie algebra. Show also that $(L_i,L_j) = c_{ij}^k L_k$ with the same structure constants c_{ij}^k.

10. Derive equation (7.58).

8

Linear transformations, rotations, and rotating frames

8.1 Overview

In this chapter we return to transformations among Cartesian frames of reference. In the first chapter we studied physics in inertial and linearly accelerated frames. Here, we formulate physics in rotating frames from both the Newtonian and Lagrangian standpoints.

Once one separates the translational motion of the center of mass then rigid body theory for a single solid body is the theory of rigid rotations in three dimensions. The different coordinate systems needed for describing rigid body motions are the different possible parameterizations of the rotation group. The same tools are used to describe transformations among rotating and nonrotating frames in particle mechanics, and so we begin with the necessary mathematics. We introduce rotating frames and the motions of bodies relative to those frames in this chapter and go on to rigid body motions in the next. Along the way, we will need to diagonalize matrices and so we include that topic here as well.

In discussing physics in rotating frames, we are prepared to proceed in one of two ways: by a direct term by term transformation of Newton's laws to rotating frames, or by a direct application of Lagrange's equations, which are covariant. We follow both paths in this chapter because it is best to understand physics from as many different angles as possible. Before using Lagrange's equations, we also derive a formulation of Newton's equations that is covariant with respect to transformations to and among rotating Cartesian frames. The resulting equation is written down in Gibbs' notation in every standard text, but usually without informing the reader that the result can be understood as a covariant equation by identifying the vectors that transform in covariant fashion. By using the same matrix/column vector notation in Cartesian coordinates to discuss Newton's law that we use to introduce and discuss the rotation group, we are aided in our identification of covariant objects.

8.2 Similarity transformations

We begin with a real Euclidean vector space of n dimensions and a tranformation from one Cartesian frame to another: $x' = Ux$ where U is an n by n matrix and x and x' are column vectors[1]. Consider any relation of the form $y = Bx$ between any two vectors in the frame. This relationship is a covariant statement so long as we restrict our discussion to coordinate systems that can be reached by linear transformation. Since $y' = Uy$, it follows that $y' = UBx = UBU^{-1}x'$, so that $y' = B'x'$ with $B' = UBU^{-1}$. The matrix U is called a similarity transformation because $y = Bx$ and $y' = B'x'$ have

---

[1] The use of bold letters to signify abstract vectors and operators, as opposed to coordinate representations of vectors and matrices is defined in section 8.3.

the same vectorial form (covariance). Note that B simply transforms like a mixed tensor B_k^i of the second rank. We restrict now to those transformations that leave the entire Euclidean space similar to itself, which means that we look only at transformations that preserve both lengths of vectors and angles between vectors. In particular, sizes and shapes of rigid bodies are left invariant under these similarity transformations. This can be summarized by saying that we use the transformation group that leaves the Euclidean scalar product *invariant*.

The Euclidean scalar product of two vectors is given by

$$\tilde{x}y = x_i y_i = |x||y|\cos\theta \qquad (8.1)$$

where $|x| = (x_i x_i)^{1/2}$ is the length of x, and θ is the angle between x and y. These transformations are called similarity transformations because the shape of a rigid body is not distorted in any way by a transformation that leaves all angles and distances within the body unchanged (invariant). In other words, the rigid body remains similar to itself under such a transformation. Since

$$\widetilde{Ux} = \tilde{x}\tilde{U} \qquad (8.2)$$

where $\tilde{x}$ is a row vector and $\tilde{U}_{ik} = U_{ki}$, it follows that

$$\tilde{x}'y' = \tilde{x}\tilde{U}Uy = \tilde{x}y \qquad (8.3)$$

is not merely a scalar but is invariant (is the *same* function of the x_i' and y_j' as it is of the x_i and y_j) whenever $\tilde{U} = U^{-1}$, that is, whenever the matrix U is orthogonal. In other words, orthogonal matrices leave rigid bodies similar to themselves because they respect the lengths and angles that are the building blocks of Euclidean geometry.

Notice also that orthogonal transformations form a group: if U_1 and U_2 are orthogonal, then $U_1 U_2$ is also orthogonal. In technical language, the restriction to real vectors and orthogonal transformations means that we restrict ourselves to single-valued representations of rotations[2]. Since the theory of rigid body motion is mathematically equivalent to the theory of time-dependent rotation matrices (as we discuss in chapters 9 and 11), it is useful to derive some facts about the rotation group before proceeding to the formulation and study of rigid body motions.

Note also that when U is a linear transformation that depends upon the time, then $dx'/dt = Udx/dt + (dU/dt)x$ is not a covariant statement, even though $x' = Ux$ is covariant. Before discussing this in the context of Newton's second law, we first develop some useful results from the theory of linear transformations.

8.3 Linear transformations and eigenvalue problems

The relation $y = Bx$ between two vectors x and y can be inverted to yield $x = B^{-1}y$ so long as $\det B \neq 0$, because $B^{-1} \propto 1/\det B$. If B can be diagonalized, then this requires that all of its eigenvalues are nonvanishing because $\det B$ is proportional to the product of its eigenvalues: if $\Lambda = U^{-1}BU$ is diagonal, then $\det \Lambda = \det B$ because $\det A \det B = \det AB$. In particular, orthogonal matrices obey $\det U = +1$ or -1 because $\det U^{-1}U = \det I = 1$, where I is the identity matrix, but also $\det U^{-1}U = \det U^{-1} \det U = (\det U)^2$, so the result follows. We restrict our discussion to proper

---

[2] Two-valued complex rotations were introduced to describe rigid body motions by Cayley and Klein and arise naturally in the quantum mechanical theory of half-integral spin (see Klein and Sommerfeld, 1897, and also Goldstein, 1980).

rotations and $\det U = 1$, since $\det U = -1$ means that there is a reflection of a coordinate axis as well as a rotation.

The rules of matrix multiplication allow us to treat a vector equation $y = Bx$ as a coordinate-free statement $\mathbf{y} = \mathbf{B}\mathbf{x}$ under similarity transformations. In other words, one can do vector and operator algebra abstractly without ever writing down a representation for the vectors and matrices in a specific coordinate system (one eventually needs a specific coordinate system in order to discuss analytic properties of solutions of the differential equations). This leads to the idea of the *same* vector x in two different coordinate systems. Therefore, we must use a notation that distinguishes the abstract vectors and operators from their *representations* in a particular coordinate system as column (or row) vectors and matrices. In this chapter we will use bold type to specify abstract vectors $\mathbf{x},\mathbf{y},\ldots$ and abstract operators $\mathbf{U},\mathbf{B},\ldots$, and normal type $x,y,\ldots$ and $U,B,\ldots$ to specify their matrix representations in a particular coordinate system. This notation distinction is in the interest of clarity because one can transform the vectors and leave the frame fixed, or one can transform the frame (transform the coordinates) and leave the vectors fixed. The results are applied to the derivation of a set of generalized coordinates called Euler angles in chapter 9.

Once we chose a Cartesian frame, we need a set of basis vectors, and it is useful to pick an orthonormal basis set $e_1,\ldots,e_n$ where $\tilde{e}_i e_k = \delta_{ik}$. In all that follows, we will use x and y to denote the components of vectors, and e and f to denote the basis vectors. Since

$$x = \begin{pmatrix} x_1 \\ \vdots \\ x_n \end{pmatrix}, \tag{8.4}$$

an obvious (but not the only) choice of orthonormal basis is given by writing $x = x_i e_i$ where

$$e_1 = \begin{pmatrix} 1 \\ 0 \\ 0 \\ \vdots \\ 0 \end{pmatrix}, e_2 = \begin{pmatrix} 0 \\ 1 \\ 0 \\ \vdots \\ 0 \end{pmatrix}, \ldots, e_n = \begin{pmatrix} 0 \\ 0 \\ 0 \\ \vdots \\ 1 \end{pmatrix}. \tag{8.5}$$

With a given orthonormal basis (not necessarily the one just written above), the components of $\mathbf{x}$ are given in that system by $x_i = (e_i, x)$ where we shall also use (x,y) to denote the scalar product $\tilde{x}y$ of any two vectors x and y. The operator $\mathbf{B}$ with respect to the same basis has a matrix representation B with matrix elements that obey $B_{ik} = (e_i, Be_k)$, by inspection[3]. We note here only that matrices can be used to represent *linear* operators, operators that satisfy $\mathbf{B}(\mathbf{x} + \mathbf{y}) = \mathbf{B}\mathbf{x} + \mathbf{B}\mathbf{y}$.

Next, we consider the eigenvalue problem because we need certain of its results in much that follows. The eigenvalue problem for the matrix B is defined by $Bx = \lambda x$ where λ is a number. In other words, an eigenvector x defines a direction that is left invariant by B. Matrices that can always be diagonalized by a transformation of coordinates, $\Lambda = U^{-1}BU$ where $\Lambda_{ij} = \lambda_i \delta_{ij}$, are symmetric matices ($B_{ij} = B_{ji}$), orthog-

---

[3] This is the same as writing $\hat{e}_i \cdot \overleftrightarrow{B} \cdot \hat{e}_k$ in Gibbs' (or 'elementary physics') notation, where $\overleftrightarrow{B}$ is a dyadic (rank two Cartesian tensor), and also the same as $\langle i | B | k \rangle$ in Dirac's 'quantum mechanics' notation, where B is any linear operator.

onal matrices ($O_{ij}^{-1} = O_{ji}$), unitary matrices ($U_{ij}^{-1} = U_{ji}$), and normal matrices. We concern ourselves here with symmetric (real self-adjoint, or real Hermitean[4]) and orthogonal matrices. Note that e^{iA} is orthogonal if A is self-adjoint ($A = A^+$), whereas iA is self-adjoint if A is real and antisymmetric ($A^* = A$ and $A^+ = -A$).

We begin by establishing two useful results. First, the eigenvalues of a symmetric matrix are real, and here we need a unitary scalar product $x^+ y = (x^*,y) = \Sigma x_i^* y_i$, where x^+ is the complex conjugate of the transpose of x (the complex conjugate of a row vector) because the eigenvectors are generally complex: with $Ax_i = \lambda_i x_i$ and $Ax_k = \lambda_k x_k$, $(x_k,Ax_i) = \lambda_i(x_k,x_i)$, but also $\lambda_k^* x_k^+ = x_k^+ A^+ = x_k^+ A$ since $A^+ = A$ (A is real and symmetric, and A^+ is the complex conjugate of the transpose of A). Therefore, we can write $(x_k,Ax_i) = \lambda_k^*(x_k,x_i)$ so that if we set $i = k$, it follows by comparison that $\lambda_i^* = \lambda_i$. Second, eigenvectors belonging to distinct eigenvalues are orthogonal: since $\lambda_i(x_k,x_i) = \lambda_k(x_k,x_i)$, it follows that $(x_k,x_i) = 0$ unless $\lambda_i = \lambda_k$. If there is degeneracy ($\lambda_i = \lambda_k$ but x_i and x_k are linearly independent) then it is possible to use the linearly independent pair to construct two new eigenvectors x_i' and x_k' which are perpendicular. Henceforth, we shall assume that if degeneracy occurs then this orthogonalization has been performed, for then we have the most important result: the eigenvectors of a symmetric matrix are a complete orthogonal set (there are n perpendicular eigenvectors if A is n by n), so that we can use the eigenvectors as a basis. We assume also that all eigenvectors have been normalized to unity, so that $e_i^+ e_k = \delta_{ik}$ where $Ae_k = \lambda_k e_k$.

Notice next that the direct product $e_i e_i^+$ is an n by n matrix, where the direct product is defined by the rule

$$xy^+ = \binom{x_1}{x_2}(y_1^* \quad y_2^*) = \begin{pmatrix} x_1 y_1^* & x_1 y_2^* \\ x_2 y_1^* & x_2 y_1^* \end{pmatrix}, \tag{8.6}$$

but also notice that we have a projection operator: with $x = \Sigma x_i e_i$, then $e_k e_k^+ x = x_k e_k$. Therefore, $\Sigma e_k e_k^+ = I$, the identity matrix. This is called the resolution of the identity, and is also called the spectral decomposition of the identity (all n eigenvalues of I are equal to unity). It follows that we have an expansion for the matrix A as well: $A = AI = A\Sigma e_k e_k^+ = \Sigma \lambda_k e_k e_k^+$, which is called the spectral resolution of the operator A in terms of its eigenvalues and eigenvectors. For any analytic function of the matrix A, it follows that $f(A)$ has the spectral resolution $f(A) = \Sigma f(\lambda_i)e_i e_i^+$.

If we transform coordinates and leave the vectors fixed in the linear space, then we must transform both basis vectors and components simultaneously. With $x = \Sigma x_i e_i = \Sigma x_k' e_k'$, the transformation matrix U in $x' = Ux$ is easy to construct: with the resolution of the identity given by $\Sigma e_k' e_k'^+ = I$, by the basis in the primed frame, we obtain

$$x = \Sigma x_k e_k = \Sigma x_k \Sigma e_i' e_i'^+ e_k = \Sigma x_i' e_i' \tag{8.7}$$

where $x_i' = \Sigma U_{ik} x_k$ with $U_{ik} = e_i'^+ e_k$. In other words, the transformation matrix has matrix elements given by scalar products of basis vectors in the two different frames. Likewise, it is easy to discover the transformation rule for the basis vectors:

$$e_i' = \Sigma e_k e_k^+ e_i' = \Sigma U_{ik}^* e_k. \tag{8.8}$$

The resolution of the identity will be used in the next chapter to show how to expand

---

[4] In an infinite dimensional space, as occurs in functional analysis and quantum mechanics, because of convergence of infinite series, the terms symmetric, self-adjoint, and Hermitean do not have identical meaning. In this case, what one usually wants is self-adjointness, which is a rather stringent condition.

mutually perpendicular angular velocities in terms of Euler's angles. The diagonaliz-ation of the moment of inertia tensor, which is real symmetric, will be used to introduce the 'principal moments of inertia' in a frame fixed in a rotating body.

8.4 Rotations form a Lie group

We know how the Cartesian coordinates x_1 and x_2 transform under a rotation clockwise through an angle θ_3 about the x_3-axis:

$$\left.\begin{array}{l} x'_1 = x_1 \cos \theta_3 + x_2 \sin \theta_3 \\ x'_2 = - x_1 \sin \theta_3 + x_2 \cos \theta_3 \\ x'_3 = x_3. \end{array}\right\} \tag{8.9}$$

Here, vectors are left fixed in space and the coordinate system is rotated. If we use matrices and column vectors, we can write $x' = R_3 x$ where

$$R_3(\theta_3) = \begin{pmatrix} \cos \theta_3 & \sin \theta_3 & 0 \\ - \sin \theta_3 & \cos \theta_3 & 0 \\ 0 & 0 & 1 \end{pmatrix}. \tag{8.10}$$

It is equally easy to write down matrices that represent clockwise rotations of frames (with vectors fixed) or counterclockwise rotations of vectors (with frames fixed) in the (x_1,x_3)- and the (x_2,x_3)-planes:

$$R_2(\theta_2) = \begin{pmatrix} \cos \theta_2 & 0 & - \sin \theta_2 \\ 0 & 1 & 0 \\ \sin \theta_2 & 0 & \cos \theta_2 \end{pmatrix} \tag{8.11}$$

$$R_1(\theta_1) = \begin{pmatrix} 1 & 0 & 0 \\ 0 & \cos \theta_1 & \sin \theta_1 \\ 0 & - \sin \theta_1 & \cos \theta_1 \end{pmatrix}. \tag{8.12}$$

Following Lie, we look at the transformations near the identity and by considering infinitesimal rotations through angles $\delta\theta_i$:

$$R_3 = \begin{pmatrix} \cos \delta\theta_3 & \sin \delta\theta_3 & 0 \\ - \sin \delta\theta_3 & \cos \delta\theta_3 & 0 \\ 0 & 0 & 1 \end{pmatrix} \approx \begin{pmatrix} 1 & \delta\theta_3 & 0 \\ - \delta\theta_3 & 1 & 0 \\ 0 & 0 & 1 \end{pmatrix} = I + \delta\theta_3 M_3 \tag{8.13}$$

where I is the identity matrix and

$$M_3 = \begin{pmatrix} 0 & 1 & 0 \\ - 1 & 0 & 0 \\ 0 & 0 & 0 \end{pmatrix} \tag{8.14}$$

is the matrix representation of the infinitesimal generator of rotations about the x_3-axis. See also chapter 7 for infinitesimal generators as differential operators, where

$$M_3 = x_1 \frac{\partial}{\partial x_2} - x_2 \frac{\partial}{\partial x_1} \tag{8.15}$$

is the corresponding infinitesimal generator for corresponding transformations of scalar functions[5]. Likewise, we obtain easily that the generators or rotations about the x_1- and x_2-axes are given by

$$M_1 = \begin{pmatrix} 0 & 0 & 0 \\ 0 & 0 & 1 \\ 0 & -1 & 0 \end{pmatrix} \qquad (8.16)$$

and

$$M_2 = \begin{pmatrix} 0 & 0 & -1 \\ 0 & 0 & 0 \\ 1 & 0 & 0 \end{pmatrix}. \qquad (8.17)$$

It is easy to show by direct calculation that we have a Lie algebra with commutation rules $(M_i, M_j) = \varepsilon_{ijk} M_k$ where ε_{ijk} is the totally antisymmetric three-index tensor ($\varepsilon_{ijk} = 1$ for cyclic permutations of indices (1,2,3), equals -1 for antisymmetric permutations, and is equal to 0 if any two indices coincide). The formation of a Lie algebra (closure of the three infinitesimal generators under commutation) is precisely the condition that orthogonal transformations depending continuously upon the three holonomic parameters $(\theta_1, \theta_2, \theta_3)$ form a Lie group. For the benefit of the reader who has postponed reading chapter 7 until later, we shall give an independent proof that the product of two or more rotation matrices is again a rotation matrix.

The most general rotation in three dimensions can be represented as a product of three successive rotations in three mutually perpendicular planes, $R = R_1(\theta_1) R_2(\theta_2) R_3(\theta_3)$. Each of the matrices R_i is orthogonal, so R is also orthogonal: $\tilde{R} = \tilde{R}_3 \tilde{R}_2 \tilde{R}_1 = R^{-1}$. The main point is that the matrix R is *also* a rotation, which means that the product of three successive rotations about three mutually perpendicular axes is equivalent to a *single* rotation through an angle η about a *single* rotation axis $\hat{n}$. This is equivalent to the statement that rotations form a three-parameter continuous group, and it is sufficient to prove that the product of any two rotations about two different (and generally not perpendicular or parallel) axes is again a rotation.

Consider two arbitrary orthogonal matrices A and B and their product $R = AB$. This product generally represents two noncommuting rotations about two different axes. Since $\det R = \det A \det B$ and $\det A = \det B = 1$ (we restrict to proper rotations with no inversions of coordinate axes), it follows that $\det R = 1$. Since $\det R = \lambda_1 + \lambda_2 + \lambda_3$ where the λ_i are the eigenvalues of R, $\det R = 1$ yields $\lambda_1 = 1$ and $\lambda_2 = \lambda_3^* = e^{i\Phi}$ because complex eigenvalues can show up only as complex conjugate pairs. This means that the eigenvector corresponding to $\lambda_1 = 1$ is an invariant axis for R, which means that R is a rotation about that axis. If we transform to a coordinate system where the invariant axis is the x_3-axis, then our matrix has the form

$$R = \begin{pmatrix} \cos\eta & \sin\eta & 0 \\ -\sin\eta & \cos\eta & 0 \\ 0 & 0 & 1 \end{pmatrix}, \qquad (8.18)$$

and it follows that $\eta = \Phi$ because $\operatorname{Tr} R = \lambda_1 + \lambda_2 + \lambda_3 = 1 + 2\cos\Phi$.

[5] One can represent continuous groups linearly by either matrices or differential operators. This is reflected in quantum mechanics by doing the dynamics in either the Schrödinger representation (differential operators) or the Heisenberg representation (matrices).

It now follows that the product of two or more orthogonal matrices is also a rotation matrix: for example, if $R = R_1 R_2 R_3$, then $R' = R_2 R_3$ is a rotation, and $R = R_1 R'$ is also a rotation. The most general rotation in three dimensions can be written as $R = R_1 R_2 R_3$, where the R_i represent rotations through angles θ_1, θ_2, and θ_3 in the three mutually perpendicular Euclidean planes, but we can obtain another form for R as follows: consider an infinitesimal rotation

$$R \approx (I + \delta\theta_1 M_1)(I + \delta\theta_2 M_2)(I + \delta\theta_3 M_3) \approx I + \delta\vec{\theta} \cdot \vec{M}. \qquad (8.19)$$

Using the idea of rotations as a group we are able to construct an arbitrary rotation matrix R, which can be thought of as the result of any number of products of rotations about different axes and through different angles, as a *single* rotation through a definite angle η about a *single* axis $\hat{n}$. Following Lie, we can build up any finite transformation R by iterating the infinitesimal transformation $(I + \eta\hat{n} \cdot \vec{M}/N)$ N times,

$$R \approx (I + \eta\hat{n} \cdot \vec{M}/N)^N, \qquad (8.20)$$

where $\delta\vec{\theta} \cdot \hat{n} = \eta$, so that

$$R = e^{\eta\hat{n} \cdot \vec{M}} \qquad (8.21)$$

follows as N goes to infinity. In other words, $R = R_1(\theta_1)R_2(\theta_2)R_3(\theta_3) = e^{\vec{\eta} \cdot \vec{M}}$, but it is important to realize that the finite angles θ_i are *not* the same as the components $\eta_i = \eta n_i$ of the single rotation angle η in the exponent of R: rotations do not commute (the infinitesimal generators do not commute), and when two matrices e^A and e^B do not commute then $e^A e^B$ is not equal to e^{A+B}. In other words, although θ_i is a canonical parameter (and holonomic coordinate) for rotations within the plane perpendicular to the axis x_i, we get entirely *different* parameters (η_1, η_2, η_3) for rotations through the *same* angles $(\theta_1, \theta_2, \theta_3)$ that are performed in different sequences. If we express the relationship by a differential form $d\eta_i = \xi_{i\sigma}d\theta_\sigma$, then this differential form is nonintegrable because the three generators M_i do not commute with each other. To say that different paths in θ_i-space give different results for the integrals of the three differential forms for the η_i is the same as saying that the parameters η_i are not holonomic variables (see chapter 12). Therefore, there is no function ('point function') of the form $\eta_i = f_i(\theta_1, \theta_2, \theta_3)$. Another way to say it is that any differential expression of the form $d\eta_i = \tau_{i\sigma}d\theta_\sigma$ will be nonexact. A rotation is a path on a Riemannian surface (a sphere), and translations on Riemannian surfaces do not commute with each other because of the intrinsic curvature of the surface.

Having proven that the matrix describing the most general result of any number of successive rotations can be written in the form $R = e^{\vec{\eta} \cdot \vec{M}}$ (M_i is antisymmetric, so R is orthogonal), we want to use this result to describe dynamics from the standpoint of an observer in a rotating frame. First it is useful to do a little tensor analysis as preparation for what follows.

Recall that if we think of a vector geometrically as anything with a magnitude and a direction, then we do not get coordinate-free statements about vectors and relationships among vectors (chapter 1 and chapter 10). Coordinate-free vectors $\mathbf{v}$ are defined as n-component objects whose components transform according to matrix multiplication rules, either as

$$V_i = \frac{\partial q^k}{\partial Q^i} v_k \qquad (8.22a)$$

or as

$$V^i = \frac{\partial Q^i}{\partial q^k} v^k \tag{8.23}$$

under general coordinate transformations $Q^i = f^i(q^1,\dots,q^f,t)$ in configuration space. This allows us to make certain coordinate-free statements.

Here, we narrow our focus and consider, *not* the most general group of coordinate transformations in configuration space, but only the rotation group. The point is that we do not need general covariance under arbitrary transformations in configuration space in order to discuss physics in rotating frames: it is enough, and is more useful, to consider only the linear transformations that leave Eucidean space similar to itself. In that case, if A is a coordinate-free Cartesian vector (which we denote as **A**) then

$$A'_i = R_{ik} A_{ik} \tag{8.22b}$$

where the matrix R is orthogonal. Under these circumstances, the position of a particle (or the center of mass of a body) transforms as a vector because $x' = Rx$ holds for all transformations R that are rigid rotations (we could also include rigid translations). However, if the transformation R takes us to a rotating frame, which means that at least one frequency ω_i is nonvanishing, then the velocity does *not* transform like a coordinate-free vector because

$$\dot{x}' = R\dot{x} + \dot{R}x \tag{8.24}$$

where

$$\dot{R} = R\Omega, \tag{8.25}$$

and where $\Omega = \omega_i M_i = \omega \hat{n} \cdot \vec{M}$ is a totally antisymmetric matrix: $\Omega_{ij} = \varepsilon_{ijk}\omega_k$, or

$$\Omega = \begin{pmatrix} 0 & \omega_3 & -\omega_2 \\ -\omega_3 & 0 & \omega_1 \\ \omega_2 & -\omega_1 & 0 \end{pmatrix}. \tag{8.26}$$

Note also that with $d\eta/dt$ defined by $dR/dt = (d\eta_i/dt)L_iR$, the three angular velocities ω_i are not the same as the velocities $d\eta_i/dt$ but are given by a transformation acting on those velocities, $(d\eta_i/dt)L_i = R\Omega\tilde{R}$. We introduce Ω instead of using the $d\eta_i/dt$ directly because Ω arises naturally in coordinate transformations: consider a vector x' that is fixed in a frame that rotates relative to an inertial frame, $x' = Rx$. Since $dx'/dt = 0$, we obtain $dx/dt = -\Omega x$ as the equation of motion of that vector in the inertial frame. Note also that $x' = Rx$ is a covariant statement whereas $dx'/dt = 0$ is not (the familiar statement of Newton's first law is not a covariant statement). Restriction to the case of rotations about the x_3-axis ($\omega_1 = \omega_2 = 0$) yields $dx_1/dt = -\omega_3 x_2$ and $dx_2/dt = \omega_3 x_1$. This shows that in the general case the velocities ω_i can be interpreted as the components of the rotation velocity vector ω projected onto the axes in the inertial frame. At any time t, the direction of ω tells us the axis of rotation, which is generally not fixed relative to the inertial frame as time goes on (a top spins about a wobbling axis).

That the time derivative of a coordinate-free vector **A** is generally not itself a coordinate-free vector leads us to ask whether we can construct a generalization of the time derivative that transforms like an abstract vector, one that transforms according to the rules of matrix multiplication. The answer is yes, and the construction follows from a general method that was introduced by Eddington (1963) to construct the covariant derivative. Consider the scalar product $(\mathbf{A},\mathbf{B})$ of any two coordinate-free vectors. When we write components in a definite Cartesian frame, then we can ignore upper indices since $g_{ij} = \delta_{ij}$ for Cartesian coordinates in Euclidean space. The Euclidean scalar product, under transformations to rotating frames, obeys the rule $d(\mathbf{A},\mathbf{B})/dt = d(\mathbf{A}',\mathbf{B}')/dt$, or

$$\frac{d}{dt} A_i B_i = \dot{A}_i B_i + A_i \dot{B}_i = \dot{A}'_i B'_i + A'_i \dot{B}'_i = \frac{d}{dt} A'_i B'_i \tag{8.27}$$

(and is also invariant under the rotation group), although the two terms in the middle, taken separately, do not obey $\dot{A}_i B_i = \dot{A}'_i B'_i$ or $A_i \dot{B}_i = A'_i \dot{B}'_i$. Consider next the special case where $\mathbf{A}$ has constant components in an inertial frame: $d\mathbf{A}/dt = 0$, and therefore defines a parallel field of constant vectors. In matrix notation, $A' = RA$ with $\dot{R} = R\Omega$ where $\tilde{\Omega} = -\Omega$. Then with $\Omega' = R\Omega R^{-1}$ we get

$$\frac{d}{dt} \tilde{A}B = \tilde{\dot{A}}B = \widetilde{\Omega'A'}B' + \tilde{A}'\dot{B}' = \tilde{A}'(\dot{B}' - \Omega'B') \tag{8.28}$$

which shows that the vector defined by

$$\frac{DB'}{Dt} = \frac{dB'}{dt} - \Omega'B' \tag{8.29a}$$

transforms by the rule of matrix multiplication and therefore is a coordinate-free vector under transformations to rotating frames (notice that $\Omega' = L_i(d\eta_i/dt)$ where the velocities $d\eta_i/dt$ are nonintegrable unless the rotation axis remains fixed in direction in inertial frames). This means: if $\mathbf{DB}/\mathbf{Dt} = 0$ in one frame then $\mathbf{DB}/\mathbf{Dt} = 0$ in all frames whereas the same is not true of either $d\mathbf{B}/dt$ or $\Omega\mathbf{B}$ taken separately because each term taken separately is coordinate-dependent. The operator $\mathbf{D}/\mathbf{Dt} = \mathrm{Id}/dt - \Omega$ is called the absolute derivative (or the Lie derivative, or the 'fishing derivative', and is the time-translation version of a covariant derivative). The interpretation of the two terms in $\mathbf{D}/\mathbf{Dt}$ is straightforward: d/dt yields the time rate of change of a vector relative to the rotating frame (the 'apparent' change of the vector, like the velocity of a fish relative to the local current), while Ω represents the time rate of change of the rotating frame relative to inertial frames (relative to 'absolute space', or like the flow of the local current relative to the bank where the fishing rod is held). The absolute derivative will be used to rewrite Newton's law in a covariant form (in a coordinate-free form) in the next section.

In order to see explicilty how the Lie derivative works as a covariant derivative we must understand how the rotation generators combine in a sequence of two or more transformations. Consider the transformation $B'' = R_2 B' = R_1 B$, where the vector B is resolved relative to an inertial frame. Then $\dot{R}_1 = R_1 \Omega_1$ and so $\dot{B}' - \Omega' B' = R_1 \dot{B}$ where $\Omega' = R_1 \Omega_1 R_1^{-1}$. In the next transformation, where $B'' = R_2 B'$, we obtain

$$R_2(\dot{B}' - \Omega'B') = R_2 \left[\frac{d}{dt}(\tilde{R}_2 B'') - \Omega'B' \right] = \dot{B}'' - R_2(\Omega_2 + \Omega')\tilde{R}_2 B'', \tag{8.30a}$$

which we can write as

$$R_2(\dot{B}' - \Omega'B') = \dot{B}'' - \Omega''B'' \tag{8.30b}$$

if we make the identification $\Omega'' = R_2(\Omega_2 + \Omega')R_2^{-1}$, which shows how the rotation rate matrices combine in two successive transformations. Let us check to see if this is correct by looking at the direct transformation $B'' = RB$ where $R = R_2R_1$, so that $\dot{B}'' - \Omega''B'$ $= R\dot{B}$. With $\dot{R} = R\Omega = \dot{R}_2R_1 + R_2\dot{R}_1 = R_2\Omega_2R_1 + R_2R_1\Omega_1$, we obtain

$$\Omega'' = R\Omega\tilde{R} = R_2\Omega_2\tilde{R}_2 + R_2\Omega'\tilde{R}_2 = R_2\Omega_2\tilde{R}_2 + R_2R_1\Omega_1\tilde{R}_1\tilde{R}_2, \tag{8.31}$$

which agrees with our conclusion drawn from two successive transformations. Note that Ω'' is a linear combination of the rotation generators L_i'' in the doubled-primed frame, and is therefore an element of the Lie algebra of the rotation group. A method for calculating transformations of generators is developed in the exercises and is used near the end of chapter 12 (if we rotate Cartesian frames rather than Cartesian vectors, then $\Omega' = R^{-1}\Omega R$, which is the form needed in chapter 12).

Before going further, it is useful to show how to write the absolute derivative in Gibbs' vector notation. Since $\Omega_{ij} = \varepsilon_{ijk}\omega_k$, it follows that the absolute derivative in Gibbs' notation is just

$$\frac{D\vec{B}}{Dt} = \frac{d\vec{B}}{dt} + \vec{\omega} \times \vec{B} \tag{8.29b}$$

because

$$(\vec{A} \times \vec{B})_i = \varepsilon_{ijk}A_jB_k. \tag{8.32}$$

8.5 Dynamics in rotating reference frames

If we begin with Newton's law $d\mathbf{p}/dt = \mathbf{F}$ for the motion of a body relative to an inertial frame with Cartesian coordinates x_i, then we can obtain the law of motion relative to a rotating Cartesian frame with coordinates x_i' directly by transformation:

$$Rdp/dt = RF = F' \tag{8.33}$$

where F' is the net force on the body in the rotating system. The point is that any force F that is the gradient of a scalar potential (like gravity) transforms like a coordinate-free vector but the velocity v and acceleration dv/dt do not: the force on the right hand side of Newton's law obeys the rule of matrix multiplication, $F' = RF$, but the left hand side transforms like a coordinate-dependent vector: $Rdv/dt \neq d(Rv)/dt$. Therefore, Newton's law $dp/dt = F$ with $p = mv$ is not a covariant statement. So far as physics goes, that does not matter: we can use Rdp/dt to derive the correct equation for a rotating frame as follows. With $x' = Rx$, $v' = dx'/dt$, and $dR/dt = R\Omega$, we have

$$v' = Rv + R\Omega x \tag{8.34}$$

(x and F transform like vectors under time-dependent rotations, but dx/dt does not) so that, with $a' = dv'/dt$,

$$a' = Ra + 2R\Omega v + R\Omega^2 x + (Rd\Omega/dt)x, \tag{8.35a}$$

which means that

$$a' = F'/m + 2R\Omega v + R\Omega^2 x + (Rd\Omega/dt)x. \tag{8.35b}$$

If the mass m is constant then $dp/dt = ma$, so that

$$Rdp/dt = Rma = dp'/dt - 2mR\Omega v - mR\Omega^2 x - m(Rd\Omega/dt)x. \tag{8.36}$$

Using $v = R^{-1}v' - \Omega x$ then yields

$$Rdp/dt = dp'/dt - 2m\Omega'v' + m\Omega'^2x' - m\dot{\Omega}'x' \qquad (8.37)$$

where $\omega_i'M_i' = \Omega' = R\Omega R^{-1}$ transforms like a second rank tensor because of the cancellation of two terms of opposite sign (one sees by differentiation that the time derivative of a coordinate-free rank two tensor is also a coordinate-free rank two tensor if and only if that tensor commutes with Ω). Note that $(\omega_m'M_m')_{ij} = \Omega_{ij}' = R_{il}R_{kj}\Omega_{lk}$ is a linear combination of the three generators M_i, and the transformed generators M_i' must also obey the closure relation $(M_i', M_j') = \varepsilon_{ijk}M_k'$. We can (and later will) also transform the rotation rates more simply according to $\omega' = R\omega$. Since $\Omega_{ij} = \varepsilon_{ijk}\omega_k$, and ε_{ijk} not only transforms like a third rank tensor but is invariant, $\Omega_{ij}' = \varepsilon_{ijk}\omega_k'$ where the ω_k' are the components of the rotation vector ω projected onto the rotating axes (x_1', x_2', x_3'). Finally, we obtain Newton's description of the motion of a body relative to a rotating coordinate system in the form

$$\dot{p}' - 2m\Omega'v' + m\Omega'^2x' - m\dot{\Omega}'x' = F'. \qquad (8.38)$$

The extra terms on the left hand side represent the effects of rotation relative to absolute Euclidean space: $2\Omega'v'$ is the Coriolis acceleration, Ω'^2x' is the centrifugal acceleration, and $(d\Omega'/dt)x'$ is sometimes called the fluctuation acceleration. Some people write $dp'/dt = F_{eff}'$, with $F_{eff}' = F' + F_{fict}$ where F_{eff}' is supposed to be (but is not) the net force on the body. In this case $F_{fict} = 2m\Omega'v' - m\Omega'^2x' + m(d\Omega'/dt)x'$ is called a fictitious force, because it is not a force: it is merely the result of acceleration relative to inertial frames that can be completely eliminated merely by transforming back to an inertial frame of reference. Forces represent the actions of energy-carrying fields, transform like coordinate-free vectors, and cannot be eliminated by a mere coordinate transformation (cannot be transformed to a frame where all of their components vanish). It is correct to write and to think of $ma' + ma_{frame} = F'$, where a' is the acceleration of the body relative to the rotating frame and a_{frame} is the acceleration of the frame relative to absolute space.

Although it is common to do so it makes no sense qualitatively to interpret the Coriolis and centrifugal accelerations as forces divided by mass. Formally, the force transforms as $F' = RF$ so that no new forces are introduced by the coordinate transformation. Physically, imagine that a ball (a force-free object) remains at rest or travels at constant velocity on ice, which we can treat as approximately frictionless. If the reader chooses to spin on ice skates and to describe the motion of the ball from his rotating frame, the spinning introduces no force, no new effect whatsoever on the motion of the ball, which continues to move at constant velocity in the inertial frame. The ball does not react at all to the spinning and therefore no new force is exerted on the ball. Seen from a deeper perspective, one should not delude oneself into believing that the second law $dp/dt = F$ is the fundamental law from which the law of inertia $dp/dt = 0$ can be 'derived'. Rather, the reverse is the case: the second law in the form $dp/dt = F$ is *only* valid in a frame where *force-free motion* is described by $dp/dt = 0$. Not only are the Coriolis and centrifugal effects not forces, Einstein also has observed that gravity, locally, is not a force either: like the Coriolis and centrifugal accelerations, gravity can be eliminated locally by a coordinate transformation, which is impossible for real forces like electromagnetic ones. The best that we can say here is that gravity behaves like an ordinary force in the Newtonian approximation to physics, providing the inertial frame is not chosen locally to be in free-fall.

There is a another way to arrive at Newton's laws of motion in rotating frames: by use of the covariant derivative. We can introduce a 'covariant momentum', a co-ordinate-free vector, by writing

$$p'_c = m \mathrm{D}x'/\mathrm{D}t = m(\mathrm{d}x'/\mathrm{d}t - \Omega'x') \qquad (8.39)$$

where $x' = Rx, R = \dot{R}\Omega$, and $\Omega' = R\Omega\tilde{R}$. As we showed in the last section, p'_c transforms according to matrix multiplication and so qualifies as an abstract vector: if the vector p'_c vanishes in one frame then it vanishes in all frames, which is not true of the velocity $v = \mathrm{d}x/\mathrm{d}t$. It is an easy exercise to show that

$$\mathrm{D}p'_c/\mathrm{D}t = F' \qquad (8.40)$$

yields equation (8.38) in rotating frames and also reduces to $\mathrm{d}p/\mathrm{d}t = F$, with $p = mv$, in inertial frames. In (8.40), we have a covariant formulation of Newton's laws, one that is valid in both rotating and nonrotating frames. That is because both sides of equation (8.40) transform in the same way, by matrix multiplication. In the absence of a net force F, we have

$$\mathrm{D}p'_c/\mathrm{D}t = 0, \qquad (8.41)$$

which is just the covariant statement of the fact that $\mathrm{d}p/\mathrm{d}t = 0$ in any inertial frame, a mere covariant rewriting of Newton's first law. Equation (8.41) means that there is a coordinate system (actually there are infinitely many) where the vector field p_c has constant components. Because $\mathrm{d}p'_c/\mathrm{d}t = \Omega'p'_c$ in force-free motion in a rotating frame, a force-free body does not move in a straight line at constant speed relative to that frame: if R is a transformation to a rotating frame from an inertial frame where $v = 0$, then the observed motion of a free body in the noninertial frame, from the viewpoint of an observer standing at the origin, is circular.

We must take care to distinguish between general covariance and a principle of relativistic invariance. The principle of relativistic invariance is in this case Galilean invariance, representing the complete equivalence of all inertial frames for performing experiments and observations. According to S. Lie and F. Klein, whose work paved the way for the generalization of geometry to topology[6] in mathematics, the laws of mechanics are dependent upon and reflect our idea of the geometry of space (here taken to be Euclidean, with the Newtonian idea of absolute time), while 'geometry' is defined by invariance under some *particular* group of transformations. In Klein's words:

Es ist eine Mannigfaltigkeit und in derselben eine Transformationsgruppe gegeben; man soll die der Mannigfaltigkeit angehörigen Gebilde hinsichtlich solcher Eigenschaften untersuchen, die durch die Transformationen der Gruppe nicht geändert werden . . . [oder] Man entwickle die auf die Gruppe bezügliche Invarianten theorie.

[Klein 1872, pp. 462–4]

The objects that are left invariant under the transformations of the group define the geometry of the space. Similar objects are transformed unchanged into each other under the transformations of the invariance group. For example, rigid bodies are transformed into themselves, with different orientations and center of mass positions,

---

[6] In topology, two geometrical objects with the same dimension (two curves, two surfaces, etc.) are regarded as similar if the one can be deformed continuously into the other by a (generally nondifferentiable) coordinate transformation. It is from this perspective, familar to us in phase space, that every simple closed curve is equivalent to a circle, and every curved trajectory is equivalent to a straight line. Clearly, Galilean invariance is not respected in topologic arguments.

under rigid rotations and translations in Euclidean space because their shapes are left invariant by those transformations (the same holds for transformations to uniformly moving frames). Rigid body shapes are left invariant because orthogonal transformations leave invariant the basic building blocks of Euclidean geometry: lengths of straight lines and angles between those lines. For classical mechanics excluding electromagnetic theory (which excludes quite a lot), the invariance group is the Galilean group, which includes translations and rigid rotations along with transformations among inertial frames. One can always enlarge the transformation group formally to permit more (like time-dependent rotations) and even arbitrary transformations (like nonlinear transformations to generalized coordinates), and such an enlargement allows us to rewrite true statements about geometry and physics in other, more general, coordinate systems. The enlargement of the group beyond the basic invariance group of Euclidean space leads to the idea of general covariance of equations of motion in configuration space (Lagrange's equations). However, if the transformation group is too large, as is the case in the general theory of relativity where the idea of covariance includes all possible differentiable (single-valued and invertible) coordinate transformations in a nonEuclidean space-time (see chapter 11), then the identity is the only function that is left invariant, and Klein's idea of how to define geometry in terms of the metric invariants of a group of transformations fails (see Bell, 1945, for a stimulating description of Klein's Erlanger Program). From the standpoint of Klein's idea, the lack of invariance is the same as lack of metric geometry, so that it is necessary to define what one means by 'geometry' in some other way than Klein's in general relativity theory. The only alternative is to concentrate on qualitative features, the topologic invariants. Noether's theorem for physicists is akin in spirit to Klein's idea in geometry: the physically important quantities are the conserved ones, which are not metric invariants (Klein's case refers only to force-free motion because the kinetic energy defines the metric for time-independent transformations (chapter 10)), but rather are quantities that leave the action invariant.

In contrast with Galilean invariance and other ideas of invariance based upon *metric* geometry (based upon the invariance of scalar products under a special transformation group), topology is a generalization of geometry whereby one forgets the metric and other quantitative details altogether, and where two or more different figures are regarded as equivalent (are *topologically* similar to each other) if one figure can be transformed into the other by any continuous deformation whatsoever, including very nonphysical and nondifferentiable deformations that get rid of sharp edges without sanding or filing (a cube and a sphere are topologically the same object, for example). In this case, one of the invariants is the number of holes in an object (a donut cannot be deformed continuously into a sphere or a cube, nor can a figure eight in two dimensions be deformed continuously into a circle). We think topologically whenever, as in the phase space formulations of chapters 3, 6, 13, 14, and 16, we argue that mechanical motions that are quantitatively different from each other are *qualitatively* similar to each other. There, we follow Lie and permit a very large class of differentiable coordinate transformations in a vector space that is not a metric space, the class of all possible differentiable, invertible, and single-valued coordinate transformations in phase space, and the resulting transformation group is so large that we end up with only *topologic* invariants, like the number of holes in a surface. In chaos theory, number theory (a subject previously ignored by physicists) is often very useful for understanding the statistics generated by whole classes of chaotic trajectories (see chapters 4 and 6 for simple examples), and more generally for understanding the topologically invariant properties of dynamical systems like the way that the periodic and nonperiodic unstable orbits arrange themselves, via the method of symbolic dynamics (computability theory and the related theory of formal grammars also come into play in symbolic dynamics). Nature is complex, and the little that we can formulate correctly in terms of nonlinear differential equations in three or more variables is often too complicated for any hope of a direct quantitative solution (fluid turbulence is a very good example of this), and so entirely different branches of mathematics are often useful in helping us to gain *some* qualitative understanding of those aspects of nature that can be quantified in the first place[7]. We use the language of the theory of continuous groups because that way of looking at things is not merely very beautiful mathematically, it also allows us to unify a lot of aspects of dynamics that would otherwise look like a conglomeration of disconnected facts. In particular, the ideas of metric and topologic invariance under transformation groups are far more useful for gaining both qualitative and quantitative understanding of nonlinear dynamics than is the idea of general covariance of equations of motion. Invariance is geometric and places constraints on *solutions*, which is an important feature for observations of nature, whereas covariance is not geometric and only places a constraint on the formal appearance of a set of differential equations (which is not important for observations of nature).

Before going on to applications, we should introduce the Lagrangian description of the equations of motion in rotating frames. We prove explicitly in chapter 10 that the Lagrangian approach automatically provides us with a covariant description of the motion, so that we can apply Lagrange's equations directly in any noninertial frame. If we start with the Lagrangian $L = T - U$ in an inertial frame, then because the Lagrangian is a scalar, $L'(x',dx'/dt,t) = L(x,dx/dt,t)$. The potential energy is also a scalar $U(x,t) = U'(x',t)$, and since

$$\tilde{v}v = \tilde{v}'v' - 2\tilde{v}'\Omega'x' - \tilde{x}'\Omega'^2x' \tag{8.42}$$

we get

$$L' = m(\tilde{v}'v' - 2\tilde{v}'\Omega'x' - \tilde{x}'\Omega'^2x')/2 - U'(x',t) \tag{8.43}$$

in Cartesian coordinates (we stick to Cartesian coordinates in all that follows). The canonical momentum is then

$$p'_i = \frac{\partial L'}{\partial v'_i} = m(v'_i - \Omega'_{ij}x'_j) \tag{8.44}$$

and is identical to the covariant momentum defined above. We show in chapter 10 that the canonical momentum transforms like a covariant vector. Therefore, Newton's law of motion takes on the form

$$\frac{dp'_i}{dt} = \frac{\partial L'}{\partial x'_i}, \tag{8.45}$$

in any frame, rotating or inertial (dp_i/dt is also a covariant vector). Writing out the details of the right hand side simply yields the law of motion in the form

---

[7] There is no justification for believing that everything in the universe can be quantified. As Wigner has stated (1967), physics is not the study of nature; physics is the study of the regularities of nature (that Newton's law generates irregular, chaotic solutions does not contradict this: Newton's law follows from Galilean invariance, which is a regularity of nature).

$$\frac{dmv'}{dt} = -\nabla'U' + 2m\Omega'v' - m\Omega'^2 x' + m\dot{\Omega}'x'. \tag{8.46}$$

Finally, we get the Hamiltonian description by the Legendre transformation

$$H' = \tilde{p}'v' - L'. \tag{8.47}$$

Replacing v' by p' and x' everywhere on the right hand side yields H' in several forms:

$$H' = m\tilde{v}'v'/2 + m\tilde{x}'\Omega'^2 x'/2 + U', \tag{8.48a}$$

$$H' = (\tilde{p}' - m\tilde{x}'\Omega')(p' + m\Omega'x')/2m + m\tilde{x}'\Omega'^2 x'/2 + U', \tag{8.48b}$$

and finally,

$$H' = \tilde{p}'p'/2m - \tilde{x}'\Omega'p' + U'. \tag{8.48c}$$

This yields Hamilton's equations:

$$\left. \begin{array}{l} v' = \dot{x}' = -\nabla_{p'}H' = p'/m - \tilde{x}'\Omega', \\ \dot{p}' = -\nabla_{x'}H' = \Omega'p' - \nabla'U', \end{array} \right\} \tag{8.49}$$

and Newton's law for rotating frames is obtained by substitution.

Note that Hamilton's equations (8.49) can be rewritten in the form

$$p' = m\frac{Dx'}{dt} \text{ and } \frac{Dp'}{Dt} = F' \tag{8.49'}$$

if we make use of the Lie derivative $D/Dt = d/dt - \Omega$. However, Hamilton's equations are not covariant under general time-dependent coordinate transformations in configuration space (or under linear transformations to rotating frames) because in that case the Hamiltonian does not transform like a scalar: $H(q,p,t) \neq H'(q',p',t)$. The reason for this is that extra terms are picked up by H' in the Legendre transformation (8.47) whenever the velocity v is not coordinate-free. In the language of chapter 15, $H' = H + \partial S/\partial t$, which is a sum of *two different* scalar functions (S is called the generating function for the transformation).

Relative to the rotating frame the Coriolis deflection, the contribution of the Coriolis acceleration to the observed trajectory, is obtained as follows. Let the x_3-axes of the rotating and inertial frames coincide and assume that there is a counterclockwise rotation of the noninertial frame relative to the inertial frame. If the rotation rate is small, then the force-free motion of a body obeys the equation

$$\frac{dv'}{dt} - 2\Omega'v' \approx 0, \tag{8.50}$$

according to an observer fixed in the rotating frame. For a constant rotation frequency $\omega_3 = \omega$, this equation is linear in v', with constant coefficients, and therefore has the solution

$$v'(t) = e^{2\Omega't}v'(0), \tag{8.51a}$$

and at short times we obtain

$$v'(t) \approx (I + 2\Omega't)v'(0). \tag{8.51b}$$

This is the Coriolis deflection, an apparent deflection of the force-free body relative to the rotating frame. Since

$$\delta v'(t) = v'(t) - v'(0) \approx 2\Omega' t v'(0) = 2\omega t \begin{pmatrix} v_2'(0) \\ -v_1'(0) \\ 0 \end{pmatrix}, \tag{8.52}$$

the apparent deflection is clockwise and perpendicular to $v(0)$ because $v(0)$ and $\delta v(t)$ are perpendicular to each other. Again, a free particle does not follow a straight line at constant speed relative to a rotating frame, but regardless of the apparent complexity of the observed path, the particle is only traveling at constant velocity relative to any inertial frame.

We now consider some applications of the above theory. For the first application, we begin with

$$H' = m\tilde{v}'v'/2 + m\tilde{x}'\Omega'^2 x'/2 + U', \tag{8.48a}$$

which shows that, from the point of view of a rotating observer, a body moves in an effective potential

$$V' = m\tilde{x}'\Omega'^2 x'/2 + U'. \tag{8.53}$$

If we choose the axis of rotation parallel to the x_3'-axis, then $\Omega^2 = -\omega_3^2 I$, and with $\omega_3 = \omega$ (8.53) reduces to

$$V' = U' - m|\omega|^2(x_1^2 + x_2^2)/2. \tag{8.54}$$

Now, suppose that a body at rest in the rotating frame is dragged by the rotation as the earth's mass is dragged by its rotation. Then the body is at rest in the rotating frame, $v' = 0$, and so equipotentials of the effective potential,

$$V' = U' - m|\omega|^2 r^2 \sin^2\theta/2 = \text{constant} \tag{8.55}$$

should describe the shape of any rotating body that is free to relax into a state of rest, a steady state, in the rotating frame. This predicts a tendency toward equatorial bulging, or flattening of the body, which is in the right direction for explaining the bulge of the earth about the equator as well as the flattened shapes of galaxies.

A second application of the same result is to Newton's famous pail experiment. If, standing on the earth (and ignoring the earth's rotation temporarily), we spin a bucket of water with the rotation axis parallel to the local direction of gravity then the equipotential surfaces are given locally by

$$V = mgz - m|\omega|^2(x^2 + y^2)/2 = \text{constant}. \tag{8.56}$$

At the first instant when the bucket is set spinning, the water does not follow but (because of no-slip boundary conditions at the walls of the bucket) is eventually dragged into rotation. When, finally the steady state is reached and the water is at rest with respect to the bucket, we expect the surface of the water to be an equipotential of V, which yields (with the constant set equal to zero) the famous parabolic shape predicted by Newton, in illustration of the physical effects of rotation relative to absolute space:

$$z = \omega^2(x^2 + y^2)/2g. \tag{8.57}$$

Note that we cannot derive these effects by setting the canonical momentum $\mathbf{p} = 0$, because that guarantees equilibrium in inertial frames, whereas we require a steady state in a particular rotating frame.

As our last application, we derive the Larmor effect, which illustrates how a transformation to a rotating frame can be used to simplify the solution of a certain class of problems. Since $-\Omega x$ represents the cross product of two vectors, $\omega \times \mathbf{x}$, it is clear that the cross product $\mathbf{v} \times \mathbf{B}$ is represented in matrix notation by $-\Omega_v B$ where B is a column vector and $(\Omega_v)_{ij} = \varepsilon_{ijk}v_k$. In a rotating frame, a charged particle in a magnetic field therefore obeys

$$m\dot{v}' = 2m\Omega'v' - m\Omega'^2 x' + m\dot{\Omega}'x' - \nabla'U' - \frac{q}{c}\Omega'_v B. \tag{8.58}$$

Since the idea of the introduction of the rotating frame is to simplify the solution of the problem, we choose ω so that

$$2m\Omega'v' - q\Omega'_v B/c = 0, \tag{8.59}$$

which means that

$$2m\varepsilon_{ijk}\omega'_k v'_j - q\varepsilon_{ijk}v'_k B_j/c = 0. \tag{8.60}$$

Rearrangement yields easily that

$$\omega' = -qB/2m. \tag{8.61}$$

Now, the equation of motion is simply

$$m\dot{v}' = -m\Omega'^2 x' - \nabla'U'. \tag{8.62}$$

Choosing the axis of rotation parallel to the x'_3-axis yields

$$m\dot{v}' = m|\omega'|^2 x' - \nabla'U'. \tag{8.63}$$

If the magnetic field is small enough, then we neglect the first term and obtain an approximate equation of motion in the rotating frame,

$$m\dot{v}' \approx -\nabla'U', \tag{8.64}$$

that is identical to the unperturbed motion in the inertial frame, where now we want to know the solution. The approximate solution follows easily: if the unperturbed orbit is a bounded motion of some shape with winding number W, then the perturbed motion in the inertial frame is simply a precession of the unperturbed orbit with frequency $\omega = qB/2m$, which means that we get an orbit with a different winding number than the unperturbed one. In particular, if we start with the Coulomb potential where $U \propto 1/r$, then the approximate solution is a precessing elliptic orbit. We leave it to the reader to determine the condition under which this precessing ellipse forms a closed orbit.

Exercises

Note: the results of exercises 7 and 8 are necessary to solve exercises 1 and 2 in chapter 12.

1. Two observers, one accelerated and one inertial, have an argument. Each claims the right to use $dp/dt = F$ to describe physics in his own frame. Suggest an experiment that can be performed to determine who is right and who is wrong.

2. If **B** is a square matrix and **A** is the exponential of **B**, defined by the infinite series expansion of the exponential,

$$\mathbf{A} \equiv e^{\mathbf{B}} = 1 + \mathbf{B} + \tfrac{1}{2}\mathbf{B}^2 + \cdots + \frac{\mathbf{B}^n}{n!} + \cdots,$$

then prove the following properties:
   (a) $e^{\mathbf{B}}e^{\mathbf{C}} = e^{\mathbf{B}+\mathbf{C}}$, providing **B** and **C** commute.
   (b) $\mathbf{A}^{-1} = e^{-\mathbf{B}}$.
   (c) $e^{\mathbf{C}\mathbf{B}\mathbf{C}^{-1}} = \mathbf{C}\mathbf{A}\mathbf{C}^{-1}$.
   (d) **A** is orthogonal if **B** is antisymmetric.
   (e) $e^{i\mathbf{B}}$ is unitary if **B** is self-adjoint.

(Goldstein, 1980)

3. A particle is thrown upward vertically with initial speed v_o, reaches a maximum height and falls back to the ground. Show that the 'Coriolis deflection', when it again reaches the ground, is opposite in direction, and four times greater in magnitude, than the Coriolis deflection when it is dropped at rest from the same maximum height.

(Goldstein, 1980)

4. Work exercise 4 of chapter 1 while taking into account the earth's rotation.

5. A projectile is fired horizontally along the earth's surface. Show that to a first approximation, the angular deviation from the direction of fire resulting from the Coriolis deflection varies linearly with time at a rate

$$\omega \cos \theta,$$

where ω is the angular frequency of the earth's rotation and θ is the colatitude, the direction of deviation being to the right in the Northern Hemisphere.

(Goldstein, 1980)

6. The Foucault pendulum experiment consists in setting a long pendulum in motion at a point on the surface of the rotating earth with its momentum originally in the vertical plane containing the pendulum bob and the point of suspension. Show that its subsequent motion may be described by saying that the plane of oscillation rotates uniformly $2\pi \cos \theta$ radians per day, where θ is the colatitude. What is the direction of rotation? The approximation of small oscillations may be used.

(Goldstein, 1980)

7. Let A and B denote two abstract operators acting on a linear vector space. Let $f(\lambda) = e^{\lambda A} B e^{-\lambda A}$.

   (a) Show that $f(\lambda) = B + \lambda(A,B) + \dfrac{\lambda^2}{2}(A,(A,B)) + \ldots$

   (Hint: prove that $f'(\lambda) = (A, f(\lambda))$, that $f''(\lambda) = (A, f'(\lambda)) = (A,(A, f(\lambda)))$, and so on.)
   (b) Show that if A commutes with the commutator (A,B), then $e^A e^B = e^{A + B + \frac{1}{2}(A,B)}$.

8. (a) Use the power series in 7(a) above to prove the following transformation rules for rotation generators M_i: if $R_3(\theta_3) = e^{\theta_3 L_3}$ then

$$\tilde{R}_3 \vec{M} R_3 = \hat{k} M_3 - \hat{k} x (M_1 \hat{j} - M_2 \hat{i}) \cos \theta_3 + (M_1 \hat{j} - M_2 \hat{i}) \sin \theta_3$$

where $(\hat{\imath},\hat{\jmath},\hat{k})$ are three orthonormal Cartesian unit vectors.

(b) If $\Omega_2 = \dot\theta_2 M_2$, show that

$$\tilde{R}_3\Omega_2 R_3 = \dot\theta_2 M_2 \cos\theta_3 + \dot\theta_2 M_1 \sin\theta_3.$$

(c) If $\Omega_2 = \dot\theta_2 M_1$, show that

$$\tilde{R}_2\Omega_2 R_2 = \dot\theta_1 M_1 \cos\theta_2 + \dot\theta_1 M_3 \sin\theta_2,$$

and

$$\tilde{R}_3\tilde{R}_2\Omega_1 R_2 R_3 = \dot\theta_1 \cos\theta_2(M_1 \cos\theta_3 - M_2 \sin\theta_3) + \dot\theta_1 M_3 \sin\theta_2.$$

9. The problem of two opposite charges interacting by a Coulomb potential is just the Kepler problem. Consider the case where $E < 0$ so that the orbit is an ellipse. Now, perturb the problem by introducing a magnetic field B perpendicular to the plane of the orbit. In the limit of weak fields B, under what condition is the perturbed orbit closed?

9
Rigid body dynamics

9.1 Euler's equations of rigid body motion

We begin with Euler's theorem: the most general motion of a rigid body about a fixed point is a rotation about some axis. This does not mean that the motion of the body is restricted to a fixed axis, only that the *net result* of any motion with one point fixed, no matter how complicated, is *mathematically equivalent* to a single rotation about a single, fixed axis. The purely rotational motion of the body can be parameterized by three independent angles representing three noncommuting rotations about three separate axes. A real motion of the body can then be described as an infinite sequence of infinitesimal transformations whose net finite result can be represented by matrices representing three independent finite rotations. Since rotations form a group, the resulting transformation is again a rotation matrix, a single matrix that describes a rotation about a single axis, the invariant axis of the matrix, through a single angle. Euler's theorem merely states that the initial and final states can always be connected by a *single* rotation matrix, no matter how complicated is the actual motion. If we allow the instantaneous rotation axis and angle to vary with time, then the rotation matrices taken at different times do not commute with each other unless they represent rotations about the *same* instantaneous axis. The representation that shows this most explicitly is given by

$$R(\eta_1, \eta_2, \eta_3) = e^{\eta(t)\hat{n}^{(t)} \cdot M}, \tag{9.1}$$

where $\hat{n}$ and η are the instantaneous rotation axis and angle at a time t, and the matrices M_i are infinitesimal generators for rotations in three mutually perpendicular planes. For different values of the angles η_i, the different matrices generally do not commute because the generators M_i of infinitesimal rotations are noncommuting.

We showed in chapter 1 that the kinetic energy divides additively into the kinetic energy of the center of mass plus the kinetic energy about the center of mass. For a rigid body, the latter is the kinetic energy of rotation about the center of mass. This decomposition corresponds to Chasles's theorem, which states that the most general motion of a rigid body is a translation of the center of mass combined with a rotation about the center of mass. This does not mean that the instantaneous rotation axis must pass through the center of mass: it means that one can treat an arbitrary motion *mathematically* as if it *were* a rotation about the center of mass, but only if one includes at the same time the required translational kinetic energy *of* the center of mass. This is illustrated better by a picture than by words: the three drawings in figure 9.1 show a double pendulum where a bob is attached to a rod and is free to rotate about the point of attachment. The first drawing shows the initial state. The second shows the center of mass in its final state, but as if the bob had *not* rotated about its point of attachment at the pivot. In the third drawing, the bob is brought into the final state by a single

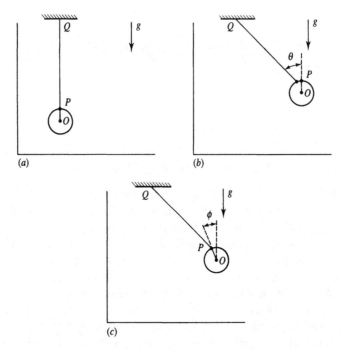

Fig. 9.1(*a*) Double pendulum: disk attached to rod at pivot point *P*.
Diagrams (*a*) and (*c*) represent configurations of the double pendulum at two
different times. Chasles's theorem is applied to the disk in (*b*) and (*c*). (*b*)
Imagine rod and disk to be disconnected: center of mass of disk *O* is
translated to new position, while rod rotates into new position about pivot
point *Q*. (*c*) Disk is rotated through angle ϕ about its center of mass *O* and
then reconnected to the rod at the pivot point *P*.

rotation about its center of mass. The point is: the real rotation is about the point of
attachment of the bob to the rod, but we can formulate the problem mathematically by dividing it artificially into a translation of the center of mass plus a rotation
about the center of mass. *The main point for the reader is that one must take the rotation
angle in the bob's kinetic energy to be ϕ, not θ in figure 9.1(c), otherwise an error will be
made.*

Summarizing, there are two possible approaches to the formulation of rigid body
problems. Whenever one point in the body is fixed, it is useful to formulate the kinetic
energy as purely rotational energy, using the fixed point as the origin of coordinates.
When no point in a body is fixed in an inertial frame, as when a top spins while sliding
on a frictionless plane, then the decomposition of the kinetic energy into translational
and rotational parts is very useful because it allows us to avoid the introduction of a
time-dependent moment of inertia tensor. The next step is to derive the moment of
inertia through the formulation of the angular momentum.

Consider the motion of any point *x* that is fixed in the rigid body that rotates about a
fixed point. In the inertial frame the motion of the point traces out a curve on the surface
of a sphere, but the point is at rest in any frame that is fixed in the rotating body. The
transformation from the inertial to the body-fixed frame is given by $x' = Rx$, so that

$$\mathrm{d}x'/\mathrm{d}t = R\mathrm{d}x/\mathrm{d}t + R\Omega x = 0 \qquad (9.2)$$

holds for the position vector x' of any and every particle in the body. Therefore, $dx/dt = -\Omega x$ is the equation of motion of a particle in the inertial frame. That is, the velocity of the pth particle in any rigid body about either the center of mass or some rotation axis is just

$$v_p = -\Omega x_p \tag{9.3}$$

where $\Omega = \omega_i L_i$, and x_p is the position of the particle relative to the chosen origin of coordinates. If we can find out how the frequencies ω_i in Ω vary with time, which is the same as solving the dynamics problem in a certain coordinate system to be introduced in section 9.2, then we can solve (9.3) and determine the curve traced out on the sphere in the inertial frame, which tells us how the rigid body moves in that frame. Toward that end, note that $\vec{r} \times \vec{v}$ is represented by $-\Omega_r v$ where

$$\Omega_r = \begin{pmatrix} 0 & x_3 & -x_2 \\ -x_3 & 0 & x_1 \\ x_2 & -x_1 & 0 \end{pmatrix}, \tag{9.4}$$

so that the total angular momentum of the rigid body is given by

$$M = \sum_{p=1}^{N} M_p = \sum_{p=1}^{N} (m\Omega_r \Omega x)_p. \tag{9.5a}$$

To go further, we use the well-known result $\varepsilon_{ijk}\varepsilon_{lmk} = \delta_{il}\delta_{jm} - \delta_{im}\delta_{jl}$, from which it follows that

$$(\Omega_r \Omega x)_i = \delta_{ij}\tilde{x}x\omega_j - x_i\tilde{x}\omega. \tag{9.6}$$

This permits us to write $M = I\omega$, where I is the 3×3 matrix

$$I_{ij} = \sum_{p=1}^{N} m_p(\delta_{ij}\tilde{x}x - x_i x_j)_p, \tag{9.7}$$

that is determined by the instantaneous distribution of mass about the chosen origin and therefore is generally time-dependent. The rotation frequencies transform according to the coordinate-free tensor rule $\Omega' = R\Omega R^{-1}$, and we leave it as an exercise for the reader to prove that this reduces to $\omega' = (\det R)R\omega$, and so to $\omega' = R\omega$ if $\det R = +1$. The column vector ω is called an axial vector. Under these conditions, $M = I\omega$ transforms like a coordinate-free axial vector under rotations, which is to say that M transforms like a coordinate-free vector under proper rotations, if the matrix I transforms like a coordinate-free second rank tensor.

A coordinate-free matrix T transforms as

$$T'_{ij} = R_{il}R_{jk}T_{lk}, \tag{9.8a}$$

or

$$T' = RT\tilde{R}, \tag{9.8b}$$

which is the same as saying that T is a coordinate-free rank two tensor. Since

$$I = \sum_{p=1}^{N} m_p(\tilde{x}xE - x\tilde{x})_p \tag{9.9}$$

where E is the identity matrix, I transforms like a second rank tensor, $I' = RI\tilde{R}$, which means that M transforms like a coordinate-free axial vector.

To the extent that it is of calculational use, we can rewrite the sum (9.7) as a Stieltjes integral over the mass distribution. In any case, the moment of inertia reflects the second moments of the mass distribution of the body.

The kinetic energy of rotation about the origin of coordinates is

$$T = \sum_{p=1}^{N} \tfrac{1}{2} m_p \tilde{v}_p v_p = - \sum_{p=1}^{N} \tfrac{1}{2} \tilde{x}_p \Omega^2 x_p, \tag{9.10}$$

and since

$$- \tilde{x} \Omega^2 x = \tilde{\omega}(\tilde{x}xE - x\tilde{x})\omega \tag{9.11}$$

we obtain

$$T = \tfrac{1}{2} \tilde{\omega} I \omega = \tfrac{1}{2} \tilde{\omega} M, \tag{9.12}$$

which is rotationally invariant because it is the scalar product of two vectors that are coordinate-free under rotations, and because rotations leave the Euclidean scalar product invariant.

From this point of view, the moment of inertia tensor looks like a metric for the rotational motion[1]. For an arbitrary rotating frame, the moment of inertia tensor is both coordinate- and time-dependent. Because the moment of inertia tensor is symmetric, it can always be diagonalized. Furthermore, if we work in any body-fixed frame then the moment of inertia is time-independent. Any frame fixed in the rotation body will do, but if we choose the frame whose axes line up with the directions of the inertia tensor's three perpendicular eigenvectors, which is called the principal axis frame, then the matrix I is both diagonal and constant (in other words the preferred coordinate system is constructed by diagonalizing the inertia tensor). This preferred frame reflects the geometric distribution of mass within the rigid body. If there is symmetry in the distribution, then it will show up as eigenvalue-degeneracy of the inertia tensor.

The rotational equations of motion are easy to write down from the standpoint of Newton's laws. Since $dM/dt = N$ in an inertial frame, where N is the net torque, the equation of motion in a frame that is fixed in the rotating body follows from $RdM/dt = RN = N'$, where R is the transformation matrix that describes the motion of any particle in the body relative to an inertial frame. Therefore, we obtain

$$\frac{DM'}{Dt} = \frac{dM'}{dt} - \Omega'M' = N' \tag{9.13a}$$

or

$$\frac{dM'_i}{dt} - \varepsilon_{ijk}\omega'_k M'_j = N'_i \tag{9.13b}$$

as the equations of motion in the body-fixed frame (the primes denote the computation of the vectors M and N and the operator Ω with respect to the body-fixed axes).

Consider next the transformation U from any body-fixed frame whatsoever to the principal axis frame: $\Lambda = UIU^{-1}$ is a diagonal matrix with elements I_1, I_2, and I_3, which are called the principal moments, and its three mutually perpendicular[2]

---

[1] However, the $\omega_i dt$ are not exact differentials of generalized coordinates – see section 12.1.
[2] If a body has symmetry then there is a degeneracy: one or more of the principle moments I_i will coincide

eigenvectors define the directions of the body's principal axes. In the principal axis frame, the equations of motion are called Euler's equations and take on their simplest form:

$$\left.\begin{aligned} I_1\dot{\omega}_1' - \omega_2'\omega_3'(I_2 - I_3) &= N_1' \\ I_2\dot{\omega}_2' - \omega_1'\omega_3'(I_3 - I_1) &= N_2' \\ I_3\dot{\omega}_3' - \omega_2'\omega_1'(I_1 - I_2) &= N_3'. \end{aligned}\right\} \tag{9.13c}$$

From the standpoint of Euler's equations we can study rigid body motion as a flow in phase space, which is very convenient. However, equations (9.13b) or (9.13c) are not a complete formulation of the problem, which generally requires a six dimensional phase space. In the most general case of no translation, the rigid body rotates about a point that is fixed in the body but is subject to an external torque due to gravity denoted by the constant force F, which is concentrated at the center of mass. The force is constant in the inertial frame but the center of mass position is not. In the inertial frame one cannot solve the problem

$$\left.\begin{aligned} \frac{dM}{dt} &= -\Omega_x F \\[2mm] \frac{dx}{dt} &= -\Omega x \end{aligned}\right\} \tag{9.13'a}$$

in the six dimensional (M,x)-phase space because the frequencies $\omega_i(t)$ are not known. Instead, one must supplement these equations by six equations of motion for the moment of inertia tensor, which requires altogether a twelve dimensional phase space.

In any body-fixed frame, one can study the problem

$$\left.\begin{aligned} \frac{dM'}{dt} - \Omega'M' &= -\Omega_x'F' \\[2mm] \frac{dF'}{dt} &= \Omega'F', \end{aligned}\right\} \tag{9.13'b}$$

where x' is the constant vector that points from the fixed point to the center of mass, in the six dimensional $(\omega_1',\omega_2',\omega_3',F_1',F_2',F_3')$-phase space because the moment of inertia tensor is constant (determining $M'(t)$ determines $\omega'(t)$, which is not true in the inertial frame). In other words, with $F' = RF$ the force components $F_i'(t)$ in the body-fixed frame are unknown until the time-dependence of $\omega'(t)$ is known, which requires solving for both ω' and F' simultaneously. There are only four known integrable cases, and we discuss them all in section 9.5.

In the body-fixed frame we have a flow in a six dimensional phase space even if we restrict to constant forces. Clearly, the M'-, x'-, and F'-subspaces are all Euclidean. The subspace of the variables (M_1',M_2',M_3'), or the qualitatively equivalent phase space $(\omega'_1,\omega'_2,\omega'_3)$ is also Euclidean because, in the body-fixed frame, the 'metric tensor' I is constant. Whenever the metric is constant then Cartesian coordinates can be defined by rescaling the variables M_i'. Therefore, the six dimensional phase space is a Cartesian space. With one or more conservation laws, the motion will generally be confined to a curved space of lower dimension.

and the corresponding eigenvectors are only linearly independent, not orthogonal. However, it is elementary to use this pair to form an orthogonal pair. Due to the symmetry of the body, it does not matter which orthogonal pair you choose.

Rigid body problems are difficult because they are very rich. All of the interesting phenomena of modern dynamical systems theory can be illustrated in the context of rigid body theory by adding dissipation to compete with the driving provided by the torque in chapter 13. We now reproduce some standard results of classical rigid body theory by using phase space analysis.

9.2 Euler's rigid body

We consider the torque-free motion of a rigid body in the principal axis frame, the frame fixed in the body where the inertia tensor is both constant and diagonal. Free-fall of any object in the earth's gravitational field is an example that satisfies the condition of absence of torque. All problems of this type are not only solvable but also are integrable, as we now show. The dynamical system where all moments of inertia are equal (a sphere or a cube with a uniform mass distribution, or a geometrically unsymmetric body with an uneven mass distribution) is trivial, so we start with the next degree of symmetry.

For a solid of revolution, where $I_1 = I_2 \neq I_3$, the frequency ω_3' is constant and we need only solve the linear equations

$$\left. \begin{array}{l} I_1 \dot{\omega}_1' - \omega_2' \omega_3' (I_1 - I_3) = 0 \\ I_1 \dot{\omega}_2' - \omega_1' \omega_3' (I_3 - I_1) = 0 \\ \dot{\omega}_3' = 0. \end{array} \right\} \tag{9.14}$$

Writing $\Omega = \omega_3'(I_1 - I_3)/I_1$ and denoting the column vector with components ω_i' by ω', we have

$$\dot{\omega}' = \Omega M_3 \omega' \tag{9.15}$$

where

$$M_3 = \begin{pmatrix} 0 & 1 & 0 \\ -1 & 0 & 0 \\ 0 & 0 & 0 \end{pmatrix} \tag{9.16}$$

is the generator of rotations about the ω_3'-axis. The solution is given by

$$\omega'(t) = e^{t\Omega M_3} \omega'(0). \tag{9.17}$$

Since $M_3^2 = -M_3$ and $M_3^4 = P_{12}$ where

$$P_{12} = \begin{pmatrix} 1 & 0 & 0 \\ 0 & 1 & 0 \\ 0 & 0 & 0 \end{pmatrix} \tag{9.18}$$

is the projection operator for projections into the plane perpendicular to x_3, it follows from resummation of the series

$$e^{\alpha M_3} = I + \alpha M_3 + \alpha^2 M_3^2 + \cdots, \tag{9.19}$$

with $\alpha = \Omega t$, that

$$\omega'(t) = (P_3 + P_{12}\cos\Omega t + M_3\sin\Omega t)\omega'(0) \qquad (9.20)$$

where $P_3 = E - P_{12}$ is the projection operator onto the x_3-axis and where E is the identity matrix.

The interpretation of the solution is simple: the vector ω rotates at constant angular frequency about the x_3-axis. This is guaranteed by the form of the kinetic energy,

$$T = \frac{I_1}{2}(\omega_1'^2 + \omega_2'^2) + \frac{I_3}{2}\omega_3'^2, \qquad (9.21)$$

because both T and ω_3 are constant: any motion that leaves $\omega_1'^2 + \omega_2'^2$ invariant is a rotation in the (x_1,x_2)-plane. From the viewpoint of an observer in an inertial frame, the angular momentum M is fixed in space in both magnitude and direction while the body rotates. Though $M = I\omega$, one cannot conclude that ω is a constant vector in the inertial frame because I varies with time there. Denoting quantities with the subscripts as those calculated relative to the space-fixed axes, with $I_s = R^{-1}IR$, where I is the tensor of the principal moments (we are denoting objects in the body-fixed frame by unprimed symbols in this section), it follows by differentiation that

$$\frac{dI_s}{dt} = (I_s,\Omega), \qquad (9.22)$$

so that I_s is constant only when the space-fixed inertia matrix commutes with the angular velocity matrix. An example is the case of a symmetric top that spins about its symmetry axis if that axis does not wobble in an inertial frame.

The result derived above has an important practical application: since the earth is approximately an oblate spheroid in free-fall about the sun, the result describes the precession of the earth's axis of rotation about the symmetry axis x_3. If we call the axis of rotation the celestial north pole and the axis x_3 the geometric north pole, then it is clear that the two are distinct.

The study of the more general problem of a completely asymmetric body is motivated by a simple experiment: wrap a rubber band around a book so that the book cannot open, and then toss the book into the air, spinning it initially and successively about each of its three principal axes. You will observe stable motion about two of the principal axes, but the motion about the third axis is unstable: the book always wobbles about one of the axes, and nothing that you do can eliminate the wobbling. Note also that the wobbling occurs about the principal axis x_2 if we order the moments in the form $I_1 > I_2 > I_3$.

In order to analyze and understand the results of the experiment described above, consider the torque-free motion of any completely asymmetric rigid body in the principal axis frame. Euler's equations in the noninertial frame are

$$\left.\begin{array}{l} I_1\dot{\omega}_1' - \omega_2'\omega_3'(I_2 - I_3) = 0 \\ I_2\dot{\omega}_2' - \omega_1'\omega_3'(I_3 - I_1) = 0 \\ I_3\dot{\omega}_3' - \omega_2'\omega_1'(I_1 - I_2) = 0. \end{array}\right\} \qquad (9.23a)$$

As we noted at the end of the last section, we have three first order differential equations for the three angular frequencies ω_i', so the phase space of this problem is either the space of the three frequencies ω_i' or the space of the three components of angular momentum M'_i:

$$\left. \begin{array}{l} \dot{M}'_1 = M'_2 M'_3 (1/I_3 - 1/I_2) \\ \dot{M}'_2 = M'_1 M'_3 (1/I_1 - 1/I_3) \\ \dot{M}'_3 = M'_2 M'_1 (1/I_2 - 1/I_1) \end{array} \right\} \tag{9.23b}$$

which we can rewrite as

$$\left. \begin{array}{l} \dot{x} = ayz \\ \dot{y} = -bxz \\ \dot{z} = cxy \end{array} \right\} \tag{9.23c}$$

where $x = M'_1$, $y = M'_2$, $z = M'_3$, and a, b, and c are positive constants if $I_1 > I_2 > I_3$. Note that this system is conservative in the general sense of chapter 3: the divergence of the phase flow velocity $V = (ayz, -bxz, cxy)$ vanishes. Note also that

$$x\dot{x} + y\dot{y} + z\dot{z} = 0 \tag{9.24}$$

so that $x^2 + y^2 + z^2 = $ constant, which is just the statement that the magnitude of the angular momentum M is conserved. This constrains the motion to the surface of a sphere of radius M in the three dimensional M'_i-phase space, and kinetic energy conservation yields the streamline equation on the sphere[3]. With two global constants of the motion, a three dimensional flow is integrable. Now, we apply the qualitative methods of analysis of chapter 3.

There are six distinct equilibria located at $(0,0, \pm 1)$, $(0, \pm 1,0)$, and $(\pm 1,0,0)$. Linear stability analysis about the first two equilibria at the points $(0,0, \pm 1)$, i.e., for rotations about the principal axis M'_3, yields the characteristic exponents $\lambda = \pm i\sqrt{ab}$, so that this pair of equilibria is elliptic. A similar result follows for rotations about the principal axis M'_1. However, rotations about M'_2 are unstable: there, $\lambda = \pm \sqrt{ac}$, so that $(0, \pm 1,0)$ is a hyperbolic point. This explains the observed results when the book is tossed into the air in rotation about the three different principal axes. The global phase portrait is shown as figure 9.2. Clearly, one need not obtain a quantitative solution of the differential equations in order to understand the qualitative nature of the motion.

We have just described the motion from the viewpoint of a hypothetical observer who is fixed in the principal axis frame. What does the motion look like from the standpoint of an inertial observer? In this case, the vector M is fixed in space and its components, the projections onto the space-fixed axes are also constants. The matrix equation of motion of the components,

$$DM/Dt = 0, \tag{9.25a}$$

is just the statement that

$$dM/dt = 0 \tag{9.25b}$$

in inertial frames. However, as the rigid body rotates, its instantaneous axis of rotation ω changes with time, and its projection onto the space-fixed axes can be studied by applying the inverse transformation R^{-1} to the principal axis solution for the angular frequency ω.

With two conservation laws that determine two of the three variables, the Euler top provides an example of an integrable flow in a three dimensional phase space. From

---

[3] Moreover, because both T and M are conserved, the orbits lie at the intersection of the sphere with the kinetic energy ellipsoid.

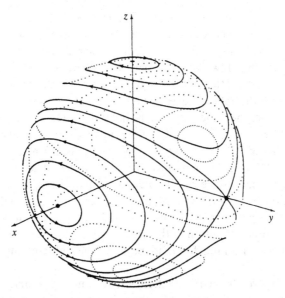

Fig. 9.2 Phase portrait for Euler's rigid body (fig. 4.31 from Bender and
Orszag, 1978).

section 3.6, we know that a flow described by the nonlinear system $dM/dt = V$ with
$V_i = \varepsilon_{ijk}M_j\omega_k$ is integrable if and only if $V_i = \varepsilon_{ijk}\partial G/\partial M_j\partial H/M_k$, where G and H are the
two conserved quantities (since the phase flow generated by $V_i = \varepsilon_{ijk}M_j\omega_k$ is conserva-
tive, the use of Jacobi's first multiplier is unnecessary). Equating the two expressions for
V_i allows us to identify G as the square of the magnitude of the angular momentum, and
H as the kinetic energy.

   If we take the first conservation law G as the square of the angular momentum, then
the motion is confined to a sphere in the M_i-phase space and conservation of kinetic
energy determines the streamlines on that sphere. If, instead, we should identify the first
integral G as the kinetic energy, then the motion in the angular momentum phase space
is confined to an ellipsoid, and constancy of angular momentum determines the
streamlines on the ellipsoid. Topologically, a sphere and an ellipsoid are identical, and
so the qualitative description of the motion is the same in both cases. In either case, the
trajectories are given geometrically by the intersections of spheres with ellipsoids
(different spheres and different ellipsoids follow from varying the magnitudes of the
angular momentum and the kinetic energy). Whittaker (1965) demonstrates the
integrability of Euler's top by using Euler angles as variables, which is more
cumbersome and less enlightening than the phase space approach used here[4].

   Integrability can be demonstrated directly by obtaining integrals for the three
angular frequencies ω_i': with two constants of the motion, T and M^2, we can eliminate
any two of the frequencies ω_k' from one of Euler's equations, say the equation for ω_i', to
obtain an equation of the form $dt = d\omega_i'/f(\omega_i')$, which is integrable by inspection, and
yields the time-development of one angular frequency $\omega_i' = \omega_i'(t)$. Therefore, the
time-dependence of each angular frequency can be found directly by integration.

   Suppose that you want to design machinery. You may then want to find out how the

---

[4] See Whittaker (1965) for solutions of most of the known integrable problems of particle and rigid body
   mechanics. Other problems, integrable and nonintegrable, are discussed in Arnol'd's *Dynamical Systems
   III*.

body moves in the inertial frame. Given the frequencies $\omega_i(t)$ in the inertial frame, the body's motion would be determined as follows: let x, a vector with one end located at the fixed point, denote the position of any point in the rigid body. As the body rotates the vector x rotates with it. If we draw a sphere that is centered on the fixed point, has a radius less than the length of the vector x, and is itself fixed in the inertial frame, then the intersection of x with the sphere traces out a curve that describes the body's motion. Now, $x' = Rx$ describes the same vector in the body-fixed frame and both x and x' have the same constant length but $dx'/dt = 0$. If the primed and unprimed axes are made to coincide at $t = 0$, then x' is the same as the initial condition $x(0)$ in the inertial frame and the equation $x(t) = R^{-1}(t)x'$ then contains the information that describes the curve that is traced out on the sphere. Since $dR/dt = R\Omega$, the frequencies ω_i determine Ω and with it the transformation matrix $R(t)$. The solution of Euler's equations in the inertial frame would determine the rigid body's motion completely, and the calculation of $x(t)$ would therefore yield a complete solution of the problem in a form that would allow one to design machinery by checking for (and perhaps then eliminating) wobbling and vibrations.

Unfortunately, as we showed in section 9.1, it is not so simple to solve for the frequencies $\omega_i(t)$ in the inertial frame. Moreover, given the solution in the form of $M'(t)$ and $\omega'(t)$ in the principal axis frame, there is not enough information to determine either $R(t)$ or $\omega(t)$. How can we solve for the motion in the inertial frame? If we could coordinatize the sphere on which the motion occurs by the parameters in our problem, which would be equivalent to a direct determination of $R(t)$, then we would reach our goal. How can we coordinatize the rotation matrix $R(t)$ in terms of parameters whose time-dependence we can calculate without so much trouble?

Since

$$\alpha_i = \int \omega_i dt, \qquad (9.26)$$

then can the time-dependent parameters $\alpha_i'(t)$ be found from the 'velocities' $\omega_i'(t)$? Can these three parameters be used to specify $R(t)$? In principle, yes, but not without practical difficulty as we show in chapter 12: there is no way to use the parameters $\alpha_i(t)$ to define a global coordinate system, although one can use them to define local Cartesian coordinate systems during an infinitesimal time interval δt at each time t. This is explained in chapter 12. Instead of digging deeper in this direction, we now introduce another way to coordinatize $R(t)$ in terms of parameters whose time-dependence we can calculate more easily from the equations of rigid body dynamics.

9.3 The Euler angles

The Euler angles (ϕ,ψ,θ) are rotation angles in three separate coordinate systems that are themselves defined by the Euler angles (see figure 9.3). Because they can be defined purely geometrically, independently of integrating the frequencies $\omega_i'(t)$, they provide a way of defining coordinates that describe motions on the sphere. In order to be able to use the Euler angles in rigid body theory, we must find a way to express the three rotation frequencies ω_k' in the body-fixed frame in terms of them and their time derivatives. What follows is an exercise in the application of results derived earlier in section 8.3 and yields a parameterization of the rotation group that is very useful for

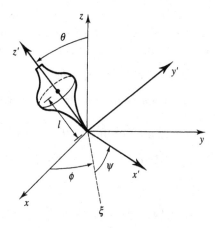

Fig. 9.3 The Euler angles defined geometrically (fig. 25 from Saletan and Cromer, 1971).

answering the question left open in the last section, and also for analyzing a certain class of dynamics problems that we study in section 9.4.

Denoting the three coordinate transformations by $\xi = D(\phi)x$, $\xi' = C(\theta)\xi$, and $x' = B(\psi)\xi'$, the transformation from the inertial to the body-fixed frame is given by $x' = Ax$ where $A(\phi,\theta,\psi) = B(\psi)C(\theta)D(\phi)$. Here, by the matrix A we mean simply the matrix $R(\theta_1,\theta_2,\theta_3)$, but expressed in terms of the Euler angles (ϕ,θ,ψ): $A(\phi,\theta,\psi) = R(\theta_1,\theta_2,\theta_3)$. By inspection of figure 9.3, the three separate transformations are given in the three separate coordinate systems by

$$D(\phi) = \begin{pmatrix} \cos\phi & \sin\phi & 0 \\ -\sin\phi & \cos\phi & 0 \\ 0 & 0 & 1 \end{pmatrix}, \tag{9.27a}$$

and

$$C(\theta) = \begin{pmatrix} 1 & 0 & 0 \\ 0 & \cos\theta & \sin\theta \\ 0 & -\sin\theta & \cos\theta \end{pmatrix}, \tag{9.27b}$$

and

$$B(\psi) = \begin{pmatrix} \cos\psi & \sin\psi & 0 \\ -\sin\psi & \cos\psi & 0 \\ 0 & 0 & 1 \end{pmatrix}. \tag{9.27c}$$

It is now clear why this representation is called degenerate: both D and B have the same infinitesimal generator, namely, M_3.

Denoting the rotation rate projected onto the body-fixed frame by

$$\omega' = \omega_1'\hat{e}_1' + \omega_2'\hat{e}_2' + \omega_3'\hat{e}_3', \tag{9.28a}$$

where $\hat{e}_i$ is the unit vector along the axis x_i', we can also resolve ω' onto three nonorthogonal axes in three of the four different coordinate systems:

$$\omega' = \dot{\phi}\hat{e}_3 + \dot{\theta}\hat{f}_1 + \dot{\psi}\hat{g}_1. \tag{9.28b}$$

Here, the f_i are three orthonormal unit vectors in the ξ-frame, and $\hat{g}_i$ are the three orthonormal unit vectors in the ξ'-frame, and the three orthonormal unit vectors $\hat{e}_i$ are in the inertial frame. It follows from taking a scalar product that

$$\omega_1' = \hat{e}_1'^+ \omega = \dot{\phi}\hat{e}_1'^+ e_3 + \dot{\theta}\hat{e}_1'^+ \hat{f}_1, \tag{9.29}$$

where x^+ is a row vector where elements are complex conjugates of the elements of the vector x.

Since the basis vectors transform according to the rule (see section 8.3)

$$\hat{e}_1' = A_{11}\hat{e}_1 + A_{12}\hat{e}_2 + A_{13}\hat{e}_3, \tag{9.30}$$

it follows that

$$\hat{e}_1'^+ \hat{e}_3 = A_{13} = \sin\psi \sin\theta. \tag{9.31}$$

And since $x' = BC\xi$, we obtain

$$\hat{e}_1' = (BC)_{11}\hat{f}_1 + (BC)_{12}\hat{f}_2 + (BC)_{13}\hat{f}_3 \tag{9.32}$$

so that

$$\hat{e}_1'^+ \hat{f}_1 = (BC)_{11} = \cos\psi. \tag{9.33}$$

Therefore, we have shown that

$$\omega_1' = \dot{\phi}\sin\theta\sin\psi + \dot{\theta}\cos\psi. \tag{9.34}$$

It is left to the reader to show by similar exercises that

$$\omega_2' = \dot{\phi}\sin\theta\cos\psi - \dot{\theta}\sin\psi \tag{9.35}$$

and

$$\omega_3' = \dot{\phi}\cos\theta + \dot{\psi}. \tag{9.36}$$

Taken together, we find that

$$\left.\begin{array}{l} \omega_1' = \dot{\phi}\sin\theta\sin\psi + \dot{\theta}\cos\psi \\ \omega_2' = \dot{\phi}\sin\theta\cos\psi - \dot{\theta}\sin\psi \\ \omega_3' = \dot{\phi}\cos\theta + \dot{\psi}. \end{array}\right\} \tag{9.37}$$

Calculating the cross-partial derivatives of the right hand side of these equations shows why we cannot use the parameters α_i defined by $d\alpha_i = \omega_i' dt$ as generalized coordinates: the velocities ω_i' are nonintegrable and so are not generalized velocities.

The Euler top provides an example of an integrable dynamical system solved in terms of variables that are themselves nonintegrable: this only means that the solutions $\omega_i'(t)$ of the Euler top cannot be integrated to define three coordinate axes α_i in *configuration space*, although the nonintegrable velocities $\omega_i'(t)$ define perfectly good Cartesian axes in phase space, in agreement with Lie's idea of a flow as the infinitesimal generator of global coordinate transformations.

9.4 Lagrange's top

Consider a solid of revolution with one point on its symmetry axis fixed in space. In the literature, this system is known both as the 'heavy symmetric top' and also as the symmetric top spinning on a rough plane. The latter phrase seems somewhat more appropriate for, without friction or something else to hold the apex of the top fixed, whether the top is heavy or light is irrelevant. We shall emphasize this below by formulating the problem where the same top spins while its apex slips on a frictionless plane.

We now consider the Lagrangian for a rigid body with one point fixed in space. It is not particularly useful to divide the kinetic energy into translational and rotational contributions, although one can always do that. It is more direct simply to compute the kinetic energy about the fixed point, which can be treated as rotational. The apex of the top, the point fixed in space, lies on the symmetry axis of the body. Therefore, if we take the apex as the origin of coordinates with the x_3-axis along the symmetry axis, then the moments of inertia are just the principal moments, the same as if we had taken the origin of coordinates to lie at the center of mass. This point is important, because it explains the difference in the way that we must formulate the kinetic energy when the top's apex is not fixed in an inertial frame but is allowed to slide frictionlessly on a plane that is perpendicular to gravity.

In terms of the Euler angles (ϕ, ψ, θ), the rotation rates in the body-fixed frame are

$$\left.\begin{array}{l} \omega_1' = \dot{\phi} \sin\theta \sin\psi + \dot{\theta}\cos\psi \\ \omega_2' = \dot{\phi} \sin\theta \cos\psi - \dot{\theta}\sin\psi. \\ \omega_3' = \dot{\phi}\cos\theta + \dot{\psi}. \end{array}\right\} \tag{9.37}$$

The symmetric top ($I_1 = I_2 \neq I_3$) with one point on the symmetry axis fixed in space then can be described by a Lagrangian $L = T - U$, and $\omega_1'^2 + \omega_2'^2 = \dot{\phi}^2 \sin^2\theta + \dot{\theta}^2$ reflects the rotational symmetry. We therefore obtain the top's kinetic energy in the form

$$T = \frac{I_1}{2}(\dot{\theta}^2 + \dot{\phi}^2 \sin^2\theta) + \frac{I_3}{2}(\dot{\phi}\cos\theta + \dot{\psi})^2, \tag{9.38}$$

and $U = -Mgl\cos\theta$ is the potential energy. From the Lagrangian L we immediately obtain two conserved canonical momenta:

$$p_\psi = \frac{\partial L}{\partial \dot{\psi}} = I_3(\dot{\phi}\cos\theta + \dot{\psi}) = M_3', \tag{9.39a}$$

is the component of angular momentum along the body-fixed x_3'-axis. The identification $p_\psi = M_3'$ follows because, as we showed in chapter 2, whenever we choose a rotation angle as a generalized coordinate then the conjugate canonical momentum is just the projection of the angular momentum vector onto the rotation axis. Therefore,

$$p_\phi = \frac{\partial L}{\partial \dot{\phi}} = I_1\dot{\phi}\sin^2\theta + I_3(\dot{\phi}\cos\theta + \dot{\psi})\cos\theta = M_3 \tag{9.39b}$$

is also constant and is the angular momentum along the inertial x_3-axis. Also, since the Lagrangian L has no explicit time-dependence H is also conserved. With three degrees of freedom and three globally analytic conserved quantities, we therefore have an integrable canonical flow in phase space. The object now, however, is to use these conserved quantities to understand the motion qualitatively in configuration space.

Since $\omega_3' = M_3'/I_3$ is constant, and M_3 is also constant, we can write the top's precession rate

$$\dot{\phi} = \frac{M_3 - M_3' \cos \theta}{I_1 \sin^2 \theta} \qquad (9.40)$$

as a function of θ alone.

Also, we can redefine $E' = H/I_1 - I_3 \omega_3'^2/2I_1$ as the Hamiltonian,

$$E' = \tfrac{1}{2}(\dot{\theta}^2 + \dot{\phi}^2 \sin^2 \theta) + Mgl \cos \theta/I_1, \qquad (9.41)$$

so that we have a one dimensional problem in the $(\theta, d\theta/dt)$-phase space with Hamiltonian

$$E' = \frac{\dot{\theta}^2}{2} + V(\theta), \qquad (9.42)$$

with the effective potential

$$V(\theta) = \frac{(M_3 - M_3' \cos \theta)^2}{2I_1^2 \sin^2 \theta} + mgl \cos \theta/I_1. \qquad (9.43)$$

Therefore, the motion is completely determined by solving the one dimensional problem for $\theta(t)$, along with specifying the six initial conditions, three of which we can think of as fixing the values of the constants E', M_3', and M_3.

For stable oscillations where θ changes by less than π, there must be two turning points $0 < \theta_1 < \theta_2 < \pi$ where $d\theta/dt$ vanishes $(V(\theta_1) = V(\theta_2) = E')$ and the θ-motion is periodic with period

$$\tau = \int_{\theta_1}^{\theta_2} \frac{d\theta}{[2(E' - V(\theta))]^{1/2}}. \qquad (9.44)$$

This means that there must also be an elliptic point somewhere in the $(\theta, d\theta/dt)$-phase space due to a minimum of the effective potential $V(\theta)$.

For the purpose of understanding the motion, it is useful to follow the trajectory traced by the intersection of the top's symmetry axis with the unit sphere, and to study the motion of that point as a function of (ϕ, θ) within the cylindrical band (θ_1, θ_2) on the sphere's surface (figure 9.4). The different ranges of values of the constant angular momenta M_3' and M_3 in the precession rate

$$\dot{\phi} = \frac{M_3 - M_3' \cos \theta}{\sin^2 \theta} \qquad (9.45)$$

yield three qualitatively different kinds of trajectories. Since $d\theta/dt$ changes sign at the turning points θ_1 and θ_2, it is only a question whether and where $d\phi/dt$ changes sign. If $M_3 > M_3'$ then $d\phi/dt$ never vanishes and we have precession with nutation, as is shown in figure 9.4(a). Although $d\phi/dt$ cannot vanish at both turning points it can vanish at θ_1 if $M_3 = M_3' \cos \theta_1$, in which case $M_3 - M_3' \cos \theta_2 > 0$. This yields the trajectory of figure 9.4(b). Finally, we can also have $d\phi/dt < 0$ at some angle θ_1 and $d\phi/dt > 0$ at θ_2 with $d\phi/dt = 0$ at some angle θ' intermediate between θ_1 and θ_2, which yields figure 9.4(c). There are no other geometrically distinct possibilities for oscillations between the two turning points. For a discussion of the functional forms of $\theta(t)$ and $\phi(t)$ we refer the reader to Whittaker (1965) and to Whittaker and Watson (1963), as each integrable problem of mechanics defines (or falls into the class of) one of the many 'special functions of mathematical physics'.

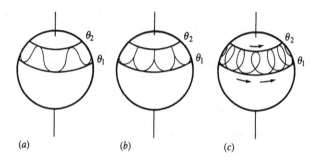

Fig. 9.4(a) Configuration space for $M_3 > M'_3$ (fig. 5.9a from Goldstein, 1980).
(b) Configuration space for $M_3 = M'_3 \cos \theta_1$ (fig. 5.9c from Goldstein, 1980).
(c) Configuration space for $M_3 = M'_3 \cos \theta'$, $\theta_1 < \theta' < \theta_2$ (fig. 5.9b from Goldstein, 1980).

Do the trajectories close within the cylindrical band or not? To answer this, we integrate $d\phi/d\theta$ one time to obtain the change $\Delta\phi = \Phi$ between any two successive turning points:

$$\Phi = \int_{\theta_1}^{\theta_2} \frac{[(M_3 - M'_3 \cos\theta)I_1^2 \sin^2\theta]^2 d\theta}{[2(E' - V(\theta))]^{1/2}}. \tag{9.46}$$

By now the reader has seen that we are studying the same kind of problem, qualitatively, that we analyzed in chapter 4 under the heading of central potentials: the configuration space trajectories close if $\Phi = 2\pi P/Q$ where P and Q are integers, but those trajectories are quasiperiodic (never close but instead made infinitely many close approaches to closure, and even fill up the θ-band ergodically (but not statistically mechanically or chaotically) as t goes to infinity) if the winding number $W = \Phi/2\pi$ is irrational. The linear twist map follows directly from this analysis, just as in chapter 4. The winding number $W = \Phi/2\pi$ is a function of the three parameters E', M'_3, and M_3. For arbitrary choices of these parameters, $W(E',M'_3,M_3)$ is irrational and the orbits are generally quasiperiodic. Closed orbits can occur only for very special choices of initial conditions. Note also that the elliptic point generally does not fall within the band $[\theta_1,\theta_2]$.

Furthermore, as is always the case with integrable motion, the orbits are marginally stable. The circle map for this problem is given by $\phi_n = f(\phi_{n-1})$ where $f(\phi) = \phi + 2\pi W$, so that $f'(\phi) = 1$ and therefore the Liapunov exponent of all orbits is $\lambda = 0$.

The analysis of the motion of a top that slips on a frictionless horizontal plane is integrable and is assigned as a set of exercises, because that problem can be formulated and solved by the same sequence of steps that we have exhibited above. There is only one difference to pay attention to: since there is no fixed point of the motion, the kinetic energy must be resolved into rotational and translational contributions about the center of mass which itself oscillates vertically.

9.5 Integrable problems in rigid body dynamics

Consider the description by Euler's equations (the Euler–Poisson equations) of a solid body that is free to rotate about a point that is fixed in the body. Let x denote the vector from the fixed point to the center of mass. In the inertial frame x varies with t while the force of gravity F is a constant. In the body-fixed frame $x' = Rx$ is a constant vector but the force $F' = RF$ in the body-fixed frame is time-dependent and is unknown until $R(t)$ is known. Therefore, we must add to the three Euler equations

$$\frac{dM'}{dt} - \Omega'M' = -\Omega_{x'}F' \qquad (9.47a)$$

the three additional differential equations

$$\frac{dF'}{dt} = \Omega'F' \qquad (9.48a)$$

where

$$\Omega = \begin{pmatrix} 0 & \omega'_3 & -\omega'_2 \\ -\omega'_3 & 0 & \omega'_1 \\ \omega'_2 & -\omega'_1 & 0 \end{pmatrix} \qquad (9.49)$$

and

$$\Omega_{x'} = \begin{pmatrix} 0 & x'_3 & -x'_2 \\ -x'_3 & 0 & x'_1 \\ x'_2 & -x'_1 & 0 \end{pmatrix}. \qquad (9.50)$$

The phase space is therefore the space of the six unknowns $(M'_1, \ldots, F'_3)$ or $(\omega'_1, \ldots, F'_3)$.

It is now easy to show by direct differentiation that the following three quantities are left invariant by the flow: first, the magnitude of the force $\tilde{F}'F' = F^2 = C_1$ is constant. Second, the total energy

$$E = \frac{\tilde{\omega}'I'\omega'}{2} + \tilde{x}'F' \qquad (9.51)$$

is conserved, and third, $\tilde{M}'F' = C_2$ is also constant. With three globally conserved quantities in a six dimensional phase space, we need one more conserved quantity in order to have an integrable problem.

Following Arnol'd (1978) and Fomenko (1988), we now list the known integrable cases:

(1) Euler's case (1750): $x' = 0$, and M is conserved.
(2) Lagrange's case (1788): $I_1 = I_2 \neq I_3$, $x'_1 = x'_2 = 0$, and ω_3 is conserved (the projection of M onto the top's symmetry axis is conserved).
(3) Kowalevskaya's case (1889): $I_1 = I_2 = 2I_3$, $x'_3 = 0$, and the extra constant of the motion is

$$(\omega'^2_1 - \omega'^2_2 - \nu F'_1)^2 + (2\omega'_1\omega'_2 - \nu F'_2)^2 \qquad (9.52)$$

where $\nu = (x'^2_1 + x'^2_2)/I_3$.
(4) Goryachev–Chaplygin's case (1900): $I_1 = I_2 = 4I_3$, $x'_3 = 0$, and in this case the scalar product of L with F vanishes.

The flow defined by the motion of a symmetric rigid body in a gravitational field with one arbitrarily chosen point in the body fixed in space is nonintegrable (there is no fourth globally conserved quantity). Kowalevskaya's case is named after the brilliant Russian mathematician of the last century, Sonya Kowalevskaya (1850–1891). Her complete solution can be found in Whittaker (1965) and is based upon the use of her discovery, local singularity analysis: she noticed that the solutions in the other two integrable cases, when extended to the complex time-plane, had singularities that were poles of finite order (there are no algebraic or logarithmic branch points of essential singularities). By requiring the solution in the general case to be of the same nature, she found the integrable case that bears her name. To understand why the absence of the spontaneous singularities other than poles in the complex time plane leads to invariants that are understood from the standpoint of symmetry, we refer the reader to the text by Fomenko (1988).

With three conservation laws we can always solve for and eliminate the three components of the force to obtain Euler's equations in the form

$$\frac{dM'}{dt} - \Omega'M' = f(M'_1, M'_2, M'_3, x'_1, x'_2, x'_3, C_1, C_2, E). \tag{9.53}$$

Therefore, the problem can always be studied in the three dimensional Euclidean phase space of the variables (M'_1, M'_2, M'_3).

If we look at Lagrange's top or any of the other three integrable rigid body cases in terms of the reduced Euler equations (9.53), then there is one conservation law and the motion is confined to a two dimensional Riemannian surface in the three dimensional Euclidean phase space (M'_1, M'_2, M'_3). If we look at the same problem in Euler angles, then there are *three* conserved quantities and the motion is confined to a three dimensional Euclidean torus in the six dimensional $(\phi, \theta, \psi, p_\phi, p_\theta, p_\psi)$-phase space. These two descriptions of the dynamics are geometrically inequivalent to each other: the canonical description cannot be transformed into the noncanonical one because the relation between the velocities $(\omega'_1, \omega'_2, \omega'_3)$ and the generalized velocities $(d\phi/dt, d\theta/dt, d\psi/dt)$ given by equations (9.34)–(9.36) is nonintegrable: there is no coordinate transformation defined by point functions from the generalized coordinates (ϕ, θ, ψ) to the nonholonomic coordinates $(\alpha_1, \alpha_2, \alpha_3)$.

In the nonintegrable case, which is the general case, one should expect to find chaotic motions in at least parts of the phase space. Any initial conditions that are far from those that yield small oscillations should provide a good starting point in the search for chaos. We continue the discussion of integrable and nonintegrable flows motivated by rigid body problems in chapter 13.

Consider next the equilibria for the general case where the body may or may not be symmetric and where the external torque is nonzero. In this case the equilibria in phase space are solutions of the six algebraic equations

$$\frac{dM'}{dt} = \Omega'M' - \Omega_x.F' = 0 \tag{9.47b}$$

and

$$\frac{dF'}{dt} = \Omega'F' = 0. \tag{9.48b}$$

Equation (9.48b) tells us that ω' and F' are colinear, $\omega' = CF'$ where C is a scalar. Since we can write

$$\Omega'_x \cdot F' = - \Omega'_F \cdot x',$$

(9.47b) can be written as

$$\Omega M' - \Omega'_x \cdot F' = - \Omega M'_1 \omega' + \Omega'_x \cdot F' = - C\Omega' M'_1 F' + \Omega'_x \cdot F' = 0 \qquad (9.54)$$

which tells us that M' and x' are colinear, that $CM' = x'$. Note that M' is constant because x' is constant, in agreement with $dM'/dt = 0$. Furthermore, with the x'_3-axis chosen to lie along the vector x' we have $M'_1 = M'_2 = 0$, which (via $\omega'_i = M'_i/I_i$) yields also that ω' is colinear with x'. Therefore, in equilibrium (which is a steady state of constant angular velocity ω'_3) the vectors $M', \omega', x',$ and F' are all colinear and there are two equilibrium points: one with x' parallel to ω' and one with x' antiparallel to ω'. We have the investigation of conditions for the stability of these two equilibria as exercises for the interested reader.

9.6 Noncanonical flows as iterated maps

It is easy to argue heuristically that a three dimensional noncanonical flow should be discussed via a two dimensional iterated map, as we did in chapter 6 for a two degree of freedom canonical flow. Here, we discuss two ways to arrive at maps from flows although there are infinitely many ways to get a map from a flow (a third method for nonautonomous systems is presented in section 14.4). This lack of uniqueness does not matter: the details of the map are irrelevant from the standpoint of topologically universal properties like the number of orbits of a given period and their periods. In other words, we expect that every map that can be derived from a particular flow belongs to a certain universality class that includes other flows as well, and that there are well-defined properties (like periods, numbers of orbits of a given period, and other abstract topologic invariants) that can be extracted from any map in the class. These properties are universal in the sense that the same properties follow from every map in the class. To develop and illustrate this 'principle of weak universality' would, however, go far beyond the scope of this text, and so we refer the interested reader to the literature on the subject, which is by now extensive (see chapter 9 in McCauley, 1993, and references therein).

We start with any noncanonical three dimensional bounded flow,

$$\left.\begin{aligned} \frac{dx}{dt} &= V_1(x,y,z) \\[2mm] \frac{dy}{dt} &= V_2(x,y,z) \\[2mm] \frac{dz}{dt} &= V_3(x,y,z), \end{aligned}\right\} \qquad (9.55)$$

and first construct a return map as follows (this is the method of E. Lorenz): pick one of the three variables, say z, and consider its time series $z(t)$. There will be one or more maxima where $dz/dt = 0$ with $d^2z/dt^2 > 0$. Denote the (unknown) times of occurrence of these maxima by t_n and denote the corresponding maxima by z_n. Since $V_3 = 0$ at these points we expect to find a functional relationship of the form $z_n = f(x_n,y_n)$ where (x_n,y_n) denotes the value of the coordinates (x,y) at time t_n. At the same time, we can think of the solution of (9.55) at any later time as a coordinate transformation on the solution at any earlier time, and for a transformation from time t_n to time t_{n+1} that transformation has the form

$$\left. \begin{array}{l} x_{n+1} = \psi_1(x_n,y_n,z_n,t_{n+1} - t_n) \\ y_{n+1} = \psi_2(x_n,y_n,z_n,t_{n+1} - t_n) \\ z_{n+1} = \psi_3(x_n,y_n,z_n,t_{n+1} - t_n). \end{array} \right\} \tag{9.56}$$

If we substitute into the condition for an extremum and write $\tau_n = t_{n+1} - t_n$, then we obtain

$$V_3(\psi_1(x_n,y_n,f(x_n,y_n),\tau_n),\psi_2(x_n,y_n,f(x_n,y_n),\tau_n),\psi_3(x_n,y_n,f(x_n,y_n),\tau_n)) = 0 \tag{9.57}$$

which we can formally solve (granted the cooperation of the implicit function theorem) to find $\tau_n = h(x_n,y_n)$, which allows us to eliminate the unknown time differences τ_n. If we substitute $z = f(x,y)$ and $\tau = h(x,y)$ formally into the first two of the three equations (9.56), then we end up with two equations that have the form of a two dimensional iterated map:

$$\left. \begin{array}{l} x_{n+1} = G_1(x_n,y_n) \\ y_{n+1} = G_2(x_n,y_n). \end{array} \right\} \tag{9.58a}$$

By instead making the elimination $x = g(y,z)$, we would obtain a different iterated map

$$\left. \begin{array}{l} y_{n+1} = H_1(y_n,z_n) \\ z_{n+1} = H_2(y_n,z_n), \end{array} \right\} \tag{9.58b}$$

which describes the distribution of successive maxima of $z(t)$. In both cases, the iterated map contains partial information about some aspects of the dynamical system.

This is not the only way to define a Poincaré map, however. Another way is to consider the (unknown) times t_n when the trajectory passes through the (y,z)-plane in phase space with positive velocity $dx/dt > 0$. In this case, $x = \psi_1$ vanishes at the times t_n and a similar formal argument leads us to expect a two dimensional map of the form (9.58a), but with different functions on the right hand side:

$$\left. \begin{array}{l} y_{n+1} = M_1(y_n,z_n) \\ z_{n+1} = M_2(y_n,z_n). \end{array} \right\} \tag{9.58c}$$

There is a one-to-one correspondence between flows and maps derived from flows. Conservative flows can have as equilibria only elliptic and hyperbolic points and the same is true of the return maps. More generally, Poincaré's recurrence theorem applies so that bounded motion can only be periodic or quasiperiodic. Regular motion follows if there is an extra conservation law that holds for all finite times. Regular motion is stable against small changes in initial conditions, whether periodic or quasiperiodic. Here, the map must have two vanishing Liapunov exponents. Nonintegrability is necessary but not sufficient for deterministic chaos, as we illustrate in chapter 17. In a chaotic region, both the periodic and quasiperiodic orbits will be sensitive to small changes in initial conditions and the map must have two Liapunov exponents $\lambda_1 = -\lambda_2 < 0$ (chapter 6).

9.7 Nonintegrable rigid body motions

We have shown above that every rigid body problem in the six dimensional phase space defined by the principal axis frame has three conservation laws, so that the motion is confined to a three dimensional bounded subspace of phase space. However, we do not know, a priori, to which subspace the motion is confined. Therefore, we perform numerical experiments to search for evidence of nonintegrability in the form of chaos by constructing two dimensional return maps in the spirit of the last section. In particular, we study the distribution of intersections of trajectories with a single Cartesian plane in the six dimensional phase space.

Figures 9.5 through 9.7 show time series, intersections of trajectories with a phase plane, and also a Poincaré section for Lagrange's top, which is an integrable flow. All trajectories are either stable periodic or stable quasiperiodic, and therefore are insensitive to small changes in initial conditions. The parameter values chosen for the numerical integrations are: $I_1 = I_2 = 106.5\,\mathrm{g\,cm^2}$, $I_3 = 17.2\,\mathrm{g\,cm^2}$, $m = 16.0\,\mathrm{g}$, $(X,Y,Z) = (0,0,2.25\,\mathrm{cm})$. The initial conditions are $\omega_1(0) = 4\,\mathrm{s^{-1}}$, $\omega_2(0) = 5\,\mathrm{s^{-1}}$, $\omega_3(0) = 2\,\mathrm{s^{-1}}$, $F_1(0) = F_2(0) = 0$, and $F_3(0) = 15\,680$ dyne.

Figures 9.8 through 9.10 show plots for a nonintegrable flow, Euler's asymmetric top with one point fixed in space, and subject to a torque due to a constant gravitational field (constant driving force). Figures 9.8 and 9.9 show the time series and a Poincaré section. Figure 9.10 shows the difference between two time series due to slightly different initial conditions. All of the figures are typical of a system that exhibits sensitivity to small changes in initial conditions. In certain other parts of the phase space, this flow is nonchaotic. The parameter values and initial conditions are (same units as above): $I_1 = 100$, $I_2 = 30$, $I_3 = 17.2$, $(X,Y,Z) = (1,2,2.25)$. The initial conditions are $\omega_1(0) = 4$, $\omega_2(0) = 5$, $\omega_3(0) = 2$, $F_1(0) = 5000$, $F_2(0) = 6000$, and $F_3(0) = 15\,680$.

There are several lessons to be learned from this work. First, one needs only a *constant* driving force in a physical system (in this case gravity) along with asymmetry (at least two unequal moments of inertia) in order to produce deterministic chaos (the symmetric top is also nonintegrable except for the Lagrange and Kowalevskaya cases). The second lesson is that rigid body motions cannot be predicted in any practical way over long time intervals when they are observed in nature: there, in comparison with computation, we know initial conditions to within only few decimal precision, and perturbations due to the environment are completely uncontrolled. This combination of circumstances limits our ability to see into the future to relatively short time intervals (see also chapter 6).

Exercises

1. Find the principal moments of inertia about the center of mass of a book with a uniform mass density.

2. A uniform right circular cone of height h, half-angle α, and density ρ rolls on its side without slipping on a uniform horizontal plane in such a way that it returns to its original position in a time τ. Find expressions for the kinetic energy and the components of the angular momentum of the cone.

    (Whittaker, 1965)

3. Use $\Omega' = R\Omega\tilde{R}$ to show that the axial-vector equation $\omega' = (\det R)R\omega$ also holds. Show also that $M' = (\det R)RM$.

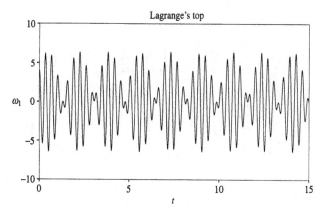

Fig. 9.5 Time series for Lagrange's top.

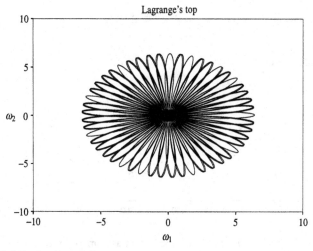

Fig. 9.6 Projection of the motion onto a phase plane for Lagrange's top.

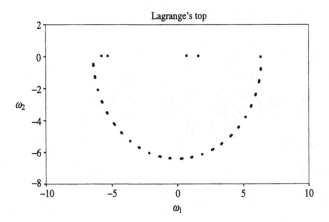

Fig. 9.7 Poincaré section for Lagrange's top.

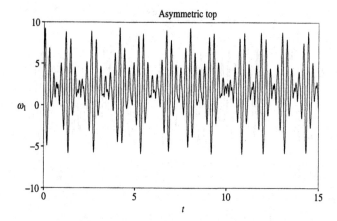

Fig. 9.8 Time series for an asymmetric top.

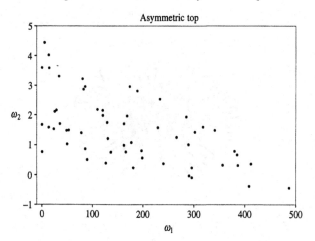

Fig. 9.9 Poincaré section for an asymmetric top.

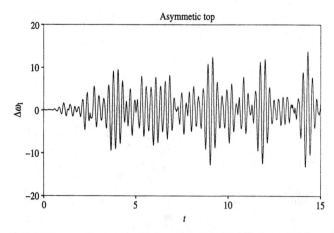

Fig. 9.10 Difference in time series for two slightly different initial conditions of an asymmetric top.

4. A plane pendulum consists of a uniform rod of length l and negligible thickness with mass m_1, suspended in a vertical plane by one end. At the other end a uniform disk of radius a and mass m_2 is attached to the rod with a pivot so that it can rotate freely in its own plane, which is the vertical plane. Set up the equations of motion in the Lagrangian formulation. (Note: use Chasle's Theorem and take care to choose the right angle for the motion of the disk!).

(Goldstein, 1980)

5. For exercise 4, locate the equilibria and determine the periods of small oscillations about any and all elliptic points. When is the motion of the pendulum periodic? Quasiperiodic?

6. An automobile is started from rest with one of its doors initially at right angles. If the hinges of the door are toward the front of the car, the door will slam shut as the automobile picks up speed. Obtain a formula for the time needed for the door to close if the acceleration f is constant, the radius of gyration of the door about the axis of rotation is r_o, and the center of mass is at a distance a from the hinges. Show that if f is 0.33 m/s² and the door is a uniform rectangle 1.33 m wide, the time will be approximately 3 seconds.

(Goldstein, 1980).

7. Using Euler's equations with two conservation laws for the torque-free case of an asymmetric rigid body, derive an integral representation of the period of oscillations about an elliptic point. (Hint: concentrate first on defining the two turning points for any one of the three variables x, y, or z.)

8. (a) Show that the components of the angular velocity along the space set of axes are given in terms of the Euler angles by

$$\omega_x = \dot{\theta}\cos\phi + \dot{\psi}\sin\theta\sin\phi,$$
$$\omega_y = \dot{\theta}\sin\phi - \dot{\psi}\sin\theta\cos\phi,$$
$$\omega_z = \dot{\psi}\cos\theta + \dot{\phi}.$$

(b) Show that the velocities ω_i are nonintegrable.

9. Show that the angular velocities ω_i' in equations (9.34), (9.35) and (9.36) are nonintegrable.

10. Express the rolling constraint of a sphere on a plane surface in terms of the Euler angles. Show that the conditions are nonintegrable and that the constraint is therefore nonholonomic.

(Goldstein, 1980).

11. For Lagrange's top, locate all equilibria of the effective potential $V(\theta)$ and determine the stability condition, the condition for an elliptic (rather than hyperbolic) point. In particular, show that near $\theta = 0$

$$V(\theta) = c + A\theta^2$$

where $A = I_3^2\omega_3'^2/8I_1 - mgl/2$ so that oscillations are confined to the region $\theta = 0$ only if $\omega_3'^2 > 4mglI_1/I_3^2$. What are the turning points θ_1 and θ_2 in this case?

12. For Lagrange's top, assume that there are two turning points $0 < \theta_2 < \theta_1 < \pi$. As the energy E' is increased $\theta_1 \to \pi$. Denote the energy by E_c' when $\theta_1 = \pi$. What is the

nature of the motion when $E' > E_c'$? (Hint: write $u = \cos\theta$ and rewrite energy conservation in the form $\dot{u}^2 = f(u) = (1 - u^2)(\alpha - \beta u) - (b - au)^2$. Analyze the problem in the context of discussing the zeros of $f(u)$, plotting $f(u)$ vs u qualitatively for both 'physical' and 'nonphysical' values of u.)

13. A symmetric top has no point fixed, as its contact point slips without friction on a horizontal plane. Formulate the Lagrangian and show that the problem is integrable by carrying out a procedure analogous to that used for Lagrange's top.

14. Consider a generally unsymmetric rigid body with one point fixed frictionlessly in space, and in a uniform gravitational field. Verify that E, $\tilde{F}F$, and $\tilde{M}F$ are always conserved quantities.

15. Consider Kowalevskaya's top. Integrate the Euler equations numerically for several different initial conditions and compute.

    (a) The time series $\omega_1(t)$.
    (b) The Poincaré section defined by the intersection of the trajectory with the (ω_1,ω_2)-plane. Compute a point (ω_1,ω_2) every time that $\omega_3 = 0$ with $\dot{\omega}_3 > 0$.

    In this exercise check your integrations for accuracy by integrating backward in time.

16. Consider the case of a completely asymmetric rigid body in the presence of gravity: $I_1 \neq I_2 \neq I_3$. Integrate the Euler equations numerically for several different initial conditions and compute.

    (a) The time series $\omega_1(t)$.
    (b) The Poincaré section defined by the intersection of the trajectory with the (ω_1,ω_2)-plane. Compute a point (ω_1,ω_2) every time that $\omega_3 = 0$ with $\dot{\omega}_3 > 0$.

    In this exercise check your integrations for accuracy by integrating backward in time.

10

Lagrangian dynamics and transformations in configuration space[1]

10.1 Invariance and covariance

We will prove explicitly that Lagrange's equations maintain their vectorial form under arbitrary differentiable coordinate transformations in configuration space, the n dimensional space of the variables q^i. Those equations have the same formal appearance

$$\frac{\mathrm{d}\partial T}{\mathrm{d}t\partial \dot{q}^k} - \frac{\partial T}{\partial q^k} = Q_k \tag{2.29'}$$

or

$$\frac{\mathrm{d}\partial L}{\mathrm{d}t\partial \dot{q}^k} - \frac{\partial L}{\partial q^k} = 0 \tag{2.35a}$$

regardless of the choice of generalized coordinates. This is equivalent to saying that we have a covariant formulation of the laws of mechanics. The Lagrangian is a scalar function under point transformations of coordinates in configuration space: expressed in terms of the new variables, $L(q,\mathrm{d}q/\mathrm{d}t,t) = L'(q',\mathrm{d}q'/\mathrm{d}t,t)$. By a *covariant* formulation of mechanics, we mean that for any other choice of generalized coordinates, $q'^k = F_k(q^1,\ldots,q^f,t)$, the Lagrangian L' also obeys Lagrange's equations in the new configuration space variables:

$$\frac{\mathrm{d}\partial L'}{\mathrm{d}t\partial \dot{q}'^k} - \frac{\partial L'}{\partial q'^k} = 0. \tag{10.1}$$

Covariance means only this formal preservation of form. It does mean that the Lagrangian is *invariant* under the transformations, and it does *not* follow that the second order differential equations for $q_i(t)$ that ones gets by substituting for T an explicit expression in terms of the generalized coordinates and generalized velocities, and for Q_i a definite function of the generalized coordinates, are the same in different coordinate systems. Invariance of the second order differential equations for the coordinates $q_i(t)$ would require *invariance* of the Lagrangian (requiring invariance of the matrix g and the force Q) under coordinate transformations, which is a much more stringent requirement than mere covariance. Lagrange's equations are, by construction, implicitly covariant, whereas invariance follows only if we restrict to transformations that are symmetries of the Lagrangian (and therefore yield conservation laws, by Noether's theorem). Covariance is implicit in the construction of chapter 2. We want to exhibit it explicitly here, term by term in Lagrange's equations.

---

[1] In this chapter (and in chapter 11) we will distinguish contravariant from covariant components of vectors by using upper and lower indices.

Newton's equations

$$\frac{d\vec{p}}{dt} = \vec{F} \tag{2.1a}$$

are covariant only with respect to transformations among inertial frames of reference, and the component form

$$\frac{dp_i}{dt} = F_i \tag{2.1b}$$

holds only for transformations among inertial frames where *Cartesian* coordinates are used.

Covariance does not require that T, U, and L are the *same* functions in two or more different coordinate systems. The Lagrangian itself is not generally *invariant* but is only a scalar function of the generalized coordinates and velocities under arbitrary configuration space transformations. An *invariant* Lagrangian would, in addition, satisfy the condition $L(q,dq/dt,t) = L(q',dq'/dt,t)$, which would mean that $L' = L$ *is the same function* in *two different* coordinate systems. In this case, the second order differential equations for the coordinates $q^i(t)$ in the two different frames are identical with each other, term by term. As we have shown in chapter 2, invariance under continuous transformations is intimately connected with symmetries and their corresponding conservation laws. Invariance and covariance are two entirely different ideas because the first is geometric while the second is only a formal mathematical rewriting of equations of motion. An example of invariance is the homogeneity and isotropy of empty Euclidean space (translation and rotation invariance), or the lack of resistance of absolute space to velocities but not to accelerations (Galilean relativity). The group of time-independent transformations leaving the kinetic energy invariant automatically leaves the matrix g invariant. The transformation group that leaves the entire Lagrangian invariant describes the symmetries of the physical system and (via Noether's theorem) yields the conservation laws of physics. The second idea, the idea of general covariance (covariance with respect to a large class of arbitrary differentiable coordinate transformations that leaves no point function invariant), is purely formal, is a mere rewriting of statements that are true from a restricted perspective (true only in inertial frames, for example) to make corresponding statements that are true from a larger perspective (true in arbitrary reference frames as well), and is devoid of any geometric or physical content. Given a set of differential equations and a group of transformations, covariance under that group is something that one generally can achieve by the definition of covariant derivatives (see chapter 8, velocity vs canonical momentum, equation (8.34) vs (8.39), and chapter 11 for specific examples).

Because our transformations in configuration space are allowed to depend explicitly upon the time t, the direct application of Lagrange's equations is permitted in noninertial frames while the same is not true of equation (2.1a). We have already used this fact in chapters 8 and 9 to discuss rotating frames and rigid body motions. One obtains exactly the same description of physics by starting with Newton's law in an inertial frame and then transforming to an accelerated frame.

Before exhibiting the general covariance of Lagrange's equations in configuration space, we first show how to express those equations in the standard form that is used in classical differential geometry and tensor analysis. In order to achieve both aims, we again divide vectors into two classes, as in chapter 8: those that transform in a certain restricted way under arbitrary configuration space coordinate transformations, and those that do not (those that are in some sense 'proper', and those whose behavior might be regarded as socially lacking in polite company[2]). We concentrate first upon defining the first-class vectors and discuss the second-class vectors afterwards, but every student of elementary physics is familiar with the second (and larger) class of vectors: they are anything with magnitude and direction. Every student of elementary linear algebra is familiar with the more sophisticated (therefore smaller) first class of vectors: they obey the rules of matrix multiplication under coordinate transformations (the first-class vectors always belong to the second class, but the reverse is not true). The main point is that the first-class vectors are coordinate-free and therefore do not change their formal appearance just because of a change of coordinates. In contrast, the coordinate-dependent vectors do not travel so well and so can look entirely different in different frames of reference. In fact, they can even disappear in certain coordinate systems, an act that is impossible for their more strictly regulated cousins.

In this section we restrict our discussion to coordinate transformations that are independent of the time t, because this permits the identification and use of a Riemannian metric in configuration space. The most general case is presented in section 10.4, based upon the results of section 10.3. In our restricted case, the kinetic energy is a scalar function of the generalized coordinates and velocities of the form $T = g_{ik}\dot{q}^i\dot{q}^k/2$, which is a matrix multiplication rule.

We can conveniently introduce an example of the contravariant components of a vector via the differential expression

$$dq'^i = \frac{\partial q'^i}{\partial q^j} dq^j, \tag{10.2}$$

so that the generalized velocity $d\mathbf{q}/dt$ transforms contravariantly,

$$\dot{q}'^i = \frac{\partial q'^i}{\partial q^j} \dot{q}^j, \tag{10.3}$$

whenever the coordinate transformations do not depend upon the time t. More generally, the components of any vector $\mathbf{A}$ that transform according to the rule

$$A'^i = \frac{\partial q'^i}{\partial q^k} A^k \tag{10.4a}$$

are called the contravariant components of that vector (as we shall see below, every coordinate-free vector must have *both* covariant and contravariant components). In agreement with our definition of first-class vectors, note that (10.4a) is merely the rule for matrix multiplication. Upper indices are used here to denote contravariant components.

The gradient of a potential energy provides the most common example of the covariant components of a vector: under an arbitrary coordinate transformation, with $U(q,t) = U'(q',t)$, we obtain

$$\frac{\partial U'}{\partial q'^k} = \frac{\partial q^l}{\partial q'^k}\frac{\partial U}{\partial q^l}, \tag{10.5}$$

---

[2] Many mathematicians are addicted to coordinate-free formulations.

and the covariant components of any vector **A** transform in exactly the same way, according to

$$A_i' = \frac{\partial q^l}{\partial q'^i} A_l. \tag{10.6a}$$

Again, this is just the rule for multiplication by an $f \times f$ matrix whose entries are $\partial q^l/\partial q'^i$. Lower indices are used to denote covariant components of vectors. Although we use upper indices on the generalized coordinates, the position q has only second-class status, as it transforms nonlinearly according to the rule $q'^k = F_k(q^1,\ldots,q^f,t)$. The position vector gains admission to first-class society only if we restrict ourselves to linear transformations.

Notice that covariant and contravariant components of first-class vectors transform under matrices that are the inverses of each other. That these matrices are inverses follows from the chain rule for differentiation:

$$\frac{\partial q^i \partial q'^j}{\partial q'^j \partial q^k} = \delta_{ik}, \tag{10.7}$$

which also allows us to define scalar products of any two vectors by combining contravariant and covariant components with each other in the following way:

$$A_i' B'^i = \frac{\partial q^l \partial q'^i}{\partial q'^i \partial q^k} A_l B^k = A_k B^k. \tag{10.8}$$

In the theory of exterior differential forms, A_k is called a vector (meaning tangent vector) while B^k is called a covector; the contraction of a vector with a covector defines the scalar product without assuming a metric: in general, we may take as our definition of scalar product the contraction rule $(\mathbf{A},\mathbf{B}) = A^i B_i$. Our idea of a first-class vector is therefore coordinate-free: given the components of the vector in one coordinate system, we use the rule of matrix multiplication to tell us the components of that same vector in any other coordinate system. The idea of a first-class vector is abstract because it is a coordinate-free idea. For example, a coordinate-free vector that vanishes in one frame must vanish in all frames because of the matrix multiplication rule. The same cannot be said of a second-class vector, like the velocity of a particle.

An $f \times f$ matrix is the representation in a given generalized coordinate system of an abstract linear operator. A linear operator **O** transforms any abstract vector in an abstract linear space into another vector in the same space and is represented by different $f \times f$ matrices in different coordinate systems but is, like the idea of a coordinate-free vector (or 'abstract' vector), a coordinate-free idea. A coordinate-free operator transforms like a tensor under our group of coordinate transformations in configuration space, and is represented in any coordinate system by a matrix whose components transform to any coordinate system like the direct product of two vectors. In other words, the covariant components of a second rank tensor form a matrix whose components T_{ij} transform in exactly the same way as the product of the covariant components of two vectors: since

$$A_i' B_i' = \frac{\partial q^l}{\partial q'^i} \frac{\partial q^k}{\partial q'^j} A_l B_k, \tag{10.6'a}$$

then the transformation rule for the covariant-component representation of **T** is

$$T'_{ij} = \frac{\partial q^l}{\partial q'^i} \frac{\partial q^k}{\partial q'^j} T_{lk}. \tag{10.6b}$$

Note that the direct product of two vectors is simply a matrix with entries $A_i B_j$. A contravariant representation of the same operator $\mathbf{T}$ is also possible: the contravariant components T^{ik} of the same tensor operator $\mathbf{T}$ transform like the product of the contravariant components of two vectors

$$T'^{ij} = \frac{\partial q'^i}{\partial q^l} \frac{\partial q'^j}{\partial q^k} T^{lk}. \tag{10.4b}$$

and mixed representations ('mixed tenors') are also possible. In a mixed representation, the components T^i_k transform like the direct product $A^i B_k$.

First-class vectors and tensors are coordinate-free ideas: they have an abstract existence independent of any particular coordinate system, and independent of whether we define coordinate systems at all. In particular, a vector equation of the form $\mathbf{A} = 0$ holds in all coordinate systems (no new terms can pop up and destroy the equality by transforming coordinates). The advantage of this point of view is: if you can rewrite the laws of physics in terms of coordinate-free vectors, in terms of abstract vectors, then the abstract vector equations of motion may be applied directly in any coordinate system, with no need whatsoever to pay any particular respect to Galilean frames of reference. This does not mean that Galilean frames and absolute space have disappeared or have lost their standing as the foundation of physics, only that they are 'second-class' ideas that are hidden in the closet in a first-class society, which shows the real danger of formalism: the most important ideas can be hidden from view in order to satisfy someone's notion of what is or is not proper. Finally, Lagrange's equations have precisely the form $\mathbf{A} = 0$, where $\mathbf{A}$ is a coordinate-free vector, as we show explicitly in section 10.4.

10.2 Geometry of the motion in configuration space

We can use the kinetic energy $T = g_{ij} \dot{q}^i \dot{q}^j / 2$ for the case of time-independent coordinate transformations to define an $f \times f$ matrix g, called the metric, by

$$g_{ij} = \frac{\partial \vec{R}_k}{\partial q_i} \cdot \frac{\partial \vec{R}_k}{\partial q_j}, \tag{10.9}$$

and where the original coordinates have been rescaled by the masses in the problem. The identification of the matrix g as a Riemannian metric is straightforward: $ds^2 = g_{ij} dq^i dq^j$ is the line element in configuration space, the square of the infinitesimal distance between two points, and $T = (ds/dt)^2/2 = g_{ij} dq^i dq^j/2dt^2$ is the kinetic energy, where $g_{ij} > 0$ because $T \geq 0$. The system is imbedded in an Nd dimensional Euclidean space, where N is the number of particles and d is the dimension of space ($d = 1, 2,$ or 3) so that the f dimensional configuration space is generally Riemannian if there are constraints (correspondingly, if the metric is nonEuclidean, then there is no global transformation to a Cartesian coordinate system[3], one where $g_{ij} = \delta_{ij}$). If there are no constraints, then the metric is Euclidean because the original Nd dimensional space is Euclidean.

[3] The ability to define a Cartesian coordinate system globally is equivalent to stating that the space is Euclidean.

Here's another way to say it. Since the matrix g is positive definite and symmetric, it can be diagonalized locally and has only positive eigenvalues, so that $ds^2 = \lambda_i d\xi^{i2}$ in a special local frame (different points $(q^1,\ldots,q^f)$ yield different local frames where g is diagonal). If the $\lambda_i = 1$ then the ξ_i are Cartesian coordinates and the space is locally (but not necessarily globally) Euclidean (one can always draw Cartesian coordinates in the tangent plane to a sphere, for example).

Motion that is constrained to the surface of the unit sphere is an example of motion on a Riemannian surface: with

$$T = \tfrac{1}{2}(\dot{\theta}^2 + \sin^2\theta\,\dot{\phi}^2),\tag{10.10}$$

we obtain from Lagrange's equations that

$$\frac{d(\dot{\phi}\sin^2\theta)}{dt} = 0 \tag{10.11}$$

so that

$$\dot{\phi}\sin^2\theta = M \tag{10.12}$$

is constant. Also,

$$\ddot{\theta} = 2\sin\theta\cos\theta\,\dot{\phi}^2, \tag{10.13}$$

so that after one integration we obtain

$$\frac{\dot{\theta}^2}{2} = \varepsilon - \frac{M^2}{2\sin^2\theta} \tag{10.14}$$

where ε is constant. We can leave it as an easy exercise for the reader to see that these combine to yield a differential equation for $\phi = f(\theta)$ that is identical to the one derived earlier in section 2.2 for geodesic motion on the surface of a sphere. The quickest way to see that we have a geodesic is to choose a coordinate system where both ϕ and its time derivative vanish initially, so that they remain zero at all future times. It then follows from $d^2\theta/dt^2 = 0$ that $\theta = \omega t + \theta_o$, which is the equation of a great circle on the sphere.

Since the kinetic energy is a scalar under transformations to and among generalized coordinate systems, and since dq^i/dt are contravariant components of the generalized velocity vector, then the operator g must transform like a second rank tensor:

$$g'_{ij} = \frac{\partial q^l}{\partial q'^i}\frac{\partial q^k}{\partial q'^j}g_{lk}. \tag{10.15}$$

Otherwise, the line element ds^2 would not be a scalar.

Given a coordinate system, there are two different ways to define the components of an abstract vector: in terms of either the contravariant components, or the covariant ones. The contravariant components V^i of a vector $\mathbf{V}$ give rise to covariant components of the same vector by the operation $V_i = g_{ij}V^j$, where g_{ij} is the metric tensor. In Cartesian coordinates, the two possible definitions coincide because in that case $g_{ij} = \delta_{ij}$. The metric not only allows us to define the idea of the distance between two points in configuration space, it also permits us to define the length of a vector: the length squared of any vector $\mathbf{V}$ is given by $(\mathbf{V},\mathbf{V}) = V^i V_i = g_{ij}V^i V^j$. Furthermore, the scalar product of two abstract vectors $\mathbf{A}$ and $\mathbf{B}$ has the form $(\mathbf{A},\mathbf{B}) = A_i B^i = g_{ij}A^i B^j$ and behaves as a scalar function of the generalized coordinates (and the generalized velocities and the time) under coordinate transformations (g_{ij} depends upon the f generalized coordinates q_i alone).

We have restricted to time-independent transformations, those where $T = g_{ij}\dot{q}^i\dot{q}^j/2$ defines the metric g_{ij} for a particular choice q_i of generalized coordinates. Since time-independent transformations guarantee that the metric g_{ij} depends upon the generalized coordinates alone, any explicit time-independence can enter into the equations of motion only through the potential U. Now we exhibit the explicit form of the equations of dynamics in configuration space.

It follows from Lagrange's equations that

$$\frac{\partial T}{\partial \dot{q}^i} = g_{ij}\dot{q}^j \tag{10.16}$$

so that

$$\frac{d\partial T}{dt\partial \dot{q}^i} = g_{ik}\ddot{q}^k + \frac{\partial g_{ik}}{\partial q^j}\dot{q}^j\dot{q}^k = g_{ik}\ddot{q}^k + \frac{1}{2}\left(\frac{\partial g_{ik}}{\partial q^j}\dot{q}^j\dot{q}^k + \frac{\partial g_{ij}}{\partial q^k}\dot{q}^j\dot{q}^k\right) \tag{10.17}$$

and with

$$\frac{\partial T}{\partial q^i} = \frac{1}{2}\frac{\partial g_{jk}}{\partial q^i}\dot{q}^j\dot{q}^k \tag{10.18}$$

we then obtain

$$\frac{d\partial T}{dt\partial \dot{q}^i} - \frac{\partial T}{\partial q^i} = g_{ij}\ddot{q}^j + [jk,i]\dot{q}^j\dot{q}^k \tag{10.19}$$

where

$$[jk,i] = \frac{1}{2}\left(\frac{\partial g_{ji}}{\partial q^k} + \frac{\partial g_{ki}}{\partial q^j} - \frac{\partial g_{jki}}{\partial q^i}\right) \tag{10.20}$$

is called a Christoffel symbol of the first kind. Lagrange's equations of the first kind then take the form

$$g_{ij}\ddot{q}^j + [jk,i]\dot{q}^j\dot{q}^k = Q_i. \tag{10.21}$$

It is an elementary exercise in tensor analysis to show that both sides of this equation transform like covariant components of vectors in configuration space, which exercise would prove that the law of motion is itself covariant. A covariant equation is one where both sides of the equation transform in exactly the same way. The idea, therefore, is to prove that

$$\left(\frac{d\partial T}{dt\partial \dot{q}'^i} - \frac{\partial T}{\partial q'^i}\right) = \frac{\partial q^l}{\partial q'^i}\left(\frac{d\partial T}{dt\partial \dot{q}^l} - \frac{\partial T}{\partial q^l}\right). \tag{10.22}$$

If, in addition, the generalized force Q_i transforms like a covariant vector, as indeed it does if it is the gradient of a scalar potential, then both sides of Lagrange's equations transform the same way. That is, they transform covariantly. However, we do not prove this here but instead prove the covariance of Lagrange's equations by the method of section 10.3 where time-dependent transformations are allowed. We shall see that we do not need a metric in order to prove the general covariance of Lagrange's equations.

Indeed, if the idea of manifest covariance were restricted to time-independent transformations, then it would be of very little interest.

Equation (10.21) has an identifiable geometric meaning: with $Q_l = 0$, it is the equation of a geodesic, the shortest distance between two points in our f dimensional configuration space. This interpretation follows easily from asking for the equation of the curve for which the arc length

$$\int_1^2 ds = \int_1^2 (g_{jk}dq^j dq^k)^{1/2} = \int_1^2 (g_{jk}\dot{q}^j \dot{q}^k)^{1/2}dt \tag{10.23}$$

is an extremum (this is an easy exercise for the reader to prove).

Defining the inverse metric tensor g^{-1} by $g^{li}g_{ik} = \delta_{lk}$, we can use the index raising operation $V^l = g^{lk}V_k$ to obtain Lagrange's equations of the first kind in the form

$$\ddot{q}^l + \{jk,l\}\dot{q}^j \dot{q}^k = Q^l, \tag{10.24}$$

where $\{jk,l\}$ is called a Christoffel symbol of the second kind:

$$\{jk,l\} = g^{lp}[jk,p]. \tag{10.25}$$

That Lagrange's equations are covariant with respect to arbitrary differentiable (single-valued and invertible) transformations in configuration space does *not* mean that all frames in configuration space are physically equivalent for doing experimental physics[4], because the *solutions* obtained in different frames generally are *not* identical in form but are related by the (generally nonlinear) coordinate transformation $q'_i = F_i(q_1,\ldots,q_n,t)$ and its inverse $q_k = f_k(q'_1,\ldots,q'_n,t)$. In transformations among Cartesian *inertial* frames, in contrast, Newton's equation of free-particle motion is not merely covariant but is invariant: *solutions of Newton's equation for free-particle motion in different inertial frames are identical for identical initial conditions.* This is the basis for the fact that one cannot perform an experiment (nor can one find a solution) that distinguishes one inertial frame from another. This, and not covariance, is what one means by 'physics', and it is Galilean invariance, not general covariance, that enforces this particular symmetry. Invariance principles form the basis for reproducibility of experiments regardless of absolute position and absolute time, and this invariance is only masked by the general covariance of Lagrange's equations under arbitrary differentiable coordinate transformations in configuration space. General covariance of equations of motion seems always to be possible because it merely reflects a mathematical rewriting of the equations of motion that neither reflects nor changes the underlying geometry of space or the physics (see Weyl, 1963, and also Wigner, 1967).

We are indebted to B. Riemann (1826–1866) for the idea of configuration space as a differentiable manifold in f dimensions that is not necessarily Euclidean. Before Riemann, there was the idea of a curved surface embedded in a three dimensional Euclidean space (Gauss (1777–1855), Bolyai (1802–1860), Lobachevskii (1793–1856)), but it was Riemann who intuitively extended this idea to higher dimensions by introducing the idea of a manifold, the generalization to f dimensions of a surface that may be neither homogeneous nor isotropic, so that there may be no translational or rotational invariance globally (in which case Cartesian coordinate systems are only possible in infinitesimal neighborhoods that are analogous to a tangent plane to a point on a sphere). The English word manifold (many-fold) comes from the German word

---

[4] General covariance does not even guarantee that all coordinate systems are available! See section 11.7 for details.

Mannigfaltikeit used by Riemann, which means simply diversity or variety. Euclidean space has no diversity and no variety because it is perfectly homogeneous and isotropic. Riemann's idea of a differentiable manifold (or of a surface imbedded in a Euclidean space) with geometric diversity can be described qualitatively as a patchwork or quiltwork of 'local' (meaning infinitesimally small) Euclidean hyperplanes. One imagines that the Riemannian surface is enveloped by a lot of very small tangent planes, each of which is Euclidean. In each of these planes, infinitesimally near to any point on the surface, we can describe the dynamics locally in Cartesian fashion. To make this useful, there must be a coordinate transformation that connects the local coordinates in any two neighboring patches with each other. The qualitative aspects of Riemannian geometry on closed, orientable surfaces of constant curvature can be understood by studying geometry and vector analysis on a two-sphere that is imbedded in three dimensional Euclidean space (one can always think of a Riemannian space of f dimensions as imbedded in a Euclidean space of $f + 1$ dimensions whenever it is convenient to do so).

On a 'flat' manifold, one where $g_{ij} = \delta_{ij}$, all Christoffel symbols vanish so that the geodesic equations reduce to the equations of a straight line in Euclidean space: $d^2 q^i/dt^2 = 0$. In general, configuration space (the Lagrangian manifold) can be 'curved' (Riemannian) due to constraints. A curved manifold is one where there is no coordinate system where the Christoffel symbols all vanish identically. In this case, there is no way to construct a Cartesian coordinate system globally, meaning over a finite portion of the manifold. Conversely, if there exists a global coordinate system where the Christoffel symbols all vanish identically, then the space is flat (Cartesian coordinates can always be constructed locally, meaning infinitesimally, in any tangent hyperplane to the manifold). In that coordinate system the g_{ij} are all constants. By a diagonaliz-ation of g and rescaling of variables, we can always construct a Cartesian coordinate system. A flat space is, by definition, one where it is possible to construct a Cartesian coordinate system.

A Cartesian manifold is any flat space with or without a metric. With a metric, a Cartesian manifold becomes a Euclidean space. In chapter 6 the torus of integrable two degree of freedom Hamiltonian systems is a Cartesian manifold (without a metric) because we were able to *define* that torus by the Cartesian axes (phase space variables are always Cartesian) $P_k = $ constant and $Q_k = \omega_k t + Q_{k_o}$. On the other hand, a sphere with a handle attached to it is not flat even though it is topologically equivalent to the torus. Topology ignores metric and curvature properties and classifies geometric objects in the most crude (most abstract) geometric way possible, which is to say purely qualitatively. The only topological invariants are the number of holes in the surface (the number of holes is called the genus of the object). Spheres with handles and tori with one hole (figure eight tori are also possible) are topologically, but not metrically, the same because they are both of genus one.

Vector algebra on a sphere (or on any curved manifold) is very different from what we are used to in the Euclidean plane. In the latter case we can freely translate vectors to form scalar and vector products, and we can rotate them without changing their lengths, but it makes no sense to compare (to add or subtract) vectors at two different points on a sphere. The reason is that the length of a vector depends on the metric g, and the metric varies from one point to another over the sphere. Geometrically, you cannot translate a vector from one place to another without changing its direction or length or both. Another way to say this is that it is impossible to construct a vector that has components that are constant over the entire sphere, or even over any finite section of it. The very idea of translating any vector parallel to itself is partly lost because this procedure would trace out a curve that is a straight line in Euclidean space, and straight lines do not exist on a sphere except infinitesimally in a tangent plane. The generalization, a geodesic or a great circle on the sphere, has some properties of a straight line in Euclidean space.

Finally, there are no metric-based geometric conservation laws in an arbitrary curved space: nothing is left invariant during the motion from one point to another because there is no underlying symmetry (differential equations that may be trivially integrable in a Euclidean space are generally nonintegrable in a curved space). Of all possible curved spaces, Helmholtz (who tried to geometrize Newtonian mechanics in configuration space) argued in favor of spaces of constant curvature because there one retains some degree of symmetry and therefore some conservation laws. Riemann pointed out that constant-curvature spaces are exactly the spaces where rigid body motions are possible (a sphere in n dimensions is an example of a space of constant curvature).

10.3 Basis vectors in configuration space[5]

It is now time to take stock of what we've been doing. We have introduced a refined or preferred idea of what constitutes a vector: in our considerations of general coordinate transformations in configuration space, a first-class vector is not merely any object with f components. Rather, in configuration space, a vector **A** is represented by any object whose components transform according to the rules of matrix multiplication either as

$$A'_i = \frac{\partial q^j}{\partial q'^i} A_j, \tag{10.26}$$

or as

$$A'^i = \frac{\partial q'^i}{\partial q^j} A^j, \tag{10.27}$$

Any f-component object that does not transform in one of these two ways does not represent a coordinate-free vector under the group of general coordinate transformations in configuration space. We repeat: an abstract vector is not merely anything with Nd configuration-space components, but is an object that transforms in one of two specific ways under a very large group of transformations, and the generalized coordinates are *not* the components of such a vector under general nonlinear coordinate transformations (see chapter 8 for an exception, that of *linear* transformations). Furthermore, the components dq^i/dt of the generalized velocity transform like the contravariant components of an abstract vector *only* when the transformations are time-independent (conversely, when the dq^i/dt transform like the components of a vector, then transformations to time-dependent reference frames are forbidden): with $q^i = f^i(q'^1, \ldots, q'^n, t)$ and $q'^k = F^k(q^1, \ldots, q^n, t)$, then

$$\dot{q}^i = \frac{\partial q^i}{\partial q'^k} \dot{q}'^k + \frac{\partial f^i}{\partial t}, \tag{10.28}$$

---

[5] The results of this section are used in section 12.1. See also Hauser (1965), Margenau and Murphy (1956), or Adler, Bazin, and Schiffer (1965) for discussions of basis vectors in configuration space.

so that the contravariant character of the generalized velocity is lost *unless* the transformation functions f^i are all independent of the time. In addition, because

$$T = \tfrac{1}{2} g_{ij} \dot{q}^i \dot{q}^j + b_i \dot{q}^i + \frac{a}{2}, \tag{10.29}$$

where

$$g_{ij} = \frac{\partial \vec{R}_k}{\partial q^i} \cdot \frac{\partial \vec{R}_l}{\partial q^j}, \quad b_i = \frac{\partial \vec{R}_k}{\partial q^i} \cdot \frac{\partial \vec{R}_k}{\partial t}, \text{ and } a = \frac{\partial \vec{R}_k}{\partial t} \cdot \frac{\partial \vec{R}_k}{\partial t}, \tag{10.30}$$

we cannot define a Riemannian metric because the condition that $ds^2 = v^2 dt^2 = g_{ij} dq^i dq^j$ does not hold any more. However, we can still define a tensor **g** that can be used to raise and lower indices, one that includes the Riemannian metric as a special case. We gain some insight by this approach because it forces us to construct basis vectors in configuration space, something that we have managed to ignore so far.

We begin once more with the transformation from Cartesian to generalized coordinates,

$$\vec{r} = (x_i, y_i, z_i) = \vec{R}_i(q_1, \ldots, q_n, t) = (x_i(q_1, \ldots, q_n, t), y_i(q_1, \ldots, q_n, t), z_i(q_1, \ldots, q_n, t)) \tag{10.31}$$

and define our Nd basis vectors by

$$\vec{e}_i = \sum_{j=1}^{N} \frac{\partial \vec{R}_j}{\partial q^i}. \tag{10.32}$$

These f basis vectors are linearly independent but are generally not orthogonal, nor are they necessarily normalized to unit length. If we expand the right hand side in terms of the f mutually orthogonal Cartesian basis vectors (there are d for the subspace of each particle, and the particle's subspaces are mutually orthogonal), then the expansion coefficients are the partial derivatives with respect to the q^i of the Nd Cartesian coordinates (x_j, y_j, z_j). Independently of the last section, it is now possible to define the contravariant components of a vector **V** by the expansion

$$\mathbf{V} = V^i \vec{e}_i. \tag{10.33}$$

Contravariance of the components V^i follows easily: first, if we transform to a new frame, then we get n new basis vectors defined by

$$\vec{b}_i = \frac{\partial q^l}{\partial q'^i} \vec{e}_l, \tag{10.34}$$

so that the basis vectors transform like covariant ones. This invariance of form of the vector **V**

$$\mathbf{V} = V^i \vec{e}_i = \bar{V}^i \vec{b}_i \tag{10.35}$$

under transformations of coordinates holds only if

$$\bar{V}^i = \frac{\partial q'^i}{\partial q^k} V^k, \tag{10.36}$$

which means that the components V^i transform contravariantly. It follows that the object g, defined now as

$$g_{ij} = \vec{e}_i \cdot \vec{e}_j, \tag{10.37}$$

transforms like the covariant components of a second rank tensor **g**:

$$g'_{ij} = \vec{b}_i \cdot \vec{b}_j = \frac{\partial q^l \partial q^k}{\partial q'^i \partial q'^j} g_{lk}. \tag{10.38}$$

Furthermore, the matrix g does the same job as a metric in the following limited sense: although we cannot define the distance between two points in configuration space in the Riemannian way, the scalar product of two vectors is given by

$$(\mathbf{V}, \mathbf{U}) = V^i U^j g_{ij} = V^i U_i = V_i U^i \tag{10.39}$$

and is a scalar function of the generalized coordinates, the generalized velocities, and the time, under arbitrary coordinate transformations. In particular, the length of a vector **U** is given by $\sqrt{(\mathbf{U}, \mathbf{U})}$.

In addition, we can define the covariant components of the vector

$$\mathbf{V} = V^i \vec{e}_i \tag{10.40}$$

by

$$V_i = g_{ij} V^j \tag{10.41}$$

so that the contravariant basis vectors in

$$\mathbf{V} = V_i \vec{f}^i \tag{10.42}$$

are given by

$$\vec{e}_i = g_{ji} \vec{f}^j, \tag{10.43}$$

which can be inverted to yield

$$\vec{f}^k = g^{ik} \vec{e}_i, \tag{10.44}$$

where $gg^{-1} = g^{-1}g = I$, and where the inverse matrix g^{-1} has elements g^{ij}.

The tensor

$$g_{ij} = \vec{e}_i \cdot \vec{e}_j = \sum_{k,l=1}^{N} \frac{\partial \vec{R}_k}{\partial q^i} \cdot \frac{\partial \vec{R}_l}{\partial q^j} = \frac{\partial \vec{R}_k}{\partial q^i} \cdot \frac{\partial \vec{R}_k}{\partial q^j} \tag{10.45}$$

(the last equality follows because the N subspaces of the N different particles are orthogonal) is not a metric whenever the transformations are time-dependent, because in that case one *cannot* make the identification

$$ds^2 = g_{ij} dq^i dq^j \tag{10.46}$$

where $T = (ds/dt)^2$ is the kinetic energy of the N bodies, due to the two extra terms in the kinetic energy. Nevertheless, we have a tensor that we can use to define both lengths and scalar products of vectors, and we can also use it to formulate the general covariance of Lagrange's equations under arbitrary point transformations in configuration space, including time-dependent ones.

10.4 Canonical momenta and covariance of Lagrange's equations in configuration space

Now, we are prepared to display the general covariance of Lagrange's equations under general linear and nonlinear time-dependent coordinate transformations.

First, note that,

$$\frac{\partial T}{\partial q^i} = \frac{\partial q'^k \partial T'}{\partial q^i \partial q'^k} + \frac{\partial T'}{\partial \dot{q}'^k}\frac{\partial \dot{q}'^k}{\partial q^i}, \tag{10.47}$$

so that the configuration-space gradient term in Lagrange's equations does not transform like a covariant vector. However, the canonical momentum

$$p_i = \frac{\partial T}{\partial \dot{q}^i} \tag{10.48}$$

always transforms like the covariant components of an abstract vector **p**, whether or not the generalized velocities dq^i/dt transform like the contravariant components of a coordinate-free vector. Since the transformation $q'^k = F^k(q^1,\dots,q^n,t)$ and its inverse are independent of the generalized velocities, then

$$\frac{\partial T}{\partial \dot{q}^i} = \frac{\partial \dot{q}'^k \partial T'}{\partial \dot{q}^i \partial \dot{q}'^k}. \tag{10.49}$$

Since

$$\dot{q}'^i = \frac{\partial q'^i}{\partial q^k}\dot{q}^k + \frac{\partial q'^i}{\partial t} \tag{10.50a}$$

where $q'^i = F_i(q^1,\dots,q^f,t)$, and since the transformation functions F_i in $q'^i = F_i(q^1,\dots,q^f,t)$ and their derivatives are independent of the generalized velocities, it follows directly from (10.50a) that

$$\frac{\partial \dot{q}'^k}{\partial \dot{q}^i} = \frac{\partial q'^k}{\partial q^i}, \tag{10.50b}$$

which states the identity of two separate $f \times f$ matrices. Insertion of (10.50b) into (10.49) yields

$$\frac{\partial T}{\partial \dot{q}^i} = \frac{\partial q'^k \partial T'}{\partial q^i \partial \dot{q}'^k}, \tag{10.51}$$

which proves that the canonical momenta

$$p_i = \frac{\partial T}{\partial \dot{q}^i} \tag{10.52}$$

are the covariant components of an abstract vector **p**. The Lagrangian L is a scalar function of generalized coordinates and velocities by construction, although the Lagrangian is not generally invariant under coordinate transformations except for the case where the transformations are time-independent *and* the transformations are restricted to those that simultaneously are symmetries of the kinetic and potential energies. In the modern language of differential forms, the canonical momentum always transforms like a covector whether or not the components dq^i/dt transform like a vector.

When $L = T - U$, where the potential U depends upon the generalized coordinates and the time alone, then we also can write

$$p_i = \frac{\partial L}{\partial \dot{q}^i}. \tag{10.53}$$

The canonical momenta play a central role in the Hamiltonian formalism, and the Hamiltonian formalism is central for understanding integrability and nonintegrability in classic and quantum mechanics.

Even if the components of a vector $\mathbf{A}$ transform by matrix multiplication rules under our large group of transformations, it does not follow that the time derivative $d\mathbf{A}/dt$ is also a coordinate-free vector. Therefore, to see how Lagrange's equations transform in configuration space we must calculate as follows: since

$$\frac{\partial T}{\partial q^i} = \frac{\partial T'}{\partial q'^k}\frac{\partial q'^k}{\partial q^i} + \frac{\partial T'}{\partial \dot{q}'^k}\frac{\partial \dot{q}'^k}{\partial q^i}, \tag{10.47}$$

and since

$$\frac{\partial T}{\partial \dot{q}^i} = \frac{\partial T'}{\partial \dot{q}'^k}\frac{\partial \dot{q}'^k}{\partial \dot{q}^i} + \frac{\partial T'}{\partial q'^k}\frac{\partial q'^k}{\partial \dot{q}^i} = \frac{\partial T'}{\partial \dot{q}'^k}\frac{\partial \dot{q}'^k}{\partial \dot{q}^i}, \tag{10.54}$$

it follows that

$$\frac{d}{dt}\frac{\partial T}{\partial \dot{q}^i} = \frac{d}{dt}\left(\frac{\partial T'}{\partial \dot{q}'^k}\right)\frac{\partial \dot{q}'^k}{\partial \dot{q}^i} + \frac{\partial T'}{\partial \dot{q}'^k}\frac{d}{dt}\frac{\partial \dot{q}'^k}{\partial \dot{q}^i}. \tag{10.55}$$

This result can be rewritten as

$$\frac{d}{dt}\frac{\partial T}{\partial \dot{q}^i} = \frac{dp'_i}{dt}\frac{\partial q'^k}{\partial q^i} + p'_k\frac{d}{dt}\frac{\partial q'^k}{\partial q^i} = \frac{dp'_i}{dt}\frac{\partial q'^k}{\partial q^i} + p'_k\frac{\partial \dot{q}'^k}{\partial q^i}. \tag{10.56}$$

Combining the above results yields

$$\left(\frac{dp_i}{dt} - \frac{\partial T}{\partial q^i}\right) = \left(\frac{dp'_i}{dt} - \frac{\partial T'}{\partial q'^i}\right)\frac{\partial q'^k}{\partial q^i}, \tag{10.57}$$

which is the desired matrix multiplication rule. Therefore, every term in Lagrange's equations, whether Lagrange's equations are written in the first form

$$\frac{dp_i}{dt} - \frac{\partial T}{\partial q^i} = Q_i \tag{10.58}$$

or in the second form

$$\frac{dp_i}{dt} = \frac{\partial L}{\partial q^i}, \tag{10.59}$$

obeys exactly the same transformation rule, that of matrix multiplication. Lagrange's equations are covariant under general time-dependent transformations in configuration space.

The converse is that when different quantities in the same equation transform in different ways (some by the matrix multiplication rule, others not), then that equation is not covariant, is not an abstract vector equation that is coordinate-free. That does not mean that the equation is wrong or useless (witness Newton's laws), only that to determine the right equation in another coordinate system you must transform each term separately (we did this for Newton's law in chapter 8 in order to show how to derive covariant equations). The worst blunder in respectable society would be to proclaim that one can write $\mathbf{F} = d\mathbf{p}/dt$ in rotating and/or linearly accelerated frames, because force is a coordinate-free idea whereas velocity and acceleration are not[6].

That each term in Lagrange's equations transforms as a covariant vector under the general point transformations $q^i = f^i(q'^1,\dots,q'^n,t)$ proves the assertion made earlier above: Lagrange's equations are covariant under general time-dependent linear and nonlinear coordinate transformations in configuration space (are abstract vector equations). A shorter way to say it is that the vector equations $\mathbf{A} = 0$ and $\mathbf{B} = 0$ hold and are coordinate-free, where in any coordinate system the f components of the vector $\mathbf{A}$ have the form $A_i = dp_i/dt - \partial L/\partial q^i$ and $\mathbf{B}$ always has the f components $B_i = p_i - \partial L/\partial \dot{q}^i$. Coordinate-free statements are sometimes convenient. If all of the components of an abstract vector vanish in one frame, then they vanish in all frames (this is not true of a second-class vector like the linear momentum or the generalized velocity). The canonical momentum is such a vector, and so the vanishing of all of the components p_i of the canonical momentum in any frame guarantees the same result in all possible frames, inertial or noninertial. In contrast, if the velocity vector vanishes in an inertial frame (or in a noninertial frame) then it does not vanish in noninertial frames (or does not vanish in inertial frames). If one wants to state coordinate-free conditions for equilibrium, or to make statements of Newton's first and second laws that are free of the restriction to inertial frames, then this can only be done by rewriting Newton's laws in a form that is acceptable in arbitrary frames by using, not the linear momentum $m\mathbf{v}$, but the *canonical* momentum $\mathbf{p}$ (see chapter 8 on rotating frames for applications of this aspect of the covariance principle).

We dedicate this chapter to the memory of Lagrange. As one reviewer noted, it was presented without the benefit of a single (badly drawn) figure.

Exercises

1. Prove that the integration of equation (10.14) yields the geodesic discussed in section 2.2.

2. By comparison of (10.11) and (10.13) with (10.24), write down the Christoffel symbols $\{jk,l\}$ for the motion of a point particle confined to a sphere of unit radius.

3. Identify and use the metric g_{ij} for spherical coordinates in three dimensional Euclidean space to write down the covariant $\vec{e}_i$ and contravariant $\vec{f}_i$ basis vectors for spherical coordinates. (Hint: start with the kinetic energy for a single particle in spherical coordinates.)

4. Construct the gradient operator

$$\nabla = \vec{e}_i \frac{\partial}{\partial q^i}$$

in spherical coordinates $(q^1,q^2,q^3) = (r,\theta,\phi)$, and compute the formula for $\nabla \cdot \vec{V}$ where $\vec{V} = (\dot{q}^1,\dots,\dot{q}^f)$ is the tangent vector in configuration space (see also exercise 15 in chapter 2 and exercise 5 in chapter 3).

---

[6] One generally cannot get rid of forces by coordinate transformations. Gravity is the only known exception: one can get rid of gravity *locally* by jumping off a cliff into a vacuum, or by falling around the earth in a satellite outside the atmosphere. In both cases there is the phenomenon of weightlessness.

11

Relativity, geometry, and gravity

11.1 Electrodynamics

We begin with Maxwell's equations of the electromagnetic field,

$$
\left.
\begin{aligned}
\nabla \cdot \vec{E} &= 4\pi\rho \\[6pt]
\nabla \times \vec{E} &= -\frac{1}{c}\frac{\partial \vec{B}}{\partial t} \\[6pt]
\nabla \cdot \vec{B} &= 0 \\[6pt]
\nabla \times \vec{B} &= \frac{1}{c}\frac{\partial \vec{E}}{\partial t} + \frac{4\pi}{c}\vec{j},
\end{aligned}
\right\}
\tag{11.1a}
$$

where static positive (or negative) electric charges are sources (or sinks) of conservative electric field lines.

A time-varying magnetic flux through a closed circuit creates a *nonconservative* electric field within the conductor, one where electric field lines loop around the wire and close on themselves. This causes free charges to move as if there were a potential difference in the circuit, even though there is none, and follows from integrating

$$
\nabla \times \vec{E} = -\frac{1}{c}\frac{\partial \vec{B}}{\partial t}
\tag{11.2a}
$$

over the area enclosed by a closed circuit and then applying Stoke's theorem to obtain the 'induced emf',

$$
\varepsilon = \oint \vec{E} \cdot d\vec{r} = -\frac{\partial \Phi_B}{\partial t},
\tag{11.2b}
$$

where Φ_B is the flux of the magnetic field B through the circuit. This result shows that the electric field that follows from a vector potential alone,

$$
\vec{E} = -\frac{1}{c}\frac{\partial \vec{A}}{\partial t},
\tag{11.3}
$$

is *nonconservative*, because the closed loop integral representing the emf ε in the circuit does not vanish. In contrast, a conservative electric field is given as the gradient of a scalar potential and therefore can only produce a field that does not close on itself but instead has electric charges as sources and sinks. Let us now look at the electrical analog of Faraday's law of induction.

Every moving charge causes source-free magnetic field lines that loop around and end on themselves (Ampère's law), but a time-varying electric flux in the absence of charge produces the same effect. If we think of a parallel-plate capacitor in an electrical circuit (it helps conceptually in what follows to let the plates be far apart from each other, with only a vacuum between them), then by integrating

$$\nabla \times \vec{B} = \frac{1}{c}\frac{\partial \vec{E}}{\partial t} \tag{11.4a}$$

over an area equal in size to a capacitor plate, and perpendicular to the electric field lines between the two plates, it follows from Stoke's theorem that

$$\oint \vec{B} \cdot d\vec{r} = \frac{1}{c}\frac{\partial \Phi_E}{\partial t} \tag{11.4b}$$

where Φ_E is the flux of the electric field between the capacitor plates. If we think of circular plates (or, equivalently, study the field far from a pair of arbitrarily shaped plates) then it follows by symmetry that the magnetic field surrounding the approximately parallel electric field lines between capacitor plates should be given by

$$\vec{B} \approx \frac{\partial \Phi_E/\partial t}{2\pi cr}\hat{\phi} \tag{11.4c}$$

during the charging process. In other words, a magnetic field loops symmetrically about the space between the two capacitor plates that contains no free charges and therefore no electric current, but only time-varying electric field lines. Magnetic field energy is located in otherwise empty space in the absence of matter or motion of matter. The magnetic field can be detected by observing the deflection of a sensitive enough compass needle that is placed within its range.

Maxwell returned to Newton's objection to the absurd idea of action at a distance between material bodies in a (Euclidean) vacuum. By including the time-varying electric field term on the right hand side of Ampère's law

$$\nabla \times \vec{B} = \frac{1}{c}\frac{\partial \vec{E}}{\partial t} + \frac{4\pi}{c}\vec{j}, \tag{11.5a}$$

action at a distance was predicted and, moreover, the propagation of electromagnetic waves at a speed equal to $c = (\mu_\circ\varepsilon_\circ)^{-1/2}$ also followed: it is easy to see that the wave equations

$$\left.\begin{array}{l} \nabla^2\vec{B} = \dfrac{1}{c^2}\dfrac{\partial^2\vec{B}}{\partial t^2} \\[4mm] \nabla^2\vec{E} = \dfrac{1}{c^2}\dfrac{\partial^2\vec{E}}{\partial t^2} \end{array}\right\} \tag{11.6}$$

follow from the source-free equations

$$\left.\begin{array}{c} \nabla \cdot \vec{E} = 0 \\[3mm] \nabla \times \vec{E} = -\dfrac{1}{c}\dfrac{\partial \vec{B}}{\partial t} \\[3mm] \nabla \cdot \vec{B} = 0 \\[3mm] \nabla \times \vec{B} = \dfrac{1}{c}\dfrac{\partial \vec{E}}{\partial t} \end{array}\right\} \tag{11.1b}$$

only because of Maxwell's 'displacement current' term $\partial\vec{E}/\partial t$ in

$$\nabla \times \vec{B} = \frac{1}{c}\frac{\partial \vec{E}}{\partial t}, \tag{11.4a}$$

which follows from Maxwell's generalization of Ampère's law. Without this term there is action at a distance, but the wave equations (11.6) still do not explain the rigid Coulomb force law as a result of wave propagation within classical Maxwell theory, although they do predict retarded and nonspherically symmetric fields for moving charges.

We know that symmetries imply conservation laws. What is the conservation law that follows from gauge invariance? By formulating a Lagrangian description and corresponding action principle for fields one can derive Noether's theorem for the field theory and then show that gauge invariance corresponds to *charge conservation* (see Mercier, 1963). We can indicate the importance of charge conservation for the theory as follows. Historically, Maxwell completed the equation

$$\nabla \times \vec{B} = \frac{4\pi}{c}\vec{j} \tag{11.5b}$$

to obtain equation (11.5a) because otherwise charge conservation is violated. To see this, we start only with a Coulomb field plus local charge conservation

$$\left.\begin{array}{l} \nabla \cdot \vec{E} = 4\pi\rho \\[2mm] \dfrac{\partial \rho}{\partial t} + \nabla \cdot \rho\vec{v} = 0. \end{array}\right\} \tag{11.7a}$$

It follows that

$$\nabla \cdot \left(\frac{\partial \vec{E}}{\partial t} + 4\pi\rho\vec{v}\right) = 0, \tag{11.7b}$$

which means that the argument of the divergence must be the curl of another (as yet unknown) vector, call it $\vec{h}$:

$$\frac{\partial \vec{E}}{\partial t} + 4\pi\rho\vec{v} = \nabla \times \vec{h}. \tag{11.8}$$

To determine $\vec{h}$, we must specify its divergence. If we do not want to introduce new field sources, for which there is no evidence, then we must choose

$$\nabla \cdot \vec{h} = 0, \tag{11.9}$$

which forces the lines of the field $\vec{h}$ to close on themselves, as they have no sources or sinks at any finite point in space. All that remains for a complete theory is to specify the curl of $\vec{E}$, because a vector is determined whenever its divergence and curl are specified. If we could do that without empirical evidence then we would arrive completely at Maxwell's equations from the standpoint of charge conservation.

Action at a distance in Newtonian gravity is consistent with Galilean relativity: a moving mass in Newtonian theory has the same gravitational field as when at rest: the field is like a rigid set of Keplerian spokes radiating outward from the body. However, rigid translation of a spherically symmetric Coulomb field is not consistent with Maxwell's theory, which takes the constancy of the velocity of light into account and leads to the special theory of relativity, where Lorentz transformations replace Galilean ones connecting inertial frames of reference. In different Lorentz frames one gets different gravitational fields, which violates the Galilean relativity principle: the static potential equation $\nabla^2 \Phi = -k\rho$ is not invariant under Lorentz transformations (see chapters 7 and 10 for the invariance of operators under coordinate transformations). A condition that the source-free Maxwell equations are invariant under Lorentz frames is that electromagnetic phenomena are transmitted at the speed of light rather than instantaneously. It was theoretical electrodynamics rather than the Michelson–Morley experiment that led Einstein (1879–1955) to suggest that the constancy of the speed of light in inertial frames is a law of nature. Gravity cannot exist merely as action at a distance, but should also travel between two macroscopic bodies at the speed of light. This leads to the expectation that gravity and electromagnetism may not be entirely independent, and may represent approximate parts of a larger field theory valid at much higher energies: if the speed of a gravitational wave is given by $c = (\mu_0 \varepsilon_0)^{-1/2}$, one can ask, then how do the basic undefined constants of electromagnetism, μ_0 and ε_0, enter into gravitational theory? In spite of this expectation, attempts earlier in the century by Einstein and Weyl (see Eddington, 1963, and Einstein, 1923), and more recently by others, have not revealed any unification that has led to unique, measurable consequences that have been empirically verified. Leaving hard questions like unification aside, we now turn to Einstein's revolutionary treatment of inertial frames of reference.

11.2 The principle of relativity

Riemann's idea of a manifold in more than three dimensions was extended to four dimensional space-time by Einstein in his special theory of relativity, which replaced Galilean relativity as the basic invariance principle of all of physics. We start by discussing Einstein's modification of the law of inertia, which follows from taking into account the invariance of the speed of light in inertial frames.

There are two invariance principles that define the special theory of relativity, or just 'relativity': (i) the results of measurement cannot depend upon the particular inertial frame in which an experiment is carried out, and (ii) the speed of light is the same universal constant c for all inertial observers. The second condition eliminates Galilean transformations as the transformations connecting different inertial frames with each other, because the Galilean law of addition of velocities $c' = c - u$ is violated by light. If we apply the second condition directly to the propagation of a light signal in an inertial frame then we can derive the correct constant velocity transformation of coordinates that connects different inertial frames with each other. If we try to connect inertial frames by transformations that are linear in position and time, then we can start by assuming that

$$\left. \begin{array}{l} x' = \gamma(x - ut) \\ t' = \gamma(t - \beta x) \end{array} \right\} \tag{11.10}$$

where γ and β are unknown constants and (x,t) and (x',t') represent two frames connected by a constant velocity transformation (both frames are inertial). At the instant that the origins of the relatively moving frames coincide (at $x = x' = 0$), two observers (each fixed at the origin of his own frame) synchronize their watches to read $t = t' = 0$. At that instant, a light signal is emitted along the x- and x'-axes. It makes no difference whether the source is fixed in either frame or in some third inertial frame: in all frames, the speed of the signal is the same, $c \approx 3 \times 10^8$ m/s. This means that

$$x^2 - (ct)^2 = x'^2 - (ct')^2, \tag{11.11}$$

which contradicts common sense. Insertion of (11.10) and comparison of coefficients yields $\beta = u/c$ and $\gamma = (1 - u^2/c^2)^{-1/2}$.

The transformations that connect different inertial frames with each other are therefore given by

$$\left. \begin{array}{l} x' = \gamma(x - ut) \\ t' = \gamma(t - ux/c^2), \end{array} \right\} \tag{11.12a}$$

where $\gamma = (1 - u^2/c^2)^{-1/2}$, and are called the Lorentz transformations after the theorist who discovered them (the transformations were also studied by Poincaré). The inverse transformation is given by

$$\left. \begin{array}{l} x = \gamma(x' + ut') \\ t = \gamma(t' + ux'/c^2). \end{array} \right\} \tag{11.12b}$$

It was left for Einstein to interpret the Lorentz transformations for us. The result is the two postulates of special relativity stated above. According to the Lorentz transformations, time is not absolute after all: clocks that are once synchronized at the same point in space in two uniformly separating frames will not be synchronized at later or earlier times. At small velocities $u \ll c$, we retrieve the Galilean transformations and the illusion of absolute time to a good first approximation:

$$\left. \begin{array}{l} x' \approx x - ut \\ t' \approx t. \end{array} \right\} \tag{11.13}$$

In addition, we can enlarge the transformations (11.12a) to include rigid spatial rotations and rigid translations, as Euclidean three dimensional space is both isotropic and homogeneous.

Consider a car of length L_o in its own rest frame (primed frame). An observer O in a moving frame can measure the car's length by marking two points x_1 and x_2 on his own moving Cartesian x-axis simultaneously, at the same time $t = t_1 = t_2$, representing the front and rear positions of the car at time t. According to (11.12b), $L = x_2 - x_1 = L_o/\gamma < L_o$. However, the car's driver O' will disagree that O marked two points on the x-axis *simultaneously*: according to O', the two points x_1 and x_2 recorded by O in the determination of the car's length L were marked at times t'_1 and t'_2 respectively (on the driver O''s watch), where (because $\Delta t = 0$ in (11.12b)) $\Delta t' = t'_2 - t'_1 = -uL_o/c^2 < 0$. In other words, the driver accuses O of marking the position of the front of the car too late. Furthermore, O' can explain the length discrepancy: note that $L_o + u\Delta t' = L_o(1 - u^2/c^2) = L_o/\gamma^2$, or $\gamma(L_o + u\Delta t') = L_o/\gamma = L$, which is the same as $L = \Delta x = \gamma(\Delta x' + u\Delta t') = L_o/\gamma$. Both observers are correct and closer agreement than this cannot be achieved: the measured length of an object depends upon the relative velocity of the measurer and the object. Disagreement between O' and O over whether both the front and rear of the car were marked by the observer O at exactly the same time is inevitable because of the invariance of the speed of light, which eliminates absolute time and, along with it, the idea of *simultaneity* in relatively moving inertial frames.

The idea of simultaneity, dear to our intuition (like absolute time), does not exist except in a highly restricted circumstance. To emphasize the point, suppose that two campfires are lit in the primed frame at the same time t' but at two spatially separated points x'_1 and x'_2. According to a moving observer O in the unprimed frame, the fires were started with a time difference $\Delta t = t_2 - t_1 = \gamma u \Delta x'/c^2$ between them, where $\Delta x' = x'_2 - x'_1$. Relatively moving observers can agree on simultaneity *if and only if two events occur at exactly the same point in space-time*. The geometry of space is still Euclidean, but the geometry of space-time is not the geometry of three dimensional Euclidean space. What does this mean?

We can understand length contraction by realizing that spatial distances and displacements are not invariant under Lorentz transformations. The idea of a rigid body, an ancient Euclidean idealization, is not invariant under Lorentz transformations. In Newtonian relativity, the geometry of space is Euclidean. According to Lie and Klein, Euclidean space is defined by the objects that are left invariant by rigid translations and rigid rotations, and the shapes of rigid bodies are left invariant by Galilean transformations. Following this idea further, in Einstein's special theory of relativity the geometry must be defined by those objects that are left invariant by the Lorentz group of transformations (11.12a) in *four dimensional space-time*. A car (an approximate rigid body) that moves at an appreciable fraction of the speed of light relative to an inertial observer, and appears to that observer to be 'squashed' in the direction of motion, is still geometrically similar via a Lorentz transformation in flat space-time to the car at rest: as we show below, the Lorentz transformations are unitary and are hyperbolic 'rotations' in space-time (the rotation angle is imaginary). The two descriptions of the car, one in the moving frame and one in the rest frame, are not invariant even though they are connected by a similarity transformation. What, is invariant if a solid object's shape is not? Maxwell's source-free equations (but not Newton's second law with an action-at-a-distance force on the right hand side) are left invariant by the Lorentz group. In other words, the analog of rigid bodies (invariant geometric structures) of space-time cannot be the rigid bodies of Euclid and Newtonian mechanics.

Lorentz transformations are similarity transformations between inertial frames in space-time: inertial frames are similar to each other. This means that the length contraction of an object in the direction of its motion is apparent in the sense that, in the rest frame of the object, there is no decrease in atomic spacings (no increase in density) due to the motion relative to other inertial frames or relative to absolute space-time. In other words, Lorentz transformations do not leave distances between objects invariant (do not leave the Euclidean metric invariant), but instead leave the metric and distances in four dimensional *space-time* invariant. The length contraction is also real in the sense that it is the result of every correctly performed measurement. There is no illusion aside from the idea of a *rigid three dimensional body*, which is a Euclidean but not a relativistic geometric idea any more than is 'rigidly ticking' or 'uniformly flowing' absolute time.

Time is not absolute, and the (x,y,z,t) space-time is a pseudo-Cartesian space with indefinite metric: the distance squared $ds^2 = (c dt)^2 - dx^2 - dy^2 - dz^2$ is left invariant by (11.12), so that g_{ik} is not positive definite. In what follows, the effect of the metric

$$g = \begin{pmatrix} 1 & 0 \\ 0 & -1 \end{pmatrix} \tag{11.14}$$

on real vectors

$$X = \begin{pmatrix} x \\ ct \end{pmatrix} \tag{11.15a}$$

is taken into account by using the Minkowski notation

$$X = \begin{pmatrix} x \\ ict \end{pmatrix},$$

(11.15b)

whereby the unitary scalar product $X^+ X = x^2 - (ct)^2$ is Lorentz-invariant. With this definition, the Lorentz transformation

$$\Lambda = \begin{pmatrix} \gamma & i\gamma\beta \\ -i\gamma\beta & \gamma \end{pmatrix}$$

(11.16)

in $X' = \Lambda X$ is unitary because $\Lambda^+ = \Lambda^{-1}$.

Since the transformation is unitary all scalar products of vectors in Minkowski space are invariant. This invariance allows us to determine the form of covariant vectors A and tensors T, objects that transform like $A' = \Lambda A$ and $T' = \Lambda^+ T\Lambda$. For example, the velocity of an object

$$V^+ V = \frac{dX^+}{dt}\frac{dX}{dt} = v^2 - c^2$$

(11.17a)

is not invariant (because v is not invariant) but the Minkowski velocity $U = dX/d\tau$, with $d\tau = dt/\gamma$ is invariant:

$$U^+ U = -c^2.$$

(11.17b)

In other words,

$$U = \begin{pmatrix} u_1 \\ u_2 \end{pmatrix} = \begin{pmatrix} \gamma v \\ \gamma ic \end{pmatrix} = \begin{pmatrix} \gamma dx/dt \\ \gamma ic \end{pmatrix}$$

(11.18)

transforms like $U' = \Lambda U$, and $d\tau = dt/\gamma$ is called the proper time. By the inverse transformation (11.12b) we get $\delta t = \gamma \delta t'$ for fixed x', so that $\delta\tau = \delta t/\gamma$ is the time recorded on a clock that travels with a body that is at rest at any point x' in the primed frame. In other words, $d\tau$ is the time interval according to a clock that travels with the body under observation. We can assume, in addition, that the two separate clocks (observer's with time t, and observed's with time τ) were synchronized to zero at the same point in space-time. The observer's time is the component of the position vector of the body under observation, whereas τ is an invariant for the observed body's motion.

The covariant momentum is therefore

$$P = mU = \begin{pmatrix} \gamma mv \\ \gamma imc \end{pmatrix}$$

(11.19)

and we expect that the law of inertia in space-time, the foundation of physics, has the form

$$\frac{dP}{d\tau} = 0.$$

(11.20)

Einstein's principle (i) stated above reflects the law of inertia, which we have now expressed in the form of Newton's first law.

Generalizing to three space dimensions, with $X = (x_1, x_2, x_3, ict)$ we have

$$\Lambda = \begin{pmatrix} 1 & 0 & 0 & 0 \\ 0 & 1 & 0 & 0 \\ 0 & 0 & \gamma & i\gamma\beta \\ 0 & 0 & -i\gamma\beta & \gamma \end{pmatrix} \tag{11.21}$$

so that Newton's second law in the three dimensional Euclidean subspace of space-time is

$$\frac{\mathrm{d}\vec{P}}{\mathrm{d}t} = \frac{\mathrm{d}}{\mathrm{d}t}\gamma m\vec{v} = \vec{F} \tag{11.22a}$$

with $\vec{P} = \gamma m\vec{v}$, and where $\vec{F}$ is a force law in classical mechanics.

Newton's second law should be covariant under the linear transformations that define the relativity principle of mechanics, as we showed in chapter 1. Covariance is easy to exhibit if we can sensibly define a fourth component F_4 of the force, so that in matrix notation (11.22a) becomes a four-component column vector equation

$$\frac{\mathrm{d}P}{\mathrm{d}\tau} = m\frac{\mathrm{d}U}{\mathrm{d}\tau} = F/\gamma = K, \tag{11.22b}$$

which defines the vector K, except for K_4. Covariance follows easily: with $P = \Lambda^+ P'$ we have

$$\frac{\mathrm{d}P}{\mathrm{d}\tau} = \Lambda^+\frac{\mathrm{d}P'}{\mathrm{d}\tau} = K, \tag{11.22c}$$

so that

$$\frac{\mathrm{d}P'}{\mathrm{d}\tau} = \Lambda K = K' \tag{11.23}$$

defines the force K' in the second inertial frame, given the force K in any other inertial frame (one proceeds in practice by choosing a convenient inertial frame in which to define a Newtonian force $\vec{F}$).

Next, we determine K_4. Again, we use the fact that all scalar products are not merely scalars but are in addition Lorentz invariants. Note that

$$U^+ K = mU^+\frac{\mathrm{d}U}{\mathrm{d}\tau} = \frac{m}{2}\frac{\mathrm{d}}{\mathrm{d}\tau}U^+U = 0, \tag{11.24a}$$

which yields directly that

$$K_4 = \frac{i\gamma}{c}\vec{F}\cdot\vec{v} = \frac{i}{c}\vec{F}\cdot\vec{u}. \tag{11.24b}$$

If we define kinetic energy T by

$$\frac{\mathrm{d}T}{\mathrm{d}t} = \vec{F}\cdot\vec{v} \tag{11.25a}$$

and use $m\mathrm{d}u_4/\mathrm{d}\tau = K_4$ with $u_4 = \gamma ic$, then (11.25a) yields the kinetic energy as

$$T = \gamma mc^2. \tag{11.25b}$$

At low velocities $\beta \ll 1$ we get $T \approx mc^2 + mv^2/2$, which is the Newtonian kinetic energy plus the 'rest energy' mc^2 of the body of mass m. Given (11.25b), the covariant momentum can be written as

$$P = mU = \begin{pmatrix} \gamma m v_1 \\ \gamma m v_2 \\ \gamma m v_3 \\ iT/c \end{pmatrix}. \tag{11.26}$$

As an example of a relativistic mechanics problem, we formulate the Kepler problem in space-time. If we take the sun to be unaccelerated by an orbiting planet of mass m, then in the sun's inertial frame the gravitational potential energy is spherically symmetric and is given by $U(r) = -k/r$. Working with the space part of Newton's second law in space-time,

$$\frac{d\vec{p}}{dt} = \frac{d}{dt}\gamma m \vec{v} = -\nabla U(r), \tag{11.27a}$$

angular momentum conservation follows immediately because we have a central force (in any other inertial frame, the gravitational force is not central, as we shall show below): $mr^2 d\theta/dt = M = $ constant so that the orbit lies in a plane in the three dimensional Euclidean subspace of space-time. The Lagrangian

$$L = -mc^2(1 - v^2/c^2) + k/r \tag{11.28a}$$

yields Newton's law (11.27a) in the sun's rest frame, and Lagrange's equations follow as usual from the action principle

$$\delta A = \delta \int_{t_1}^{t_2} L \, dt = 0. \tag{11.28b}$$

As Bergmann (1942) points out, the form (11.28a) of the Lagrangian follows from assuming that the covariant momentum $P_i = \gamma m v_i$ is the canonical momentum in Cartesian coordinates, and then integrating

$$P_i = \gamma m v_i = \frac{\partial L}{\partial v_i} \tag{11.27b}$$

once to yield $L = -mc^2\gamma^{-1} + $ constant. We have set the integration constant equal to zero.

With cylindrical coordinates (r,θ), and canonical moment (p_r, p_θ) calculated from L, one easily finds the Hamiltonian to be

$$H = T + U(r) = \gamma m c^2 - k/r, \tag{11.29}$$

which is constant because L has no explicit time-dependence. Noting that

$$\gamma^2 = \left(\frac{H + k/r}{mc^2}\right)^2 = \frac{1}{1 - (\dot{r}^2 + M^2/\gamma^2 m r^2)} \tag{11.30}$$

we can solve for γ as a function of dr/dt:

$$\gamma^2(1 - \dot{r}^2/c^2) = 1 + M^2/m^2 c^2 r^2. \tag{11.31}$$

This allows us to find a differential equation for the orbit equation $\theta = f(x)$ where $x = 1/r$. Using $d\theta/dt = -f'(x)(dr/dt)/r^2$ and angular momentum conservation to get $dr/dt = -M/\gamma m f'(x)$, we can combine our results and write

$$\left(\frac{H + kx}{mc^2}\right) = 1 + \frac{M^2 x^2}{m^2 c^2} + \left(\frac{M}{mcf'(x)}\right)^2, \tag{11.32}$$

which is the nonlinear orbit equation for the relativistic Kepler problem. Next, we try to analyze the problem qualitatively.

To study the radial problem as a flow in the (r,p_r)-subspace of phase space, one must first prove that there are no singularities at any finite time. We look only at the problem from the standpoint of bounded orbits, ignoring the scattering problem. For bounded orbits, there must be two turning points $r_{min} < r < r_{max}$ where $dr/dt = 0$ defines the turning points. Therefore, the equation of the turning points is

$$\left(\frac{H + k/r}{mc^2}\right)^2 = 1 + \left(\frac{M}{mcr}\right)^2. \tag{11.33}$$

With $x = 1/r$ we get a quadratic equation for x whose solution is

$$x = \frac{Hk/M^2 c^2 \pm \{(Hk/M^2 c^2) - 4[1 - (k/Mc)^2](m^2 c^4 - H^2)/M^2 c^2\}^{1/2}}{2[1 - (k/Mc)^2]}. \tag{11.34}$$

For $k/Mc = 1$ there are no turning points. For $k/Mc > 1$ and large enough there are no turning points. For $k/Mc < 1$ there are always two turning points if $0 < H < mc^2$; in this case there is a phase flow with bounded orbits.

For the case $k/Mc < 1$ and $0 < H < mc^2$ we can study the problem as a flow in the (r,p_r)-phase space: Lagrange's equations for p_r and dp_r/dt yield

$$\dot{r} = p_r/\gamma m \tag{11.35}$$

and

$$\dot{p}_r = \frac{\partial L}{\partial r} = \frac{M^2}{\gamma m r^3} - \frac{k}{r^2}. \tag{11.36}$$

Since

$$\gamma = \frac{H + k/r}{mc^2} \tag{11.37}$$

we can write the phase flow equations in the form

$$\left. \begin{aligned} \dot{r} &= \frac{c^2 p_r}{H + k/r} \\[2mm] \dot{p}_r &= \frac{M^2 c^2}{(H + k/r)r^3} - \frac{k}{r^2} \end{aligned} \right\} \tag{11.38}$$

By setting the right hand sides equal to zero, we find that there is an equilibrium point (meaning a circular orbit in configuration space) at

$$r = \left(\frac{M^2 c^2}{k} - k\right)\frac{1}{H}. \tag{11.39}$$

In addition, the orbit equation (11.32) can be integrated exactly in closed form by a transformation. As we have foreseen, the solutions differ qualitatively depending upon the parameter k/Mc. If $k/Mc < 1$ and $H > 0$ there is a precessing ellipse that yields a closed orbit in the limit $c = \infty$. When $k/Mc = 1$ there is escape to infinity at a finite value of θ, indicating a spontaneous singularity. If $k/Mc > 1$ then gravitational collapse is possible, but r goes to zero as θ goes to infinity. With $M \approx h/2\pi$ as the smallest possible angular momentum (h is Planck's constant), and $k = Ze^2$ for the analogous Coulomb problem (e is the electronic charge and Z is the net charge on the nucleus), the fine structure constant is $2\pi e^2/hc \approx 1/137$ so that $Z > 137$ for collapse. Corresponding- ly, the Dirac equation in relativistic quantum mechanics has no solution for $Z > 137$.

This formulation of the Kepler problem illustrates that you do not have to use equations that are explicitly covariant in order to study physics (finding solutions requires the use of symmetry, which means abandoning manifest covariance). When- ever the equations of motion are not covariant (our Lagrangian formulation is not covariant, but it could be replaced by a covariant one), then you have to carry out the transformations explicitly to see what physics looks like in other coordinate systems. In any other inertial frame than the 'heavy sun's', the relativistic Kepler problem is mathematically more difficult because the sun's gravitational field is neither spherically symmetric nor time-independent in a frame where the sun moves. In fact, the relativistic Kepler problem for two bodies requires a time-dependent potential energy and may not conserve energy locally because of gravitational radiation. We now show why we expect this order of difficulty from the relativistic two-body problem of gravity.

Consider the gravitational field due to a body moving with constant velocity. The gravitational potential Φ satisfies $U = m_1\Phi$ where m_1 is the mass of any body that experiences the gravitational field of the moving body. In the rest frame of the body mass m_2 that creates the field, both the potential and field $\vec{g} = \nabla\Phi$ are spherically symmetric: $\Phi = Gm_2/r$, meaning that the field lines radiate out like spokes from the body and the equipotential surfaces are concentric spheres. As the body moves, the radial field lines that are dragged along with it cannot appear radial in the eyes of an observer who is not in the body's rest frame. Newtonian gravity makes the approxi- mation of assuming action at a distance, which is like setting $c = \infty$ in a relativistically covariant wave equation like the scalar equation

$$\frac{1}{c^2}\frac{\partial^2\Phi}{\partial t^2} - \nabla^2\Phi = 4\pi\rho, \tag{11.40}$$

where ρ is the density of matter creating the field and vanishes outside the source body. This equation predicts that gravity travels at the speed of light, in accordance with the requirement that no signal can exceed the speed of light in an inertial frame, so that any motion of the source will create a change in the field lines that will start locally and propagate outward from the source at the speed c. The condition, in relativity theory, that the field lines are dragged rigidly along and are spherically symmetric *in the body's rest frame* is that

$$\frac{d\Phi}{dt} = \frac{\partial\Phi}{\partial t} + \vec{v}\cdot\nabla\Phi = 0, \tag{11.41}$$

where v is the body's speed. With a constant velocity v along the x-axis, we get

$$(1 - \beta^2)\frac{\partial^2\Phi}{\partial x^2} + \frac{\partial^2\Phi}{\partial y^2} + \frac{\partial^2\Phi}{\partial z^2} = 0 \tag{11.42}$$

which, by a change of variables, yields the gravitational potential as seen by an observer located at the field point (x,y,z), and who sees the source of the field moving at constant speed v along the x-axis:

$$\Phi = \frac{GM}{[(x - v^2t^2)^2 + (1 - \beta^2)(y^2 + z^2)]^{1/2}}. \tag{11.43}$$

Clearly, this field does not have spherical symmetry. Unfortunately, this is not an extension to space-time of the theory of gravity either: the equation determining the gravitational field

$$\vec{g} = \nabla\Phi \tag{11.44}$$

is not compatible with the wave equation (11.40) ($\nabla \cdot \vec{g} = 4\pi\rho$ and (11.44) do not yield the wave equation (11.40)). The attempt to repair the internal contradiction via a vector field leads to a Maxwell-type theory where like masses repel. The only alternative is a theory based upon a higher rank tensor, which one could attempt to construct by requiring that the gravitational field should satisfy the equation for local mass–energy conservation analogous to the way that the existence of the magnetic field can be deduced from the existence of an electric field by the requirement of charge conservation.

Maxwell's equations are not covariant under Galilean transformations (and Newton's static gravitational field is not a vector under Lorentz transformations). Lorentz discovered that Maxwell's equations are covariant under the one parameter transformations (11.12), and that the source-free equations are invariant. However, he did not interpret their meaning for physics correctly because he and a lot of other physicists were still ensmogged by the pre-medieval idea of an ether. Even a genius like Maxwell believed in the ether. Einstein, like Newton before him, eliminated the reign of confusion by discarding the ether and assuming that the velocity of light is a Lorentz invariant, is the same universal constant c for all *inertial* observers. Like electromagnetism, the force of gravity should be a wave propagating at the speed of light. According to $T = \gamma mc^2$, no body with finite rest mass m can travel at a speed as large as c whereas massless energy, like light and gravity, must travel with the speed of light.

Before going further, we exhibit Maxwell's equations in manifestly covariant form. Starting with the Maxwell equations (11.1a), which are valid in any Lorentz frame, we construct the electromagnetic field tensor F,

$$F = \begin{pmatrix} 0 & B_3 & -B_2 & -iE_1 \\ -B_3 & 0 & B_1 & -iE_2 \\ B_2 & -B_1 & 0 & -iE_3 \\ iE_1 & iE_2 & iE_3 & 0 \end{pmatrix}, \tag{11.45}$$

where the 3×3 magnetic submatrix has the form of the infinitesimal generator of a rotation. The two Maxwell equations with sources then become

$$\frac{\partial F_{\mu\nu}}{\partial x_\nu} = \frac{4\pi}{c} j_\mu \tag{11.46}$$

where $j = (\mathbf{j}, ic\rho)$ is the electric current density and satisfies charge conservation locally in the form

$$\frac{\partial j_{\mu}}{\partial x_{\mu}} = 0. \tag{11.47}$$

The two sourceless Maxwell equations are given by

$$\frac{\partial F_{\mu\nu}}{\partial x_{\lambda}} + \frac{\partial F_{\nu\lambda}}{\partial x_{\mu}} + \frac{\partial F_{\lambda\mu}}{\partial x_{\nu}} = 0. \tag{11.48}$$

This equation requires that

$$F_{\mu\nu} = \frac{\partial A_{\mu}}{\partial x_{\nu}} - \frac{\partial A_{\nu}}{\partial x_{\mu}} \tag{11.49}$$

where $A = (\mathbf{A}, i\Phi)$, and represents the fields in terms of the potentials:

$$\left. \begin{aligned} \vec{E} &= -\nabla\Phi - \frac{1}{c}\frac{\partial \vec{A}}{\partial t} \\[2mm] \vec{H} &= \nabla \times \vec{A}. \end{aligned} \right\} \tag{11.50}$$

In addition, gauge invariance (demanded by charge conservation) can be formulated covariantly in the Lorentz gauge where

$$\frac{\partial A_{\mu}}{\partial x_{\mu}} = 0. \tag{11.51}$$

The differential equation expressing magnetic induction locally,

$$\nabla \times \vec{B} = \frac{1}{c}\frac{\partial \vec{E}}{\partial t} + \frac{4\pi}{c}\vec{j}, \tag{11.52}$$

makes an interesting prediction: for an observer who travels along with a steadily moving charge, there is only a steady and radially symmetric electric field, the Coulomb field. However, an observer who watches the charge from any other inertial frame can prove that the moving charge generates a magnetic field given by $\vec{H} = \vec{v} \times \vec{E}/c$, where $\vec{E}$ is the field measured by that observer (and is *not* spherically symmetric about the moving source even though it is radial). In other words, you can eliminate the magnetic field (but not the electric field) by a Lorentz transformation. This tells us that electric and magnetic phenomena are not two separate ideas: linear combinations of both fields define the new fields under a Lorentz transformation. To be completely coordinate-free, we can discuss the properties of 'the electromagnetic field' as defined by the abstract tensor **F**.

A four dimensional space-time manifold also is used in the general theory of relativity, a geometric theory of gravity based upon the principle of equivalence. The latter is a local invariance principle. Special relativity also ends up as a local (as opposed to global) invariance principle in Einsteinean gravity. A global invariance principle is one that is valid over finite regions of space-time where gradients of fields vary nontrivially. A local one is rigorously valid only over finite but small enough regions of space-time that fields can be linearized because gradients are ignored (on the scale of the universe, one can sometimes treat the solar system over observationally long time spans as a local region). The sections that follow approach gravity as geometry from the standpoint of integrability. We also continue to develop the tensor analysis that we could easily have continued in chapter 10.

11.3 Coordinate systems and integrability on manifolds

Gravity, as we have noted in section 1.4, is peculiar in a way that other forces (electrostatic, magnetic, farmer pushing a cart, etc.) are not. No other force obeys the principle of equivalence, the equivalence, resulting from the strict equality of the inertial and gravitational definitions of mass, between the effects of linear acceleration and a local gravitational field. Furthermore, free-fall along the gravitational field lines locally *eliminates* the effect of the gravitational force completely and constitutes a local inertial frame. So long as the gradient of the gravitational field is negligible, a freely falling frame is inertial: you would become weightless if you could jump off a cliff into a vacuum, and a ball placed at rest beside you would remain there during the free-fall. The equivalence is complete: because of the equality of inertial and gravitational mass, an observer in an accelerated frame feels a noninertial effect that is indistinguishable from the effect of 'weight' in the inertial frame of a nonaccelerated planet. The latter follows because a force $F' = MA$ is required to hold an observer at rest in a noninertial frame, so that the acceleration A of the frame relative to absolute space produces an effect equivalent to that of a local gravitational field where $g = A$.

First Mach, and later Einstein, wanted to explain the Newtonian noninertial effects of acceleration relative to absolute space as having their true origin in acceleration relative to very distant matter in the universe. Newton himself was aware that the prediction of observable physical effects due to acceleration relative to absolute space is open to philosophic objection. The physical effects attributed to acceleration relative to inertial frames may be philosophically undesirable in some quarters[1], but they are present in Newtonian theory *because of the law of inertia*, and are present in general relativistic theories as well[2]. No one has yet managed to construct a working theory that explains noninertial effects systematically, free of objection, as global effects that are due to the presence of very distant matter in the universe.

Mach's principle is philosophically attractive but is also unpractical (which does not mean that it's wrong). Physics has historically been the attempt to describe the regularities and some irregularities of nature universally as if they were the consequences of a simple impersonal machine whose behavior is describable using a few essential local variables and local interactions that obey universally valid equations of motion. Physics has, from the time of Galileo, been based upon simplification and idealization, upon ignoring details of interactions not only with far-away bodies but even of nearby ones in order to understand how relatively simple, ideal dynamical systems behave *in principle*. Any theory that attempts to take all of the matter in the universe into account will be likely to be too complicated to be of any use in answering very many questions. Another way to say it is that a theory that tries to take too many details into account, 'a theory of everything', would be a theory that could account in practice for relatively few observations, if any, and would therefore effectively be a theory of nearly nothing. The reason for this is there would be no possibility of solving the equations of such a theory by any method other than by ignoring all 'unnecessary details'. Even though we expect the Newtonian three-body problem to be considerably

---

[1] In quantum field theory there is no such thing as empty space due to vacuum fluctuations (zero-point energy). Perhaps distant matter is conceptually unnecessary: that acceleration relative to the local distribution of zero-point energy should produce physical effects might make sense.

[2] Einstein first abolished absolute time and later geometrized gravity, but he did not manage to abolish the inertial effects of acceleration locally. That is, Mach's principle has not yet successfully been derived from or built into general relativity.

less complex than fluid turbulence, with its nonlinear partial differential equations and infinitely many variables and boundary conditions, *we cannot even solve the three-body problem in the Newtonian theory of gravity except by making approximations that are so drastic that we still do not understand the full three-body problem mathematically.* Therefore, physicists usually accept the idea of absolute space implicitly in practice, if not philosophically in principle, just as they accept that classical mechanics must be used to formulate theories of fluid turbulence in spite of the (effectively irrelevant) fact that every atom and molecule of a fluid obeys the laws of quantum mechanics (turbulence in superfluids may be different). Newton's relativity principle represents the law of inertia, the idealization of an essentially empty universe as an infinite Euclidean space, and is useful for describing local motions where all relative velocities are small compared with the speed of light, and where gravitational effects are not too strong.

Newton's first law depends upon Galilean invariance between inertial frames of reference and presumes that empty absolute space is Euclidean: perfectly long straight lines are presumed to exist, so that Cartesian coordinates can be used globally. Cartesian axes are homogeneous. Another way to state the first law is: in inertial frames, the momentum of a body is conserved so long as no net force acts on that body. Homogeneity of space means simply that, along a particular direction, empty space is completely uniform, is 'flat' as relativity theorists and geometers would say. If, instead, empty space were curved or bumpy, then the linear momentum of a force-free body could not be conserved (the effect of curvature of configuration space in the destruction of linear momentum conservation is implicit in the geometric formulation of Lagrange's equations in chapter 10). *Conservation of momentum reflects the translational symmetry of empty Euclidean space.* Newton's first law and the corresponding Newtonian relativity principle could not be true in the absence of the uniformity of empty space. Euclidean geometry alone does not force us to the conclusion that the law of inertia is right (translational invariance is consistent with the Aristotelian equation of motion (1.21)), but the law of inertia and the corresponding conservation law for momentum cannot hold globally unless empty space is globally uniform.

The special relativity principle did not solve the problem of gravity and indicates that Newton's law of gravity is only a static approximation. A better theory would yield gravity as a signal that propagates at the speed of light rather than instantaneously as action at a distance, but gravity is not an ordinary force because it can be eliminated by free-fall in a uniform field. The purpose here is not to discuss all possible technical details but instead to show how gravity was understood as geometry by Einstein by using only the principle of equivalence and the special relativistic law of inertia ($d^2x^\mu/d\tau^2 = 0$ *locally* for a force-free body) on a curved space-time manifold that is locally flat. In general relativity, the Lorentz transformations, the law of inertia, and the principle of equivalence hold only locally, which means over small enough regions of the space-time manifold (in a tangent plane) that gradients of the gravitational field can be neglected.

Globally, space-time in general relativity is intrinsically curved by the presence of mass and energy. Euclidean and pseudo-Euclidean space-times are called flat. We will consider smooth curved manifolds that are locally flat (locally Lorentzian) because we assume them to be differentiable, so that Cartesian (x^1, x^2, x^3, ct) and all other curvilinear coordinate systems can be defined over a small enough finite region in the tangent plane to any point on the manifold; then, we can always describe free-body motion locally by $d^2x^\mu/d\tau^2 = 0$ in a tangent hyper-plane. The attempt to define coordinate systems globally in a curved space poses a nontrivial and very interesting integrability problem that we will discuss because it is central to understanding general relativity theory and also has application to the geometry of invariant manifolds in phase space. In what follows, ds represents the arc-length along an arbitrary curve on an arbitrary space-time manifold. In a local Lorentz frame we can use $ds = cd\tau$ where τ is the proper time of the observed body. More than this, we cannot yet say.

Einstein took the principle of equivalence, a local invariance principle, as his starting point. It suggested to him that gravity might be accounted for geometrically in a nonEuclidean space-time. Locally, the Lorentz transformations connect nearby frames that are inertial because they are in free-fall. The principle of inertia, which states that a force-free body moves at constant speed with respect to an inertial frame of reference, holds locally in any inertial frame in space-time and is simply

$$\frac{d^2x^\mu}{d\tau^2} = 0, \tag{11.53}$$

where $(x_1, x_2, x_3, x_4) = (x, y, z, ct)$. On a curved manifold, the law of inertia is not a global law of force-free motion. It represents the approximation whereby a force-free body travels in a straight line at constant speed in a tangent plane near a point on a nonEuclidean manifold, and is valid locally over small enough space-time intervals (the size of the interval is determined by the field gradients that are neglected).

The formal generalization to the global equation of free-particle motion is simple: the shortest distance between two points in flat space-time is a straight line, the extremum of a variational principle of the integral of the arc-length ds along a path evaluated between two fixed points in space-time. The condition that a body travels along the shortest path between two globally separated points along the space-time path is that we get a minimum for the integral

$$\int_1^2 ds = \int_1^2 (g_{jk}dq^j dq^k)^{1/2} = \int_1^2 (g_{jk}\dot{q}^j \dot{q}^k)^{1/2}ds \tag{11.54a}$$

which requires that

$$\frac{d^2q^\mu}{ds^2} + \{\nu\lambda,\mu\}\frac{dq^\nu}{ds}\frac{dq^\lambda}{ds} = 0. \tag{11.55}$$

This free-body equation of motion is the obvious generalization to nonflat spaces of the idea of force-free motion at constant speed along a straight line in a flat space. Next, let (q^1, q^2, q^3) denote any admissible set of generalized coordinates in space (holonomic coordinates) and let $q^4 = ct$. Usually, the q^i are curvilinear coordinates that reflect the symmetry of some finite portion of the manifold where we want to study the motion, and in a tangent plane we will normally take the generalized coordinates q^i to be Cartesian. If gravity can be understood as geometry on a curved four dimensional manifold, then the *force-free* equations (11.55) should describe how a body of relatively very small mass moves in the presence of a second and relatively massive body that creates the gravitational field described geometrically by the metric g, which must reduce approximately to the diagonal Lorentz metric $(-1,1,1,1)$ in any local Cartesian coordinate system. In this approximation the body that is the source of the gravitational field is treated here to zeroth order as unaccelerated (static gravitational field).

What determines the Christoffel symbols $\{v\lambda,\mu\}$ that are supposed to reproduce the motions that we identify as caused by gravity? In a purely mechanical theory, only the distribution of matter and energy are allowed to do that. Einstein had to use trial and error to arrive at gravitational field equations that allow matter to determine the geometry in the simplest possible way. We now follow the path that leads to Einstein's field equations, which determine the geometry and thereby allow the geodesic equation for a given distribution of mass–energy in space-time to determine the motion of a 'test mass'.

We begin with the idea of parallelism in flat and curved spaces by starting with an example of a manifold that indicates that global parallelism generally does not exist on curved manifolds. Then, we go on to a local approach on an arbitrary manifold and explain why global parallelism is impossible on curved manifolds.

If you should transport the tail of a vector around any closed curve in Euclidean space so that the vector is always moved parallel to itself locally, then the vector will still be parallel to its original direction when the tail is returned to its starting point. The corollary is that parallelism is path-independent, so that constant vector fields are possible in Euclidean space over finite and even infinite distances. To visualize the creation of a constant vector field via 'parallel transport', imagine translating the vector parallel to itself, along a path that is locally parallel to the vector, all the way to infinity. Then, returning to the vector's initial position, translate it transversely over a finite distance to some other starting point in the space (which changes neither its length nor direction), and repeat the process of parallel transport to infinity. Do this for a large enough number of points in space and you will have a globally parallel and constant vector field, a lot of arrows of the same length flying in the same direction over a semi-infinite region of space.

Global parallelism exists trivially in a flat space; the existence of global parallelism can even be taken as the definition of a flat space. Correspondingly, the system of differential equations $dx^\mu/ds = v^\mu =$ constant is defined and integrable everywhere in a flat space and defines Cartesian axes $x^\mu = v^\mu s + x^\mu_\circ$ globally. In contrast, the analogous equations on a sphere require polar coordinates $(q^1,q^2) = (\theta,\phi)$ but the required flow equations $d\theta/dt = 0$ and $d\phi/dt = 1$ are not defined at $\theta = 0$ and π, and the constant components $(v_\theta,v_\phi) = (0,1)$ also do not define a constant vector field on the sphere because the unit vector in the ϕ direction is not constant along an orbit. Global parallelism and the construction of Cartesian coordinates on a manifold are equivalent to the integrability of the equations $dx^\mu/ds = v^\mu =$ constant at *every* point on the manifold. The construction of an orthogonal (θ,ϕ) coordinate system on a sphere is equivalent to the integrability of the equations $d\theta/ds = 0$ and $d\phi/ds = 1$ on the sphere, except at the elliptic points $\theta = 0$ and π. In general, the global construction of a set of orthogonal coordinates $\{q^i\}$ (curvilinear coordinates) on a manifold requires the integrability of the generating set of differential equations $dq^i/ds = v^i$ with constant right hand sides on that manifold. Spherical coordinates cannot be defined globally on a saddle-surface: ellipsoidal coordinates cannot be defined globally on a sphere with a handle attached to it, and Cartesian coordinates do not exist globally on a sphere. A flow field $dq_i/dt = v_i$ with constant velocity components v_i that is trivially integrable on one surface is nonintegrable (perhaps because not everywhere defined) on metrically inequivalent surfaces. If geometry determines gravity, and if geometry is that of a curved space-time manifold that is itself determined by the distribution and motion of matter, then we can expect that we may have a lot of trouble finding admissible global coordinate systems in general relativity. The reason is simply that the number of coordinate systems that can be defined globally on arbitrary curved manifolds is very small compared with the number that can be defined in Euclidean space, where every conceivable coordinate system is compatible with the (flat) manifold.

The impossibility of global parallelism on a curved manifold corresponds to a general kind of nonintegrability that corresponds geometrically to a certain lack of path-independence. Consider the case of a Riemannian surface of constant positive curvature, the sphere: if a ship follows a great circle on the oceans, a geodesic, then it can arrive back at its starting point parallel to itself after a trip around the world. Consider next any other hypothetical closed path on the globe where, at the starting point, the ship's fore–aft direction does not line up with the tangent vector to the path. If the ship is transported (by its motor or sail, or abstractly in a geometry calculation) around the loop while, in each very small but finite displacement, maintaining the ship's orientation parallel to itself ('parallel transport'), then the ship will not line up with it's initial orientation after a complete 2π-circuit around the loop. The amount of the angular discrepancy depends on the choice of path. This is a kind of nonintegrability and it arises because the ship's fore–aft direction is not initially parallel to the tangent vector to the curve along which the ship moves on the curved manifold. In general, parallel transport is globally integrable on a curved manifold only if the vector is transported parallel not only to itself but also parallel to the tangent to the curve along which it moves. Whenever the vector is itself the velocity vector along the path, then this kind of parallel transport defines a geodesic (see below).

In what follows, we shall not rely upon the global definition of geodesics by the variational principle, because we will not assume a metric. We start instead with the less restrictive idea of local parallelism as expressed by a set of 'affine connections' in an inner product space (see section 10.3 for the construction of the tensor **g** that allows the transformation of covectors into vectors and vice versa). Whenever we can use a metric along with holonomic coordinates, then the affine connections can be calculated from the metric. If we were to use, instead, a local description in terms of nonintegrable velocities (see section 12.2), then the connections are the structure coefficients of a Lie algebra.

Finally, note also (see chapter 10, for example) that the geodesic equations (11.55) generated by (11.54a) above also follow from Hamilton's principle

$$\delta \int T \mathrm{d}s = 0 \qquad (11.54\mathrm{b})$$

if we take $T = g_{ik}\mathrm{d}x^i\mathrm{d}x^k$, which is the square of the integrand in (11.54a). Note that the same geodesic equations fall out if we replace the integrand T of (11.54b) by any differentiable function $F(T)$.

Einstein's idea is to understand the two laws of nature represented by the Galilean law of inertia and the principle of equivalence as purely *local* consequences of a *single global* law of nature represented by the geodesic equation (11.55) along with a geometric condition that determines the gravitational field. Einstein wants us to understand the Keplerian orbits of planets as the consequence of a globally valid *law of inertia*, an idea that reminds us superficially of Galileo's notion that inertia alone is enough to cause a circular orbit.

11.4 Force-free motion on curved manifolds

We need not, and do not, assume that the tensor **g** defined by the method of section 10.3 defines a metric: we do not assume that the space-time manifold is a metric space. Instead, we assume that space-time is an affine manifold locally, which means that comparisons of vectors at slightly separated points is possible by means of a formula that is linear in the vector and the displacement, with local coefficients that are called the affine connections, or just 'connections'. Whenever space-time is a metric space and holonomic coordinates are used, then the connections coincide with the Christoffel symbols defined by the metric in chapter 10.

Our starting point is the definition of parallelism locally, the idea of the translation of a vector parallel to itself, and parallel to the tangent vector to a curve, over a small but finite distance along any curve on the manifold. Parallel transplantation along a curve is always possible locally in the tangent plane of a curved differentiable manifold (spheres and saddles are examples of such manifolds). Consider a vector **A** that is transported parallel to itself locally so that it has the same components A_μ at q and at $q + dq$ (locally, the vector **A** has constant components in the q-frame). Seen from another coordinate system $q'^\mu = f^\mu(q)$, the components A'^ν of the vector **A** in the q'-frame are

$$A'^\mu = \frac{\partial q'^\mu}{\partial q^\nu} A^\nu. \tag{11.56}$$

The condition for a constant vector locally, meaning local parallelism in the q-frame, is that

$$dA'^\mu = \frac{\partial^2 q'^\mu}{\partial q^\nu \partial q^\sigma} A^\nu dq^\sigma = \frac{\partial^2 q'^\mu}{\partial q^\nu \partial q^\sigma} \frac{\partial q^\nu}{\partial q'^\eta} dq^\sigma A'^\eta = \Gamma'^\mu_{\sigma\eta} dq^\sigma A'^\eta, \tag{11.57}$$

where $\Gamma'^\mu_{\nu\eta}$ is called the affine connection in the q'-frame (affine because it transforms like a tensor under affine transformations, which are linear transformations (see chapter 1), but does not transform like a tensor under nonlinear coordinate transformations).

Although we have not yet explained how the connections are to be calculated, the condition for parallel transport of *any* vector **A** in the q-frame is just

$$dA_\mu = \Gamma^\mu_{\nu\eta} dq^\nu A^\eta, \tag{11.58}$$

or

$$\frac{\partial A^\mu}{\partial q^\nu} = \Gamma^\mu_{\nu\eta} A^\eta. \tag{11.59}$$

The condition for parallel transport over a finite distance along a curve $q^\mu = F^\mu(s)$, where s is the arc-length, is that the partial differential equations (11.59) are integrable. The condition for global parallelism, as we shall see, is that there is a global coordinate system where all of the connections vanish identically. In what follows, we assume that the connections are symmetric in the lower indices, which rules out the use of nonintegrable velocities as local coordinates (see Dubrovin et al., 1984).

Notice that if we consider a tangent vector $A^\mu = dq^\mu/ds$, then we can define a special class of curves on the manifold by the equations

$$\frac{\partial}{\partial q^\nu} \frac{dq^\mu}{ds} - \Gamma^\mu_{\nu\varepsilon} \frac{dq^\varepsilon}{ds} = 0, \tag{11.60}$$

so that

$$\frac{dq^\nu}{ds} \frac{\partial}{\partial q^\nu} \frac{dq^\mu}{ds} - \Gamma^\mu_{\nu\varepsilon} \frac{dq^\varepsilon}{ds} \frac{dq^\nu}{ds} = 0, \tag{11.61}$$

which states that

$$\frac{d^2 q^\mu}{ds^2} - \Gamma^\mu_{\nu\varepsilon} \frac{dq^\varepsilon}{ds} \frac{dq^\nu}{ds} = 0. \tag{11.62a}$$

In other words, the condition for parallel transport of the tangent to a curve along its own direction yields the differential equation of a curve that describes 'the straightest' path between two points on the manifold. This defines geodesics on an affine manifold. The manifold, through the connections, determines the 'straightest' possible curves on itself. Whenever the connections are not identically zero in some coordinate system on the manifold, then the equations $d^2 q^\mu/ds^2 = 0$ are not integrable at some points on that manifold. Cartesian coordinates and global parallelism cannot exist on the manifold in this case (the sphere provides the simplest case).

Note that if V^σ is an arbitrary tangent vector to a curve, then the equation

$$\frac{dV^\sigma}{ds} - \Gamma^\sigma_{\alpha\beta} V^\beta \frac{dq^\alpha}{ds} = 0 \tag{11.62b}$$

expresses the idea that the tangent vector is displaced parallel to itself along the curve. Although dq^σ/ds is an obvious tangent vector, any vector of the form $V^\sigma = \lambda(s)dq^\sigma/ds$ is also a tangent vector so that the most general geodesic equation on an affine manifold has the form

$$\frac{d}{ds}\left(\lambda(s)\frac{dq^\sigma}{ds}\right) - \lambda(s)\Gamma^\sigma_{\alpha\beta} V^\beta \frac{dq^\alpha}{ds} = 0. \tag{11.62c}$$

However, by a new choice of parameter given by $dp = ds/\lambda(s)$ we can always bring this equation into the form (11.62b).

We can now offer a more analytic definition of global flatness. Flat manifolds are manifolds where the equations $d^2 q^\mu/ds^2 = 0$ are defined and integrable everywhere on the manifold, *for all points* $-\infty < q^\mu < \infty$, without excluding even one or two points from the manifold. In this case, the coordinates q^μ are Cartesian coordinates x^μ: flat manifolds are those where Cartesian coordinates can be constructed globally all the way to infinity and are called Cartesian manifolds. Euclidean space is a Cartesian manifold with a metric. Curved manifolds are those where the dynamical system $d^2 q^\mu/ds^2 = 0$ is not everywhere integrable because the connections do not vanish identically in any global coordinate system on the manifold. On a sphere, for example, in polar coordinates $(q^1, q^2) = (\theta, \phi)$ the equation $d^2\phi/ds^2 = 0$ is undefined at $\theta = 0$ and π because ϕ is undefined at those two points.

We now introduce the idea of the covariant derivative, which allows us to organize our ideas more systematically. Consider the scalar product of any covector A_μ with the vector $dq^\nu/d\tau$ that defines the tangent to some curve on the manifold. The derivative with respect to s of the scalar product is also a scalar along the curve

$$\frac{d}{ds} A_\mu \frac{dq^\mu}{ds} = \frac{\partial A_\mu}{\partial q^\nu} \frac{dq^\mu}{ds} \frac{dq^\nu}{ds} + \Gamma^\sigma_{\mu\nu} \frac{dq^\mu}{ds} \frac{dq^\nu}{ds} A_\sigma = A_{\mu,\nu} \frac{dq^\mu}{ds} \frac{dq^\nu}{ds}, \tag{11.63}$$

where

$$\frac{\partial A_\mu}{\partial q^\nu} + \Gamma^\sigma_{\mu\nu} A_\sigma = A_{\mu,\nu} \tag{11.64}$$

is called the covariant derivative because it transforms like a rank two tensor under arbitrary coordinate transformations, so long as A_μ transforms like a covector. Note that

$$A_{\mu,\nu} - A_{\nu,\mu} = \frac{\partial A_\mu}{\partial q^\nu} - \frac{\partial A_\nu}{\partial q^\mu} \tag{11.65}$$

follows from $\Gamma^\lambda_{\mu\nu} = \Gamma^\lambda_{\nu\mu}$. This means that

$$A_{\mu,\nu} - A_{\nu,\mu} = \frac{\partial A_\mu}{\partial q^\nu} - \frac{\partial A_\nu}{\partial q^\mu} = 0 \tag{11.66}$$

guarantees that the differential form $A_\mu dq^\mu$ is exact. In other words, $A_{\mu,\nu} = A_{\nu,\mu}$ is the condition that each component A_μ of the vector $\mathbf{A}$ is a function of the q_ν, defined independent of path, on the manifold. The parallel transport condition

$$\frac{\partial A^\mu}{\partial q^\nu} = \Gamma^\mu_{\nu\eta} A^\eta \tag{11.67}$$

is therefore the infinitesimal statement that the covariant derivative of the vector $\mathbf{A}$ vanishes.

As a check on our definition, note that the covariant derivative of the velocity vector $A^\mu = dq^\mu/ds$ is

$$A^\mu_{,\nu} = \frac{\partial}{\partial q^\nu} \frac{dq^\mu}{ds} + \Gamma^\mu_{\nu\varepsilon} \frac{dq^\varepsilon}{ds}, \tag{11.68}$$

and so the contraction of this derivative with the velocity vector $dq^\mu/d\tau$ yields

$$\frac{dq^\nu}{ds} A^\mu_{,\nu} = \frac{d^2 q^\mu}{ds^2} + \Gamma^\mu_{\nu\varepsilon} \frac{dq^\varepsilon}{ds} \frac{dq^\nu}{ds} = 0, \tag{11.69}$$

which is the equation that defines the curve.

The vanishing of the covariant derivative of a vector $\mathbf{A}$ guarantees that the vector is translated along a curve on the manifold with its direction maintained locally parallel to itself. The allowed curves on the manifold are determined by the geometry of the manifold through the connections. A flat space accepts everything, but not all curves are allowed on every curved manifold because the defining differential equations may be nonintegrable on a given manifold. Now for the main question.

When is global parallelism generated by the integration of the equations of local parallelism? In other words, when does parallel transport of a vector $\mathbf{A}$ about an arbitrary path lead to global parallelism, to the possibility of an everywhere constant vector field? This reduces to the question of integrability of the arbitrary differential form

$$dA^\nu = dq^\mu \frac{\partial A_\nu}{\partial q^\mu}, \tag{11.70}$$

of a vector field $\mathbf{A}$ whose covariant derivative vanishes. Such a differential form would define a constant vector in a flat space, because with $\Gamma = 0$ in Cartesian coordinates it would follow that

$$A_{\mu,\nu} = \frac{\partial A_\mu}{\partial q^\nu} = 0 \tag{11.71}$$

so that the A_μ are globally constant in a Cartesian frame (in contrast, the components of a constant vector **A** vary whenever $\Gamma \neq 0$ in a nonCartesian frame), defining a constant vector field. The differential form (11.70) is integrable if

$$dA_\mu = \frac{\partial A_\mu}{\partial q^\nu} dq^\nu = -\Gamma^\sigma_{\mu\nu} A_\sigma dq^\nu \tag{11.72}$$

is closed, which requires that

$$\frac{\partial}{\partial q^\alpha} \Gamma^\sigma_{\mu\nu} A_\sigma = \frac{\partial}{\partial q^\nu} \Gamma^\sigma_{\mu\alpha} A_\sigma. \tag{11.73}$$

Direct differentiation, along with substitution of (11.72), yields the integrability condition as

$$A_\sigma R^\sigma_{\mu\alpha\nu} = 0 \tag{11.74}$$

where

$$R^\sigma_{\mu\alpha\nu} = \frac{\partial}{\partial q^\alpha} \Gamma^\sigma_{\mu\nu} - \frac{\partial}{\partial q^\nu} \Gamma^\sigma_{\mu\alpha} + \Gamma^\tau_{\mu\alpha}\Gamma^\sigma_{\tau\nu} - \Gamma^\tau_{\mu\nu}\Gamma^\sigma_{\tau\alpha} \tag{11.75}$$

is called the Riemann curvature tensor. Whenever this tensor vanishes identically over all of space-time, with no points excepted (delta-function singularities might represent point particles), then the differential form

$$dA_\nu = dq^\mu \frac{\partial A_\nu}{\partial q^\mu} \tag{11.76}$$

is exact, is globally integrable. When (11.76) is exact, then

$$\int_1^2 dA_\mu = A_\mu(2) - A_\mu(1) \tag{11.77}$$

is path-independent, so that we can define the components A_μ at a second arbitrary point in space by starting at point 1 and translating A_μ parallel to itself along any curve that connects the two points. In particular, since the integral (11.77) vanishes for arbitrary closed paths, if we transport the vector about a closed loop then it must line up with its former direction after a 2π-revolution.

In other words, if the Riemann curvature tensor vanishes identically then the differential equations

$$\frac{\partial A^\mu}{\partial q^\nu} = \Gamma^\mu_{\nu\eta} A^\eta \tag{11.78}$$

can be integrated independently of path to produce a globally constant and parallel vector field **A**, which means that the manifold is flat. In this case, the differential equations $d^2 x^\mu/ds^2 = 0$ (meaning $dx^\mu/d\tau = v^\mu = $ constant) can be integrated globally to construct mutually perpendicular Cartesian axes defined by $x^\mu = v^\mu s + x^\mu_o$. In this Cartesian coordinate system all of the connections $\Gamma^\lambda_{\mu\nu}$ vanish, and so $R^\alpha_{\beta\gamma\delta}$ vanishes as well. The vanishing of Γ in a Cartesian coordinate system is equivalent to the vanishing of $R^{\alpha}_{\beta\gamma\delta}$ in all coordinate systems because $R^{\alpha}_{\beta\gamma\delta}$ transforms like a fourth rank tensor, which means that we should write **R** in order to remind ourselves that the Riemann curvature tensor is defined completely coordinate-free.

What does the integrability following from $\mathbf{R} = 0$ mean geometrically? We have seen from the discussion above that it means that we can define parallelism globally by freely translating vector fields around on the manifold independently of path, an idea that we took for granted in the study of elementary physics. The Riemann tensor **R** is called the curvature tensor because a space with nonvanishing curvature tensor is not flat, and distant parallelism (with corresponding global Cartesian axes) does not exist on such a manifold. A space with a nonvanishing Riemann tensor is called a curved space. There, integrability fails and differential forms that would be trivially integrable in a flat space are generally nonintegrable.

For example, differentials dx^{μ} of Cartesian coordinates that are defined locally in a tangent plane cannot be integrated to define a global Cartesian coordinate system unless the space-time is flat (and therefore has no mass at all): the simplest differential equation of special relativity, $d^2x^{\mu}/d\tau^2 = 0$, in Cartesian coordinates is nonintegrable in a curved space-time, and so the parameters x^{μ} that we would like to identify as Cartesian coordinates are nonholonomic rather than generalized coordinates (see also section 12.1 for nonintegrability and nonholonomic coordinates). In other words, the very point that we normally take for granted in physics, the construction at will of a global coordinate system, rests on the question of whether a locally defined velocity vector $dq^{\mu}/d\tau$, where the q^{μ} are any set of curvilinear coordinates defined locally in a tangent plane, can be integrated independently of path, but this is only possible if the metric has enough symmetry to admit global conservation laws. Arbitrary curvilinear coordinate systems are not guaranteed to exist globally in an arbitrary curved space-time. In other words, the search for solutions of the equations of general relativity is tied much more strongly to the search for a coordinate system that reflects the symmetry of the system than it is in classical mechanics, where one can write down *any* coordinate system and integrate numerically or make approximations. Even worse, unless we know the symmetry of the manifold we cannot know in advance if a global coordinate system exists. This is the main geometric point that is missed in discussions that concentrate upon the purely formal idea of general covariance.

11.5 Newtonian gravity as geometry in curved space-time

We now present Cartan's beautiful explanation of Newtonian gravity as the lack of parallelism in a curved space-time where space is Euclidean and time is absolute. Consider a body of mass m_1 in the gravitational field of a much heavier body of mass m_2. The idea is to rewrite Newton's law of motion

$$\frac{d^2x^i}{dt^2} + \frac{\partial\Phi}{\partial x^i} = 0 \tag{11.79}$$

as a geodesic equation in space-time, where $U = m_1\Phi$ is the gravitational potential energy. This is possible only because of the principle of equivalence: Φ is independent of the moving body's mass. Cartan's geometric reformulation of Newtonian physics is simple: fixed-time snapshots of three dimensional space are Euclidean (the three (x^i, x^k)

planes with $i = 1,2,3$ have no curvature), time is absolute, but the three (x^k,t) manifolds are all curved. In particular, the time is affine, $s = at + b$, so that each body's clock ticks uniformly from the standpoint of any other body's clock, where s is now the time as read by a clock that travels along with our body of mass m. In what follows we set $x^4 = t$ and we can rescale the time variable for the motion of any planet about the sun so that, effectively, $a = 1$, $b = 0$.

Working in the Cartesian system of coordinates[3] centered on the source-body of mass M ('the sun'), the geodesic equations are

$$\left. \begin{array}{c} \dfrac{d^2x^i}{ds^2} + \dfrac{\partial\Phi}{\partial x^i}\left(\dfrac{dt}{ds}\right)^2 = 0, \ i = 1,2,3 \\[3mm] \dfrac{d^2t}{ds^2} = 0. \end{array} \right\} \tag{11.80}$$

That last equation tells us that a global Cartesian time axis exists, and it is trivial to read off the affine connections:

$$\Gamma^i_{\infty} = \frac{\partial\Phi}{\partial x}, \ \Gamma^i_{kl} = 0 \text{ otherwise,} \tag{11.81}$$

where $i,k,l = 1,2,3$. Therefore, the Riemann tensor in the 'sun's' Cartesian coordinate system is given by

$$R^i_{\circ k\circ} = \frac{\partial\Phi}{\partial x^i}, \ R^i_{jkl} = 0 \text{ otherwise.} \tag{11.82}$$

Defining the Ricci curvature tensor as the contraction ('trace') of the Riemann curvature tensor

$$R_{kl} = R^i_{ikl}, \tag{11.83}$$

we find that

$$R_{\infty} = \nabla^2\Phi, \ R_{kl} = 0 \text{ otherwise,} \tag{11.84}$$

so that the gravitational field equations

$$\nabla\cdot\vec{g} = \nabla^2\Phi = 4\pi\rho \tag{11.85}$$

become

$$R_{\infty} = 4\pi\rho. \tag{11.86}$$

What Hertz (who tried to geometrize forces) found to be impossible in configuration space, Cartan (with the hindsight of Einstein's correct formulation of gravity) has managed for gravity: the complete geometrization of Newtonian gravity in absolute space and absolute time.

To abandon the Cartesian coordinates of the field-source, one can get a (coordinate-free) description of gravity and motion in a gravitational field only by first making a suitable replacement for the mass density on the right hand side of equation (11.86). A static point mass is represented by a delta function density, and the singularity of the

---

[3] Misner et al. (1973) and Fock (1959) systematically call Cartesian coordinates in an inertial frame 'Galilean coordinates'. Galileo introduced and emphasized the idea of inertial frames, but he certainly did not invent Cartesian coordinates.

Ricci curvature tensor determines the connections and therefore the static gravitational field in that case. The underlying invariance principle, 'the physics', is simply the Newtonian principle of equivalence, first discovered by Galileo: replacement of gravity by a geodesic equation is possible because all bodies fall freely at the same rate, independent of mass, in a local (or uniform) gravitational field.

Because space is Euclidean global parallelism exists there, but not in space-time cuts: the amount by which a four-vector (A^1, A^2, A^3, A^4) fails to line up with itself after parallel transport about any small-enough closed curve (an ε-sized rectangle) in an (x^i, t) manifold is described by the covariant derivative and determined by the Riemann curvature (local nonintegrability):

$$\left. \begin{array}{l} \delta A^4 = 0 \\ \delta A^i = - R^i_{ook} A^\circ (\delta t) \delta x^k. \end{array} \right\} \tag{11.87}$$

Kepler's elliptic orbits about the sun are due to free-fall of a planet in the sun's gravitational field, are geodesics in absolute space-time, and are curved because of the curvature of the three space-time manifolds (x^k, t). In general, two geodesics that start very near to each other will separate spatially from each other (will not remain parallel to each other), and the deviation δx^k is given to first order by

$$\left. \begin{array}{c} \dfrac{\mathrm{d}^2}{\mathrm{d}t^2} \delta x^k + \dfrac{\partial \Phi}{\partial x^k \partial x^1} \delta x^l = 0, \; k = 1,2,3 \\[2ex] \dfrac{\mathrm{d}^2}{\mathrm{d}t^2} \delta x^4 = 0, \end{array} \right\} \tag{11.88}$$

an idea that leads to the formulation of Liapunov exponents for neighboring trajectories (but no necessarily to nonintegrability and chaos in phase space – see chapter 13). Of course, trajectories that separate from each other more slowly than exponentially have vanishing Liapunov exponents. Stated covariantly, equation (11.88) reads

$$\frac{D}{Ds} \frac{D \delta x^i}{Ds} = R^i_{jkl} \frac{\mathrm{d}x^j}{\mathrm{d}s} \frac{\mathrm{d}x^k}{\mathrm{d}s} \delta x^l \tag{11.89}$$

where D/Ds denotes the covariant derivative. This suggests that exponentially fast separation of nearby trajectories may arise from the negative curvature of a manifold, a point that brings the geometry of a two dimensional manifold to bear on our discussion of the integrability of phase space flows where $n \geq 4$ (chapter 13).

We can now treat gravity, Coriolis and centrifugal accelerations, and also the noninertial effect of a linearly accelerated frame from a unified geometric point of view. Abandoning the inertial frame of the sun, let us transform to a frame that is both rotating and linearly accelerated relative to the sun's inertial frame: $x' = Rx + a$ where both the orthogonal matrix R and the translation vector a are time-dependent (the primed frame could be a frame that rotates with the sun, for example). Combining the results of chapters 1 and 8, the equation of motion is

$$\frac{\mathrm{d}^2 x'}{\mathrm{d}t^2} - 2\Omega' \frac{\mathrm{d}x'}{\mathrm{d}t} + (\Omega'^2 - \Omega')x' + \ddot{a} + \nabla' \Phi = 0. \tag{11.90}$$

Understood as a geodesic (11.90) representing *force-free* motion in absolute space-time, this equation yields the connections

$$\left.\begin{array}{l} \Gamma'^i_{jk} = 0,\ i = 1,2,3 \\ \Gamma'^i_{0k} = \Gamma'^i_{k0} = -2\Omega'_{ik} \\[2ex] \Gamma'^i_{00} = \dfrac{\partial \Phi'}{\partial x'^i} + \Omega'_{ij}\Omega'_{jk}x'^k - \dot{\Omega}'_{ij}\dot{x}'^j + \ddot{a}. \end{array}\right\} \tag{11.91}$$

The connections defined by Ω are called gauge potentials and are not symmetric in the lower indices (the space is not a metric space). Gauge potentials are the generalization to noncommuting potentials defined by matrices (or operators) of the idea of the 'Abelian' gauge potential **A** of classical electromagnetic theory: as is implicit in section 8.6, the gauge potential connections above can be rewritten in terms of the structure constants of the Lie algebra of the rotation group. Note that if we define $\Phi' = \Phi + \tilde{x}\ddot{a}$ then the transformation to a frame that is accelerated linearly, with constant acceleration $\ddot{a}$, can be treated as an abelian gauge transformation on the gravitational potential Φ. Here, the gravitational potential Φ does not transform like a scalar function, and correspondingly the gravitational force does not transform like a covariant vector. So is the principle of equivalence built into Cartan's reformulation of Newtonian gravity!

It is now easy to write down Einstein's field equations for curved space-time. Those equations motivated Cartan's ingenious geometric description of Newtonian gravity.

11.6 Einstein's gravitational field equations

What are Einstein's field equations, the equations that determine the connections? Clearly, we can not determine them by setting the Riemann tensor $R^\alpha_{\eta\pi\gamma}$ equal to zero because then space would be flat, and Einstein's idea is that space-time, including the three dimensional 'spatial' subspace, is curved by mass and energy. If we take the trace over the two indices α and β and set the result equal to zero,

$$R^\alpha_{\eta\alpha\gamma} = 0 \tag{11.92}$$

then space is flat. This tensor also characterizes a curved space (does not vanish identically in a curved space because there is then no global frame where the Γ's vanish). Einstein identified (11.92) as the equations of the gravitational field, the equations that determine the connections, by assuming that $R^\alpha_{\eta\alpha\gamma} = 0$ except where matter is located. In particular, this tensor must be singular at the location of a point mass, just as is the Newtonian gravitational potential. More generally, the Einstein field equations are given by

$$R^{\mu\nu} - \tfrac{1}{2}g^{\mu\nu}R = -\kappa T^{\mu\nu} \tag{11.93}$$

where $R = g_{\mu\nu}R^{\mu\nu} = T$, $T^{\mu\nu}$ is the mass–energy tensor (the generalization to moving mass–energy of the mass density ρ) and κ is a constant to be determined. This turns out to be the simplest mathematics ansatz that completely determines the connections Γ and simultaneously yields the right answer in the Newtonian limit. For a point singularity at the origin, it yields connections that, in the geodesic equations, reproduces Newton's second law with the Newtonian inverse square law force as a zeroth order approximation (we expect this from the standpoint of Cartan's geometric reformulation of Newtonian gravity). More generally, and in contrast with Cartan's formulation but in agreement with special relativity, the Einstein field equations describe the propagation of nonlinear waves at the speed of light ('gravitational waves').

The Newtonian limit is realized in weak fields as follows. To keep the analysis easy, we assume that the connections are given as the Christoffel symbols determined by a metric g where $ds^2 = g_{ik}dq^idq^k$,

$$\Gamma_{bc}^a = \tfrac{1}{2}g^{ad}\left(\frac{\partial g_{bd}}{\partial q^c} + \frac{\partial g_{cd}}{\partial q^b} - \frac{\partial g_{bc}}{\partial q^d}\right). \tag{11.94}$$

In the case of connections defined by a metric, the connections are symmetric: $\Gamma_{bc}^a = \Gamma_{cb}^a$, a restriction that does not apply to affine connections defined by rotating frames in Euclidean space. A slowly varying weak field g, to lowest order and in local Cartesian coordinates, must reduce to $g_{ab} \approx -\delta_{ab}$ for the spatial indices and $g_{oo} \approx +1$ for the time index. In this case $ds/c \approx d\tau \approx dt$ and the spatial variation coordinates along the geodesic obey

$$\frac{d^2q^a}{dt^2} + \Gamma_{oo}^a c^2 + \Gamma_{bc}^a \frac{dq^b dq^c}{dtdt} \approx 0. \tag{11.95}$$

If we neglect $dq^a/dt \ll c$ we also get

$$\frac{d^2q^a}{dt^2} + \Gamma_{oo}^a c^2 \approx 0. \tag{11.96}$$

With index $a \neq 0$ this reduces to

$$\frac{d^2q^a}{dt^2} - \tfrac{1}{2}c^2\frac{\partial g_{oo}}{\partial q^a} \approx 0. \tag{11.97}$$

Therefore, we identify $g_{oo} \approx 1 + 2\Phi/c^2$ as the approximate matrix element where Φ is the Newtonian gravitational potential due to a source fixed at the origin of coordinates. With $d\tau$ denoting the time interval measured by a clock that is fixed in the x-frame at the spatial point (x^1, x^2, x^3), we have

$$cd\tau = c(g_{oo})^{1/2}dt \tag{11.98}$$

so that

$$d\tau \approx dt(1 + 2\Phi/c^2)^{1/2}. \tag{11.99}$$

In (11.99) $d\tau$ is the proper time of a clock at rest in the gravitational field Φ. Therefore, a clock at rest in the gravitational field ticks at a slower rate the lower the potential. A clock fixed in the gravitational field is effectively accelerated: to be at rest in a gravitational field is locally equivalent to being fixed in an accelerated frame of reference.

From a broader point of view, Newtonian physics represents the zeroth order solution to the Einstein field equations, and can be derived from a systematic perturbation expansion in powers of $1/c^2$ (see Fock, 1959, or Dubrovin et al., 1984).

Schwarzschild, who also introduced the action-angle variables (see chapter 16) around 1916, solved Einstein's free-space gravitational field equations (11.91) to determine the space-time manifold surrounding a mass distribution represented by a fixed singularity of mass m at the origin of a set of local coordinates. Global coordinates that extend over all of space-time do not necessarily exist for an arbitrary manifold, but the Schwarzschild problem has spherical symmetry so that spherical coordinates can be used to find that solution. The result, plugged into (11.97), reproduces Kepler's elliptic orbits and Newton's second law with the Newtonian gravitational force, to lowest order in perturbation theory, indicating that gravity can be geometrized. Experimental tests of some low order corrections to Newtonian predictions can be found in the literature (see Bergmann, 1942, Adler et al., 1965, Misner et al., 1973, or Eddington, 1963). Next, we display the Schwarzschild solution (see Adler et al., pp. 164–74 for details).

With spherical symmetry we can take

$$ds^2 = g_{oo}(r)c^2dt^2 - g_{rr}(r)\,dr^2 - r^2d\theta^2 - r^2\sin^2\theta d\phi^2. \tag{11.100}$$

A straightforward three- to four-page solution of Einstein's field equations $R_{\alpha\beta} = 0$ using the Christoffel symbols yields

$$ds = \left(1 - \frac{2\alpha}{r}\right)c^2dt^2 - \frac{dr^2}{1 - 2\alpha/r} - r^2d\theta^2 - r^2\sin^2\theta d\phi^2. \tag{11.101}$$

The coefficient of dt^2, with no modification of the spatial part of the metric, would reproduce Newtonian mechanics with the choice $\alpha = Gm/c^2$. However, the coefficient of dr^2 suggests that we have something more than a Newtonian result. In particular, we have the famous Schwarzschild singularity: g_{rr} is infinite when $r_s = 2\alpha = 2Gm/c^2$. For the sun, the Schwarzschild radius r_s is on the order of a kilometer. Solutions obtained from (11.101) are valid only if $r > r_s$. Note that the radius R of a planet whose mass is m and whose escape speed is the speed of light is given in Newtonian theory by $R = r_s$.

Denoting the derivative d/ds by a dot, the geodesic equations are given by the Euler–Lagrange equations of Hamilton's principle (11.54b)

$$\delta\int\left[\left(1 - \frac{2\alpha}{r}\right)c^2\dot{t}^2 - \left(1 - \frac{2\alpha}{r}\right)^{-1}\dot{r}^2 - r^2(\dot{\theta}^2 + \sin^2\theta\dot{\phi}^2)\right]ds = 0 \tag{11.102}$$

and read

$$\left.\begin{array}{c} \dfrac{d}{ds}(r^2\dot{\theta}) = r^2\sin\theta\cos\theta\dot{\phi}^2 \\[2ex] \dfrac{d}{ds}(r^2\sin^2\theta\dot{\phi}) = 0 \\[2ex] \dfrac{d}{ds}\left(1 - \dfrac{2\alpha}{r}\right)\dot{t} = 0. \end{array}\right\} \tag{11.103}$$

It is easy to see, as expected from spherical symmetry, that the orbit is planar. This means that we can choose $\theta = \pi/2$. With this choice, $r^2d\phi/ds = h$ is an integration constant as is also expected, so that $d\phi/dt$ can be eliminated as in the Newtonian problem. Also, we have $(1 - 2\alpha/r)dt/ds = l = $ constant so that dt/ds can also be eliminated for $r > r_s$. The radial equation is given by

$$l = \left(1 - \frac{2\alpha}{r}\right)^{-1} c^2 l^2 - \left(1 - \frac{2\alpha}{r}\right)^{-1} \dot{r}^2 - \frac{h^2}{r^2}, \qquad (11.104)$$

which is also a conservation law. With $r = 1/x$ the turning points are determined by the solutions of the cubic equation $x^3 - x^2/2\alpha + x/h^2 + (c^2 l^2 - 1)/2\alpha h^2 = 0$. As in the classical problem, there must be exactly two positive turning points r_1 and r_2 where $dr/ds = 0$, representing the case of bounded motion. Note that if $r < r_s$ there are no turning points.

The equation for the orbit $d\theta/dr$ can be found and integrated to yield a result analogous to the angular displacement (4.10b).

$$\Phi = \int_{x_1}^{x_2} \frac{dx}{[2(E - w(x))]^{1/2}}, \qquad (11.105)$$

where $E = (c^2 l^2 - 1)/2h^2$ and $w(x) = x^2/2 - \alpha x/h^2 - \alpha x^3$. We know from chapter 4 that this integral can represent a rational number multiplied by 2π only for a very sparse and special set of values of the constants α, l, and h. Therefore, the bounded orbits are generally quasiperiodic, which differs from Newtonian gravity. Perturbation theory, for the case where the bounded orbit is approximately Newtonian (approximately elliptic) yields a precessing ellipse. The computed precession rate can be found in the literature and compares well with the known precession rate of the orbit of the planet Mercury, although this is not the only possible source of orbital precession. The reader who has understood the method of this chapter and that of chapter 10 will have no difficulty working out the details missing in the above analysis by following the very readable chapter 6 of Adler et al. (1965).

In the Schwarzschild solution, the connections representing a static point mass are static, and the curved space-time near the mass is spherically symmetric. A particle of small enough mass that the source of the field does not accelerate appreciably (and thereby generate a time-dependent metric) obeys the force-free equation (11.62a) and so follows a geodesic on the curved space-time manifold that is caused by the massive, static point source. The two-body problem in general relativity is still unsolved. The metric may be time-dependent because of gravitational radiation (energy could not be strictly conserved within the solar system in that case).

In the Schwarzschild solution action at a distance (which Newton could not explain by a mechanism) has been replaced by the steady warping of empty Euclidean space due to the presence of a mass. Far from concentrations of mass, the Einsteinean space-time manifold is flat, but the static Schwarzschild solution does not clarify for us the wave-propagation mechanism whereby flat space-time is caused to become curved by the sudden introduction of matter.

Of all papers and books written as introductions to special and general relativity, none are easier or more enjoyable to study than Einstein's original work, including the English translation. Integrability is discussed in Bergmann, (1942), Adler et al., (1965) and Eddington (1963). The book by Misner et al. (1973) is hard to read but is encyclopedic and constitutes a comprehensive and invaluable source of information.

11.7 Invariance principles and laws of nature

It is not necessary to look deeper into the situation to realize that laws of nature could not exist without principles of invariance. This is explained in many texts of elementary physics even though only few of the readers of these texts have the maturity necessary to appreciate these explanations. If the correlations between events changed from day to day, and would be different for different points of space, it would be impossible to discover them. Thus the invariances of the laws of nature with respect to displacements in space and time are almost necessary prerequisites that it be possible to discover, or even catalogue, the correlations between events which are the laws of nature.

E. P. Wigner (1967)

An arbitrary nonEuclidean manifold has no global symmetry. Accelerating bodies radiate gravitational waves and therefore lose energy. Local conservation laws (partial differential equations that are continuity equations) therefore generally do not give rise to global conservation laws over arbitrary finite submanifolds of space-time, although one expects mass–energy to be conserved if all of the mass–energy in the universe were taken into account. Given local conservation laws, some degree of integrability is required for global conservation laws, but the required differential forms in a curved space-time are generally nonintegrable (see section 13.2 for the analogous problem in phase space). If the metric has spherical symmetry, then global spherical coordinates may fall out, as they do in the Schwarzschild solution described incompletely above, but if a metric has no symmetry at all then there is no guarantee that a global coordinate system can be constructed (nonintegrability). In general relativity Klein's attempt to define geometry by transformations that leave the metric invariant (the Erlanger Program) fails. It's hard to understand what one can possibly mean by a 'geometric object' in a space-time where there are no four dimensional analogs of rigid bodies because there is no invariance. In a space-time without adequate symmetry the search for metric invariants can be replaced by the search for purely qualitative invariants, namely, topologic invariants. In the spirit of Klein's program, topologic invariants can be used to distinguish qualitatively different curved spaces from each other. The geometric objects that are called 'wormholes' in geometrodynamics are called spheres with one or more handles in dynamical systems theory. Very little is known about the behavior of dynamical systems on such manifolds.

While the general two-body problem remains unsolved, Fock (1959) has shown how to solve it as a systematic perturbation expansion to second order in v^2/c^2. There, with assumptions that yield symmetry, all of the conservation laws of classical mechanics hold. The dynamics problem is integrable and yields a precessing ellipse for the bound-state motions.

The equations of general relativity can be proven to be covariant for arbitrary differentiable transformations in space-time. As his starting point for the geometric theory of gravity, Einstein argued (unconvincingly, for this author) that the principle of equivalence means that inertial frames should play no special role in the formulation of physics, that any reference frame, accelerated or not, should be as good as an inertial frame for the formulation of the laws of physics. By stating that one frame should be as good as any other for the formulation of the laws of physics, Einstein seemed to elevate covariance of equations (coordinate-free descriptions of geometry and motion) to the status of a physical principle. But what can it possibly mean to say that an arbitrary frame is as good as an inertial frame for the formation of the laws of physics?

In local regions of the universe where matter is irregularly distributed and rapidly evolving, the space-time manifold may have no global symmetry and may change rapidly. In such a locale, inertial frames, Lorentz transformations, and uniform gravitational fields might be good approximations only over space-time regions that are too small to be of any macroscopic physical significance. In that case, what we have identified in chapter 1 as universal laws of nature do not hold macroscopically, although we expect Lorentz invariance to hold microscopically. As Wigner has pointed out (1967), it is only an untested article of faith, a local dogma of present-day science, to believe that physics (meaning the universal laws that follow from the *regularities* of nature) is the same everywhere and at all times in the universe.

In an arbitrary curved space-time, there is no guarantee that a global coordinate system can be constructed, except far enough away from matter that space-time is approximately flat. In that case, we know what we mean by 'physics', and approximate inertial frames are certainly preferred there for experiments and observations of nature. It seems impossible to apply Einstein's statement in any meaningful way to a space-time region with an arbitrarily rapidly varying metric with no symmetry. We conclude that Einstein's statement about covariance makes sense only in a restricted sense in the context of local frames, and even then only to space-time regions where local frames are a good approximation for describing phenomena over macroscopic space-time regions, like the solar system over thousands of years. There, inertial frames are again preferred because identical experiments performed in different inertial frames give identical numbers, and the laws of the regularities of nature take on their simplest form in those frames. But another class of frames is not far behind in theoretical value, and is even better from a practical standpoint: a uniformly accelerated frame is just as good as a nonaccelerated frame in a uniform gravitational field, because in either of these two cases an identically prepared and performed experiment will yield the same numbers. It seems that the latter either is or at least should have been what Einstein meant by 'general relativity'.

Global nonintegrability rules out any possibility that general covariance could provide us with a strong principle, even if it could be granted that general covariance could be regarded as a physical principle (Kretschmann pointed out in 1917 that it cannot be). Transformations to arbitrarily curvilinear (more generally, holonomic) coordinates are possible globally in the $3Nd$ Euclidean space of an unconstrained classical Newtonian system, but most of those curvilinear coordinate systems are possible only locally in general relativity or in a constrained classical Lagrangian system because of the general nonintegrability (or lack of definition at singular points on a manifold) of the velocity fields needed to generate them. In other words, if a manifold is curved, then stating that a dynamics theory is 'generally covariant' does not guarantee that there are a lot of different global reference frames available for the discussion of large-scale, long-time motions. The number and type of global coordinate systems that are available for use is a question of symmetry and integrability (invariance). Cartesian and Euclidean spaces have maximal symmetry, so that all possible curvilinear coordinate systems can be constructed in those spaces.

As Wigner has pointed out (in the context of general relativity and the idea of the expanding universe), it is impossible to give examples that prove that the laws of regularities of nature are the same in every corner of the universe and at all times, because time translational invariance and also spatial translational and rotational symmetries do not hold globally in such a universe. In other words, we cannot have confidence that at early or late times, and in far corners of the expanding universe, the laws of nature are guaranteed to be always the same. In order to identify laws of nature empirically, there must be *some* sort of motion whose description is free of the initial conditions of time and location. Otherwise, every set of observations would look like a law unto itself (as seems to be the case in observations of social and individual behavioral phenomena). Covariance of *equations* does not guarantee an experimenter the existence of coordinate-free *regularities*, phenomena and correlations that are reproducible independently of time and place because they reflect an underlying invariance principle. Only invariance principles can produce true (as opposed to formal vectorial) reference frame independence.

Geometric gravity theory has no *relativity principle* other than the principles of equivalence and Lorentz invariance, both of which are strictly local. In principle, there need be no global invariants on a finite portion of any arbitrary manifold where bodies are continually and even chaotically in motion.

Finally, if one uses holonomic coordinates q^i and introduces a metric according to $ds^2 = g_{ij}dq^i dq^j$, then the connections Γ are the Christoffel symbols of section 10.2 and are symmetric in the two lower indices. The derivations of this chapter after section 11.2 hold for an affine space with or without a metric. For example, the connections Γ^i_{jk} defined by using rotating frames in Newtonian mechanics are metric-free and need not be symmetric in the two lower indices. In the absence of a metric the tensor g that is needed for raising and lowering of indices can be constructed from local basis vectors by the method of section 10.3. In general, commuting translations of locally orthogonal coordinates indicate a flat manifold: the coordinates can be chosen both locally and globally to be Cartesian. The statement that the transport of an arbitrary vector parallel to itself along a very small path (x,y) to $(x + \delta x,y)$ to $(x + \delta x,y + \delta y)$ in the Cartesian plane yields the same result as parallel transport along the alternative path (x,y) to $(x,y + \delta y)$ to $(x + \delta x,y + \delta y)$ is equivalent to the statement that the infinitesimal generators of translations d/dx and d/dy commute with each other. Commutation of operators can be used to express the idea of parallelism locally, just as the Cartesian axes that are generated by the iteration of the infinitesimal transformations express the idea of parallelism globally. In other words, flatness of a manifold and commutation of translations on the manifold are not independent ideas.

In contrast, noncommuting translations in locally orthogonal coordinates indicate a curved manifold: that parallel transport of a vector along a path defined by (θ,ϕ) to $(\theta + \delta\theta,\phi)$ to $(\theta + \delta\theta,\phi + \delta\phi)$ yields a different result than the transport of a vector parallel to itself along the path (θ,ϕ) to $(\theta,\phi + \delta\phi)$ to $(\theta + \delta\theta,\phi + \delta\phi)$ is not different than saying that the generators of rotations do not commute with each other. It would be both natural and interesting to go on and discuss Berry's geometric phase, noncommuting gauge potentials, and metric-free connections defined by the structure constants of a Lie algebra, as well as geodesic equations expressed in terms of non-integrable velocities, but as every reasonable discussion must eventually end we shall refer the reader instead to the literature (see Shapere and Wilczek, 1989, or Dubrovin et al., 1984).

Utiyama (1956), has derived a nonvanishing Riemann tensor from a locally Lorentz invariant microscopic theory. There, one starts with quantum mechanics and noncommuting gauge fields in a locally flat space-time, and the passage from the local picture to the global one is nonintegrable, which gives rise to a Riemann curvature tensor in terms of the gauge fields (see also Kaempffer, 1965).

For an entertaining account of cosmology as the marriage of particle physics with general relativity, and the Platonism of both cosmology and ultra-high energy particle physics for lack of adequate empirical data, see the popular book *The End of Physics* by Lindley.

Exercises

1. Show that Lorentz transformations leave the metric g defined by $ds^2 = c^2 dt^2 - dx^2 - dy^2 - dz^2$ invariant.

2. Show that Newton's law in space-time $dP_\mu/d\tau = K_\mu$ is covariant under the Lorentz transformations if $P = \begin{pmatrix} \gamma\vec{p} \\ \gamma mic \end{pmatrix}$ and $K = \begin{pmatrix} \gamma\vec{F} \\ \gamma_t\vec{F}\cdot\vec{v}/c \end{pmatrix}$.

3. Work exercise 7.5.

4. Show for a charge traveling with constant velocity $\vec{v}$ that $\vec{H} = \vec{v} \times \vec{E}/c$. (Hint: use Maxwell's equations directly.)

5. Show that $[\mu v,\sigma] + [\sigma v,\mu] = \dfrac{\partial g_{\mu\sigma}}{\partial x^v}$.

6. Prove that the covariant derivative of $A^\mu = g^{\mu v}A_v$ is given by

$$A^\mu{}_{,v} = g^{\mu\sigma}A_{\sigma,v} = \frac{\partial A^\mu}{\partial x^v} + \{\varepsilon v,\mu\}A^\varepsilon.$$

7. Show that $g_{\mu v,\sigma} = 0$.

8. Following Adler et al., (1965), work out the details of the Schwarzschild solution to Einstein's field equations.

12

Generalized vs nonholonomic coordinates

Die Untersuchung nichtholonomer, d. h. nicht integrabler Bedingungsgleichungen in der Mechanik hat die Mathematiker in den letzten 20 Jahren vielfach beschäftigt. . . . das Studium der Lagrange–Eulerischen Gleichungen führte mich immer tiefer in die Gruppentheorie.

G. Hamel (1904)

12.1 Group parameters as nonholonomic coordinates

We return to rigid body theory in order to illustrate the difference between holonomic (generalized) and nonholonomic (quasi) coordinates. If we begin with the description of the rotational motion of a rigid body from the standpoint of an observer in an inertial frame then transformation to a rotating frame is given by

$$x'_p = Rx_p, \tag{12.1}$$

where the entries in the column vector x_p are the Cartesian coordinates of the pth particle of the body in the inertial frame, x'_p denotes the position vector of the same particle in the rotating frame, which is any frame fixed in the rigid body, and R is an orthogonal matrix of the form

$$R = R_1(\theta_1)R_2(\theta_2)R_3(\theta_3) \tag{12.2a}$$

where $R_i(\theta_i) = e^{\theta_i M_i}$. The angles θ_i are clockwise rotation angles in three different planes (are the canonical group parameters corresponding to the infinitesimal generators M_i), and the generators M_i are 3×3 matrices obeying the commutation rules $(M_i, M_j) = \varepsilon_{ijk} M_k$ where ε_{ijk} is the completely antisymmetric three-index symbol and summation over the index k is implicit. More compactly, we can write

$$R = e^{\eta \hat{n} \cdot \vec{M}} = e^{\vec{\eta} \cdot \vec{M}} \tag{12.3a}$$

where η and $\hat{n}$ are a *single* rotation angle and axis. That equations (12.2a) and (12.3a) are equivalent is the content of Euler's theorem, the statement that rotations in three dimensions form a group: the product of any number of rotations about *different* axes can be accomplished by a *single* rotation through a *single* axis. The angle θ_i is the canonical parameter for the one dimensional subgroup of rotations $R_i(\theta_i)$ confined to one particular Cartesian plane, while the three angles η_i provide the 'canonical parameterization' (see Eisenhart, 1961, or Campbell, 1966) of the full rotation group $SO^+(3)$. Since the operators M_i do not commute, it is clear that the parameters η_i cannot be globally additive: in the product of two successive transformations $e^{\eta_i M} e^{\eta'_i M} = e^{\gamma_i M_i}$, the three parameters γ_i are connected with the six parameters η_i and η'_i by the Baker–Campbell–Hausdorf formula, an advanced result from linear operator theory.

Next, we define the completely antisymmetric matrix Ω by writing $dR/dt = R\Omega$ where, $\Omega_{ij} = \varepsilon_{ijk}\omega_k$. The Cartesian entries ω_i of the column vector ω represent the components of the instantaneous rotation frequencies projected upon the inertial axes; $\omega' = R\omega$ is then the rotation vector in the body-fixed frame. It follows that the three angles defined by

$$\alpha_i = \int \omega_i dt \qquad (12.4)$$

also provide a possible parameterization of rotations in three dimensions. Hence, we have at least four different possible parameterizations of rotations: via (i) the canonical parameters $(\theta_1,\theta_2,\theta_3)$ of the three noncommuting one dimensional subgroups, (ii) the Euler angles, (iii) the canonical parameters (η_1,η_2,η_3) of the full three dimensional rotation group, and (iv) the time integrals $(\alpha_1,\alpha_2,\alpha_3)$ of the three rotation frequencies $(\omega_1,\omega_2,\omega_3)$. The parameters $(\theta_1,\theta_2,\theta_3)$ and the Euler angles are holonomic and can be used as generalized coordinates, but the parameters η_i are nonholonomic. According to the above analysis, we suspect that the parameters α_i also are nonholonomic. If the α_i are not holonomic, then there is no way to rewrite the six differentials $d\alpha_i = \omega_i dt$ combined with the three velocities ω_i given by Euler's equations (9.13c) in Hamiltonian form, but an Euler–Lagrange formulation of the equations of motion is still possible.

Observe that the transformation R that takes one from space-fixed to body-fixed axes at the *same* instant t has an inverse R^{-1} that describes the time-evolution of the rigid body. Each particle in the rigid body is fixed relative to the body, so that $dx'_p/dt = 0$ holds for all N particles, including the pth one. Transforming back to the space-fixed frame by using $dx'_p/dt = Rdx_p/dt + R\Omega x_p = 0$ then yields

$$dx_p/dt = -\Omega x_p, \qquad (12.3b)$$

which is the same as equation (9.3) of rigid body theory. Once one solves the dynamics equations to find the time-dependence of Ω, then (12.3b) can be integrated to yield the time-development of (each particle in) the rigid body, that is, the solution, in the form

$$x_p(t) = R^{-1}x_p(0), \qquad (12.5a)$$

where $x_p(0) = x'_p$ is the pth particle's initial position at time $t = 0$ and

$$R^{-1} = e^{-\hat{n}\vec{n}\cdot\vec{M}}. \qquad (12.5b)$$

The reason for the result (12.5b) is that the inverse R^{-1} of the matrix R satisfies the matrix differential equation

$$dR^{-1}/dt = -\Omega R^{-1}. \qquad (12.5c)$$

Combination of these results and direct differentiation shows that the matrix R^{-1} is the time-evolution operator for the dynamical system. By this approach, we have *implicitly* parameterized the motion of (every one of the N particles in) the rigid body by any one of the four possible sets of group parameters (i)–(iv) that we should choose to work with. *This is the same as treating the three group parameters as the coordinates of the rigid body*. If one chooses the parameters α_i, the time integrals (12.4) of the Cartesian rotation rates ω_i, then the continuation of this procedure leads to the derivation of Euler's equations of rigid body motion in terms of the 'velocities' ω_i directly from Newton's equations of motion.

We derived Euler's equations from Newton's second law rather than from Lagrange's equations of the first kind because the parameters α_i *do not qualify as generalized* coordinates. The reason for this is that *the rotation frequencies ω_i are not generalized velocities.* In particular, $\omega_i dt$ is not an exact differential unless the motion is restricted to a single plane, in which case α_i and M_i are a canonically conjugate coordinate- and momentum-pair, where M_i is the angular momentum about the Cartesian axis x_i.

Furthermore, the parameters α_i and η_i are not identical. This follows from a result of chapter 8 where we showed that the matrices Ω and $\vec{\eta} \cdot \vec{M}$ are not identical but are related by the time-dependent transformation matrix R. A qualitative argument indicates that the parameters η_i are generally nonholonomic: concentrating upon the matrix R as expressed by both (12.2a) and (12.3a), it is clear that we cannot assign a single triplet of parameters (η_1, η_2, η_3) to a single triplet of coordinates $(\theta_1, \theta_2, \theta_3)$, because if we replace (12.2a) by

$$R' = R_1(\theta_1)R_3(\theta_3)R_2(\theta_2),\tag{12.2b}$$

then the result, for the same values of $(\theta_1, \theta_2, \theta_3)$ as before, is

$$R' = e^{\vec{\eta}' \cdot \vec{M}}\tag{12.3c}$$

with a *different* set of parameters $(\eta_1', \eta_2', \eta_3')$. In other words, *the parameters η_i cannot be defined independently of the specific path traced out by the angles θ_i*. We show by direct calculation in the next section that the θ_i are holonomic coordinates for precisely the same reason that the Euler angles are: they uniquely define different geometric orientations of a rigid body independently of the motion of the body. Another way to say it is that one does not have to integrate a set of differential equations in order to define the variables θ_σ because those variables are defined *globally* geometrically rather than merely locally.

In order to simplify the notation, we can define the purely rotational kinetic energy of a rigid body by

$$T = \tfrac{1}{2}v^2 = \tfrac{1}{2} \sum_{i=1}^{N} v_i^2\tag{12.6a}$$

where

$$\vec{v} = \sum_{i=1}^{N} \frac{d\vec{r}_i}{dt} = \sum_{i=1}^{N} \vec{v}_i,\tag{12.7}$$

which means that the masses of the particles have been absorbed into the definition of the velocities by rescaling each velocity. It follows that

$$2T = v^2 = \sum_{i,k=1}^{N} \vec{v}_i \cdot \vec{v}_k = \sum_{i=1}^{N} v_i^2,\tag{12.8}$$

because velocities of different particles belong to orthogonal subspaces in the Nd dimensional configuration space, where d is the dimension of space.

Using a matrix notation that is close to Hamel's notation, and beginning with any set of $f = Nd$ generalized coordinates q_i for the position of a particle in a rigid body, we write

$$\dot{q}_i = V_{\sigma i}\omega_\sigma\tag{12.9}$$

where

$$\vec{v}_i = \frac{\partial \vec{r}_i}{\partial q_l} \dot{q}_l \qquad (12.10)$$

is the velocity of the ith particle in the rigid body, and inversely that

$$\omega = \pi dq/dt \qquad (12.11a)$$

where $\pi V = V\pi = I$ is the identity matrix. We can therefore introduce basis vectors $\vec{e}_\lambda$ in configuration space via

$$\vec{v} = \sum_{i=1}^{N} \frac{\partial \vec{r}_i}{\partial q_k} \dot{q}_k = \vec{e}_\lambda \omega_\lambda, \qquad (12.12a)$$

in agreement with the definition in chapter 10, and so

$$\vec{e}_\lambda = V_{\lambda\mu} \sum_{i=1}^{N} \frac{\partial \vec{r}_i}{\partial q_\mu}. \qquad (12.13)$$

Next, we define the virtual change

$$\delta \alpha_\rho = \pi_{\rho k} \delta q_k \qquad (12.14)$$

where δq_k is any variation whatsoever.

The idea of virtual changes is familiar from our study of variational principles in chapter 2, but δq_k here is taken to represent any change whatsoever. In particular and in contrast with the variational derivation of Lagrange's equations via variations at fixed time and with fixed end points (as in chapter 2), we consider variations δ that can be parameterized by the time t. The central question is: when can one treat the parameters α_k defined by (12.4) above as the generalized coordinates of a rigid body, for the velocities ω_k of equation (12.11a)? The condition for this is simply that the expressions $\omega_i dt$ are exact differentials, which is also the condition that the velocities ω_i can serve as generalized velocities.

To analyze the question, we write

$$d\alpha_\rho = \pi_{\rho k} dq_k, \qquad (12.15)$$

where '$d\alpha_k$' $= \omega_k dt$ will eventually have to be understood as a nonexact differential, and to lowest order it follows that

$$d\delta\alpha_\rho - \delta d\alpha_\rho = \left(\frac{\partial \pi_{\rho k}}{\partial q_l} - \frac{\partial \pi_{\rho l}}{\partial q_k} \right) \delta q_k dq_l = \beta_{\mu\nu\rho} \delta \alpha_k d\alpha_l \qquad (12.16a)$$

where we define

$$\beta_{\mu\nu\rho} = \left(\frac{\partial \pi_{\rho k}}{\partial q_l} - \frac{\partial \pi_{\rho l}}{\partial q_k} \right) V_{\mu k} V_{\nu l}. \qquad (12.17)$$

In order that the α_i can be used as generalized coordinates ('holonomic' coordinates), it is necessary and sufficient that $\beta_{\mu\nu\rho} = 0$ holds as an identity, because this is equivalent to the *integrability condition*

$$\frac{\partial^2 \alpha_k}{\partial q_\sigma \partial q_\lambda} = \frac{\partial^2 \alpha_k}{\partial q_\lambda \partial q_\sigma} \qquad (12.18a)$$

for the existence of the angles α_i as *path-independent functions* ('point functions') of the original generalized coordinates q_k, where

$$\frac{\partial \alpha_k}{\partial q_l} = \pi_{kl}. \tag{12.19}$$

Therefore, when $\beta_{\mu\nu\rho} \neq 0$, then the parameters α_i, the time integrals of the velocities ω_i in (12.11a), are 'nonholonomic coordinates', which means that the functionals (12.4) depend upon the path followed by the rigid body during its motion.

But what is $\beta_{\mu\nu\rho}$, and what is the relation between the $d\theta_i/dt$, which are exact derivatives of holonomic coordinates θ_i, and the nonintegrable velocities ω_i? To answer these questions, we follow Hamel (1904).

We ask now for the meaning of the three-index symbol $\beta_{\mu\nu\rho}$. We know from chapter 7 that there is another place in mathematics where a three-index symbol arises in a question of integrability: in the theory of groups of coordinate transformations depending continuously upon several parameters.

Mathematically, we shall interpret the differential form

$$dq = V\delta\theta, \tag{12.11b}$$

written using matrix notation, as an infinitesimal transformation with n group parameters θ_i ($n = 3$ for the motion of simple rigid bodies, but we can generalize to include groups other than $SO^+(3)$ that describe more complicated motions, like two wheels coupled by an axle, a problem discussed by Hamel). Equation (12.11b) follows from (12.11a) and the reason why we must write $\delta\theta$ in (12.11b) rather than $d\theta$ is explained by equations (12.22)–(12.26) below. In addition, we write $\delta\theta$ rather than $\delta\alpha$ because we parameterize our transformations via the holonomic variables θ that are analogous to the variables θ in equation (12.2a) above rather than directly following Hamel, who considered only the parameterization analogous to that given by equation (12.4).

We begin by summarizing the necessary results from the transformation theory of chapter 7. The conditions for a set of transformations

$$q_i(\theta_1,\ldots,\theta_n) = f_i(q_1,\ldots,q_f;\theta_1,\ldots,\theta_n), \tag{12.20}$$

depending continuously and differentiably upon n parameters $(\theta_1,\ldots,\theta_n)$ to form a group are: (i) that the group combination law, written symbolically as $f(f(q,\theta),\beta) = f(q,\gamma)$, is possible because there is a combination law for the parameters

$$\gamma_\sigma = \phi_\sigma(\theta_1,\ldots,\theta_n;\beta_1,\ldots,\beta_n), \tag{12.21}$$

(ii) that there is an identity, $q = f(q,\theta_e)$; (iii) that each transformation has a unique inverse $q = f(q(\theta),\theta')$, and (iv) the group combination law must be associative.

Lie studied the transformations locally in the tangent space near the identity, which reduces to a study of the n vector fields with components $V_{i\sigma}$ in the infinitesimal coordinate transformation

$$dq_i = V_{i\sigma}\delta\theta_\sigma, \tag{12.22}$$

which is the same as equation (12.11b). Correspondingly, the closure condition for the infinitesimal transformation becomes

$$\theta_\sigma + d\theta_\sigma = \phi_\sigma(\theta_1,\ldots,\theta_m;\delta\theta_1,\ldots,\delta\theta_n) \tag{12.23}$$

and guarantees that *two successive* transformations with parameters θ and $\delta\theta$ are equivalent to a *single* transformation with parameter $\theta + d\theta$. The reader will note that we have introduced an exact differential $d\theta_\sigma$ and also a different variation $\delta\theta_\sigma$. To obtain the connection between the two, we expand (12.23) to lowest order in $\delta\theta_\sigma$ to obtain

$$d\theta_\sigma = v_{\sigma\rho}\delta\theta_\rho, \tag{12.24a}$$

which by definition is an exact differential.

Consider next the matrix v with entries $v_{\sigma\rho}$ and let λ denotes its inverse. Both v and λ depend upon $\theta = (\theta_1,\dots,\theta_n)$. Then

$$\delta\theta_\rho = \lambda_{\rho\sigma}d\theta_\sigma, \tag{12.25a}$$

and so it follows that

$$dq_i = V_{i\sigma}\lambda_{\sigma\tau}(\theta)d\theta_\tau. \tag{12.26}$$

The transformation that is induced in any scalar function $F(q_1,\dots,q_f)$ is therefore given by

$$dF = \delta\theta_\sigma X_\sigma F \tag{123.27}$$

where

$$X_\sigma = V_{i\sigma}(q)\frac{\partial}{\partial q_i} \tag{12.28}$$

is the infinitesimal generator determined by the velocity field V_σ whose components are given by $V_{i\sigma}(q)$.

As is emphasized above, we have begun with the transformations in the form $q_i(\theta_1,\dots,\theta_n) = f_i(q_1,\dots,q_f,\theta_1,\dots,\theta_n)$, which correspond in rigid body theory to the formula $x' = R_1(\theta_1)R_2(\theta_2)R_3(\theta_3)x$, although the canonical parameterization in terms of the η_i, where $x' = e^{\eta_i M_i}x$, is always possible. Instead of the θ_σ, we could try to think of the transformation functions f_i as functions of the Euler angles, the α_σ, or the η_σ. Hamel regarded f_i as a function of the parameters α_σ without considering the other three possibilities explicitly, but the parameters that Hamel denotes in his 1904 paper as θ_σ are denoted here by α_σ.

We have now arrived at our main point: $d\theta_\sigma$ is exact, but there is absolutely no guarantee, and no reason to believe, that the right hand side of equation (12.25a) for $\delta\theta_\sigma$ is an exact differential. We can therefore introduce a new, generally nonholonomic variable α_σ whose inexact derivative is denoted as '$d\alpha_\sigma$' and is defined by setting '$d\alpha_\sigma$' equal to $\delta\theta_\sigma$, the right hand side of (12.25a): '$d\alpha_\rho$' $= \lambda_{\rho\sigma}d\theta_\sigma$. That '$d\alpha$' $= \omega_\sigma dt$ is the correct identification follows from comparing (12.11a) and (12.11b) (see also equation (12.22)).

It takes only a few more manipulations to derive the condition that infinitesimal transformations X_σ generate a group. The result, Lie's integrability condition for the global existence of the coordinate transformations f_i over all of configuration space, is just the closure condition

$$(X_\rho,X_\sigma) = c_{\rho\sigma}^\tau X_\tau, \tag{12.29}$$

for the infinitesimal generators, which is simply the closure condition guaranteeing that the n operators X_σ form a Lie algebra. If the Lie algebra can be integrated to define a group then the coefficients $c^\tau_{\rho\sigma}$ are constants, the structure constants of the group, and are independent of both the group parameters and the generalized coordinates. If the $c^\tau_{\rho\sigma}$ are not constants then the integrability condition is not satisfied and the Lie algebra does not give rise to a Lie group.

Our next step is to show that $\beta_{\mu\nu\rho} = c^\rho_{\mu\nu}$. The proof is as follows: note that

$$q_l + dq_l + \delta(q_l + dq_l) = q_l + d\theta_\nu X_\nu q_l + \delta\theta_\nu X_\nu q_l + \delta d\theta_\nu X_\nu q_l + \delta\theta_\mu d\theta_\nu X_\mu X_\nu q_l, \quad (12.30a)$$

whereas

$$q_l + \delta q_l + d(q_l + \delta q_l) = q_l + \delta\theta_\nu X_\nu q_l + d\theta_\nu X_\nu q_l + d\delta\theta_\nu X_\nu q_l + \delta\theta_\mu d\theta_\nu X_\mu X_\nu q_l, \quad (12.30b)$$

Next, we can write $d\delta q_l = \delta dq_l$ because we arrive at the same point $q_l + dq_l$ starting from the point q_l in configuration space no matter whether we get there by a single transformation via one parameter $\theta_\sigma + d\theta_\sigma$, starting from $\theta_\sigma = 0$, or by way of two successive transformations with two parameters θ_σ and $\delta\theta_\sigma$. Setting the left hand sides of (12.30a) and (12.30b) equal to each other then yields

$$0 = (X_\nu, X_\mu)q_l d\theta_\nu \delta\theta_\mu + X_\nu q_l(d\delta\theta_\nu - \delta d\theta_\nu). \quad (12.31a)$$

Substituting the commutation rule $(X_\rho, X_\sigma) = c^\tau_{\rho\sigma}X_\tau$ yields

$$0 = c^\rho_{\nu\mu}\delta\theta_\nu d\theta_\mu X_\rho q_l + X_\nu q_l(d\delta\theta_\nu - \delta d\theta_\nu). \quad (12.31b)$$

Since $X_\rho q_l = V_{\rho l}$, we finally obtain

$$d\delta\theta_\rho - \delta d\theta_\rho = c^\rho_{\mu\nu}d\theta_\mu\delta\theta_\nu, \quad (12.32)$$

which shows by comparison with equation (12.16a) that $\beta_{\mu\nu\lambda} = c^\lambda_{\mu\nu}$. Note that, applied to the parameters θ_ρ, the operations d and δ commute, $(d,\delta) = 0$, if and only if all of the n infinitesimal operators commute, $(X_\mu, X_\nu) = 0$.

The structure constants $c^\tau_{\kappa\lambda}$ fail to vanish whenever the operators X_σ do not commute, and in this case there is, by (12.32), a set of nonholonomic variables. But which differential expression is nonintegrable when $d\delta\theta_\sigma \neq \delta d\theta_\sigma$? The differential coefficients $\pi_{\rho k}$ in

$$\beta_{\mu\nu\rho} = \left(\frac{\partial\pi_{\rho k}}{\partial q_l} - \frac{\partial\pi_{\rho l}}{\partial q_k}\right)V_{\mu k}V_{\nu l} \quad (12.18b)$$

from section 12.1 follow from the equation

$$d\alpha_\rho = \pi_{\rho k}dq_k, \quad (12.16b)$$

so that the noncommutation condition $c^\tau_{\kappa\gamma} \neq 0$ guarantees that the differentials $\omega_\rho dt$ are nonintegrable, hence that the variables α_ρ are nonholonomic. Finally, if one asks why it is $\omega_\rho dt$ that is the left hand side of (12.16a) and not $d\theta_\rho/dt$, then the answer is provided by the expressions (12.22)–(12.26) above. The differential form for $d\alpha_\sigma$ in (12.16a) is integrable and *then* becomes an exact differential $d\alpha_\sigma = d\theta_\sigma$ if and only if all of the rotations are confined to a *single* plane perpendicular to the Cartesian axis x_σ.

By starting with holonomic variables θ_σ, variables with exact derivatives $d\theta_\sigma$, we have arrived quite naturally at a set of nonintegrable variations $\delta\theta_\rho$ that are precisely what we mean by the inexact differentials '$d\alpha_\sigma$'. In this way, for a noncommuting group, does the holonomic give rise to the nonholonomic. One can do the same with any set of generalized coordinates: an arbitrary differential form $a_{ik}(q_1,\dots,q_f)dq_k$ is generally nonintegrable.

Granted that the right hand side of equation (12.25a) is generally nonintegrable, if we next parameterize those equations by the time t, then we obtain

$$d\theta_\sigma/dt = v_{\sigma\rho}\omega_\rho \qquad (12.24b)$$

and

$$\omega_\rho = \lambda_{\rho\sigma}d\theta_\sigma/dt, \qquad (12.25b)$$

which show explicitly how the nonintegrable velocities ω_σ are related to the exact time derivatives $d\theta_\sigma/dt$. In section 12.3 we provide two explicit examples of equation (12.25b) in the context of rigid body motion. In fact, the usual expressions for the rotation rates ω_σ for rigid body motions in inertial and rotating frames in terms of the time derivatives of the Euler angles would provide examples of (12.25b) were it not for the fact that two of the generators X_σ corresponding to the Euler angles are equal, so that those three generators do not close under commutation (do not generate a Lie algebra).

In the next section, we show how the structure constants of the Lie algebra enter Euler's equations of rigid body motion by deriving Euler's equations by a coordinate transformation within a variational equation rather than directly from Newton's law of motion.

12.2 Euler–Lagrange equations for nonintegrable velocities

We begin as earlier with the kinetic energy of a rigid body in the form

$$T = \tfrac{1}{2}v^2 = \tfrac{1}{2}\sum_{i=1}^{N} v_i^2 = \tfrac{1}{2}g_{ij}\dot{q}_i\dot{q}_j, \qquad (12.6b)$$

with arbitrary variations given by

$$\delta T = \sum_{i=1}^{N} \vec{\ddot{r}}_i \cdot \delta\vec{r}_i, \qquad (12.33)$$

and with the Euler–Lagrange equations given by

$$\left(\frac{d\partial T}{dt\partial\dot{q}_i} - \frac{\partial T}{\partial q_i}\right)\delta q_i = Q_i\delta q_i \qquad (12.34)$$

with no further immediate simplification because the generalized coordinates of the particles of the body are fixed at any time t by the three parameters α_i. With

$$\vec{v} = \vec{e}_\lambda\omega_\lambda \qquad (12.12b)$$

we get

$$T = \tfrac{1}{2}\tilde{\omega}I\omega = \tfrac{1}{2}I_{\mu\nu}\omega_\mu\omega_\nu \qquad (12.35)$$

where $I_{\mu\nu}$ is the metric tensor for the group-parameter description of the motion. It follows that

$$J_\lambda = \frac{\partial T}{\partial\omega_\lambda} = I_{\lambda\sigma}\omega_\sigma \qquad (12.36)$$

and that

$$Q_\sigma \delta q_\sigma = N_\lambda \delta \alpha_\lambda. \tag{12.37}$$

To go further, we note that with

$$\delta T = \sum_{i=1}^{N} \ddot{\vec{r}}_i \cdot \delta \vec{r}_i, \tag{12.38}$$

and

$$\vec{v} \cdot \delta \vec{v} = \frac{\mathrm{d}}{\mathrm{d}t} \dot{\vec{r}}_i \cdot \delta \vec{r}_i - \ddot{\vec{r}}_i \cdot \delta \vec{r}_i = \frac{\mathrm{d}}{\mathrm{d}t} \dot{\vec{r}}_i \cdot \delta \vec{r}_i - \vec{F}_i \cdot \delta \vec{r}_i \tag{12.39}$$

we obtain

$$\frac{\mathrm{d}}{\mathrm{d}t} \dot{\vec{r}}_i \cdot \delta \vec{r}_i = \delta T + N_\rho \delta \alpha_\rho. \tag{12.40}$$

But

$$\dot{\vec{r}}_i \cdot \delta \vec{r}_i = I_{\mu\nu} \omega_\mu \delta \alpha_\nu = J_\mu \delta \alpha_\mu \tag{12.41}$$

so that the Euler–Lagrange equations now read

$$\frac{\mathrm{d}}{\mathrm{d}t} J_\lambda \delta \alpha_\lambda = \delta T + N_\lambda \delta \alpha_\lambda. \tag{12.42}$$

But

$$\mathrm{d}\delta\alpha_\lambda - \delta \mathrm{d}\alpha_\lambda = \beta_{\mu\nu\lambda} \delta \alpha_\mu \omega_\nu, \tag{12.43}$$

so that

$$\frac{\mathrm{d}}{\mathrm{d}t} \delta \alpha_\lambda - \delta \omega_\lambda = \beta_{\mu\nu\lambda} \delta \alpha_\mu \omega_\nu, \tag{12.44}$$

and so

$$\frac{\mathrm{d}J_\lambda}{\mathrm{d}t} \delta \alpha_\lambda + J_\lambda \frac{\mathrm{d}}{\mathrm{d}t} \delta \alpha_\lambda = J_\lambda \delta \omega_\lambda + \frac{\partial T}{\partial q_k} \delta q_k + N_\lambda \delta \alpha_\lambda \tag{12.45}$$

finally yields

$$\left(\frac{\mathrm{d}J_\lambda}{\mathrm{d}t} + \beta_{\lambda\mu\nu} \omega_\mu J_\nu - \frac{\partial T}{\partial q_k} \pi_{\lambda k} - N_\lambda \right) \delta \alpha_\lambda = 0. \tag{12.46}$$

Since the variations in the three parameters α_λ are unconstrained, we obtain the Euler–Lagrange equations for rigid body motion in the form

$$\frac{\mathrm{d}J_\lambda}{\mathrm{d}t} + \beta_{\lambda\mu\nu} \omega_\mu J_\nu - \frac{\partial T}{\partial q_k} \pi_{\lambda k} = N_\lambda \tag{12.47}$$

where the constants $\beta_{\lambda\mu\nu} = c_{\lambda\mu}^\nu$ are the structure constants of the Lie algebra. The condition that the α_λ are generalized coordinates and that the J_λ are conjugate canonical momenta is that the structure constants all vanish, which is the requirement that all of the infinitesimal operators of the Lie algebra commute with each other:

$(\mathbf{X}_\mu, \mathbf{X}_\nu) = 0$. In this case, and only in this case, do we retrieve Lagrange's equations in the standard holonomic form of chapter 2. If the structure coefficients $c_{\lambda\mu}^\nu$ are not all constants, then we still have a Lie algebra in terms of the nonintegrable velocities J_λ, but the infinitesimal transformations $\omega dt = \Pi dq$ then cannot be integrated independently of path in configuration space to yield a Lie group.

We started in chapter 2 with a coordinate transformation to an arbitrary set of coordinates, called generalized coordinates or holonomic coordinates, and now we have found that there are definite conditions that must be satisfied before a set of parameters can serve as generalized coordinates, or just 'coordinates'. On the other hand, a Lagrangian description of the equations of motion is still possible even in the nonholonomic case. Euler's equations of rigid body motion provide an example: with $\beta_{ijk} = \varepsilon_{ijk}$, we have the $SO^+(3)$ group, and since the kinetic energy of a body has no explicit dependence upon the generalized coordinates q_k but depends only upon the nonintegrable velocities ω_λ, we have

$$\frac{dJ_\lambda}{dt} + \varepsilon_{\lambda\mu\nu}\omega_\mu J_\nu = N_\lambda \tag{12.48}$$

where $J_\mu = I_{\mu\nu}\omega_\nu$ is the angular momentum, $I_{\mu\nu}$ is the moment of inertia, and N_λ is the torque. Other applications of the Euler–Lagrange equations (12.47), especially to motions of rigid bodies with more than three nonintegrable velocities ω_λ, can be found in Hamel (1904), in Whittaker (1965), and in Kilmister (1964) (see also Marsden, 1992, and Arnol'd, 1993).

12.3 Group parameters as generalized coordinates

We begin with the connection of the parameter η_i to the canonical parameters θ_i for rotations in the three mutually independent planes. With $\Omega = \tilde{R}\dot{R}$ or $\tilde{R} = \Omega R$ where $\tilde{\Omega} = -\Omega$ and $\Omega_{ij} = \varepsilon_{ijk}\omega_k$, we can also write $R = e^{\eta_i M_i}$. Therefore, we can also write

$$R(\eta_1, \eta_2, \eta_3) = R_1(\theta_1)R_2(\theta_2)R_3(\theta_3) \tag{12.49}$$

where the θ_i are clockwise rotation angles about three different axes in three different coordinate systems (see figure 12.1) and $R_i = e^{\theta_i M_i}$ (no summation convention here). In order to get a relationship between the nonintegrable frequencies ω_i and the exact derivatives $d\theta_i/dt$, we start with

$$\dot{R} = \dot{R}_1 R_2 R_3 + R_1 \dot{R}_2 R_3 + R_1 R_2 \dot{R}_3, \tag{12.50}$$

or

$$R\Omega = R_1\Omega_1 R_2 R_3 + R_1 R_2 \Omega_2 R_3 + R_1 R_2 R_3 \Omega_3, \tag{12.51}$$

from which it follows that

$$\Omega = \tilde{R}_3 \tilde{R}_2 \Omega_1 R_2 R_3 + \tilde{R}_3 \Omega_2 R_3 + \Omega_3. \tag{12.52}$$

It follows from the sequence of problems given as exercises 1 and 2, based upon the transformation rule for generators that

$$e^{-\vec{\theta}\cdot\vec{L}}\vec{L}e^{\vec{\theta}\cdot\vec{L}} = \hat{n}\hat{n}\cdot\vec{L} - \hat{n}\times\hat{n}\times\vec{L}\cos\theta + \hat{n}\times\vec{L}\sin\theta, \tag{12.53}$$

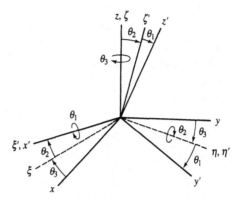

Fig. 12.1 The holonomic coordinates θ_i defined geometrically by three successive rotations of Cartesian axes.

where $\vec{\theta} = \theta\hat{n}$, that

$$\left.\begin{aligned}
\omega_1 &= \dot{\theta}_1 \cos\theta_2 \cos\theta_3 + \dot{\theta}_2 \sin\theta_3 \\
\omega_2 &= -\dot{\theta}_1 \cos\theta_2 \sin\theta_3 + \dot{\theta}_2 \cos\theta_3 \\
\omega_3 &= -\dot{\theta}_1 \sin\theta_2 + \dot{\theta}_3.
\end{aligned}\right\} \tag{12.54a}$$

These ω_i are the projections of the rotation frequency vector onto the space-fixed axes. An alternative to the method used here is to derive (12.54a) by the method of section 9.3.

Checking the second cross-partials shows immediately that the $\omega_i dt$ are not exact differentials. From this standpoint it is also interesting to note that the attempt to treat the α_i as coordinates is essentially like trying to use the components n_i of the instantaneous rotation axis as coordinates, because $\alpha_i = \alpha n_i$.

We cannot take the α_i as generalized coordinates, but if we interpret the angles θ_i as rotations about *three axes in three different coordinate systems* (see figure 12.1), according to $\xi = R_3 x$, $\xi' = R_2 \xi$, and $x' = R_1 \xi'$, then we can take the θ_i as generalized coordinates even though the three corresponding rotations still do not commute. In fact, we can interpret the x'-frame as any body-fixed frame, in particular, we can take it to be the principal axis frame of the rigid body. The frequencies in the principal axis frame follow from $\omega' = R\omega$, where

$$R = R_1 \begin{pmatrix} \cos\theta_2 \cos\theta_3 & \cos\theta_2 \sin\theta_3 & -\sin\theta_2 \\ -\sin\theta_3 & \cos\theta_3 & 0 \\ \sin\theta_2 \cos\theta_3 & \sin\theta_2 \sin\theta_3 & \cos\theta_2 \end{pmatrix}, \tag{12.55}$$

and are given by

$$\left.\begin{aligned}
\omega'_1 &= \dot{\theta}_1 - \dot{\theta}_3 \sin\theta_2 \\
\omega'_2 &= \dot{\theta}_3 \sin\theta_1 \cos\theta_2 + \dot{\theta}_2 \cos\theta_1 \\
\omega'_3 &= \dot{\theta}_3 \cos\theta_1 \cos\theta_2 - \dot{\theta}_2 \sin\theta_1,
\end{aligned}\right\} \tag{12.54b}$$

which is an example of the group theoretic equation (12.25b). As an application of the generalized coordinates $(\theta_1, \theta_2, \theta_3)$, we can study the symmetric top (with principal moments $I_1 = I_2 \neq I_3$) with one point on the symmetry axis fixed in space. The Lagrangian is $L = T - U$, where $U = -Mgl\cos\theta_1 \cos\theta_2$, and the kinetic energy in the principal axis frame is given by

$$T = \frac{I_1}{2}(\dot{\theta}_1 - \dot{\theta}_3 \sin\theta_2)^2 + \frac{I_1}{2}(\dot{\theta}_2 \cos\theta_1 + \dot{\theta}_3 \sin\theta_1 \cos\theta_2)^2 \qquad (12.56)$$

$$+ \frac{I_3}{2}(\dot{\theta}_3 \cos\theta_1 \cos\theta_2 - \dot{\theta}_2 \sin\theta_1)^2.$$

Of the three canonical momenta

$$p_i = \frac{\partial L}{\partial \dot{\theta}_i}, \qquad (12.57)$$

only p_3 is conserved, where

$$p_3 = \frac{\partial L}{\partial \dot{\theta}_3} = I_1(\dot{\theta}_1 - \dot{\theta}_3 \sin\theta_2)\sin\theta_2 + I_1(\dot{\theta}_2 \cos\theta_1 + \dot{\theta}_3 \sin\theta_1 \cos\theta_2)\sin\theta_1 \cos\theta_2$$
$$+ I_3(\dot{\theta}_3 \cos\theta_1 \cos\theta_2 - \dot{\theta}_2 \sin\theta_1)\cos\theta_1 \cos\theta_2, \qquad (12.58)$$

If M is the angular momentum in the space-fixed frame then $M' = RM$ is the angular momentum in the body-fixed frame. We can compute $M = R^{-1}M'$ to obtain

$$M_3 = I_1(\dot{\theta}_1 - \dot{\theta}_3 \sin\theta_2)\sin\theta_2 + I_1(\dot{\theta}_2 \cos\theta_1 + \dot{\theta}_3 \sin\theta_1 \cos\theta_2)\sin\theta_1 \cos\theta_2 \qquad (12.59)$$
$$+ I_3(\dot{\theta}_3 \cos\theta_1 \cos\theta_2 - \dot{\theta}_2 \sin\theta_1)\cos\theta_1 \cos\theta_2.$$

showing that $p_3 = M_3$, as expected, where p_3 is the canonical momentum conjugate to θ_3. Because neither of the other two canonical momenta p_i is conserved (the Lagrangian L depends explicitly upon the angles θ_1 and θ_2), this choice of generalized coordinates does not provide the shortest path to the solution of the top problem. Since the corresponding component of angular momentum is the canonical momentum conjugate to the rotation angle about a given axis taken as a generalized coordinate, we expect that p_2 and p_1 are angular momenta about the two axes ξ_2 and ξ'_1 respectively.

Exercises

1. If $\dot{R} = R\Omega$ and $R = R_1(\theta_1)R_2(\theta_2)R_3(\theta_3)$, use the results of exercises 7 and 8 in chapter 8 to show that $\Omega = \tilde{R}_3\tilde{R}_2\Omega_1 R_2 R_3 + \tilde{R}_3\Omega_2 R_3 + \Omega_3$ (compare this result with equation (12.53)).

2. (a) Derive equations (12.55) and (12.54b).
   (b) Prove directly by differentiation that the angular velocities ω_i and ω'_i are non-integrable.

3. Derive the Lagrangian of section 12.3 where the kinetic energy is given by equation (12.56).
   (a) Prove by a coordinate transformation that $p_3 = M_3$ where M_3 is the projection of the angular momentum onto the x_3-axis in the inertial frame.
   (b) Prove by a coordinate transformation that p_2 and p_1 are the components of angular momentum about the two axes ξ_2 and ξ'_1, respectively.

13

Noncanonical flows

Die Mechanik ist das Rückgrat der mathematischen Physik.

A. Sommerfeld (1964)

13.1 Flows on spheres

We begin with the symmetric top in the absence of a torque (section 9.2). The motion is confined to a sphere (or an ellipsoid) in the three dimensional phase space. On the sphere there are four elliptic points bounded by the separatrices that connect two hyperbolic points with each other (figure 9.2). We can generalize this to other integrable flows that are confined to a spherical surface, or to a flow confined by a conservation law to any closed surface that is topologically equivalent to a sphere (any closed surface that has no holes in it, is not toroidal, is topologically and differentiably equivalent to a sphere).

Consider first any flow that is confined by conservation laws to a closed two dimensional surface. Since all streamlines are bounded and cannot intersect each other (the solutions are unique), every streamline must either close on itself or else must start and end on hyperbolic points or, for a dissipative flow, on sources and sinks. If all streamlines close on themselves (the first family-defining case), then the flow is topologically equivalent to one where the streamlines are geodesics on a sphere. In that case there are two elliptic points where the velocity field vanishes. If not all of the streamlines close on themselves and the flow is conservative (the second general case), then there must be hyperbolic points, but it is geometrically clear from imagining all possible smooth distortions of the streamlines of the Euler top that hyperbolic points can only come in pairs combined with four elliptic points: every such flow is therefore qualitatively equivalent to the flow described by Euler's top. Flows on two dimensional surfaces that can be transformed differentiably into a sphere are integrable[1].

Euler's equations for a solid of revolution

$$
\left.
\begin{aligned}
I_1 \dot{\omega}_1 - \omega_2 \omega_3 (I_1 - I_3) &= 0 \\
I_1 \dot{\omega}_2 - \omega_1 \omega_3 (I_3 - I_1) &= 0 \\
\dot{\omega}_3 &= 0
\end{aligned}
\right\}
\qquad (9.14a)
$$

can be used to illustrate the first family-case mentioned above, that of closed streamlines with two elliptic points on the sphere. We can rewrite these equations in the form

---

[1] The concept of a differentiable manifold was introduced intuitively and nonrigorously by Riemann in his development of his original ideas about the geometry intrinsic to nonEuclidean spaces. See section 10.2 for a description of Riemann's idea. According to Arnol'd (1979–81), all orientable closed two dimensional surfaces are topologically equivalent to a sphere or to a sphere with one or more handles.

$$\left.\begin{array}{l} \dot{x} = \Omega y \\ \dot{y} = -\Omega x \\ \dot{z} = 0, \end{array}\right\} \qquad (9.14b)$$

along with conservation of angular momentum $x^2 + y^2 + z^2 = 1$. There are two equilibria at $(0,0, \pm 1)$ and both are elliptic points. Note that if we transform to angular variables then the differential equations on the sphere are simply $\dot{\theta} = 0$ and $\dot{\phi} = 1$, which describe a constant speed translation due to the conservation law $\theta = $ constant (figure 13.1). The transformation from (x,y) to (θ,ϕ) is the Lie transformation. The Lie transformation defines an orthogonal coordinate system on the sphere except at the two elliptic points where ϕ is undefined. Each component of the velocity vector on the sphere is constant but the direction of that vector is not constant. This case is 'generic', which means that it defines a topologically distinct family for all flows on closed surfaces where there are two elliptic points: other nonlinear flows with two elliptic points on a different closed surface must be a differentiable transformation of the streamlines and locations of the equilibria of the simplest case.

Topological arguments based upon differentiable coordinate transformations are very powerful because one can use the simplest dynamical system in the class to understand and characterize the entire class, so long as one can identify the topologically invariant quantities. The number and different types of equilibria are the simplest examples of topological invariants if we exclude degeneracy (the case where two elliptic points coalesce is degenerate and produces a dipolar flow pattern on the sphere). We have achieved more than we expected by using the Eulerian description of rigid body motion: we have discovered that spheres in phase space are typical for integrable noncanonical motion where there are two or more elliptic points.

To show that the above geometric arguments also apply in the dissipative case, we give an example of a source–sink flow on the sphere that is an easy generalization of the conservative flow with two elliptic points. Equations (9.14b) translate into the equations of geodesics on the sphere with elliptic points at both poles, $\dot{\theta} = 0$ and $\dot{\phi} = 1$ (we have rescaled the time by Ω), and so if we write instead that

$$\left.\begin{array}{l} \dot{\theta} = \sin\theta \\ \dot{\phi} = 1 \end{array}\right\} \qquad (13.1)$$

then we obtain a source at $\theta = 0$ and a sink at $\theta = \pi$. The streamlines are spirals about the sphere (figure 13.2). Integration of the exact differential $d\phi = d\theta/\sin\theta$ yields the nontrivial conservation law $h(\theta,\phi) = \ln[\tan(\theta/2)] - \phi = C_2$. The lack of continuity of this conservation law at the source and sink prevents its triviality: different streamlines define different constants C_2. The same is not true of the system $\dot{\theta} = \dot{\phi} = \sin\theta$: in that case, $\theta = \phi$ and $C_2 = 0$ for all streamlines, by continuity of the flow into the source or sink. Our conclusion is again very general: any velocity field on the sphere $v = (v_\theta, v_\phi)$ where v_θ vanishes linearly at a source/sink (or as θ^α with $\alpha > 1$) will yield a nontrivial conservation law because of a logarithmic singularity (or because of a $\theta^{1-\alpha}$ divergence) at the point-source/sink. Note also that Lie's reduction of the problem to a single constant speed translation still makes sense: with $v = h(\theta,\phi) = C_2$ and $w = f(\theta,\phi) = t + D = g(\theta, C_2)$, the time t goes to minus infinity as any streamline approaches the source (due to the singularity in h), and to plus infinity as the same streamline approaches the sink. Infinite time is required to leave an unstable

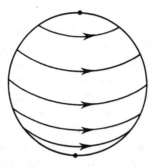

Fig. 13.1 Translations on a sphere.

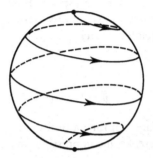

Fig. 13.2 Spiraling flow from a source into a sink.

equilibrium point or to reach a stable one because ϕ goes to minus infinity as θ approaches the source, and to plus infinity as θ approaches the sink.

Finally, we rewrite this flow as a three dimensional one in Cartesian coordinates: with $x = \sin\theta\cos\phi$, $y = \sin\theta\sin\phi$, and $z = \cos\theta$, which amount to assuming that $G(x,y,z) = x^2 + y^2 + z^2 = C_1$ with $C_1 = 1$, we obtain from (13.1) the nonconservative flow

$$\left.\begin{aligned} \dot{x} &= xz - y \\ \dot{y} &= yz + x \\ \dot{z} &= -x^2 - y^2, \end{aligned}\right\} \tag{13.2a}$$

which has a source at $(0,0,1)$ and a sink at $(0,0,-1)$. This flow is confined to the unit sphere because $dr/dt = 0$. The corresponding conservative velocity field $\tilde{V} = MV = \nabla G \times \nabla H$ is easily constructed and has singularities at $z = 1$ and $z = -1$. We discuss singular conservative flows on the sphere from a hydrodynamics viewpoint below.

If we make the substitution $-x^2 - y^2 = z^2 - 1$ in the equation for dz/dt in (13.2a) then we obtain

$$\left.\begin{aligned} \dot{x} &= xz - y \\ \dot{y} &= yz + x \\ \dot{z} &= z^2 - 1. \end{aligned}\right\} \tag{13.2b}$$

In contrast with (13.1), the motion is no longer confined to the unit sphere: because $r\,dr/dt = (r-1)z$, the upper hemisphere repels any streamline that starts at a point where $r > 1$ and $0 < z < 1$ (streamlines that start at $z > 1$ or $z < -1$ cannot cross the planes $z = 1$ and $z = -1$). Such a streamline must spiral about the sphere with $dz/dt < 0$ and therefore eventually crosses the equator at $z = 0$ in finite time, whereafter it continues to spiral downward as it is attracted by the lower hemisphere. In modern language, the sphere is an invariant set for the flow: any streamline that starts on the sphere stays on the sphere and obeys the integrable equations (13.2a). For streamlines that begin outside the sphere in the region $-1 < z < 1$ (with $r < 1$ or $r > 1$), the upper hemisphere repels and the lower hemisphere attracts.

Singular flows where the velocity field diverges at two or more points are possible on the surface of a sphere and are important in ideal fluid hydrodynamics. In fact, there is also a singular type of conservative flow. We need only consider two qualitatively distinct possibilities: flow from a source to a sink, with source and sink located at opposite poles, and 'parallel-transport' on a sphere. Both flows can be studied as planar problems that are mapped onto the sphere by stereographic projection (see Klein, 1963), and the two flows are related to each other by a conformal mapping but we take the direct approach instead.

With

$$\vec{v} = \nabla \times \hat{r}\psi \tag{13.3}$$

where $\psi(\theta,\phi)$ is the stream function for motion on a sphere, and (θ,ϕ) are spherical coordinates on the unit sphere, an incompressible, steady flow satisfies

$$\nabla \cdot \vec{v} = 0. \tag{13.4}$$

From $\nabla \times \nabla \times \hat{r}\psi = -\hat{r}\nabla^2\psi$ (proven in Cartesian coordinates), we obtain Laplace's equation on the unit sphere for the stream function

$$\nabla^2\psi = \frac{1}{\sin\theta}\left(\frac{\partial}{\partial\theta}\sin\theta\frac{\partial\psi}{\partial\theta} + \frac{1}{\sin\theta}\frac{\partial^2\psi}{\partial\phi^2}\right) = 0 \tag{13.5}$$

except at the singular points (the right hand side is really a delta-function, in other words).

Consider first the source–sink flow, where $\psi = \psi(\theta)$. The solution is

$$\psi = \frac{J}{2\pi}\ln\tan\frac{\theta}{2}, \tag{13.6}$$

where J is the flow rate out of the source and into the sink, and this yields

$$v_\theta = \frac{J/2\pi}{\sin\theta} \tag{13.7}$$

which diverges at both poles, at $\theta = 0$ and π. Hence, steady nonsingular flow is not possible in this case. In general, the divergence-free condition satisfied by $\vec{V} = \nabla \times \vec{A}$ does not yield integrable motions for an arbitrary three dimensional flow where there is no conservation law, but the regularity of the streamlines in this case tells us that this flow is integrable. The flow depends upon only two variables θ and ϕ and is integrable.

In the second case, that of the generalization of 'parallel flow' to a sphere (the flow lines are nonintersecting circles), $\psi = \psi(\phi)$ and here we obtain

$$\psi = \frac{\kappa}{2\pi}\phi, \tag{13.8}$$

a multivalued potential representing a vortex of strength κ at $\theta = 0$ and one of opposite strength at $\theta = \pi$. In this case,

$$v_\phi = \frac{\kappa/2\pi}{\sin\theta} \tag{13.9}$$

which also blows up at $\theta = 0$ and π. The flow is qualitatively like that of figure 13.1.

Our hydrodynamics is linear: we can superpose stream functions to construct other flows. If we write $-J = \kappa = -2\pi$ and then add (13.6) and (13.8) together, we get $\psi = \ln[\tan(\theta/2)] - \phi$ which is the conserved quantity $h(\theta,\phi)$ for the dissipative flow (13.1) on a sphere. The stream function for a singular conservative flow on a sphere is simultaneously the conservation law for a corresponding nonsingular dissipative flow on the same sphere! In the hydrodynamic case where $\psi = \ln[\tan(\theta/2)] - \phi$, the streamlines swirl out of a vortical source at $\theta = 0$ and into a vortical sink at $\theta = \pi$. This is a symmetry-breaking solution, as the streamlines do not have the symmetry of Laplace's equation which is just the symmetry of the sphere.

13.2 Local vs complete integrability

We pointed out in chapter 7 that there are at least three different meanings of the word integrability in the context of differential equations. A very general kind of integrability that guarantees that holonomic coordinates are generated by integrating differential forms globally was defined in chapter 12 and is satisfied by every flow: the solutions of the differential equations *exist globally as functions of the initial conditions and the time* and define the n variables $x_i(t) = U(t)x_{i_\circ}$ as holonomic coordinates by a global coordinate transformation $U(t)$ on the initial conditions $(x_{1_\circ},\ldots,x_{n_\circ})$, which are Cartesian holonomic coordinates. Complete integrability, as defined by Jacobi and Lie, requires in addition that the streamlines can be 'straightened out' globally (excluding sources and sinks) by a coordinate transformation that is determined by $n-1$ conservation laws, which means that the time-evolution operator $U(t)$ is equivalent for all finite times t (by a coordinate transformation that is itself determined by the conservation laws) to a pure time-translation $U(t) = e^{td/dy_n}$, a constant speed motion on a two dimensional manifold. This corresponds to 'integrability by quadratures' in the older literature. In addition to complete integrability of a flow, there is also the idea of 'local integrability', that is restricted to small but finite regions in phase space and is satisfied by arbitrarily chosen flows, including chaotic ones.

We remind the reader that 'global' means a finite (or infinite) region of phase space where the velocity field varies nontrivially (where nonlinearity may be important in determining the flow). By local, we mean a finite but small enough region of phase space that gradients of the velocity field can be neglected. The size of any local region can be estimated from the neglected gradients of the velocity field.

Before discussing local integrability, we use certain of the results of chapters 7, 11, and 12 to give a more precise definition of phase space by starting with an n dimensional manifold rather than with the idea of a coordinate system in an a priori undefined space. We can then use integrability on the manifold to derive any desired coordinate system *if* the differential equations that generate the coordinate system infinitesimally are globally integrable. In particular, we found in chapter 11 that one does not need a metric in order to describe the geometry of an affine manifold. Here we need only the connections $\Gamma^\sigma_{\mu\nu}$, so we assume only an affine manifold. A flat manifold is one where the Riemann tensor vanishes. In that case the n differential equations $dx_i/dt = v_i =$ constant are globally integrable ($d^2x_i/dt^2 = 0$ is globally integrable for $i = 1,\ldots,n$) with no points of the manifold excluded and can be used to define a global Cartesian coordinate system via $x_i(t) = v_i t + x_{i_o}$ for all points $-\infty < x_i(t) < \infty$ on the manifold, starting from an arbitrary point $(x_{1_o},\ldots,x_{n_o})$ on the manifold.

On a manifold of positive curvature, the n local differential equations $dx_i/dt = v_i =$ constant defining local parallel flow are nonintegrable unless singular points like equilibria are excluded from the manifold. In that case, constant components of velocity do not represent a constant vector field because the vectors that define the different directions locally vary on the manifold, but a remnant of the idea of global parallelism survives in the form of integrability of a vector that is translated along a curve with its direction always maintained parallel to the curve's tangent vector. For example, the constant components of velocity on a sphere defined in spherical coordinates by $d\theta/dt = 0$, $d\phi/dt = 1$ exclude the elliptic points at $\theta = 0$ and π but the global velocity field, while locally parallel to itself, is not globally parallel on the sphere. Phase space, as defined in chapter 3, was implicitly assumed to be a flat differentiable manifold, one with a Cartesian volume element but with no corresponding requirement of a metric, so long as we stick to purely qualitative analysis and ignore Picard's approximation method. Global parallelism and Cartesian coordinates follow from the assumption that the flow $dx_i/dt = v_i =$ constant is defined, and therefore is integrable, for $-\infty < x_i < \infty$ with *no* points on the manifold excluded. With this definition, phase space can always be called a Cartesian space. If there is a metric then Cartesian space coincides with Euclidean space.

We have used the Lie–Jacobi method (see section 3.6) to show that a completely integrable flow

$$\frac{dx_i}{dt} = V_i(x_1,\ldots,x_n) \tag{13.10a}$$

requires that there are $n - 1$ global isolating integrals $G_i(x_i,\ldots,x_n) = C_i$, and that the global coordinate transformation (excluding all equilibrium points)

$$\left.\begin{array}{l} y_i = G_i(x_1,\ldots,x_n) = C_i, \quad i = 1,\ldots,n-1 \\ y_n = F(x_1,\ldots,x_n) = t + D \end{array}\right\} \tag{13.11}$$

to a parallel flow is possible via the determination of $n - 1$ variables through the $n - 1$ isolating integrals and holds for all times $-\infty < t < \infty$. Clearly, whenever the conservation laws are multivalued functions of the x's, then some care is necessary in the application of (13.11) – see the discussion of radial flow in section 3.6 for a simple example where $G = \phi$ is a phase angle. The solution for the nth coordinate x_n then has the form $t + D = f(x_n, C_1,\ldots,C_{n-1}) = F(x_1,\ldots,x_n)$. Again, 'global' need not mean the whole phase space but can mean a finite part of it between any two equilibrium points, or within or outside a separatrix or limit cycle: the Lie transformation (13.11) sends sources and sinks off to infinity, and so when more than one equilibrium point (or other differentiable attractor or repeller) is present then the transformation (13.11) transforms the streamlines within the basin of attraction of one equilibrium point, or inside the separatrix of a conservative problem. For example, for the problem of the simple pendulum the functional form of the conservation law (energy conservation) is the same everywhere in phase space, but the solutions inside and outside the separatrix (representing the limit of a 2π-rotation) are qualitatively different from each other, and so there are two different Lie transformations for the two qualitatively different regions in phase space. The Lie transformation is 'global' in the sense that it is not restricted merely to a local region in phase space and holds for all finite times. By using the transformation (13.11) based upon $n - 1$ global isolating integrals, an integrable flow is transformed into a flow that is locally parallel to itself over some finite (or possibly infinite) portion of phase space.

Through $n - 2$ isolating integrals an integrable flow in the n dimensional flat space is confined to a two dimensional surface that may be curved (unless $n = 2$). The $(n - 1)$th isolating integral, if it exists, is the streamline equation on that surface. If the surface has curvature then any two global coordinates defined on the surface and describing the integrable flow are not Cartesian but are curvilinear and orthogonal. An example is given by the torque-free symmetric top, where the Lie transformation is simply the transformation from Cartesian to spherical coordinates: $d\theta/dt = 0$ and $d\phi/dt = 1$ except at the two elliptic points $\theta = 0$ and π.

Now for the third idea of integrability: *all* flows are trivially integrable in a local sense. Local integrability is represented by what Arnol'd calls 'The basic theorem of the theory of ordinary differential equations': in a small enough neighborhood of any nonequilibrium point, every smooth flow can be transformed into a flow that is locally parallel to itself. The smoothness requirement eliminates from consideration velocity fields like $V(x) = x^2 \sin(1/x)$ near $x = 0$. We now discuss Arnol'd's basic theorem.

In a small enough neighborhood of any nonequilibrium point x, where $V_i(x) \neq 0$, for at least one index i, and with $\delta x_i(t) = x_i(t + \delta t) - x_i(t)$ where $x(t) = x$, we can write

$$\delta x_i(t) \approx V_i(x_1, \ldots, x_n)\delta t \tag{13.10b}$$

so that the solution locally (neglecting field gradients) is just

$$x_i(t + \delta t) \approx x_i(t) + V_i(x_1, \ldots, x_n)\delta t. \tag{13.10c}$$

Here, the local velocity field $V(x)$ is a constant for δt small enough. Local integrability requires the knowledge of only n trivial constants of the motion, the n initial conditions $x_i(t)$. Next, we show how to replace this local flow by a locally parallel flow via a smooth local coordinate transformation.

Assume that a rotation (a transformation that is linear in the phase space coordinates) has already been performed so that at the point x_o the velocity field $V(x_o)$ is parallel to the x_n-axis. In other words, the flow passes through the point x_o perpendicular to the $(x_1, \ldots, x_{n-1})$ hyperplane. In this case $dx_i/dt = 0$ with $i \neq n$ at x_o but $dx_i/dt \neq 0$ with $i \neq n$ at nearby points $x_o + \delta x_o$ in the hyperplane. We next transform from the coordinates x at time t to another coordinate system $(y_1, \ldots, y_n)$ at the same time t, and where the vector $V(x_o)$ is also parallel to the axis y_n and perpendicular to the $n - 1$ hyperplane axes y_i with $i \neq n$. Concentrating for the moment on the state $x + \delta x$ of the flow at time $t + \delta t$, we choose for our local transformation the pure translation

$$x_i + \delta x_i = e^{\delta y_n d/dy_n} y_i, \tag{13.12}$$

where $(x_1,\ldots,x_n)$ and $(y_1,\ldots,y_n)$ both lie in the hyperplane and represent states at time t. Notice that only parallel lines transverse to the hyperplane connect the point $(y_1,\ldots,y_{n-1},y_n + \delta y_n)$ with the point $(y_1,\ldots,y_{n-1},y_n)$. With $\delta y_n = \delta t$, this transformation therefore defines a flow where $dy_i/dt = 0$ if $i \neq n$ but $dy_n/dt = 1$. Note that both flows arrive at the same point $(x_1 + \delta x_1,\ldots,x_n + \delta x_n) = (y_1,\ldots,y_{n-1},y_n + \delta y_n)$ in phase space at the same time $t + \delta t$, but from different initial conditions at time t: the first flow starts at a point $x_o + \delta x_o$ where $V(x_o + \delta x_o)$ has a component that lies in the hyperplane, while the second flow starts at the transformed point y_o that is the projection of the tail of the velocity vector $V(x + \delta x)$ at the point $x + \delta x$ onto the hyperplane. Equation (13.12) therefore transforms the flow $dx/dt = V(x)$ locally into a purely parallel local flow transverse to the hyperplane.

Every n dimensional flow has n 'local constants of the motion'. In the transformed system there are n trivially conserved quantities: the $n - 1$ locally defined coordinates y_i for $i = 1,\ldots,n - 1$, and the constant D in $y_n = t + D$. However, local integrability based upon initial conditions as constants of the motion generally does not imply global integrability of the flow for all times t where $n - 1$ nontrivial isolating integrals are required: local integrability holds for all flows at nonequilibrium points whether the flow is integrable or nonintegrable, including chaotic flows. When is a flow completely integrable (when can locally defined constants of the motion be extended analytically to hold for all finite times t)? This question can be reformulated in the following way: when are the locally defined coordinates $(y_1,\ldots,y_n)$ holonomic? This is the same as asking when the locally defined velocity field $v(x) = (v_1,\ldots,v_n)$ is globally integrable.

We can summarize the two transformations described above in the form

$$\left. \begin{array}{l} v_i = U_{ik}V_k(x) = 0, \ i \neq n \\ v_n = U_{nk}V_k(x) = 1 \end{array} \right\} \tag{13.13a}$$

where the n local coordinates y_i are defined in an infinitesimal region near x_o by the locally defined velocities v_i by $dy_i = v_i dt$. The transformation matrix U includes a linear transformation (a rotation of coordinates) and an infinitesimal translation, and was constructed above for points x in the neighborhood of an arbitrary nonequilibrium point x_o where $V(x_o)$ is parallel to v_n.

The axes $(y_1,\ldots,y_n)$ where local integrability holds have been constructed only in a small neighborhood of a point x_o and generally do not define a global coordinate system in phase space. We ask next for the condition that the locally defined coordinates y_i can be globally extended. Since the matrix U is constant only in the neighborhood of a single point x_o, we must replace (13.12) by the more general statement

$$\left. \begin{array}{l} v_i(x) = U_{ik}(x)V_k(x) = 0, \ i \neq n \\ v_n(x) = U_{nk}(x)V_k(x) = 1 \end{array} \right\} \tag{13.13b}$$

where the velocity field $v(x)$ is generally nonintegrable. Any n local coordinates y_i defined by n arbitrarily chosen differential forms

$$dy_i = v_i(x)dt = U_{ik}(x)dx_k \tag{13.13c}$$

are typically nonholonomic, which is why there is no global extension of Arnol'd's 'basic theorem' for an arbitrarily chosen flow in phase space.

Consider next an arbitrary transformation $U(x)$ of the velocity field $V(x)$ given by (13.13b) that is not restricted to the local transformation (13.12). The condition for n holonomic coordinates (n globally defined, path-independent coordinates) y_i is that the velocities $v_i(x)$ are integrable, that

$$\frac{\partial U_{ik}}{\partial x_l} = \frac{\partial U_{il}}{\partial x_k} \tag{13.14}$$

for all indices $l \neq k$ and for all indices i, which allows us to rewrite (13.13c) to read

$$dy_i = v_i(x)dt = dF_i(x_i, \ldots, x_n) \tag{13.13d}$$

so that $y_i = F_i(x, \ldots, x_n)$ for $i = 1, \ldots, n$ is just the statement that a global transformation of coordinates exists. The differential equations in the new globally defined coordinates are given by

$$\dot{y}_i = v_i(F^{-1}(y)) = \frac{\partial F_i(F^{-1}(y))}{\partial y_k} V_k(F^{-1}(y)) \tag{13.10d}$$

where $x_i = F^{-1}(y_1, \ldots, y_n)$ is the inverse transformation.

If, now, we ask when local integrability near an arbitrary nonequilibrium point x reflects complete integrability of the flow, then we must restrict transformations $U(x)$ of the velocity field to those that are generated locally by (13.12),

$$\left.\begin{array}{l} dy_i = v_i(x)dt = U_{ik}(x)dx_k = 0, \ i \neq n \\ dy_n = v_n(x)dt = U_{nk}(x)dx_k = dt. \end{array}\right\} \tag{13.13e}$$

Complete integrability then requires that the integrability condition (13.14) holds for (13.13e) for all finite times t, which yields

$$\left.\begin{array}{l} dy_i = v_i(x)dt = dG_k(x_1, \ldots, x_n) = 0, \ i \neq n \\ dy_n = v_n(x)dt = dF(x_1, \ldots, x_n) = dt. \end{array}\right\} \tag{13.15}$$

This is just the infinitesimal statement of Lie's global transformation

$$\left.\begin{array}{l} y_i = G_i(x_1, \ldots, x_n) = C_i, \ i = 1, \ldots, n-1 \\ y_n = F(x_1, \ldots, x_n) = t + D, \end{array}\right\} \tag{13.16}$$

which holds for any completely integrable flow.

The question whether a flow is completely integrable is therefore equivalent to the question whether the locally defined velocity field $v(x)$ is integrable and yields a set of holonomic coordinates. If a flow is completely integrable, then integrability of the first, second, and third kinds coalesce into a single idea of integrability (which was assumed implicitly in some of the older literature). Conversely, a nonintegrable flow is one where the coordinates $x_i(t)$ are holonomic, but the variables y_i defined locally by (13.12) are not holonomic over arbitrarily long time intervals. The unification of the separate ideas of integrability, the reduction of the idea of an integrable flow to the construction of a special set of holonomic coordinates, is implicit in Lie's construction of his transformation to constant speed translations. Confusion of trivial with nontrivial conserv ation laws led some earlier writers to ignore the possibility of nonintegrability of flows, meaning the neglect of flows that are not integrable by quadratures (see

Eisenhart, 1961, pp. 33–4, Duff, 1962, p. 26, and also Whittaker, 1965, pp. 53 and 275; see also the discussion of Goldstein and Landau–Lifshitz in section 16.1 below).

In the theory of partial differential equations it is possible to prove that $n - 1$ functionally independent solutions of a linear partial differential equation $V \cdot \nabla G = 0$ exist. Locally, these are just the constants $(y_1, \ldots, y_{n-1})$. Palmore (1996) has argued that these local constants can always be extended analytically until one meets a singularity. In other words, by analytic continuation, every 'nonintegrable' flow in phase space has $n - 1$ time-independent global conservation laws $y_i = C_i = G_i(x_1, \ldots, x_n)$, but they are singular: during a finite time $0 \leq t < t_1$ the flow can be parallelized via a Lie transformation (13.11), but at time t_1 one of the conservation laws has a singularity and so there is no parallelization by that particular transformation for $t \geq t_1$.

According to Arnold's basic theorem, however, there is nothing special about the time t_1: the flow can again be parallelized locally for $t \geq t_1$, and the same argument applies again. This means that the singularities of the conservation laws of a nonintegrable flow must be like branch cuts or phase singularities: moveable and arbitrary, like the international date line in the attempt to impose linear time on circular time.

This leads to the following picture of a nonintegrable flow: integrability holds piecewise within infinitely many time intervals $0 \leq t < t_1$, $t_1 \leq t < t_2, \ldots$, but at the times t_i a singularity of a conservation law is met and 'the clock must be reset'.

13.3 Globally integrable noncanonical flows

Questions of integrability inevitably reduce to the question whether a given differential form or set of differential forms is exact. We call a differential form $X_1 dx_1 + \ldots + X_n dx_n$ integrable if it is a closed differential or else has an integrating factor that makes it closed, and we call it nonintegrable otherwise. The condition for a closed differential form is that all of the cross-partials are equal, that

$$\frac{\partial X_j}{\partial x_k} = \frac{\partial X_k}{\partial x_i}, \tag{13.17}$$

in which case there is a point function $u(x_1, \ldots, x_n)$ such that

$$du = X_1 dx_1 + \ldots + X_n dx_n, \tag{13.18}$$

because $X_k = \partial u / \partial x_k$. When this is true, then all line integrals of the differential form are independent of path so long as isolated singular points where (13.17) is violated are not crossed. For nonclosed differential forms, line integrals of the form are generally path-dependent. For a Hamiltonian system, the differential form in question is $Ldt = \Sigma p_i dq_i - Hdt$, where L is the Lagrangian, the integrand of the action. We show in chapter 16 that the condition for a Hamiltonian flow to be integrable is that Ldt is the closed differential form of a multivalued q-space potential, a state of affairs that was not foreseen in the action principle of chapter 2.

If the X_k are the covariant components of a coordinate-free vector, one that transforms according to the matrix multiplication rule

$$X_k = \frac{\partial y_i}{\partial x_k} Y_i \tag{13.19}$$

where $y_i = F_i(x_1, \ldots, x_n)$ and therefore $dy_i = \partial y_i/\partial x_k dx_k$, then $X_k dx_k = Y_i dy_i$ holds but

$$\delta X_k dx_k = \left(\frac{\partial X_k}{\partial x_i} - \frac{\partial X_i}{\partial x_k} \right) dx_k \delta x_i = \left(\frac{\partial Y_k}{\partial y_i} - \frac{\partial Y_i}{\partial y_k} \right) dy_k \delta y_i \tag{13.20}$$

holds as well, showing that integrability (or lack of it) is an invariant property: if (13.17) holds then the variational equation (13.20) vanishes in every coordinate system and $Y_k dy_k$ is also integrable. Integrability is not coordinate-dependent: like the effect of a positive Liapunov exponent, integrability cannot be eliminated by differentiable coordinate transformations (the Lie transformation is generally singular and can change the Liapunov exponent, however). The 'two form' (13.20) is called the exterior derivative of the 'one form' $X_i dx_i$ in the modern literature (see Buck, 1956, Dubrovin et al., 1984, 1985, or Arnol'd 1983, 1989).

By Lie's theorem (section 3.6), every one-parameter transformation group on two variables has an invariant, so that flows in the Cartesian plane are generally completely integrable by way of an isolating integral. Correspondingly, flows in the Cartesian plane cannot be chaotic: they are equivalent by a differentiable change of coordinates to a translation at constant speed, representing a global idea of perfect order that is the opposite of deterministic chaos. The Lie transformation to time translations is singular at a sink or source of a dissipative flow: in the original coordinates the nearby streamlines are repelled by the source and also by each other, but in Lie's translation variables the sink is sent off to infinity so that the (singularly-) transformed flow lines are all parallel to each other and are nonattracting and nonrepelling. We discuss this more carefully in section 13.6, where we observe that there are open flows where positive Liapunov exponents and integrability coexist harmoniously (there can be local 'streamline-repulsion' without chaos).

We now discuss further some general aspects of integrability where $n \geq 3$, where

$$\frac{dt}{1} = \frac{dx_1}{V_1} = \cdots = \frac{dx_n}{V_n}. \tag{13.21}$$

The condition that a function $G(x,t)$ is conserved is that G satisfies the first order partial differential equation

$$\frac{dG}{dt} = \frac{\partial G}{\partial t} + V \cdot \nabla G = 0, \tag{13.22}$$

which is just the condition that G is constant along streamlines in phase space (G is equal to different constants along different streamlines). The streamlines are determined by (13.21), which are called the characteristic curves (or just characteristics) of (13.22). Here, $G(x_1, \ldots, x_n, t)$ is generally a time-dependent constant of the motion. The linear, homogeneous partial differential equation (13.22) may have more than one solution, which is just the statement that a given dynamical system may have more than one conserved quantity.

Examples of time-dependent conservation laws for five dimensional flows are given in Sneddon's treatment of the solution of nonlinear partial differential equations in two variables by Cauchy's method of characteristic strips (1957).

A completely integrable noncanonical system with a time-dependent right hand side (representing a flow in an $n + 1$ dimensional phase space) is one where n functionally independent isolating integrals that hold for all finite times can be found, and it is easy to see in principle how this works (here, we follow in part the discussion in Kilmister,

1964). Suppose that there are $n + 1$ conserved quantities $G_\alpha(x_1, \ldots, x_n, t)$, $n + 1$ solutions of the equation

$$\frac{\partial G_\alpha}{\partial t} + V_i \frac{\partial G_\alpha}{\partial x_i} = 0, \tag{13.23}$$

where we sum over the repeated index i. Considered all together, we have a system of $n + 1$ linear equations in $n + 1$ variables. By Cramer's rule, the determinant of the coefficients must vanish,

$$\frac{\partial(G_1, \ldots, G_{n+1})}{\partial(x_1, \ldots, x_n, t)} = 0, \tag{13.24}$$

which means that at least one of the constants G_α is not functionally independent of the other n (it is easy to prove by differentiation that any function of one or more constants of the motion is also a constant of the motion, a functionally dependent one). It follows by the same reasoning that if we look for time-independent conservation laws of an autonomous system, then there can be at most $n - 1$ of them.

Mathematically, a bounded invariant set is the closure of any region of phase space where a bounded orbit is confined as the time goes to infinity. Every point in an invariant set is a possible initial condition, and if an initial condition is chosen from within that set then the resulting trajectory is confined to the invariant set forever. In a dissipative system, invariant sets may be either attracting or repelling in the sense that they attract or repel nearby orbits. Attraction/repulsion of orbits outside an invariant set is impossible in a conservative system on account of Liouville's theorem. Equilibria and periodic orbits are examples of zero and one dimensional invariant sets. Any integrable three dimensional flow has exactly two functionally independent isolating integrals, and so bounded orbits can also be confined to two dimensional invariant sets, as we have discussed in section 13.1 above and also in chapter 3 for the case of flow on an attracting torus.

Consider the case of an integrable flow in three dimensions where there are two time-independent isolating integrals. In this case, as we showed in section 3.6, the vector field has the form $MV = \nabla G \times \nabla H$ where $G(x, y, z) = C_1$ and $H(x, y, z) = C_2$ are the two invariants of the flow, and the motion is confined to the two dimensional surface $G = C_1$. If we can construct new variables (u, ξ_1, ξ_2), where $u = G$ and (ξ_1, ξ_2) are local coordinates defined on the surface $G = C_1$, then $H(x, y, z) = h(\xi_1, \xi_2) = C_2$ depends only upon those two coordinates (is the streamline equation). We have also shown that if the vector field has equilibria on the surface $G = C_1$, then those equilibria can appear only in certain compatible combinations (excluding degeneracy, at least two elliptic points are necessary for a bounded velocity field on the closed surface) and the surface has the topology of a sphere, a closed Riemannian surface with no holes. We now ask: what is the geometry of a surface $G(x, y, z) = C_1$ for the case where the flow velocity V *never* vanishes on that surface? We consider only the case where the motion in the variables (x, y, z) is uniformly bounded for all times t, so that the surface G is closed.

In this case, the differential equations on the surface G must have the form

$$\left. \begin{array}{l} \dot{\xi}_1 = v_1(\xi_1, \xi_2) \\ \dot{\xi}_2 = v_2(\xi_1, \xi_2) \end{array} \right\} \tag{13.25}$$

where the vector field $v = (v_1, v_2)$ has no zeros. In order to understand the different possible flows qualitatively, it is enough to understand the simplest possible case, the case where the velocity vector v is a constant, where both components v_i are constants. This means that the manifold is Cartesian and that we can transform to a constant velocity translation. Denoting the coordinates (ξ_1, ξ_2) by (θ_1, θ_2) for this special case, and the constant velocities by ω_i, the solution is simply

$$\left. \begin{aligned} \theta_1(t) &= \omega_1 t + \theta_{1_o} \\ \theta_2(t) &= \omega_2 t + \theta_{2_o} \end{aligned} \right\} \tag{13.26}$$

Before going further, we should make contact with Lie's form of the solution as a *single translation* in the coordinates $w + t + D$, with both u and v constant (to show that the two translations combine to form a single translation on the confining surface). The axis of the coordinate $u = G(x,y,z)$ is locally normal to the surface G and is constant, $u = C_1$. If we write

$$\theta_2 = \omega_2 \theta_1 / \omega_1 + C_2, \tag{13.27}$$

then this is just the streamline equation

$$h(\theta_1, \theta_2) = \theta_2 - (\omega_2/\omega_1)\theta_1 = C_2 \tag{13.28}$$

and defines the coordinate $v = H(x,y,z) = h(\theta_1, \theta_2)$, which is perpendicular to the coordinate $w = t + \theta_{1_o}/\omega_1 = \theta_1/\omega_1 = t + D = F(x,y,z)$. Now, what is the topology of a confining surface G where the velocity field cannot vanish and that admits translations at constant velocity?

Since the motion is bounded in the original variables (x,y,z), G is a closed surface. Therefore, the variables (θ_1, θ_2) must be periodic with period 2π. This means that the points 0 and 2π on the axis θ_1 must be identified as the same point, and likewise 0 and 2π are the same point on the θ_2-axis. The surface G is therefore a circular torus with two radii r_1 and r_2 fixed by u and v (see figure 3.8b).

For the general case of bounded motion where $v_1(\xi_1, \xi_2)$ and $v_2(\xi_1, \xi_2)$ are not constants but do not vanish, then the coordinates (ξ_1, ξ_2) on G must be periodic modulo two different lengths, if we think of ξ_i as a position variable. Here, we can transform to the variables $(v,w) = (C_2, t + D)$, and then to the constant velocity variables (θ_1, θ_2) by a rotation and a rescaling. The point is that the more general flow where v is not constant is equivalent, by a differentiable transformation of coordinates, to the constant velocity case so that the surface $G(x,y,z) = C_1$ cannot be other than a smooth deformation of the circular torus. Note that a sphere with a handle is topologically the Riemannian analog of a torus. This corresponds to the case where there are no equilibria, but where a constant velocity field on the confining surface is impossible because of curvature.

From the standpoint of the theory of oscillations, the problem of flow on a sphere is fundamentally a *one*-frequency problem because the periodic orbits enclose elliptic points (for separatrix motion, that period is infinite). In contrast, the flow on a Euclidean torus is irreducibly a *two*-frequency problem: ω_2/ω_1 determines the winding number W of a constant speed translation $w = t + D$ on the torus, and the orbits on the torus close on themselves only if the frequency ratio is rational. Otherwise, the orbits do not close in finite time and are quasiperiodic on the torus (as we discussed in sections 4.3 and 6.2). In other words, the form of the solution on a Cartesian torus is irreducibly of the form $x_i(t) = \psi_i(x_{1_o}, x_{2_o}, x_{3_o}, \omega_1(t - t_o), \omega_2(t - t_o))$ whenever the frequency ratio ω_2/ω_1 is irrational.

If we consider an integrable flow in n dimensions, then there are $n - 1$ isolating integrals $G_i = C_i$ and a single translation $t + D = F$. The flow is restricted to a two dimensional surface by $n - 2$ of the conservation laws which determine $n - 2$ of the coordinates x_i as functions of the other two x_i and the $n - 2$ constants C_k. The remaining $(n - 1)$th isolating integral then describes the streamlines on the confining two dimensional surface. Consider only the case where the motion in the variables $(x_1, \ldots, x_n)$ is uniformly bounded, so that the confining two dimensional surface is closed and has no boundaries (the surface closes on itself). If the flow has equilibria on that surface, then the surface has the topology of a two-sphere imbedded in n dimensions. If the velocity field does not vanish anywhere on the surface, then the surface has the topology of a 2-torus in the n dimensional phase space, and the trajectories are either stable periodic or stable quasiperiodic on the torus. In the case of dissipative flow on a sphere there must be at least two equilibria, just as in the conservative case (in the conservative case, there is also the degenerate possibility that two elliptic points that are not separated by a separatrix may coincide, giving rise to a dipolar flow pattern). Two geometrically consistent possibilities are: (i) source–sink flow, and (ii) two stable foci separated by an unstable limit cycle, or two unstable foci separated by a stable limit cycle. One can also replace the two stable foci in (ii) by two unstable foci surrounded by two stable limit cycles that are themselves separated by a third unstable limit cycle, and so on.

There are conditions that must be met for the above-stated conclusions to be correct (the same restrictions hold in the case of time-dependent conservation laws). In order that the $n - 2$ conservation laws $G_i(x_1, \ldots, x_n) - C_i = 0$ determine $n - 2$ variables x_i, $i = 3, 4, \ldots, n$, in terms of two variables x_1 and x_2 the $n - 2$ constants $C_1, \ldots, C_{n-2}$, the $n - 2$ conservation laws must yield a nonsingular two dimensional surface. In this case, the $(n - 2) \times n$ matrix $\partial G_i / \partial x_k$ has rank $n - 2$, and by the implicit function theorem we can then define local coordinates in the neighborhood of any point on the surface where the normal to that surface does not vanish. If one or more of the invariants G_i is such that this procedure fails, then the motion is not necessarily isolated on a two dimensional surface in the n dimensional phase space but may only be confined to a surface of higher dimension and is therefore nonintegrable. Motion on Möbius strips, Klein bottles, and other nonorientable surfaces is possible. Even in the case where the reduction procedure works, the two-sphere and the 2-torus may not be the only possible closed, boundary-less two-surfaces to which integrable motion need be confined. Another topologically distinct possibility is a toroidal two-surface with two or more holes in the shape of a figure eight (see Armstrong, 1983, or Fomenko, 1988), or the Riemannian case of a sphere with two or more handles attached (Arnol'd (1993) and Fomenko point out that analytic conservation laws cannot exist on the latter type of surface). Now for the main point.

We have not yet asked the extremely interesting and pregnant question what happens if one perturbs a conservative integrable flow defined by a velocity field V by adding a 'small' vector field $\varepsilon \delta V$ that makes the resulting flow driven-dissipative. The Hamiltonian case is discussed in chapter 17. Here, we discuss the noncanonical case. For the cases $n = 2$ and 3 do all of the two-spheres or 2-tori disappear immediately, or do some of them survive the perturbation in distorted form? The analysis of this question leads to the study of attractors in phase space.

Complete integrability is essential for understanding nonchaotic Hamiltonian systems but is difficult to prove for most nonchaotic dissipative systems whenever $n \geq 3$. For example, we can prove (so far) that the Lorenz model

$$\left.\begin{array}{l} \dot{x} = \sigma(y - x) \\ \dot{y} = \rho x - y - xz \\ \dot{z} = -\beta z + xy \end{array}\right\} \qquad (13.29)$$

has enough time-dependent conservation laws to be completely integrable only for three sets of parameter values where $(\sigma,\beta,\rho) = (1/2,1,0)$, $(1,2,1/9)$, or where $(1/3,1,\rho$ is arbitrary). In spite of the lack of proof of complete integrability, this model is known to have nonchaotic motions for $\rho < \sigma(\sigma + \beta + 3)/(\sigma - \beta - 1)$, with $\sigma - \beta + 1 > 0$. This is a much larger parameter range for the occurrence of orbitally stable bounded motions than can be explained on the basis of our ability to prove complete integrability. In a driven-dissipative system we can often short-circuit the question of complete integrability in order to rule out chaos everywhere in phase space. It turns out that *local* integrability due to a zero, or differentiable one or two dimensional *attractor* is enough to keep the motion regular, and so we turn to the discussion of how self-confinement and attractors appear through nonconservative perturbations of integrable systems with bounded integrable motion. Examples of attracting closed curves and tori were given in chapter 3.

13.4 Attractors

In order that a nonconservative system has a periodic orbit, Bendixson's 'negative criterion' must be satisfied: the divergence of the flow cannot be of a single sign over all of phase space. The proof is easy and follows from an application of the divergence theorem in phase space. Let C be any periodic solution of a flow with velocity field V; C is therefore a closed curve. Let $\hat{n}$ denote the outer normal to that curve. Since V is everywhere perpendicular to $\hat{n}$ it follows that

$$\int \nabla \cdot V d^n x = \oint V \cdot \hat{n} ds = 0, \qquad (13.30)$$

where the first integral is over the area bounded by C and the second is the closed line integral around C. Therefore, if V has a divergence that does not change sign in some region of phase space, there are no periodic solutions within that region.

The idea of foliation of the phase space of completely integrable systems by 'confining surfaces' is useful. If we use Euler's top as an example of complete integrability, then we can understand how global conservation laws can 'foliate' a phase space: for each different value of the angular momentum, the top's motion is confined to a sphere in the three dimensional angular momentum phase space (kinetic energy conservation then yields the streamlines on each sphere). In the phase space, these spheres are all concentric with each other and there is no motion, for any set of initial conditions, that is *not* confined to one of these spheres. Consequently, the spheres are said to 'foliate' the phase space, which can be thought of geometrically as consisting of the concentric system of infinitely thin spherical 'foils'. For the one dimensional simple harmonic oscillator, the phase space is foliated by concentric ellipses, which are one dimensional tori. When dissipation and driving are added to such a system (as in the generalization of Euler's top in section 13.1), is all of the foliation immediately destroyed or does some part of it survive?

To answer that question in part, we now give an example of a periodic solution of a driven-dissipative equation. Consider as an example the Newtonian system

$$\ddot{x} + \varepsilon(x^2 + \dot{x}^2 - 1)\dot{x} + x = 0, \tag{13.31}$$

which we can think of as a particle with unit mass attached to a spring with unit force constant, but subject also to a nonlinear dissipative force. In what follows, we take $\varepsilon > 0$. The energy $E = (\dot{x}^2 + x^2)/2$ changes according to

$$dE/dt = -\varepsilon\dot{x}^2(x^2 + \dot{x}^2 - 1). \tag{13.32}$$

From Newton's law it follows that $\dot{E} = -f\dot{x}$, where the nonconservative force is given by $f = \varepsilon\dot{x}(x^2 + \dot{x}^2 - 1)$. This force represents contributions from both external driving and dissipation, and the effect that dominates depends upon where you look in the phase plane. To see this, note that there is an energy flow through the system corresponding to positive damping (represented by a net dissipation) whenever $r^2 = x^2 + \dot{x}^2 > 1$, but the system gains energy in the region $r^2 < 1$. Therefore, the phase plane is divided into two distinct regions of energy gain and energy loss by a circle at $r = 1$, because $\dot{E} = 0$ on that circle. On the unit circle, we have a steady state where energy gain is balanced by dissipation so that the energy E is maintained constant. By inspection, the unit circle $x^2 + \dot{x}^2 = 1$ is a solution of (13.31), and this solution illustrates what we mean by a 'limit cycle' when we show below that nearby initial conditions have this solution as their (periodic) limit (see figure 3.8(a)), which is the same as saying that the limit cycle is a one dimensional attractor for nearby initial conditions (in this special case, the limit cycle is a global attractor for all possible initial conditions).

When $\varepsilon = 0$ then the solutions of the conservative system (the harmonic oscillator) are concentric circles and the phase plane is 'foliated' by one dimensional tori. When $\varepsilon > 0$, all of those tori except one are destroyed: the only surviving torus is the one at $r = 1$, where periodicity through energy balance is maintained because the dissipation rate matches the energy input rate at $r = 1$.

If, in our model, we transform to polar coordinates $r^2 = x^2 + \dot{x}^2$, $\tan\theta = \dot{x}/x$, it follows that

$$\left.\begin{array}{l} \dot{r} = r(1 - r^2)\sin^2\theta \\ \dot{\theta} = (1 - r^2)\sin\theta\cos\theta - 1, \end{array}\right\} \tag{13.33}$$

so that the cycle is stable: $\dot{r} > 0$ if $r < 1$, $\dot{r} < 0$ if $r > 1$, $\dot{\theta} \sim -1$ when $r \sim 1$, and so the direction of motion on the limit cycle is clockwise. Note that there is (by necessity) an unstable spiral at the origin. Note also that the amplitude and period of oscillation on the limit cycle are completely independent of the initial conditions, in contrast with the case of linear oscillations. More to the point, the amplitude and period of the limit cycle are actually determined by one particular solution of the corresponding conservative system. In other words, we have a one dimensional periodic attractor that is defined by a conservation law. Because the motion on the limit cycle is stable periodic, it is therefore integrable.

In general, a limit cycle of a set of driven-dissipative nonlinear equations is a periodic solution that divides the phase plane into regions of net average energy gain and energy loss. For Newtonian flows the generalization of the above example is easy, and we state the analysis for the case of a single particle with energy $E = (dx/dt)^2 + U(x)$, although one can make a more general statement by starting with a Hamiltonian flow and perturbing it dissipatively. With $x = x$ and $y = dx/dt$, a planar Newtonian flow is defined by

$$\left. \begin{array}{l} \dot{x} = y \\ \dot{y} = -U'(x) + f(x,y) \end{array} \right\} \tag{13.34}$$

so that $dE/dt = -fy$. Suppose that f changes sign in an open region of phase space consistent with the requirement of a periodic solution. If we integrate around any closed curve C we get

$$E(T) - E(0) = -\int_t^{t+T} y(s)f(x(s),y(s))ds, \tag{13.35a}$$

and if C is a limit cycle, a periodic solution of the Newtonian flow (13.35) with period T, then this requires that $E(T) = E(0)$, which means that

$$\int_t^{t+T} y(s)f(x(s),y(s))ds = 0 \tag{13.35b}$$

on C. Stability would require $E(T) - E(0) > 0$ for any curve C' enclosed by C, and also that $E(T) - E(0) < 0$ for any curve C'' that encloses C. For weakly dissipative Newtonian flows, this leads to a practical method for locating limit cycles.

Let $f(x,y) = -\varepsilon h(x,y)$ where ε is small. In this case, we require a potential $U(x)$ such that, for some range of energies $E_{\min} < E < E_{\max}$, the motion of the conservative system (the $\varepsilon = 0$ system) is bounded and is therefore periodic with period

$$\tau = 2\int_{x_1}^{x_2} \frac{dx}{[2(E - U(x))]^{1/2}}, \tag{13.36}$$

where x_1 and x_2 are the turning points of $U(x)$. To lowest order in ε, we expect that the limit cycle is approximately, very roughly, given by one of these periodic solutions (we assume that we can construct the limit cycle from the conservative solution by a perturbation approach). The first question is: which unperturbed periodic orbit should we start with? The second question is: how can we use that unperturbed orbit to compute the limit cycle approximately to first order in ε? Both questions are analyzed in the easily readable text by Jordan and Smith (1977), but we address only the first question here.

Let A be the amplitude parameter in the solution $(x_o(t),y_o(t))$ of the conservative solution with period τ that is nearest to the limit cycle. If in (13.35a) we replace the exact solution by this nearby unperturbed solution (whose period and amplitude we have not yet determined), then A is fixed by the condition that $E(T) - E(0) = 0$, or by the condition

$$g(A) = \int_t^{t+T} y_o(s,A)f(x_o(s,A),y_o(s,A))ds \approx 0, \tag{13.37}$$

which determines the amplitude A_o of the conservative cycle that best approximates the limit cycle for small ε. The stability condition can then be formulated as $g'(A_o) < 0$, which means energy gain inside the cycle and energy loss outside it, causing solutions off the cycle to spiral toward it. The method introduced here is the method of energy balance, or 'the method of averaging'.

When an integrable conservative system with stable periodic orbits is perturbed weakly by nonlinear dissipative and driving that are able to satisfy energy balance ('averaging'), then a limit cycle is born through the destruction of global periodicity, leaving behind one particular distorted periodic orbit as an attractor or repeller. In a nonlinear driven-dissipative system that is far from (unperturbed) integrable conservative ones, a limit cycle can be born from a certain kind of bifurcation as a control parameter is varied: a stable limit cycle can follow from the loss of stability of a spiral sink through a Hopf bifurcation. A Hopf bifurcation occurs when the real parts of two complex conjugate eigenvalues λ and λ^* pass through zero as a control parameter μ is varied, so that $\mathrm{Re}\,\lambda < 0$ if $\mu < \mu_c$ (spiral flow into a sink) and $\mathrm{Re}\,\lambda > 0$ if $\mu > \mu_c$ (spiral flow out of a source) with $d\mathrm{Re}\,\lambda/d\mu > 0$ at μ_c. As an example of a Hopf bifurcation, consider again our model (13.31) but with a control parameter μ,

$$\left.\begin{aligned} \dot{x} &= y \\ \dot{y} &= (\mu - x^2 - y^2)y - x. \end{aligned}\right\} \tag{13.38a}$$

The origin is an equilibrium point, so consider the linear flow

$$\left.\begin{aligned} \dot{x} &= y \\ \dot{y} &= \mu y - x \end{aligned}\right\} \tag{13.38b}$$

near the origin. The eigenvalues, for $-2 < \mu < 2$, are complex:

$$\lambda = \frac{\mu \pm i(4 - \mu^2)^{1/2}}{2}. \tag{13.39}$$

When $\mu = 0$ the origin is elliptic and the phase plane of the linear flow is foliated by 1-tori. If $\mu < 0$, all tori are destroyed and the origin is a sink due to over damping: the only remnant of the tori is the spiraling motion into the sink. When $\mu > 0$, the motion spirals out of a source at the origin, but in the full nonlinear system (13.38a) a 1-torus at $r = \sqrt{\mu}$ survives as a stable limit cycle that attracts all other trajectories. As μ passes through 0 we have a Hopf bifurcation that describes the change of stability from spiraling flow into a sink to spiraling flow out of a source and onto the limit cycle.

Regardless of the details of how a limit cycle is born, it is always a closed streamline where the flow is both integrable and periodic: a closed curve is topologically equivalent to a circle, and so there is a continuous transformation that takes us from the dynamics on the limit cycle to dynamics on a circle where the differential equations have the explicitly integrable form

$$\left.\begin{aligned} \dot{r} &= 0 \\ \dot{\phi} &= \omega. \end{aligned}\right\} \tag{13.40}$$

We have seen how periodicity follows as a consequence of an average energy balance in Newtonian systems: the net energy is conserved for each cycle of the motion although energy is not conserved pointwise along the limit cycle.

After one dimensional attractors, the next smooth case is that of a two dimensional attractor, a sphere or a torus (or a topologically equivalent surface). We consider next the case of an attracting Cartesian torus. Three dimensional tori in four dimensions occur for two simple harmonic oscillators, but two dimensional tori can also occur in three and higher dimensional nonNewtonian flows.

On a circular torus the differential equations have the explicitly integrable form

$$\left.\begin{aligned} \dot{r} &= 0 \\ \dot{\phi}_1 &= \omega_1 \\ \dot{\phi}_2 &= \omega_2. \end{aligned}\right\} \tag{13.41}$$

In a Newtonian system, local periodicity should follow as a consequence of local energy-flow balance. A three dimensional model of an attracting 2-torus is given by the flow

$$\left.\begin{aligned} \dot{\phi}_1 &= \omega_1 - \beta r \\ \dot{\phi}_2 &= \omega_2 \\ \dot{r} &= -\beta r. \end{aligned}\right\} \tag{13.42}$$

With $\theta_n = \phi_1(t_n)$ and $r_n = r(t_n)$ where $t_n = n\tau_2 = 2\pi n/\omega_2$, we obtain the linear twist map

$$\left.\begin{aligned} \theta_n &= \theta_{n-1} + 2\pi\Omega - br_{n-1} \\ r_n &= br_{n-1} \end{aligned}\right\} \tag{13.43a}$$

with $b = e^{-\beta\tau}$ and $\Omega = \omega_1/\omega_2$. Since $0 < b < 1$ we obtain the linear circle map

$$\theta_n \approx \theta_{n-1} + 2\pi\Omega \tag{13.43b}$$

for large n, which represents motion on a circular 2-torus. Attraction onto the torus is represented by the contraction to zero of r_n, which can be regarded geometrically as a variable perpendicular to the surface of the torus.

With one and two dimensional attractors, one does not need to prove complete integrability in order to rule out the possibility of chaos in dimensions $n \geq 3$. The motion is integrable on a smooth attractor which is either a differentiable curve or surface. Attractors can only occur in dissipative systems. In either the Newtonian or nonNewtonian case the first step in the search for an attractor is to neglect driving and dissipation and search for a conservation law $G(x_1,\ldots,x_n) = C$ that is analogous to the energy of a Newtonian flow on the angular momentum of an Euler top. If a conservation law is found that globally foliates the phase space by an infinite hierarchy of smooth Euclidean surfaces, then the next step is to see whether the inclusion of damping and driving leads to circumstances whereby $dC/dt < 0$ at large distances but $dC/dt > 0$ at small scales, which suggests a region of phase space where the motion is trapped due to the competition between driving and dissipation. Inside any trapping region we expect to find at least one attractor. We turn now to an interesting class of Newtonian flows where the idea of an attractor can occur in much more general form than that of a limit cycle, a sphere, or a torus.

13.5 Damped–driven Euler–Lagrange dynamics

Writers of purely mathematical texts do not feel compelled to restrict themselves to the study of mathematical models that reflect laws of nature, but Newton's laws of motion (and their relativistic and quantal generalizations) are the only universal laws of motion that are known to man: the empirically based law of inertia is the fundamental basis for the laws of motion. We want to show how the interesting recent discoveries of nonlinear dynamics are reflected in mathematical models that are inspired by physics, but in order to study self-confinement and attractors of autonomous flow in three dimensions (or in any odd dimensional phase space) we cannot start with Newtonian particle mechanics. There, except for time-dependent external driving (which we consider explicitly via two models in the next chapter), the phase space dimension is always even. There is also a reason of convenience to choose flows that follow from Newtonian mechanics as our take-off point for the study of attractors in phase space: we have just seen in the discussion of limit cycles that it is useful to start with a conserved quantity that confines the motion of an unperturbed conservative flow in order to arrive at the idea of an attractor with dimension one. In that case, we had no need to distinguish between a trapping region (a region where, after long times, the flow is self-confined) and an attractor. In what follows we use conservation laws that are broken by dissipation and driving to introduce trapping regions in phase space. Because the trapping region has the dimension n of the phase space, attractors of any dimension less than n may in principle occur in a trapping region of a dissipative flow.

A Newtonian starting point for the self-confinement of a damped-driven nonlinear flow to a definite finite region in a flat n dimensional phase space is provided by the Euler–Lagrange equations

$$\frac{dJ_\lambda}{dt} + c_{\lambda\mu}^\nu \frac{J^\mu J^\nu}{I_\mu} = N_\lambda, \tag{13.44a}$$

which are conservative. Here, the inertia matrix I is taken to be diagonal and constant and the c's are the structure constants of some Lie algebra (of rigid body systems more complicated than a top, for example) and so must obey the antisymmetry condition $c_{\mu\nu}^\lambda = - c_{\nu\mu}^\lambda$. The symbols J, I, and N can, but need not, denote the angular momentum, moment of inertia, and torque of a rigid body problem: for the time being, all that we require is the mathematical structure of the Euler–Lagrange equations (13.44a), so it does not matter whether those equations are taken to represent rigid body motions or some other model. We consider only the case where the vector N is a constant vector in what follows. That is, we study an autonomous system with constant forcing N in the n dimensional phase space with Cartesian coordinates $(J_1, \ldots, J_n)$ (because the metric $I_{\mu\nu}$ is diagonal and constant, the J_λ are Cartesian). In the geometric language of chapter 11, (13.44a) with vanishing torque is the equation of a geodesic, described in nonholonomic coordinates, in the tangent plane to a point on the manifold. The affine connections are simply $\Gamma_{\mu\lambda}^\nu = c_{\mu\lambda}^\nu/I_\mu$. A similar relationship can be derived for use in the pseudo-Cartesian tangent planes of general relativity theory (see 'tetrads' in Dubrovin et al., 1984).

There are always at least two quantities that are conserved whenever $N_\lambda = 0$, and the study of the nonconservation of either of those quantities in the presence of damping can lead to the deduction of a trapping region in the $(J_1, \ldots, J_n)$-phase space. A trapping region is more general than the idea of an attractor: it is a region within which any and all trajectories become self-confined after a long enough finite time (the trapping time is trajectory-dependent). If there is an attractor, then it lies within the trapping region but the trapping region (as we shall see) generally has a larger dimension than the attractor. For example, the trapping region always has dimension n, but fixed points and limit cycles have dimension 0 and 1 respectively.

If $n = 3$ with $N_\lambda = 0$, and there is no damping, then the motion is integrable (we prove this below), but this feature is not essential in what follows, and whenever we assume that $n > 3$. To obtain a model that can have an attractor of dimension greater than zero, we must generalize the Euler–Lagrange dynamics (13.45a) by adding damping, which we take to be linear:

$$\frac{dJ_\lambda}{dt} + b_{\lambda\mu} J_\mu + c_{\mu\nu}^\lambda \frac{J_\mu J_\nu}{I_\mu} = N_\lambda. \tag{13.45}$$

Since

$$\nabla \cdot V = \frac{\partial \dot{J}_\sigma}{\partial J_\sigma} = - \operatorname{Tr} b, \tag{13.46}$$

if we require that $\operatorname{Tr} b > 0$ then we expect to find that very large scale motions will be damped, which is a necessary condition for an attractor to exist somewhere in the phase space. To have an attractor that is not a zero dimensional sink then the energy input must dominate dissipation locally at small scales to make the smallest scale motions unstable. That is, to have an attractor the large scale motions must be unstable against further growth but the motions at small enough scales must also be unstable against further decay: these circumstances lead to attractors of dimension higher than zero in phase space, and it is helpful to use a quantity that is conserved whenever $b = 0$ in order to formulate this notion quantitatively. We show next that the conservative flow with $N = 0$ always has two algebraic invariants, so that the $n = 3$ flow without forcing is always integrable.

The conservative flow (13.45a) with $N_\lambda = 0$ always leaves the kinetic energy $T = (J_1^2 I_1 + J_2^2/I_2 + \ldots + J_n^2/I_n)/2$ invariant. The proof of a second conserved quantity C is really only a disguised form of the proof that a Lie algebra has at least one Casimir operator (see Hammermesh, 1962, or Racah, 1951), an operator that commutes with all of the infinitesimal generators of the algebra, and this conserved quantity is a generalization of the square of the angular momentum.

We can show directly by differentiation that the quantity

$$C = g^{\rho\sigma} J_\rho J_\sigma \tag{13.47}$$

is left invariant by the unforced conservative flow

$$\frac{dJ_\lambda}{dt} + c_{\lambda\mu}^\nu \frac{J_\mu J_\nu}{I_\mu} = 0, \tag{13.44b}$$

where by $g^{\rho\sigma}$ we mean the inverse of the matrix

$$g_{\mu\nu} = c_{\mu\alpha}^\beta c_{\nu\beta}^\alpha. \tag{13.48}$$

Hammermesh states that the matrix g provides a metric for defining inner products in linear representations of Lie algebras, but that g is such a metric is not used at all in what follows. Instead, from the differential equations (13.44b) we obtain

$$\frac{dC}{dt} = - 2g^{\rho\sigma} \frac{c_{\rho\mu}^\nu}{I_\mu} J_\mu J_\nu J_\sigma \tag{13.49}$$

so that $dC/dt = 0$ follows if the rank three tensor $g^{\rho\sigma} c_{\rho\sigma}^\nu$ is antisymmetric under exchange of the two indices σ and ν. In fact, this tensor is antisymmetric under exchange of any two of the three indices (σ,μ,ν). Therefore, $dC/dt = 0$ and C is left invariant by the conservative flow (13.44b).

We have two conservation laws that describe two ellipsoids in phase space, or two hyperellipsoids if $n > 3$. Denote the kinetic energy ellipsoid by E_1 and the 'angular momentum' ellipsoid described by the conserved quantity C by E_2. In the conservative flow (where $b = 0$), the invariant curves (the solutions of (13.44b)) lie in intersections of E_1 and E_2 and both these ellipsoids foliate the phase space. In this case, each streamline of the flow is self-confined to an $n - 2$ dimensional hypersurface in the n dimensional phase space.

When $b \neq 0$, then (even if $N = 0$) the foliations of phase space are destroyed (invariant curves of (13.44a) quite generally do not lie on intersections of E_1 with E_2). Next, we state what we expect to happen when dissipation and steady driving are added, and then we prove that the expectations are indeed borne out by the model under consideration.

Consider the surface E_2 for large values of J, and assume that $\text{Tr}\, b > 0$. Think of any point on E_2 as an initial condition for the flow so that the ellipsoid changes rigidly in size as the single point on E_2 follows the streamline: then for large enough J we expect that $dC/dt < 0$, while E_2 contracts rigidly according to the law

$$\frac{dC}{dt} = -\tilde{J}g^{-1}bJ + \tilde{N}J. \tag{13.50a}$$

Since this expression is generally not positive definite, we cannot conclude that C decreases eventually to zero as t goes to infinity. In fact, in order to get a trapping region that contains an attractor with dimension greater than zero, the action of the matrix $g^{-1}b$ combined with that of the constant vector N in (13.50a) must be such as to divide the phase space into regions of net dissipation at large enough J and regions of net energy input at small enough J. Let us assume that the matrix b has been constructed so that this is the case. Then as E_2 contracts rigidly while one point on E_2 follows a streamline, E_2 will eventually intersect a third surface E_3 that is defined by the condition that $dC/dt = 0$ at all points along the intersection. Inside the surface E_3, $dC/dt > 0$ and so any trajectory that enters E_3 tends to be repelled outward again. If these expectations are borne out, then we can define the trapping region optimally as the smallest ellipsoid E_{2c} that contains the (still unknown) surface E_3. Every streamline must enter the n dimensional region E_{2c} in finite time, and having entered cannot escape again (this is self-trapping, or self-confinement). For each set of parameter values we also expect to find at least one attracting point set somewhere within the ellipsoidal trapping region E_{2c}.

To find E_3 and E_{2c} explicitly, to show that the expectations just described are borne out, we follow Lorenz and define next a vector e with components e_ν that solve the linear equations

$$(b' + \tilde{b}')e = N, \tag{13.51}$$

where $b' = g^{-1}b$. In this case, direct comparison with (13.50a) yields

$$\frac{dC}{dt} = \tilde{e}b'e - (J - e)b'(J - e). \tag{13.50b}$$

This means that $dC/dt = 0$ on the surface E_3 defined by

$$\tilde{e}b'e = (J - e)b'(J - e). \tag{13.52}$$

To get a trapping region (a *closed* surface E_3) it is sufficient to have a positive definite matrix b', which explicitly guarantees that $\text{Tr}\, b' > 0$ and also makes E_3 an ellipsoid or hyperellipsoid. It follows from (13.50b) that $dC/dt < 0$ outside the ellipsoid E_3 (attraction toward E_3) but $dC/dt > 0$ inside the ellipsoid E_3 (expulsion from E_3), but the ellipsoid E_3 usually does not contain an invariant curve (invariant set, or just 'solution') of the flow (13.45). We can define the trapping region E_{2c} to be the smallest ellipsoid E_2 that completely contains the smaller ellipsoid E_3.

Note also that if $N_\lambda = 0$ in equations (13.45), then a set of equations with finite, constant values of N_λ can always be obtained by a translation $L_i' = L_i + a_i$ in phase space. This translation is carried out explicitly for the Lorenz model in the exercise 7.

That definition of trapping regions is not unique (not necessarily optimal) because they need only enclose all of the attractors: had we started with the kinetic energy rather than the 'angular momentum' C, then we would have arrived at a quantitatively different (but qualitatively similar) trapping region.

It is possible that there are integrable attractors inside the trapping region E_{2c}, but it is also possible that a trajectory that is confined within E_{2c} continually enters, leaves, and re-enters the ellipsoid E_3 in a very complex way that does not correspond to any sort of global integrability at all. With $\text{Tr } b < 0$, all finite-sized volume elements in phase space are always contracted so that $\delta\Omega(t) \to 0$ as t goes to infinity. An attractor represents the closure of an infinite time orbit and therefore always defines a point set with size $\delta\Omega = 0$. An attractor is therefore a point set with dimension less than n, whereas the trapping region always has precisely the dimension n. If the attractor is a stable equilibrium point, then it has dimension 0. A stable limit cycle has dimension 1, and a quasiperiodic orbit on a stable torus has, asymptotically, dimension 2. We shall see in the next chapter that chaotic orbits can define attractors and repellers where nonintegrable motions occur and that can have nonintegral dimensions (we introduce Hausdorff's generalization of the intuitive idea of dimension in chapter 14 and show there that it is easy to find invariant sets of chaotic dynamical systems that have nonintegral Hausdorff dimensions). Numerical studies of the Lorenz model indicate that chaotic motions can occur within a trapping region. This leads to the idea of 'strange attractors', which may or may not have a nonintegral dimension. Let us now return to the details of our class of models.

As a point on an initial ellipsoid $E_2(0)$ taken as initial condition at $t = 0$ for (13.45a) follows a trajectory, then E_2 contracts rigidly in size according to (13.50b). That trajectory is therefore uniformly bounded at all times $t \geq 0$ by the initial ellipsoid $E_2(0)$. Since this is true for any and all trajectories, this proves that (13.45a) defines a flow in phase space (under these circumstances the solutions can have no finite-time singularities). The converse, however, is not true: an arbitrary flow is not necessarily *uniformly* bounded for all negative times.

To give an example, we return to a specific class of damped-driven Euler–Lagrange flows. If $n = 3$ and $c_{\mu\nu}^\lambda = \varepsilon_{\lambda\mu\nu}$, which represents the motion of a rigid body with three moments of inertia in three dimensions, then $C = J_1^2 + J_2^2 + J_3^2$ is just the square of the angular momentum. More generally, we can assume that $n \geq 3$ but still require that $c_{\mu\nu}^\lambda J_\mu J_\nu J_\lambda = 0$ (this condition defines the class of models considered in section 2 of Lorenz's paper, 1963). In this case, it still follows that $C = J_1^2 + \ldots + J_n^2$, and so the trapping region E_{2c} is the smallest hypersphere, the sphere with radius $\sqrt{C_c}$, that contains the entire ellipsoid E_3.

With different labeling of terms, we can see by inspection that our class of models includes the model discussed by Lorenz in his original paper, namely, the model

$$\left.\begin{array}{l} \dot{x} = \sigma(y - x) \\ \dot{y} = \rho x - y - xz \\ \dot{z} = -\beta z + xy. \end{array}\right\} \tag{13.53}$$

Normally, textbook writers will advise the reader to ignore all nonlinear terms in order to identify a specific linear system. To the contrary, a conservative Euler top flow is obtained by deleting all *linear* terms in (13.53), yielding

$$\left.\begin{array}{l} \dot{x} = 0 \\ \dot{y} = -xz \\ \dot{z} = xy. \end{array}\right\} \tag{13.54}$$

This flow represents a torque-free symmetric top with $I_1 = I_2 \neq I_3$. For example, if we write the equations of Euler's top in the form

$$\left.\begin{array}{l} \dot{x} = ayz \\ \dot{y} = -bxz \\ \dot{z} = cxy, \end{array}\right\} \tag{9.23c}$$

then $x^2 + y^2 + z^2 = \text{constant}$ and with $a = 0$, $b = c = 1$ (symmetric top with $I_1 = I_2 \neq I_3$), and with linear damping-driving added on corresponding to the matrix

$$\mathbf{b} = \begin{pmatrix} \sigma & -\sigma & 0 \\ 0 & 1 & -\rho \\ 0 & 0 & \beta \end{pmatrix}, \tag{13.55}$$

then we get exactly the Lorenz model. The Lorenz model is therefore formally equivalent to a certain damped, driven symmetric top, and to get a trapping region in phase space we need only $\operatorname{Tr} b = -\sigma - \beta - 1 < 0$ along with the condition that σ, β, and ρ are all positive. The latter condition is specifically needed in order to get the ellipsoid E_3, and the trapping region is therefore the smallest sphere that contains that ellipsoid. In general, truncations to finitely many modes of the Navier–Stokes equations can give rise to equations of motion of the form (13.55) of the damped, driven Euler–Lagrange equations of rigid body motion.

We now have a generalization of the circumstances that led, in the simpler case of the last section, to the deduction of a stable limit cycle. We cannot determine, without further analysis, the nature of any attractor that we may find within the trapping region E_{2c} but we do expect to find at least one attractor there for a given set of parameter values (an expectation is, however, never a proof).

The different attractors of the Lorenz model (the damped symmetric Euler top) include sinks and limit cycles, for different parameter values, but are not restricted either to those geometric objects or to tori or to any other smooth Euclidean surfaces that would limit the flow to complete integrability.

We have explained heuristically (chapter 9) how three dimensional flows can lead to uniquely invertible two dimensional iterated maps. Numerical computations on the Lorenz model for $\rho > \rho_c$ (see Schuster, 1988, or McCauley, 1993, chapter 2) lead to an apparently one dimensional hat-shaped map $z_{n+1} = f(z_n)$ that is superficially reminiscent of the tent map, which is continuous and has a two-valued inverse. In order to preserve uniqueness of the inverse $z_{n-1} = f^{-1}(z_n)$, the numerically indicated Lorenz map must therefore be nondifferentiable: discontinuous and highly fragmented[2].

---

[2] The speculation that the Lorenz map may be fractal was made by Jan Frøyland in 1985 (unpublished).

It is commonplace to emphasize that the addition of nonlinear terms to linear equations can lead to deterministic chaos, but a viewpoint that implicitly takes linear differential equations as the starting point for discussions of nonlinear dynamics is not useful: most of the textbook problems of classical mechanics are conservative and integrable but are *nonlinear*. A better starting point is to begin with nonlinear integrable systems, especially with nonlinear conservative ones, and then add dissipation and driving. The addition of linear damping and driving to nonlinear conservative systems can lead to attractors and even to chaos, as is illustrated by the Lorenz model.

13.6 Liapunov exponents, geometry, and integrability

The one dimensional flow $dx/dt = -x$ has a negative Liapunov exponent $\lambda = -1$ in forward integration and a positive Liapunov $\lambda = 1$ in backward integration. Dissipative flows toward zero and one dimensional sinks have analogous Liapunov exponents, which explains why *regular* flows like the Lorenz model with $\rho < \rho_c$ *cannot* be integrated backward to recover initial conditions except over excessively short time intervals, whenever floating-point arithmetic is used. The abstract Lorenz model is *precisely* time reversible: $x_{i_0} = U(-t)x_i(t)$, as is true of every flow, whether conservative or driven-dissipative.

Positive Liapunov exponents alone are not sufficient for chaos and can even be compatible with integrability, as the example above illustrates. Sensitivity with respect to small changes in initial conditions leads to chaos if (i) the motion is uniformly bounded for all times and (ii) there are no regular attractors in the bounded region of phase space.

The usual recipe given in chaos cookbooks for computing Liapunov exponents by long-time floating-point integrations is misleading. A typical flow cannot even be characterized by a single Liapunov exponent (see chapter 14 for easy examples). For an arbitrary flow, Jacobi's equation describing the initial separation of any two nearby trajectories (see Whittaker, 1965, or Jordan and Smith, 1977) is

$$\frac{d}{dt}\delta x_i \approx \delta x_k \frac{\partial V_i(x)}{\partial x_k}. \tag{13.57}$$

At large times one might expect to find that $\delta x_i(t) \approx \Sigma \delta x_i(0)f_k(t)e^{\lambda_k t}$, where $f_k(t)$ is uniformly bounded and the λ_k are Liapunov exponents, but times large enough that $t \gg \lambda_i$ violate the linearization approximation (13.56) severely. The Jacobi equation is *only* valid at times $t \ll \lambda_i$ and initial separations $\delta x_i(0)/x_i(0) \ll 1$ where the ignored nonlinear terms are *negligible*. Liapunov exponents, defined in this way, are at best a *local*, not a global idea. We show in the next chapter that Liapunov exponents generally vary strongly with changes in initial conditions, and that for an arbitrary flow or map, whether chaotic or not, a Liapunov exponent can be constant, at best, for a restricted class of initial conditions.

The question of integrability of a flow with $n \geq 4$ that is confined to a two dimensional manifold imbedded in $n \geq 4$ dimensions cannot be decided independently of the geometry and topology of the manifold to which it is confined. Geodesics on two dimensional surfaces that have enough symmetry to admit global, orthogonal, *holonomic* coordinates q_1 and q_2 are globally describable by a Lagrangian

$L = g_{ik}\dot{q}_i\dot{q}_k/2$. The Lagrangian configuration space is generally curved with metric g. The corresponding Hamiltonian is $H = p_i(g^{-1})_{ik}p_k/2$ and is conserved, and phase space has four dimensions. For integrability, we need a second conservation law. For motion on a two-sphere, a surface of constant positive curvature, the variables are $(\theta,\phi,p_\theta,p_\phi)$ and p_ϕ is conserved. In the case of a surface of constant negative curvature, the negative curvature yields a positive Liapunov exponent. If the surface is compact then the trajectories are bounded and there is no analogous second conservation law that is analytic. The motion is both nonintegrable and chaotic, and was first studied by the French mathematician Hadamard. The initial separation of two neighboring trajectories on a surface of Riemannian curvature K is described locally, but not globally, by the simple differential equation $\ddot{y} + Ky = 0$, which follows from equation (11.89). If $K < 0$ then there is one positive and one negative Liapunov exponent (see appendix 1 in Arnol'd, 1989, and also Fomenko, 1988).

Consider the motion on a billiard table with one ball fixed and another free to move. Removal of the convex region corresponding to the fixed ball introduces the effect of negative curvature in configuration space. The idealized problem can be studied mathematically as a limiting case of geodesic motion on a surface of constant negative curvature (see Sinai, 1970).

Exercises

1. For Euler's top show that $\vec{V} = k\nabla T \times \nabla \vec{M}^2$ gives the right flow, where T and $\vec{M}$ are the kinetic energy and angular momentum. What is k?

2. Verify the argument of the text that, for

$$\dot{x} = xz - y$$
$$\dot{y} = yz + x$$
$$\dot{z} = z^2 - 1,$$

   the unit sphere is an attractor if $\pi/2 < \theta < \pi$, a repeller if $0 < \theta < \pi/2$.

   (a) Show that any solution that starts in the region $r > 1$, $-1 < z < 1$ (or $r < 1$) is confined to that region forever.
   (b) By rewriting the differential equations in cylindrical coordinates, show that the streamlines in the region $-1 < z < 1$ are spirals. Show that the upper hemisphere with $r = 1$ is a repeller and the lower hemisphere with $r = 1$ is an attractor for all initial conditions in $-1 < z < 1$.

3. Show that the transformation

$$x = (R - \cos\phi_2)\cos\phi_1, \quad -\pi \leq \phi_1 \leq \pi$$
$$y = (R - \cos\phi_2)\sin\phi_1, \quad -\pi \leq \phi_2 \leq \pi$$
$$z = \sin\phi_2, \quad R > 1$$

   from Cartesian coordinates (x,y,z) to angular coordinates (ϕ_1,ϕ_2) defines a circular torus (with radius R corresponding to the angle ϕ_1). Derive the differential equations that describe motion on the torus in Cartesian coordinates and also in torus coordinates.

4. For the Van der Pol oscillator

$$\ddot{x} + \varepsilon(x^2 - 1)\dot{x} + x = 0, \, \varepsilon > 0,$$

find approximately the amplitude A of the limit cycle (assume that $T \approx \tau = 2\pi$ if we set $\varepsilon = 0$). Compute also $g'(A)$ and show that the limit cycle is an attractor.

5. Show that the Lorenz model has three equilibria at $(0,0,0)$ and at $(\pm \sqrt{[\beta(\rho - 1)]}, \pm \sqrt{[\beta(\rho - 1)]}, \rho - 1)$.

6. Show that the last equilibrium point to become unstable does so via a Hopf bifurcation at $\rho_n = [\sigma(\sigma + \beta) + 3\sigma]/(\sigma - \beta - 1)$ (see Schuster, 1988, or McCauley, 1993 for the method).

7. Show that, with $R^2 = x^2 + y^2 + z'^2$ where $z' = z - \rho - \sigma_1$ then $\dot{R} = 0$ on the intersection of the sphere with an ellipse given by

$$\sigma x^2 + y^2 + \beta[z' + (\sigma + \rho)/2]^2 = \beta[(\sigma + \rho)/2]^2,$$

and that $\dot{R} > 0$ inside the ellipse but that $\dot{R} < 0$ outside it. (See McCauley, 1993, chapter 2.)

8. Show that the flow defined by $\ddot{y} - y = 0$ is (a) integrable and (b) has Liapunov exponents $\lambda_1 = -\lambda_2 = 1$.

9. Even in the regular regime where $\rho < \rho_c$, the Lorenz model and other regular dissipative systems cannot be integrated backward in time over intermediate to long time intervals to recover initial conditions if floating-point arithmetic is used. To understand this, observe that the simplest flow with a positive Liapunov exponent in forward integration is $\dot{x} - x = 0$ (the flow $\dot{x} = -x$ has a positive Liapunov exponent in backward integration). Integrating forward and backward in time numerically, and using floating-point arithmetic, how far forward in time can you integrate (a) $\dot{x} = x$ and (b) $\dot{x} = -x$ and still recover the first digit of the initial condition? Compare the results of both floating-point integrations with tables of e^t and e^{-t} that are known to several decimal places.

10. Show that the initial conditions $x_{i0} = \psi_i(x(t), t_0 - t) = U(t_0 - t)x_i(t)$ trivially solve equation (13.22).

14

Damped-driven Newtonian systems

14.1 Period doubling in physics

We now give an example of a simple mechanical system where, by making approximations along with a single simplifying assumption, we can derive the algebraic form of the return map. Although the resulting iterated map looks deceptively simple algebraically it yields chaotic motions at long times for large enough values of the control parameter. Depending upon the strength of a certain control parameter, there is only one frequency in the driven system but we shall find that the response to the periodic driving force may be either periodic (frequency locking without a second frequency) or nonperiodic.

Consider the problem of a partially inelastic ball that bounces vertically in a uniform gravitational field, colliding successively with a vibrating table top. Using the y-axis to measure positions of both ball and table top, let y denote the ball's position at time t and let $Y = A \sin \omega t$ describe the location of the table top, whose equilibrium position is at the origin $Y = 0$. In other words, we assume that the table is massive and is rigidly driven: its reaction to the ball is completely ignored. We also assume that $h \gg A$, where $h_n = (y_n)_{\max}$ is the maximum height achieved by the ball after the nth collision with the table. In this case, we can make the approximation that the time between two successive collisions is given by

$$t_n - t_{n-1} \approx \frac{2v'_n}{g} \qquad (14.1)$$

where v'_n is the velocity of the ball at $y \approx 0$ just after the nth collision and v_n is the ball's velocity just before the nth collision. With this approximation, we have $v_{n+1} \approx -v'_n$, and we shall model the inelasticity by following Newton: assume that

$$v'_n - V_n = -\alpha(v_n - V_n) \qquad (14.2)$$

where $V_n = A\omega \cos t_n$ is the table's position at the time t_n of the nth collision and α is Newton's coefficient of restitution (section 1.3). The parameter α, with $0 < \alpha < 1$, measures the degree of inelasticity of each collision: $\alpha = 0$ yields $v'_n = V_n$, which represents a totally inelastic collision where the ball sticks to the table, and $\alpha = 1$ yields the limit of a totally elastic collision where $v'_n = -v_n + 2V_n$. There are two competing effects. With $\alpha < 1$, the ball tends to lose kinetic energy in a collision but if it hits the table as the table rises then there is the tendency of the ball's kinetic energy to increase, especially as the table moves upward and is near to $Y = 0$, where the table's velocity is a maximum. The interaction between these two competing tendencies causes chaotic orbits of the ball in a definite range of control parameter values. In order to determine the control parameter we next write the map in dimensionless form.

Our two dimensional map has the form

$$t_n = t_{n-1} + \frac{2v'_{n-1}}{g} \left.\vphantom{\begin{array}{c} \\ \\ \end{array}}\right\}$$

$$v'_n = \alpha v'_{n-1} + (1 + \alpha)\omega A \cos \omega t_n. \tag{14.3a}$$

Using dimensionless variables $u = 2\omega v'/g$ and $\phi = \omega t$, the map takes on the deceptively simple looking form

$$\phi_n = \phi_{n-1} + u_{n-1} \left.\vphantom{\begin{array}{c} \\ \\ \end{array}}\right\}$$

$$u_n = \alpha u_{n-1} + \gamma \cos \phi_n, \tag{14.3b}$$

so that γ and α are dimensionless control parameters. In what follows, we fix α and vary γ. Taking the (u,ϕ)-space as phase space, we see that, with J the Jacobi matrix, then

$$\det J = \frac{\partial(\phi_n,u_n)}{\partial(\phi_{n-1},u_{n-1})} = \begin{vmatrix} 1 & 1 \\ -\gamma \sin \phi_n & \alpha - \gamma \sin \phi_n \end{vmatrix} = \alpha \tag{14.4}$$

so that elastic collisions are area-preserving while inelastic collisions lead to contractions of blobs of initial conditions in phase space as the discrete time n increases, characteristic of a dissipative system. Next, we deduce the existence of an attractor for the inelastic case. The method of deduction is very similar to that carried out for the Lorenz model (see McCauley, 1993).

Taking $0 < \alpha < 1$ to be fixed in all that follows, note that if

$$|u_{n-1}| < \frac{\gamma}{1-\alpha} \tag{14.5a}$$

then

$$|u_n| < \frac{\gamma}{1-\alpha} \tag{14.5b}$$

as well, so that orbits that start inside the strip $[-\gamma(1-\alpha),\gamma(1-\alpha)]$ remain there forever (self-confinement, or self-trapping through nonlinearity). Note also that if

$$|u_{n-1}| > \frac{\gamma}{1-\alpha} \tag{14.5c}$$

then

$$|u_n| < |u_{n-1}|, \tag{14.5d}$$

which means that iterations that start outside the strip $[-\gamma(1-\alpha),\gamma(1-\alpha)]$ must approach the strip as the time n is increased. The iterations must enter the rectangular strip in finite time and then remain there forever. This suggests the presence of one or more attractors within the strip $[-\gamma(1-\alpha),\gamma(1-\alpha)]$. The main problem in what follows is to discover the number and nature of those attractors, and so we begin with attractors of the simplest sort, those with dimension equal to zero. In this case, the attractor must be a point, a finite number of points, or else a countable set of points.

To start our discovery of attractors, we begin by defining cycles of the map. A 1-cycle is just a fixed point of the map,

$$\phi' = \phi + u \text{ where } u = 2\pi n, \ n = 0, \pm 1, \pm 2,\ldots \left.\vphantom{\begin{array}{c} \\ \\ \end{array}}\right\}$$

$$u = \alpha u + \gamma \cos \phi, \tag{14.6}$$

so that the fixed points are confined to a cylinder in phase space. We shall define cycles of higher order below as the need arises. In a dissipative system a fixed point represents a steady state of the driven system (energy must be continually pumped through the system and dissipated to maintain the steady state) while a cycle of order $n > 1$ is the discrete analog of a limit cycle, which also requires energy flow through the system. A stable n-cycle is an attractor. An unstable one is a repeller. Stable fixed points are the simplest attractors and so we start by looking for them. The physics follows from the balance of the energy input rate with the dissipation rate. To deduce the consequences of this condition requires mathematics, and the mathematical consequences of a steady state can be nontrivial, both algebraically and number theoretically, as we shall soon discover.

The fixed point condition is

$$\left.\begin{array}{c} \phi' = \phi + u \text{ where } u = 2\pi n \\[2mm] \cos\phi = \dfrac{2\pi n(1 - \alpha)}{\gamma} \end{array}\right\} \tag{14.7}$$

and where $u = 2\pi n$ guarantees that $\phi' = \phi$ modulo $2\pi n$, where n is allowed to be any positive or negative integer or zero, so far. However, because $-1 < \cos\phi < 1$ we also obtain the condition that

$$\gamma > \gamma_{\min} = 2\pi n(1 - \alpha), \tag{14.8}$$

otherwise there is no fixed point. In other words, when $\gamma < \gamma_{\min}$ then it is impossible to find a *periodic* orbit ($\gamma \ge 0$ here and in all that follows), because a 1-cycle is an orbit with period zero. We shall see that orbits with higher period follow for higher values of γ. An example of the motion for $0 < \gamma < \gamma_{\max}$ is shown as figure 14.1.

Note also that the integer n cannot be arbitrarily large: the condition $-1 < \cos\phi < 1$ determines $N = n_{\max}$ whenever α and γ are fixed. In what follows, we always will assume that α is fixed while γ is varied by varying the table's frequency ω or amplitude A. Having located all fixed points, we must now determine when they are stable and when they are unstable.

A stable fixed point is an attractor for the motion of the ball whereas an unstable fixed point is a repeller, and fixed points ($u = 2\pi n$) exist only if $\gamma > 2\pi n(1 - \alpha)$, for each value of n. Next, we perform linear stability analysis for fixed points of two dimensional maps. The method applies to any two dimensional map although we stick to the notation of our particular map for the time being.

Consider the motion near, but not on, any fixed point with $u = 2\pi n$ and $\cos\phi = 2\pi n(1 - \alpha)/\gamma$. Writing

$$\left.\begin{array}{c} \phi_n = \phi + \delta\phi_n \\ u_n = u + \delta u_n \end{array}\right\} \tag{14.9}$$

and expanding the map to first order in the small changes about the fixed point (corresponding to one value of n) yields the linear equations

$$\left.\begin{array}{c} \delta X_n = J\delta X_{n-1}, \\ \delta X_n = J^n \delta X_{\circ} \end{array}\right\} \tag{14.10}$$

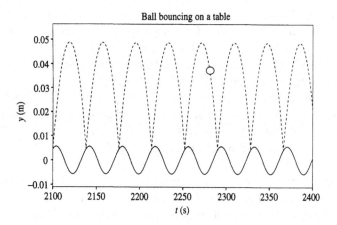

Fig. 14.1 1-cycle for $0 < \gamma < \gamma_{max} = 3.81$.

where δX_o is the initial condition,

$$\delta X_n = \begin{pmatrix} \delta\phi_n \\ \delta u_n \end{pmatrix}, \tag{14.11}$$

and where

$$J = \begin{pmatrix} 1 & 1 \\ -\gamma\sin\phi & \alpha - \gamma\sin\phi \end{pmatrix} \tag{14.12}$$

is the Jacobi matrix evaluated at the equilibrium point under consideration.

If we diagonalize J by solving the eigenvalue problem $Je = \lambda e$, then stability can be discussed in terms of the two eigenvalues μ_1 and μ_2 where $\det J = \mu_1\mu_2 = \alpha < 1$. In particular, stability of a fixed point requires that both eigenvalues have magnitudes less than unity. If either eigenvalue has magnitude greater than unity then the fixed point is unstable. Writing $\mu_i = r_i e^{i\theta}$, there are three possibilities: either both exponents are real, in which case $\mu_1 = \alpha/\mu_2$ with $\mu_1 > 1$ and $\mu_2 < 1$ (or vice versa) and we have a hyperbolic point, or else with $\mu_2 < 1$ we also have $\mu_1 = \alpha/\mu_2 < 1$, and we have a sink. Or, both eigenvalues are complex, in which case $r_1 = r_2 = \alpha^{1/2} < 1$, and we have spiraling flow into a sink. Whenever γ increases enough that one eigenvalue lands on the unit circle, then we have a bifurcation because a further increase in γ will yield an unstable fixed point.

In our model the area contraction condition $\mu_1\mu_2 = \alpha < 1$ rules out all possibilities for 1-cycles except hyperbolic points and sinks. With $\beta = \sin\phi = [\gamma^2 - (2\pi n)^2(1-\alpha)^2]^{1/2}/\gamma$, the eigenvalues are given by

$$\mu_{1,2} = \frac{1 + \alpha - \gamma\beta \pm [(1+\alpha-\gamma)^2 - 4\alpha]^{1/2}}{2} = \frac{x \pm (x^2 - 4\alpha)^{1/2}}{2} \tag{14.13}$$

where $x = 1 + \alpha - \beta\gamma$ and we take λ_1 to have the plus sign. Note that $\mu_2 = \alpha/\mu_1 < \mu_1$, and note also that both eigenvalues are increasing functions of x. Recalling that, for each value of n, the condition that $-1 < \cos\phi < 1$ requires that $\gamma \geq 2\pi n(1-\alpha) = \gamma_{min}$, there are no fixed points for $\gamma < \gamma_{min}$ so that $\gamma = \gamma_{min}$ corresponds to $\phi = 0$ which is the configuration of maximum momentum transfer from the table to the ball. This corresponds to $0 < \mu_2 = \alpha < \mu_1 = 1$. For a slightly greater value of x we therefore have $0 < \mu_2 < \mu_1 < 1$, and the ball has stable periodic oscillations. In other words, for a fixed range of the variable x there are two real eigenvalues in the range $-1 < \mu_2 < \mu_1$ (there is also an intermediate range of γ-values where both eigenvalues are complex) so that the fixed point attracts all iterations. For $\gamma > \gamma_{\min}$ we have $\phi > 0$ and $\mu_1 < 1$, but there is (for each fixed n) a limit: when x is decreased enough (by increasing γ) so that $\mu_2 = -1$, then we find that $\gamma = 2[(2\pi n)^2(1 - \alpha)^2 + (1 + \alpha)^2]^{1/2} = \gamma_{\max}$. Therefore, for $\gamma > \gamma_{\max}$ there is a hyperbolic point rather than a sink.

In other words, the stability region for each fixed point is defined by $\gamma_{\min} < \gamma < \gamma_{\max}$, which corresponds to $0 < \sin\phi < [1 - (2\pi n(1 - \alpha)/\gamma_{\max})^2]^{1/2}$. We can summarize this stability range of the fixed point by the diagram of figure 14.2: stability requires that both eigenvalues lie within the unit circle in the complex λ-plane. Whenever one of the two eigenvalues reaches the circle, as a result of changing the dimensionless control parameter γ, then a further increase causes the sink to disappear. Physically, if you drive the system too weakly there is no fixed point at all, but if you drive it too strongly then every fixed point that arises from a bifurcation as the driving strength γ is increased eventually becomes unstable because the table contacts the ball too much out of phase with the table's point of maximum velocity on the upswing.

The ball has no natural frequency, but in the range $\gamma_{\min} < \gamma < \gamma_{\max}$ the ball's motion is periodic with a period that is an integral multiple of the table's frequency. In other words, we have a kind of frequency locking with only one frequency on hand initially, the frequency ω of the table. For example, the return time of the ball (between successive collisions with the table) is $\tau_n = t_n - t_{n-1} = 2v'_{n-1}/g$, so that the fixed-point condition (14.7) can be rewritten as $\tau = 2\pi n/\omega$, which says that the ball's motion is locked into periodicity with a period τ that is n times the table's period. However, this locked motion is stable only if the ball's motion is controlled by momentum transfers that occur sufficiently near to the maximum possible momentum transfer, otherwise the locked motion is unstable. Next, we look for the stable cycles of higher order that arise through bifurcations whenever $\gamma \geq \gamma_{\max}$.

For any two dimensional map

$$\left.\begin{array}{l} x_n = f(x_{n-1}, y_{n-1}) \\ y_n = g(x_{n-1}, y_{n-1}) \end{array}\right\} \tag{14.14}$$

the condition for a 2-cycle is

$$\left.\begin{array}{l} x_2 = f(x_1, y_1) \\ y_2 = g(x_1, y_1), \end{array}\right\} \tag{14.15a}$$

and the same with the indices 1 and 2 interchanged, which is just a cycling of the system between two separate points (x_1, y_1) and (x_2, y_2). Another way to say it is that the 2-cycle is a fixed point of the second iterate of the map:

$$\left.\begin{array}{l} x_i = f(f(x_i, y_i), g(x_i, y_i)) \\ y_i = g(f(x_i, y_i), g(x_i, y_i)) \end{array}\right\} \tag{14.15b}$$

holds for both of the two points. A stable 2-cycle is a zero dimensional attractor, an unstable one is a zero dimensional repeller. To determine stability or lack of same, we must look at orbits sufficiently near to the 2-cycle.

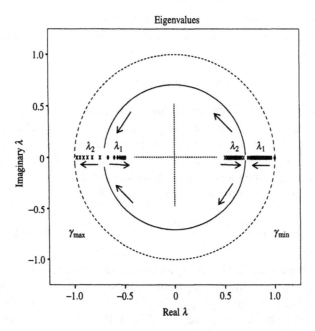

Fig. 14.2 Eigenvalues in the complex γ-plane as a function of γ.

Setting $x_n = x + \delta x_n$ and $y_n = y + \delta y_n$, where (x,y) is either of the two points on the 2-cycle yields

$$\left.\begin{aligned}\delta X_n &= J\delta X_{n-1}\\ \delta X_n &= J^n \delta X_\circ\end{aligned}\right\} \tag{14.16}$$

where

$$\delta X_n = \begin{pmatrix} \delta\phi_n \\ \delta u_n \end{pmatrix} \tag{14.17}$$

and the matrix J is given by

$$J = J_1 J_2 = \begin{pmatrix} 1 & 1 \\ -\gamma\sin\phi_1 & \alpha - \gamma\sin\phi_1 \end{pmatrix}\begin{pmatrix} 1 & 1 \\ -\gamma\sin\phi_2 & \alpha - \gamma\sin\phi_2 \end{pmatrix} \tag{14.18}$$

and is just the product of two Jacobi matrices evaluated at the two points on the 2-cycle. Denoting the eigenvalues of the matrix J by μ_1 and μ_2 where $\mu_1\mu_2 = \alpha^2$, we have a stable 2-cycle if neither eigenvalue has a magnitude greater than unity, instability if either eigenvalue has magnitude greater than unity, and bifurcations occur whenever either eigenvalue reaches the unit circle as a result of changing γ. We now look for the conditions that a 2-cycle exists and is stable.

We can write the 2-cycle equations in the form

$$\left.\begin{aligned}\phi_2 &= \phi_1 + u_1 = \phi_\circ + u_1 + u_2\\ u_1 &= \alpha u_2 + \gamma\cos(\phi_\circ + u_2)\\ u_2 &= \alpha u_1 + \gamma\cos(\phi_1 + u_1)\end{aligned}\right\} \tag{14.19}$$

where ϕ_o and ϕ_2 are the same modulo 2π. This requires that $u_1 + u_2 = 2\pi k$, where, at this stage, k can be either zero or any positive or negative integer. We can simplify these equations a bit more: substituting the u_1 equation into the u_2 equation yields

$$\gamma \cos \phi_o = u_2(1 + \alpha) - 2\pi\alpha k. \tag{14.20}$$

Next, we note that the 2-cycle equations must hold trivially for the 1-cycle. If we insert the fixed point solutions $u = 2\pi n$ and $\cos \phi_o = 2\pi n(1 - \alpha)/\gamma$ into (14.20), then we obtain $k = 2n$, which says that k can only be a positive or negative integer with magnitude $2,4,6,\ldots$. In particular, if we should find that a nontrivial 2-cycle grows out of the bifurcation at $\gamma = \gamma_{max}$, then $k = 2n$ must hold for that 2-cycle.

Finally, if we substitute into the equation for u_1 above, then we obtain

$$\gamma \cos\left(\phi_o + \frac{\gamma \cos \phi_o + 4\pi n\alpha}{1 + \alpha}\right) = -(\gamma \cos \phi_o + 4\pi n\alpha) + 4\pi n, \tag{14.21}$$

which can be solved numerically for ϕ_o. At this stage, we must begin to quote results rather than derive them. Taking $\cos \phi_o$ as the variable to be solved for, there is only one solution, the stable fixed point if $\gamma_{min} < \gamma < \gamma_{max}$. If $\gamma > \gamma_{max}$, then there are three solutions: one corresponding to the unstable cycle or order one, and two more corresponding to a nontrivial 2-cycle (i.e., to $\cos \phi_o$ and also to $\cos \phi_1$). Further algebra shows that the 2-cycle is stable if $\gamma_{max} < \gamma < \gamma'$. A stable 2-cycle is shown in figure 14.3.

In the literature (see Holmes, 1982), it is shown that the bouncing ball map has an infinite sequence of pitchfork bifurcations, also called period-doubling bifurcations: for each fixed value of n, at corresponding parameter values $\gamma_{m-1} < \gamma < \gamma_m$, a stable cycle of order 2^m exists and cycles of orders $1,2,\ldots,2^{m-1}$ persist but are all unstable. At $\gamma = \gamma_m$, the 2^m cycle also becomes unstable and a stable cycle of order 2^{m+1} appears (in this notation, $\gamma_{min} = \gamma_1, \gamma_{max} = \gamma_2$, and $\gamma' = \gamma_3$). This bifurcation sequence continues until, at a critical point γ_∞, all cycles of order 2^m are unstable, for $m = 1,2,3,\ldots,\infty$. Each branch in the bifurcation diagram (figure 14.4) gives rise to a complete binary tree qualitatively like the one shown in figure 14.5 for the period-doubling sequence of the logistic map. For $\gamma > \gamma_\infty$ chaotic orbits with positive Liapunov exponents can occur. Results of some numerical computations for the bouncing ball map are shown as figure (14.6).

Universal scaling near the period-doubling critical point was discovered for a class of one dimensional maps by Feigenbaum (1980), who argued (i) that period doubling is ubiquitous in mechanical systems and (ii) that, with enough dissipation, universal behavior shown by a class of one dimensional maps may also appear in higher dimensional systems, like the bouncing ball model.

In order to explain period doubling and how it can lead to chaos, we fall back on one dimensional maps $x_n = f(x_{n-1})$ that peak once in the unit interval (unimodal maps). Note that self-confined motion occurs whenever $0 < f_{max} < 1$, because this condition guarantees that $0 < x_n = f(x_{n-1}) < 1$ so long as $0 < x_{n-1} < 1$. In particular, if we consider the logistic map $f(x) = Dx(1 - x)$, where $f_{max} = D/4$, then self-confined motion on the unit interval (the phase space) occurs so long as $0 \leq D \leq 4$. Like the tent map (chapter 6), the logistic map has a two-valued inverse; the reason why multivaluedness of the map's inverse is necessary for chaos in one dimension is shown implicitly in the next section by the method of backward iteration.

The logistic map can be understood as a simple-minded population growth model based upon two competing effects: if x_n denotes the fraction of some species living at

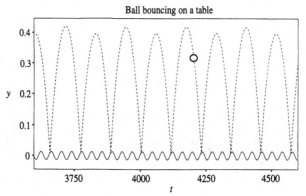

Fig. 14.3 Stable 2-cycle for $k = 3$, $\alpha = 0.8$, and $\gamma > 5.21$ ($A = 0.013$).

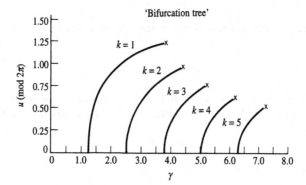

Fig. 14.4 First branches of the bifurcation trees.

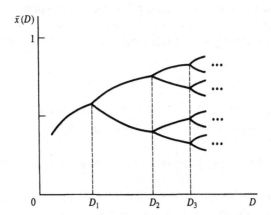

Fig. 14.5 Period-doubling bifurcation diagram.

time n, then births are represented by the term Dx_n and deaths by $- Dx_n^2$. As we shall see, even this simple-minded model leads to chaotic dynamics, dynamics where all orbits are unstable and unstable nonperiodic orbits with 2^∞ points appear. The point to be understood is that the orbits of the logistic map have the complexity of the orbits of the bouncing ball model, but some aspects of the logistic map are a little easier to handle both algebraically and numerically.

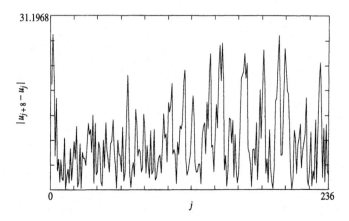

Fig. 14.6 Time series for chaotic bouncing ball.

The condition for a cycle of order N is that the mth iterate of the map has a fixed point of order N:

$$x = f^{(N)}(x) = f(f(\ldots f(x)\ldots)), \tag{14.22}$$

where $f^{(n)}$ denotes the nth iterate of the map f. Since the study of motion near a fixed point of order N (near an N-cycle of points $(\bar{x}_2,\ldots,\bar{x}_N)$) yields the linear equation

$$\delta x_n \approx f'(\bar{x}_1)\ldots f'(\bar{x}_N)\delta x_{n-1}. \tag{14.23}$$

Stability of the N-cycle therefore requires that

$$-1 < f^{(N)'}(\bar{x}_1) = f'(\bar{x}_1)\ldots f'(\bar{x}_N) < 1 \tag{14.24}$$

where $\bar{x}_i$ is any one of the N points on the N-cycle. By a stable N-cycle, we therefore mean one that attracts nearby trajectories whereas by an unstable one we mean an N-cycle that repels nearby trajectories. Bifurcations occur as the control parameter D is varied whenever

$$|f'(x_1)\ldots f'(x_N)| = 1, \tag{14.25}$$

which is the condition that the N-cycle lies at a border of stability (is marginally stable).

Since the Liapunov exponent of an N-cycle is given by

$$\lambda = \frac{1}{N}\ln|f^{(N)'}(x_1)| = \frac{1}{N}\ln|f'(x_1)\ldots f'(x_N)| = \frac{1}{N}\sum_{i=1}^{N}\ln|f'(x_i)|, \tag{14.26}$$

a cycle is stable if $\lambda < 0$, marginally stable if $\lambda = 0$ (defining the bifurcation points), and unstable if $\lambda > 0$. A stable cycle is an attractor; every attractor has a negative Liapunov exponent because the negative Liapunov exponent quantifies the mechanism through which nearby orbits are attracted toward the N-cycle. Conversely, $\lambda > 0$ means simply that the N-cycle is a repeller, that orbits that lie near but not on the cycle are repelled by the N-cycle. Positive Liapunov exponents are necessary but not sufficient for chaos: if there are one or more (necessarily unstable) cycles with positive Liapunov exponents, but also at least one stable cycle, a cycle with a negative Liapunov exponent, then arbitrary initial conditions that lie within the basin of attraction of the stable cycle may yield orbits that are not chaotic but rather are attracted by the stable cycle. For the logistic map, the basin of attraction of a stable cycle is almost the entire unit interval. We turn now to the analysis of the logistic map.

It is easy to calculate the first few bifurcations analytically: there are two fixed points f the logistic map, $x = 0$ and $x = 1 - 1/D$, and stability is determined by the condition $- 1 < D(1 - 2x) < 1 : x = 0$ is stable for $0 < D < 1$, unstable otherwise, with Liapunov exponent $\lambda = \ln|D|$. The second fixed point at $x = 1 - D^{-1}$ comes into existence at $D = 1$, is stable for $1 < D < 3$ but persists unstable if $D > 3$. Note that this 1-cycle has a different Liapunov exponent given by

$$\lambda = \ln|D - 2|. \tag{14.27}$$

The Liapunov exponent is negative whenever the 1-cycle is stable, when $1 < D < 3$, and is positive for $D > 3$. In other words, both 1-cycles persist in unstable form beyond $D = 3$ as repellers, a repeller being any orbit with a positive Liapunov exponent. Here, again, these positive Liapunov exponents do not produce any chaos because as soon as D is increased slightly beyond 3 there is another attractor available.

According to Feigenbaum (1980), initial conditions near but not on the second 1-cycle will not quite repeat after one cycle of the motion but will slightly overshoot or undershoot. However, after two cycles of the motion the orbit may close exactly, yielding an attracting 2-cycle.

Therefore, to determine the attractor when $D > 3$, let us search systematically for a 2-cycle, whose equations are given by

$$\left.\begin{array}{l} x_1 = Dx_2(1 - x_2) \\ x_2 = Dx_1(1 - x_1). \end{array}\right\} \tag{14.28}$$

The 2-cycle is stable whenever $|f'(x_1)f'(x_2)| < 1$, which is the same as saying that the Liapunov exponent

$$\lambda = \tfrac{1}{2}\ln|f'(x_1)f'(x_2)| \tag{14.29}$$

for the 2-cycle is negative. An easy calculation of a few pages yields

$$\left.\begin{array}{l} x_1 = [1 + 1/D + (1 - 2/D - 3/D^2)^{1/2}]/2 \\ x_2 = [1 + 1/D - (1 - 2/D - 3/D^2)^{1/2}]/2, \end{array}\right\} \tag{14.30}$$

and therefore that

$$\lambda = \tfrac{1}{2}\ln|2D + 4 - D^2|. \tag{14.31}$$

Note that both the 1-cycles and the 2-cycle all have entirely different Liapunov exponents, which illustrates the very general fact that Liapunov exponents depend strongly upon initial conditions (except in the very special cases of symmetric tent maps and Bernoulli shifts). Note also from (14.30) that the 2-cycle does not exist for $D < 3$ but is born of a pitchfork bifurcation at $D = 3$, just where the fixed point becomes marginally stable (the Liapunov exponents of both the 1-cycle and the 2-cycle vanish at $D = 3$). However, for $3 < D < 1 + \sqrt{6}$ the 2-cycle is stable because it has a negative Liapunov exponent. For $D > 1 + \sqrt{6}$, $\lambda > 0$ for the 2-cycle, and both the 1-cycles and the 2-cycle are repellers rather than attractors.

Further analysis shows that, at $D = 1 + \sqrt{6}$ an attracting 4-cycle is born, and this period doubling continues systematically until a critical parameter value D_c is reached: denoting $D_1 = 3$ and $D_2 = 1 + \sqrt{6}$, there are period-doubling bifurcations at further parameter values $D_3, D_4, \ldots$ where, at D_n a stable 2^n-cycle is born but it becomes unstable at D_{n+1}. In other words, in the 'window of stability' $D_n < D < D_{n+1}$ all cycles of order 2^{n-1} are repellers with (different) positive Liapunov exponents, but there is also an attractor: a 2^n-cycle with a negative Liapunov exponent. That 2^n-cycle is born of a pitchfork bifurcation at D_n but persists as a repeller for $D > D_{n+1}$. At a critical parameter value D_c, all cycles of order 2^n with n finite have become repellers with positive Liapunov exponents, and the 'infinite cycle' with 2^∞ points, the first nonperiodic orbit, has a vanishing Liapunov exponent (this nonperiodic orbit is actually a multifractal, but we are not yet prepared to explain what this means).

It is via period doubling that chaos, in the form of unstable periodicity and unstable nonperiodicity is created systematically in a one-frequency problem: for D slightly greater than D_c, not only do all of the periodic orbits with periods 2^n have positive Liapunov exponents, but the nonperiodic orbit that was marginally stable (with $\lambda = 0$) at D_c now also has a small positive Liapunov exponent. The condition for deterministic chaos is satisfied at D slightly greater than D_c: there is bounded motion with no attractors. All periodic and nonperiodic orbits are now repellers, so that sensitivity with respect to small changes in initial conditions follows for every choice of initial conditions.

The reason why a 4-cycle occurs before a 3-cycle is explained by Sarkovskii's theorem: if a continuous map of the unit interval has a 3-cycle, then cycles of all other integer orders also occur in the following sequence, corresponding to starting around $D = 4$ with the logistic map and then decreasing D systematically to 0 (see Devaney, 1986, for the proof):

$$3 > 5 > 7 > \ldots > 2.3 > 2.5 > 2.7 > \ldots > 2^m.3 > 2^m.5$$

$$> 2^m.7 > \ldots > 2^\infty > \ldots > 2^m > \ldots > 2^2 > 2 > 1. \tag{14.32}$$

This means: as D is increased in the logistic map (or in any other unimodal map), then 1-cycles must occur first, cycles of all orders 2^n follow, and therefore a nonperiodic with 2^∞ points must occur before a cycle of odd order can occur, the 3-cycle occurring last as D is increased. For an arbitrary continuous map of the interval, there need not be even a single cycle (nothing is said here about stability, either), but if a cycle of any order at all occurs, then all cycles listed to the right of that cycle in the Sarkovskii sequence (14.32) must occur as well (the proof of the theorem assumes only that the map is continuous but is not necessarily differentiable). We leave it as a homework exercise for the reader to show that the logistic map has a 3-cycle. Devaney also gives a nice example of a map with piecewise constant slope that has a 5-cycle but has no 3-cycle.

Near the period-doubling critical point (the transition to chaos), where the logistic map's orbit is nonperiodic with 'period' 2^∞, there are universal exponents that describe the transition to the region where unstable nonperiodicity is possible. Those exponents are analogous to universal critical exponents in the theory of phase transitions in equilibrium statistical mechanics and quantum field theory. As Feigenbaum has shown (1980), all one dimensional maps with self-confined motion and a quadratic maximum (for the logistic map, $f \approx f_{max} - c(x - 1/2)^2$ near $x = 1/2$) have the same critical exponents. Maps with maxima of other orders define other universality classes, and each class has its own critical exponents, in qualitative agreement with the idea of universality classes in the theory of phase transitions. However, this idea of universality relies upon fixed points of maps at the transition to chaos but does not describe any of the topologically universal properties of chaotic orbits of the map that occur for $D > D_\infty$. The treatment of universality properties at the transition to chaos is based upon the renormalization group idea and is beyond the scope of this text. For the study of universality at the transition to chaos, we refer the reader to the extensive literature on the subject, in particular, see the informative Los Alamos Science paper by Feigenbaum, and also the very readable review article by Hu (1987) (see also Schuster, 1988, or McCauley, 1993).

For the logistic map, just beyond the period-doubling critical point D_∞, nonperiodic orbits with positive Liapunov exponents appear. However, as D is increased 'windows of stability' must also occur: the periodic orbits in the Sarkovskii sequence (14.32) that are not present for $D < D_\infty$ appear via bifurcations, and are stable in a small range of parameter values of D. All of these 'windows of stability' first appear and then disappear via bifurcations as D is increased from D_∞ to $D = 4$. The last periodic orbit to appear and then go unstable is the 3-cycle. The case where $D = 4$ is very special because the dynamics there again become simple: at $D = 4$, orbits of all possible integer periods exist and are unstable. The dynamics at $D = 4$ are simple because the logistic map with $D = 4$ turns out to be a simple pseudo-random number generator that is conjugate, by a coordinate transformation, to the binary tent map.

Consider the binary tent map,

$$x_{n+1} = \begin{cases} 2x_n, & x_n < 1/2 \\ 2(1 - x_n), & x_n > 1/2. \end{cases} \tag{14.33}$$

If we transform variables according to

$$x_n = \frac{2}{\pi} \sin^{-1} \sqrt{z_n}, \tag{14.34}$$

then the equation $x_{n+1} = 2x_n$ becomes

$$\frac{2}{\pi} \sin^{-1} \sqrt{z_{n+1}} = \frac{4}{\pi} \sin^{-1} \sqrt{z_n}, \tag{14.35}$$

or

$$z_{n+1} = [\sin(2 \sin^{-1} \sqrt{z_n})]^2. \tag{14.36}$$

Using $\theta = \sin^{-1} \sqrt{z}$, $\sin\theta = \sqrt{(z/1)}$, and $\cos\theta = \sqrt{(1 - z)/1}$, yields $\sin 2z = 2 \sin z \cos z$, or

$$z_{n+1} = 4z_n(1 - z_n), \tag{14.37}$$

which is a logistic map. Periodic orbits of the tent map occur with all possible periods for rational initial conditions, and these orbits transform into period orbits of the logistic map with irrational initial conditions (the period of an orbit is an example of a topologic invariant). Nonperiodic orbits of the tent map occur with every irrational initial condition, and these orbits transform into nonperiodic orbits of the logistic map with both rational and irrational conditions (every periodic orbit of the logistic map starts at an irrational number, and every rational initial condition yields a nonperiodic rbit). Since the Liapunov exponent is invariant under differentiable coordinate transformations, almost every orbit of the logistic map has the same Liapunov exponent $\lambda = \ln 2$ even though the slope of the logistic map varies, and even though the logistic map has different Liapunov exponents for different initial conditions for every other value of D (one sees by substitution of $D = 4$ into the formulae above for the Liapunov exponents for one of the 1-cycles and the 2-cycle that $\lambda = \ln 2$ in all three cases). That the 1-cycle $x = 0$ has an exponent $\lambda = \ln 4$, rather than $\ln 2$ at $D = 4$, is allowed by the fact that the transformation from the tent map to the logistic map is not differentiable at $x = 0$.

It was first noted by von Neumann that the logistic map would make a good pseudo-random number generator on a computer were it not for the fact that roundoff/truncation error in floating-point arithmetic destroys that possibility in practice. In general, the forward iteration of a chaotic one dimensional map on a computer in the floating-point mode does not generate correct orbits of the map because a chaotic map magnifies the roundoff/truncation error exponentially with each iteration. The answer to the question of how to compute chaotic orbits correctly, with controlled precision, was described in chapter 6 and is described briefly again in this chapter. In contrast to the numerically unstable procedure of forward iteration of chaotic maps on a computer, there are universal properties and universality classes *within the chaotic regime* that follow from topological invariance. The topologic properties can be discovered by a combination of analytic and numerical methods, especially by discovering the hierarchy of unstable periodic orbits.

Again, we stress that while the 1-cycles and 2-cycle have different Liapunov exponents for arbitrary values of D, at $D = 4$ all orbits have $\lambda = \ln 2$, as is required by conjugacy with the binary tent map except for the singular orbit at $x = 0$. This equality of exponents for (nearly) all values of initial conditions is a singular phenomenon that is not reflected by typical chaotic dynamical systems, where Liapunov exponents generally vary from orbit to orbit with changes in initial conditions. In other words, a dynamical system does not have merely 'a' Liapunov exponent, but has instead an entire spectrum of exponents. We will emphasize this aspect of chaotic dynamics again in the next section.

Summarizing, period doubling represents one possible route to chaos via a definite bifurcation sequence: for $D_n < D < D_c$ all orbits with periods 2^{n-1} are unstable with positive Liapunov exponents, but there is no chaos at long times there because stable attractors with periods from 2^n to 2^∞ still exist. At $D = D_c$ the Liapunov exponent for the first orbit with 2^∞ points vanishes, and for $D > D_c$ all orbits with periods $2^n, n = 1, 2, 3, \ldots, \infty$ have positive Liapunov exponents, as do the nonperiodic orbits that are created for $D \geq D_c$, so that the motion is necessarily chaotic because it is bounded, but with no attractor available to capture the iterates (except when D falls into a window of stability of an orbit with period other than 2^n). We expect that period doubling is the typical path to chaos in an oscillatory system with only one frequency. In a flow defined by differential equations period doubling can take the form of successive doublings of the period of the limit cycles of a dissipative equation (see Feigenbaum's discussion of the Duffing equation (1980) as an example).

We did not discuss the critical orbit of the logistic map, the nonperiodic orbit with vanishing Liapunov exponent that occurs at $D = D_\infty$. That orbit is multifractal but the multifractal property requires a study of dynamics near the transition point. We turn next to simpler dynamical systems that generate fractal and multifractal orbits in phase space in order to show how different classes of initial conditions generate both spectra of Liapunov exponents and corresponding spectra of fractal dimensions. At the end of that discussion, we shall return briefly to the fractal orbit generated by the logistic map at criticality.

14.2 Fractal and multifractal orbits in phase space

If we were to continue with either the bouncing ball model or the logistic map then in the limit of infinitely many period doublings we would arrive at an orbit consisting of 2^∞ points, as would be indicated by the continuation of the bifurcation diagram shown as figure 14.5. This orbit, we call the critical orbit, because it occurs for the limiting value D_∞ of D and is the first nonperiodic orbit to occur as D (or γ) is increased. At D_∞ the complete invariant set (the closure of all possible orbits at infinite time) includes the critical orbit of infinitely many period doublings along with all periodic orbits with periods 2^n, and organizes itself naturally onto a complete binary tree (orbits with periods $2, 2^2, 2^3, \ldots, 2^\infty$ are all present). By writing the natural numbers between zero and one as binary strings, the continuum of the unit interval can also be organized onto a complete binary tree. Therefore, the critical orbit has 'as many' points as does the continuum in the sense of 'cardinality' defined by the mathematician Cantor (1845–1918). In fact, the critical orbit defines a Cantor set, a set of 2^∞ points that can be approximated, to within any desired finite precision, by an unstable 2^n cycle at D_c. As we shall later see, this possibility to describe the critical orbit, to within finite precision, by the hierarchy of unstable orbits with periods 2^n, gives us a way to estimate the fractal dimension of the critical orbit. First, we must define 'fractal dimension'.

We can think of a Cantor set as any set that has 'as many' points as the continuum (meaning the same cardinality), but such a set does not necessarily take up any space on the real line. To explain this, and to introduce the idea of fractal dimension in its simplest terms, we now turn to uniform Cantor sets that are generated by much simpler dynamical systems than the bouncing ball model or the logistic map (as we shall see below, nonuniform Cantor sets are also generated by the logistic map whenever $D > 4$).

Examples of Cantor sets can be constructed algorithmically by starting with the continuum and systematically deleting finite parts of it. The simplest example is the middle thirds Cantor set: start with the unit interval and remove the open middle third interval. The Cantor set is covered by the two closed intervals $[0,1/3]$ and $[2/3,1]$ and the interval end points 0, $1/3$, $2/3$, 1 belong to the Cantor set. Next, remove the open middle thirds of the two remaining closed intervals: the Cantor set is still covered by the four closed intervals $[0,1/9]$, $[2/9,1/3]$, $[2/3,7/9]$, and $[8/9,1]$, and consists of the end points $0, 1/9, 2/9, 1/3, 2/3, 7/9, 8/9, 1$ as well as infinitely more points that are still covered by the four closed intervals. At the nth stage of recursive construction, the Cantor set is covered by 2^n closed intervals, each of length $l_n = 3^{-n}$, and consists of all end points of those intervals plus infinitely many points that still lie beneath the 2^n intervals.

Notice at this stage that we can write down an exact scaling law: if we denote as N_n the number of intervals of size l_n in generation n needed to cover the Cantor set, then $N_n = l_n^{-D}$ where $D = \ln 2/\ln 3$, for $n = 1, 2, 3, \ldots$. The number D is called the fractal dimension (Hausdorff dimension) of the Cantor set, which is the set of all end points and limits of infinite sequences of end points that follow from taking the limit of infinitely many open interval removals as prescribed above.

The middle thirds Cantor set has no length: $N_n l_n = l_n^{1-D}$ goes to zero as n goes to infinity because $D < 1$. However, there is a sense in which this set has 'as many' points as the continuum: consider the numbers from 0 to 1 written as binary strings. These numbers can be thought of as occupying all of the branches of a complete binary tree. Next, note that the intervals that cover the middle thirds Cantor set also lie on a complete binary tree; there are 2^n intervals in each generation n, and these intervals give rise to 2^∞ points in the Cantor set as n goes to infinity (Cantor set points are represented by the end points of the closed intervals shown as solid bars in figure 14.7). Both the Cantor set and the continuum consist of '2^∞' points (this is Cantor's idea of cardinality: both sets have the same cardinality, but a set with cardinality 10^∞ is not qualitatively different from one with cardinality 2^∞, because numbers in the continuum can be expanded in either base ten or base 2, or in any other base). In contrast, the rational numbers can be put into one-to-one correspondence with the integers and so have only a cardinality of ∞. The dimension of each rational is zero (the dimension of a point is zero because a point has no size), and the dimension of the rationals is also zero because the dimension of any countable point set is zero. It is easy to prove that any countable point set (like the rationals) has no size: simply cover each rational by an interval of size $\varepsilon/2^n$, where $n = 0,1,2,\ldots$ represents the ordering of the countable set according to the integers (the set can be put into one-to-one correspondence with the nonnegative integers $n = 0,1,2,3,\ldots$). The size of the set is therefore less than

$$\sum_{n=0}^{\infty} \varepsilon/2^n = 2\varepsilon, \tag{14.38}$$

which can be made as small as we like since ε is arbitrary. A countable point set like the rationals takes up no space at all in the continuum although between any two numbers on the real line, no matter how closely spaced, there are always infinitely many rational numbers. The middle thirds Cantor set is nonintuitive: unlike the rationals, it has as many points as the continuum but, like the rationals, it has no size at all. The mathematical idea of the continuum is itself nonintuitive, except perhaps to a mind like Cantor's, and is not the same as the physicist's intuitive idea of continuity based upon geometry.

In physics we are not concerned with the mathematical continuum because infinite precision in nature (in observation and in experiment) is both impossible and meaningless (which is why we cannot trust the predictions of 'measure theory'), so what does a physicist mean by a fractal? By fractal, we mean that there is a scaling law whereby, when you measure (for example) the size of an object in nature (or try to cover a point set representing a Poincaré section of an orbit in phase space) with precision l_n, then the number of rulers (bins, boxes) of size l_n needed to measure the size of the object optimally (or to cover the point set optimally) obeys the scaling law $N_n \approx l_n^{-D}$ where D is not an integer. Roughly speaking, the object looks, at resolution l_n, as if it consists of N_n fragments, but the number of fragments that it appears to consist of increases exponentially whenever you increase the resolution l_n of the measurement. Furthermore, this kind of scaling cannot go on forever in nature: both in observations and experiments, fractal scaling can hold at best between two extreme limits $l_{min} \ll l_n \ll l_{max}$ where, in the crudest case, the limits are $l_{min} \approx$ an interatomic spacing and $l_{max} \approx$ the physical extent of the object. An example is provided by the coastline of Norway, whose perimeter size $L = Nl = l^{1-D}$ obeys fractal scaling with $D \approx 1.28$ for a range of 'ruler sizes' $4\,\text{km} < l < 370\,\text{km}$ (imagine estimating the perimeter of the coastline by placing rulers end to end on a large map).

The middle thirds Cantor set can be generated systematically by backward iteration of the ternary tent map (similar to figure 14.7, but with equal slopes of magnitude $a = b = 3$). Any unimodal map that peaks above unity, like the logistic map $f(x) = 4D(1 - x)$ with $D > 4$ or like any asymmetric version of the tent map or logistic

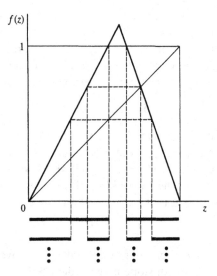

Fig. 14.7 Backward iteration of the unit interval by a tent map with slope magnitudes $a < b$, and $ab > a + b$, generates a Cantor set that organizes itself naturally onto a complete binary tree.

map, will also generate a fractal (a Cantor set) in backward iteration. For any of these maps, we can divide the unit interval into two nonoverlapping sets, the escape set and the invariant set. The escape set consists of all initial conditions x_o such that, for some finite n, $x_n > 1$ so that $x_n = f^{(n)}(x_o)$ goes off to minus infinity as n goes to infinity (each map is extended continuously to minus infinity). For example, for the ternary tent map, if $1/3 < x_o < 2/3$ then $x_1 > 1$ and the remaining iterations escape to infinity. In contrast, the invariant set consists of all possible initial conditions x_o such that $x_n = f^{(n)}(x_o)$ remains in the unit interval as n goes to infinity.

The ternary tent map's invariant set can be constructed recursively by backward iteration of the entire unit interval by the tent map. In one backward iteration, we obtain the two closed intervals $[0,1/3]$ and $[2/3,1]$. Any initial condition x_o that is covered by these two intervals will therefore yield a first iterate that obeys $0 \leq x_1 \leq 1$, but the resulting x_2 is not necessarily still in the unit interval. If we perform two backward iterations of the unit interval then we obtain four closed intervals $[0,1/9]$, $[2/9,1/3]$, $[2/3,7/9]$, and $[8/9,1]$. Any initial condition x_o that lies beneath any of these four intervals will yield $0 \leq x_2 \leq 1$, but the third iterate is not necessarily still confined to the unit interval. Generalizing, if we want to choose an x_o so that x_n is still inside the unit interval after n or fewer iterations, then we can and must choose that initial condition to be any number that lies beneath any one of the $N_n = 2^n$ intervals of size $l_n = 3^{-n}$ that are given by n backward iterations of the whole unit interval by the ternary tent map. The limit where there would be infinitely many backward iterations leads us to infer that the invariant set of the ternary tent map, the set of initial conditions that do not escape to infinity, is the middle thirds Cantor set. In other words, bounded motion of the ternary tent map (or of any unimodal map peaking above unity) takes place on a repelling fractal. The fractal is a repeller because any initial condition that is chosen near to, but not on, the fractal will escape from the unit interval in finite time. The escape time depends upon how near to the fractal one starts the iterations. Another way to see that the fractal is a repeller (a 'strange repeller') is to say that each orbit on (and off) the fractal invariant set has a positive Liapunov exponent $\lambda = \ln 3$, so that particular bounded orbit is a repeller.

Note that we can write the interval sizes as $l_n = e^{-n\lambda}$ where $\lambda = \ln 3$ is the Liapunov exponent of every orbit of the ternary tent map. This formula can be deduced from a more general standpoint. Consider the error propagation condition

$$\delta x_n \approx e^{n\lambda(x_o)}\delta x_o \qquad (14.39)$$

where $\lambda(x_o)$ is the Liapunov exponent for the trajectory starting at x_o. The maximum possible error (where the phase space is the unit interval or a subset thereof) is $\delta x_n \approx 1$ (more precisely, the maximum error is $\delta x_n \approx 1/\mu$ if we compute in base μ arithmetic). If the initial condition lies under the interval l_n, and is known or specified to precision l_n, then this is the same as saying that $\delta x_o \approx l_n$. The result is

$$l_n \approx e^{-n\lambda(x_o)}, \qquad (14.40)$$

which is exact for all symmetric tent maps that peak at or above unity ($l_n = a^{-n} = e^{-n\lambda}$ with $\lambda = \ln a$ for a tent map with slope magnitude $a \geq 2$).

It is impossible to compute bounded trajectories near a repelling fractal using forward iterations with floating-point arithmetic on a computer. Determining the intervals by backward iteration is a convergent process, however, because backward iterations contract intervals. Another way to say it is that one dimensional maps that expand intervals in forward iteration have a negative Liapunov exponent in backward iteration. Therefore, one can always determine the interval sizes and positions as accurately as one wishes by using floating-point arithmetic and backward iteration. We illustrate next that it is also possible to compute exact orbits, to within any desired precision, on a repelling fractal.

If we study the ternary tent map in ternary arithmetic, expanding all iterations in the form

$$x_n = \sum_{i=1}^{\infty} \frac{\varepsilon_i(n)}{3^i} = 0.\varepsilon_1(n)\varepsilon_2(n)\cdots\varepsilon_N(n)\cdots, \qquad (14.41)$$

where $\varepsilon_i(n) = 0, 1,$ or 2, then the ternary tent map can be rewritten as an automaton for the ternary digits:

$$\varepsilon_i(n) = \begin{cases} \varepsilon_{i+1}(n-1) & \text{if } \varepsilon_1(n-1) = 0 \\ 2 - \varepsilon_{i+1}(n-1) & \text{if } \varepsilon_1(n-1) = 1. \end{cases} \qquad (14.42)$$

This follows easily from using $1 - x_n = 0.(2 - \varepsilon_1(n))(2 - \varepsilon_2(n))\cdots(2 - \varepsilon_N(n))\cdots$ because $1 = 0.22222\cdots2\cdots$ in ternary arithmetic, just as $1 = 0.999\cdots9\cdots$ in decimal arithmetic. The advantage of using this procedure is that one sees precisely how the 'far away' digits in the initial condition are propagated in finite time into the leading digits of higher iterations, which property is the vary hallmark of deterministic chaos (sensitivity with respect to small changes in initial conditions).

The next point is: by using the replacement rules $0.1000\cdots = 0.02222\cdots$, $0.0100\cdots = 0.00222\cdots$, and so on, the points on the middle thirds Cantor set can be seen to consist of all possible ternary numbers that can be constructed using only 0's and 2's (1's are not allowed, except by the replacement rules). It is easy to see by examples that all of the end points can be written as strings using only 0's and 2's: $1/9 = 0.001 = 0.000222\cdots$, $7/9 = 2/3 + 1/9 = 0.021 = 0.020222\cdots$, and so on. Hence, end points of the Cantor set consist of finite binary strings with only 0's and 2's. Note that with only 0 and 2, we have a binary rather than a ternary alphabet, corresponding geometrically to the fact that all points in the Cantor set are covered either by the left interval $[0,1/3]$ or by the right interval $[2/3,1]$.

However, all of the orbits of the map starting from end points taken as initial conditions hop onto the unstable fixed point at $x = 0$ in finite time. Unstable periodic orbits with all possible integer periods follow from rational numbers that are infinite periodic binary strings using 0's and 2's. For example, $x_\circ = 0.0202\cdots$ yields $x_1 = 0.2020\cdots$ and $x_n = 0.2020\cdots$ for $n \geq 2$, so that the string $0.2020\cdots$ must be the ternary expansion for the unstable fixed point $x = 3/4$. However, the choice $x_\circ = 0.002002\cdots$ yields $x_1 = 0.02002002\cdots, x_2 = 0.200200\cdots, x_3 = 0.220220\cdots, x_4 = 0.0200200\cdots = x_2$, which is an unstable 3-cycle. Cycles of all possible orders can be constructed systematically by inventing different algorithms for period binary strings (the binary expansion of every rational number provides just such an algorithm).

The periodic orbits with cycle numbers greater than one are the points of the Cantor set that are rational limits of infinite sequences of end points. For example, $2/9 = 0.02000\cdots$, and every truncating string in 0 and 2 is an end point, so that an end point sequence starting with 0.02, $0.022000\cdots$, $0.0220200.$, $0.0220202000.$, $0.02202022000\cdots$, gives rise to the periodic limit $0.022022022022\cdots$, which is a rational number and is also a periodic point (belongs to a 3-cycle) for the ternary tent map on the Cantor set.

Limits of sequences of end points also give rise to irrational numbers on the Cantor set (like every set with 2^∞ elements, most of the points on the Cantor set are irrational), and these numbers lie on (and define) the unstable periodic orbits of the ternary tent map. For example, the initial condition (constructed by an obvious algorithm: if n 0's precede a 2, then $n + 1$ 0's follow that 2 before the next 2 occurs) $x_\circ = 0.02002000200002000002\cdots$ belongs to the Cantor set and gives rise to an unstable periodic orbit of the tent map. By computing this initial condition to high enough precision, we can compute the tent map's corresponding orbit to any desired precision on the repelling fractal via the automaton (14.42). In order to know the tent map's orbit to one decimal precision for n iterations, for example, we must compute the first n ternary digits of $x_\circ$. This is the same as saying that we have one digit per iteration information transfer in base three because the Liapunov exponent is $\lambda = \ln 3$. The generalization of this idea to one dimensional maps with variable slope that peak at or above unity is easy: one computes slopes and symbol sequences by backward iteration (see McCauley, 1993). This method leads to the study of universal properties of classes of chaotic maps. For higher dimensional maps, or for maps of the unit interval that peak beneath unity, instead of iterating a map forward in time on a computer, which only leads to error buildup after very short times, the right approach is to extract the unstable periodic orbits of the dynamical system systematically. This method also leads to the study of universal properties of classes of dynamical systems, because the periods of orbits and their symbol sequences are topologic invariants (see the references by Cvitanovic et al., 1985).

The fractal dimension $D = \ln 2/\ln 3$ of the middle thirds Cantor set is the dimension of the set of all possible initial conditions in $[0,1]$ that yield bounded motion of the ternary tent map. In other words, $D = \ln 2/\ln 3$ is the dimension of a set that consists of all possible bounded orbits, both periodic and nonperiodic, of the tent map. We have a single dimension because there is only one Liapunov exponent $\lambda = \ln 2/\ln 3$ for every orbit. The logistic map $f(x) = 4x(1 - x)$ at $D = 4$ has only a single Liapunov exponent because of conjugacy to the binary tent map. In general, as we saw above for the logistic map with D unequal to 4, a map with variable slope does not have merely one Liapunov exponent but has instead different Liapunov exponents for different classes of initial conditions. We show next that, whenever there is a spectrum of Liapunov orbits, then there is also a corresponding spectrum of fractal dimensions.

First, as a prelude to what follows, notice that the middle thirds Cantor set exhibits an exact geometric self-similarity. If you pick out of figure 14.5 any nth generation sub-tree for the case where both slopes are equal to the same number $a > 2$, the one that grows from a single interval of size $l_n = a^{-n}$, and then rescale all intervals on that sub-tree by multiplying them by the same factor a^n, you obtain an exact replica of the entire original binary tree of 2^∞ intervals. This is the same as saying that the relation $N_n l_n^D = 1$ is exact for all $n = 1, 2, \ldots$. *It is also the same as saying that the entire fractal is generated recursively by the iteration of a single largest length scale* $l_1 = 1/3$. Or, it is the same as saying that the ternary tent map has only one Liapunov exponent, independently of the choice of initial conditions.

Whenever (as is typical of both dynamical systems and nature) we encounter a fractal that does not possess exact geometric self-similarity, but a more complicated sort of self-similarity, then the relation $N_n l_n^D = 1$ must be generalized to allow for nonuniformity of the hierarchy of intervals that, in each generation, cover the fractal optimally,

$$\sum_{i=1}^{N_n} l_i^{D_H} \approx 1, \tag{14.43}$$

where $\{l_i\}$ is the optimal partitioning, the partitioning generated by the dynamical system (backward iteration for one dimensional maps) in generation n, and D_H is called the Hausdorff dimension ($D = \ln 2/\ln 3$ is therefore the Hausdorff dimension for the middle thirds Cantor set). If you choose a partitioning other than the optimal one, then you will always overestimate D_H.

However, the Hausdorff dimension that follows from an optimal partitioning hierarchy that is made up of fundamentally nonuniform intervals is only one point in an entire spectrum of fractal dimensions. Returning to equation (14.40) where $l_i \approx e^{-n\lambda}$, we see that nonuniform partitionings are generated by variable Liapunov exponents. Any fractal that is generated recursively by the iteration of two or more largest length scales is called a multiscalefractal, or just 'multifractal' (there are also multifractal probability distributions, but we do not consider them here). The Cantor set at the period-doubling critical point is not so simple as a two-scale multifractal. It is a (more complicated) multifractal, but there is a way, in a fairly good approximation (see McCauley, 1993), to estimate the fractal dimension by using a two-scale approximation to that Cantor set.

To see how one gets a spectrum of fractal dimensions from of a simple example, we need look no further than any asymmetric tent map that peaks above unity (figure 14.7):

$$f(x) = \begin{cases} ax, & x < a/(a + b) \\ b(1 - x), & x > a/(a + b) \end{cases}. \tag{14.44}$$

This map has slopes a and $-b$, and there is an escape set and a complementary invariant set that is a repelling fractal whenever $ab/(a + b) > 1$, which is just a disguised form of the condition $1/a + 1/b < 1$ on the two first generation intervals $l_1 = 1/a$ and $l_2 = 1/b$. In the case where $1/a + 1/b = 1$, then the map peaks at unity and the

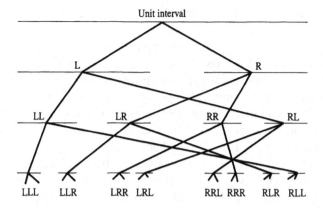

Fig. 14.8 Complete binary tree of symbol sequences generated by backward iteration of unimodal maps peaking above unity.

condition $1/a + 1/b = 1$ is just the condition that backward iteration of the map partitions the unit interval without generating an escape set: every point in the unit interval belongs to the invariant set. If, with $l_1 = 1/a$ and $l_2 = 1/b$, $1/a + 1/b < 1$, then there is an escape set and also a corresponding repelling fractal invariant set. However, in contrast with the middle thirds Cantor set, this fractal is not geometrically self-similar (the fractal is not uniform): if you pick out any infinite sub-tree of the complete binary tree of intervals (see figure 14.7), say, the sub-tree that grows from the second generation interval $1/b^2$, then merely by rescaling every interval on the infinite tree by the same number b^2 it is not possible to obtain an exact replica of the entire original tree. The reason for this is the nonuniformity of the fractal. The fractal is nonuniform because it is constructed recursively (by backward iteration of the tent map) by the iteration of *two* different largest scales: $l_1 = 1/a$ and $l_2 = 1/b$.

Because the tent map has only two slopes a and $-b$ and generates a complete binary tree of intervals in backward iteration, we still have binary combinatorial simplicity: in the nth generation of description of the fractal there are $N_n = 2^n$ intervals and $N_m = n!/m!(n-m)!$ have the same size $l_m = a^{-m}b^{-n+m} = e_m^{-\lambda}$, where

$$\lambda_m = \frac{m}{n}\ln a + \frac{n-m}{n}\ln b \qquad (14.45)$$

is the Liapunov exponent for a trajectory that starts under any interval with size l_m. By necessity, such a trajectory visits the left first generation interval $l_1 = 1/a$ (to which we assign the letter L) m times, and the right first generation interval $l_2 = 1/b$ (denoted by R) $n - m$ times in n forward iterations of the map (different intervals l_m with different LR orderings generate the same exponent, because the exponent is independent of the orderings), where the meaning of the 'symbol sequences' (binary LR-sequences) in figure 14.8 is: if you start with a specific interval l_m and then iterate n times forward, the map will visit the two first generation intervals $l_1 = 1/a$ (L) and $l_2 = 1/b$ (R) in exactly the sequence shown in figure 14.8. In other words, in generation $n = 3$, LRL means that x_o lies under l_1, x_1 lies under l_2, and x_2 lies under l_1, where $l_1 = 1/a$ and $l_2 = 1/b$ denote the two first generation intervals (symbolic dynamics, which we are introducing here, is the generalization to maps with arbitrary slope of the method of studying the ternary tent map in ternary arithmetic).

The symbol sequence orderings shown in figure 14.8 are universal for all unimodal maps that peak at or above unity and contract intervals to generate a complete binary tree in backward iteration (the same maps, peaking beneath unity, generate incomplete trees, define other universality classes, and are a lot harder to analyze). More generally, the allowed symbol sequences of any given map are universal: they are not properties of a specific map but rather are universal topologic properties characteristic of an entire class of maps. This is an extremely important point whose further discussion would take us well beyond the scope of this introductory book (see McCauley, 1993, for the easy case, Cvitanovic et al., 1985, for the harder cases).

We have seen already that two scales means a spectrum of Liapunov exponents: for the asymmetric tent map, $\lambda(x) = x \ln a + (1 - x)\ln b$ with $x(= m/n)$ varying from 0 to 1 has the range $\lambda_{min} \leq \lambda \leq \lambda_{max}$ where (with $a < b$) $\lambda_{min} = \ln a$ corresponds to $x = 0$ and to the symbol sequence $LLL\ldots L\ldots$, and $\lambda_{max} = \ln b$ corresponds to $x = 1$ and the symbol sequence $RRR\ldots$. We show next how this multiplicity of scales gives rise to a spectrum of fractal dimensions.

First, because the two first generation scales determine everything, the condition (14.43) for the Hausdorff dimension deduces to $a^{-D_H} + b^{-D_H} = 1$ for the tent map:

$$1 = \sum_{i=1}^{2^n} l_i^{D_H} = \sum_{m=0}^{n} \frac{n!}{m!(n-m)!}(a^{-m}b^{-n+m})^{D_H} = (a^{-D_H} + b^{-D_H})^n. \qquad (14.46)$$

If $1/a + 1/b = 1$, then $D_H = 1$, but $0 < D_H < 1$ follows if $1/a + 1/b < 1$. Now, very important to note is that, although there is no geometric self-similarity of the entire nonuniform fractal, there are special subsets of our multiscalefractal that do scale exactly because of perfect geometric self-similarity: consider the infinite hierarchy of intervals defined by $l_m = a^{-m}b^{-n+m}$ where the ratio $x = m/n$ is fixed, but where n varies from 1 to infinity (therefore, m also varies to maintain fixed x). For each fixed n, there are N_m such intervals, and these intervals define a fractal dimension $D(\lambda_m)$ according to

$$N_m = l_m^{-D(\lambda_m)}, \qquad (14.47)$$

where λ_m is the Liapunov exponent given by $l_m = e^{-n\lambda_m}$. Remember, the idea is to fix $x = m/n$, while varying m and n. Also, borrowing an idea from Boltzmann, we can define an entropy S by writing

$$S(\lambda_m) = \ln N_m. \qquad (14.48)$$

If we combine the Mandelbrot equation (14.47) with the Boltzmann equation (14.48) by writing $l_m = e^{-n\lambda_m}$, and also define $s(\lambda) = S(\lambda)/n$, then we arrive at the very general relationship

$$D(\lambda) = s(\lambda)/\lambda \qquad (14.49)$$

as the fractal dimension of the set of initial conditions with the same Liapunov exponent λ. What is the meaning of the entropy in this expression? It is simply the logarithm of the number of different symbol sequences that have the same ratio $x = m/n$ regardless of the orderings of the L's and R's within the sequence (to each infinite precision initial condition there corresponds a unique LR-sequence but many different LR-sequences may have the same ratio m/n). So, whenever we write $D = \ln 2/\ln 3$ for the ternary tent map, we can now understand that $\ln 3$ in the denominator represents the Liapunov exponent, while $\ln 2$ in the numerator represents

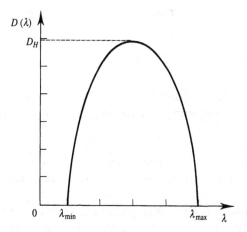

Fig. 14.9 Spectrum of fractal dimensions $D(\lambda)$ vs spectrum of Liapunov exponents λ.

the entropy of binary completeness: there are 2^n ways to form symbol sequences of length n from 2 letters L and R.

Going further, if we make Stirling's approximation whereby $\ln m! \approx m\ln(m/n) + (n - m)\ln[(n - m)/n]$, then we obtain the expression

$$D(\lambda) \approx \frac{- x\ln x - (1 - x)\ln(1 - x)}{x\ln a + (1 - x)\ln b} \tag{14.50}$$

where $\lambda = x\ln a + (1 - x)\ln b$, so that both the D- and λ-spectra are parameterized by x, which varies from 0 to 1. There is only one initial condition of the tent map that generates the symbol sequence LLL...L..., so that $D = 0$ when $x = 0$. Likewise, $D = 0$ for $x = 1$ corresponds to the single sequence RRR...R.... The spectrum of fractal dimensions has the general form shown as figure 14.9, and the maximum of $D(\lambda)$ is given by the condition $x = \ln(a/b^2)/\ln(a/b)$ that follows directly from $D'(\lambda) = 0$. Next, we use a more general argument to show that the maximum value of $D(\lambda)$ is equal to the Hausdorff dimension D_{H}, as $l_m \to 0$.

Writing down the generating function

$$Z_n(\beta) = \sum_{i=1}^{N_n} l_i^\beta \tag{14.51}$$

where β is a free parameter that we can vary from minus to plus infinity at will, we see that whenever $Z_n(\beta) = 1$ then $\beta = D_{\mathrm{H}}$. For the tent map we get

$$Z_n(\beta) = \sum_{m=0}^{n} N_m a^{-m\beta} b^{-(n-m)\beta} = (a^{-\beta} + b^{-\beta})^n = e^{-ng(\beta)} \tag{14.52a}$$

where $g(\beta) = -\ln(a^{-\beta} + b^{-\beta})$. We can also make a largest term approximation as follows:

$$Z_n(\beta) = \sum_{\lambda_m} e^{ns(\lambda)} e^{-n\beta\lambda} \approx e^{n(s - \lambda\beta)_{\mathrm{min}}} \tag{14.52b}$$

where the extremum condition required to minimize the exponent in (14.52b) is given by

$$\beta = \frac{ds}{d\lambda}. \tag{14.53}$$

For a given value of β, this condition determines a value $\lambda(\beta)$ such that we can approximate the entire generating function by a single term. Hence,

$$g(\beta) = \lambda\beta - s(\lambda) \tag{14.54}$$

holds approximately with λ determined by (14.53), and where $g(\beta) = -\ln(a^{-\beta} + b^{-\beta})$. Note that $\lambda = dg/d\beta$ follows from (14.53) and (14.54). Note also that, since $g = 0$ at $\beta = D_H$, it must be true that $D_H = \beta = s(\lambda)/\lambda = D(\lambda)$ for the particular value of λ that determines D_H. Which part of the $D(\lambda)$ spectrum is this? To answer this question, we maximize $D(\lambda) = s(\lambda)/\lambda$. The extremum condition $D'(\lambda) = 0$ yields the condition that $s'(\lambda) = s(\lambda)/\lambda = D(\lambda)$, but since $\beta = s'(\lambda)$ we can then conclude that $\beta = D_H = D(\lambda)_{max}$. In other words, we expect that the peak of the $D(\lambda)$ spectrum is the same as the Hausdorff dimension of the fractal repeller. This argument can be extended to maps other than the tent map, but we do not provide the proof here.

The idea propagated in some texts that a chaotic dynamical system can be characterized by a single Liapunov exponent is very misleading: only the simplest systems, systems with constant slope and systems that are differentiably conjugate to systems with constant slope, can be characterized by a single Liapunov exponent. In general, Liapunov exponents of trajectories of chaotic systems depend strongly on classes of initial conditions, as we have shown for the logistic map in the last section and in this section for the asymmetric tent map. Correspondingly, there is a spectrum of exponents, and this spectrum of exponents generates a spectrum of fractal dimensions of which the Hausdorff dimension is the largest dimension.

As a prelude to devil's staircases in the two-frequency problems of section 14.4, we end this section with the construction of the devil's staircase for the middle thirds Cantor set. In the first stage of iterative construction of the middle thirds Cantor set, we discovered that the end points 3^{-1} and $1 - 3^{-1}$ belong to the Cantor set. For $n = 2$, we discovered in addition the end points 3^{-2}, $3^{-1} - 3^{-2}$, $1 - 3^{-1} + 3^{-2}$, and $1 - 3^{-2}$. These end points and all other end points of the Cantor set can be written as finite length ternary strings using only two symbols, $\varepsilon_i = 0$ or 2. For example, $3^{-1} = 0.1 = 0.022\cdots$ while $1 - 3^{-2} = 0.2200\cdots$, and so on. All of the elements of the Cantor set can be written using the binary symbols $\varepsilon_i = 0$ or 2, and all possible combinations of 0's and 2's describe all the elements in the set. Hence, the Cantor set is in one-to-one correspondence with all possible binary strings, so that the cardinality of the set is 2^∞, which is the same as the cardinality of the continuum. This tells us that 'almost all' of the points in the Cantor set are irrational numbers.

We construct a function called the Cantor function as follows: with $\varepsilon_i = 0$ or 2 we interpret the function defined by

$$P(x) = 0.\frac{\varepsilon_1}{2}\frac{\varepsilon_2}{2}\cdots\frac{\varepsilon_N}{2}\cdots \tag{14.55}$$

as a binary number ($\varepsilon_i/2 = 0$ or 1), where $x = 0.\varepsilon_1\varepsilon_2\cdots\varepsilon_N\cdots$ is an element of the Cantor set ($\varepsilon_i = 0$ or 2). It is easy to see that if x_1 and x_2 are the end points of a removed open interval (part of the escape set of the tent map), then $P(x_1) = P(x_2)$. For

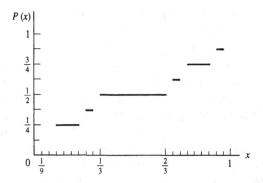

Fig. 14.10 A few steps in the devil's staircase of the ternary tent map.

example, with $x_1 = 3^{-1} = 0.022\cdots$ we get $P(x_1) = 0.01111 = 0.1 = 1/2$. If $x_2 = 1 - 3^{-1}$, then $P(x_1) = 0.1 = 1/2$. If $x_1 = 1/3^2 = 0.01 = 0.00022\cdots$, then $P(x_1) = 0.00111\cdots = 0.01\cdots = 1/4$, whereas $x_2 = 3^{-1} - 3^{-2}$ also yields $P(x_2) = 0.01\cdots = 1/4$. Having $x_1 = 1 - 3^{-1} + 3^{-2}$ yields $P(x_1) = 0.1011\cdots = 0.11\cdots = 3/4$, whereas $x_2 = 1 - 3^{-2}$ also yields $P(x_1) = 0.11\cdots = 3/4$. If we extend the definition of $P(x)$ to include $P(x) =$ constant on the closure $[x_1,x_2]$ of every removed open interval, then $P(x)$ has the following properties: P is continuous but nondecreasing and has a slope that vanishes almost everywhere. The plot of $P(x)$ vs x is called the devil's staircase, and the first few staircase steps are plotted in figure 14.10.

The Cantor function $P(x)$ can be understood as a certain probability distribution for the iterates of the ternary tent map on the middle thirds Cantor set. Consider any nonperiodic orbit of the symmetric tent map that generates, as its distribution of iterates, a uniform probability distribution on the Cantor set: $P_i = 2^{-n}$ on any nth generation interval $l_n = 3^{-n}$ of the tent map (think of the distribution $\{P_i\}$ as a histogram). That the Cantor function represents the uniform distribution can be seen as follows: $P(x) = 1/2$ for $x = 3^{-1}$ means that half of the iterates of the map fall into the interval $[0,3^{-1}]$ and half in $[1 - 3^{-1},1]$. That $P(x) = 1/4$ for $x = 1/3^{-2}$ means $1/4$ of the iterates fall into the interval $[0,1/3^{-2}]$, while $P(x) = 3/4$ if $x = 3^{-1} + 3^{-2}$ means that $3/4$ of the iterates fall into the three intervals $[0,3^{-2}]$, $[3^{-1} - 3^{-2},3^{-1}]$, and $[1 - 3^{-1}, 1 - 3^{-1} + 3^{-2}]$. The remaining $1/4$ land in the interval $[1 - 3^{-2},1]$. The method of construction of the very special (and hard to find) class of initial conditions of the ternary tent map that generates even histograms on the fractal is discussed in McCauley (1993).

Next, note that if you should choose a covering that is too large, then by virtue of $N_n l_n^D \approx 1$, you will overestimate the fractal dimension D, because $l_n = 1/N_n$ yields $D = 1$. Similarly, an underestimate of the covering size will underestimate D: in the extreme case we have $D = 0$ for a countable point set, a set that can be covered by a countable infinity of arbitrarily small intervals. The same argument holds for nonuniform coverings where we must use the generalization of $N_n l_n^D \approx 1$, whereby

$$\sum_{i=1}^{N_n} l_i^{D_H} \approx 1. \tag{14.56}$$

We now return to the nonperiodic orbit of the logistic map at the transition to chaos. That orbit has 2^∞ points and is multifractal. Although the nonperiodic orbit at criticality (at $D = D_c \approx 3.57$) has a vanishing Liapunov exponent, it is born as a result of the orbits with periods $2, 2^2, \ldots, 2^n, \ldots$, all of which have positive Liapunov exponents at $D = D_c$. We can try to use the periodic orbits systematically to estimate the fractal dimension of the critical nonperiodic orbit. Consider the 4-cycle at $D_c \approx 3.57$. We can locate that cycle approximately as follows: if we start the iterations near the unstable fixed point $x = 1 - 1/D_c \approx 0.72$, they will stay there for a while and will then diverge, but the next path that they follow is one near the 2-cycle. After landing near the 2-cycle for a while, the iterates will diverge and will land near the unstable 4-cycle. In this way, we can locate approximately the points on the 4-cycle, and also the points on the 8-cycle and all cycles of higher order. We find the 4-cycle is given approximately at D_c by $x_1 = 0.89$, $x_2 = 0.35$, $x_3 = 0.81$, and $x_4 = 0.55$. We know, by completeness of the binary tree of period doubling, that we need 2^n intervals in generation n to cover the Cantor set of the nonperiodic orbit at criticality. Another way to say it is: if you measure the points on the critical orbit with crude enough resolution, then the observed motion will look like a nonperiodic bouncing around in 2^n bins (you can see the nonperiodicity corresponding to 2^∞ points only by resolving the orbit with infinite precision over infinite times, both of which are completely impossible). In particular, we can take the two first generation intervals to be defined by the 4-cycle: $l_1 = x_4 - x_2 \approx 0.2$ and $l_2 = x_1 - x_3 \approx 0.08$. Insertion of these two scales into

$$l_1^{D_H} + l_2^{D_H} \approx 1 \tag{14.57}$$

yields $D_H \approx 0.34$, which is much less than the known estimate of $D_H \approx 0.54$ that follows from a higher order analysis. This tells us that the two intervals yielded by the 4-cycle are too small to cover the Cantor set, that the points on the 2^∞-orbit are much more dispersed than our crude estimate of D_H would indicate.

It is instructive to see how well we can estimate the fractal dimension $D = \ln 2 / \ln 3$ of the middle thirds Cantor set by looking at low order periodic orbits. The 4-cycle of the ternary tent map is easily found by digital experimentation to be given by the rational numbers $x_o = 0.22202220\cdots$, $x_1 = 0.00200020002\cdots$, $x_2 = 0.0200020002\cdots$, $x_3 = 0.2002002\cdots$, $x_4 = 0.22202220\cdots = x_o$. The two intervals that cover this 4-cycle are $l_1 = x_2 - x_1 \approx 4/27$ and $l_2 = x_o - x_3 \approx 8/27$, so that equation (14.43) yields $D_H \approx 0.4$, which is much smaller than $D = \ln 2 / \ln 3 \approx 0.63$. Here it is geometrically clear why the 4-cycle underestimates the fractal dimension: the 4-cycle, along with all other cycles, is covered by the optimal choice of two intervals of length $l_1 = 1/3$ whose end points are transients on the path to an unstable 1-cycle at $x = 0$: because a point on a cycle cannot be an end point in the optimal covering, cycle coverings will by necessity always underestimate D_H.

Finally, while the critical nonperiodic orbit is a multifractal it is not merely a 2-scale multifractal. However, there is a particular 2-scale approximation to the fractal that yields a very good approximation to the fractal dimension. Feigenbaum (1980) and Cvitanovic have shown that the period-doubling limit near D_c is described by the functional equation (universal for maps with quadratic maxima) $g(x) = -\alpha g(g(-x/\alpha))$, where $l_1 = 1/\alpha$ and $l_2 = 1/\alpha^2$ can be understood as two first generation length scales. With the universal result $\alpha \approx 2.50\cdots$, and

$$l_1^{D_H}(1 + l_1^{D_H}) \approx 1, \tag{14.58}$$

we can solve this quadratic equation for $l_1^{D_H}$ to find

$$D_H \approx \frac{\ln[2/(\sqrt{5}-1)]}{\ln \alpha} \approx 0.5245 \cdots, \tag{14.59}$$

which is fairly close to the standard numerical estimate of $D_H \approx 0.537 \cdots$.

Let us not give up so easily in our quest to deduce this dimension from the hierarchy of unstable periodic orbits. We return to the 4-cycle of the logistic map at D_c, but instead of thinking like a naive theorist let us approach the problem from the standpoint of an empiricist who only knows that, at resolution l_n one needs 2^n boxes of that size to cover the orbit. That is the key: we cover the 4-cycle with 2 boxes of size $l_1 = 0.2$, the smaller interval $l_2 = 0.08$ being too small for resolution. This procedure yields $2 \approx l_1^{-D}$ with $D \approx 0.43$, which is slightly better but still too small, the same procedure applied to the 4-cycle of the ternary tent map yields $2 \approx l_1^{-D}$ with $l_1 \approx 8/27$, so that $D \approx 0.57$, which is not too bad for the crudest level of empiricism.

Theorists will not be satisfied with this last method of estimation, widely known as 'box-counting', but many researchers use the 'box-counting method' in order to look for evidence of the scaling law $N_n \approx l_n^{-D}$ over a finite, usually small, range of different box sizes (a range of experimental resolutions) $l_1, l_2, \ldots, l_N$. Evidence for fractal scaling of raw data is usually not accepted by the community unless (among other things) one can produce strong evidence that the scaling holds over a range $l_1/l_N \geq 10^3$.

A theoretically more respectable way to approach the problem of determining D_H (and also one that can be applied to experimental data) is via the transfer matrix method (adapted from statistical mechanics), which compares the sizes of the intervals in generation n with those of generation $n+1$. The use of the resulting interval contraction rates (the transfer matrix elements) leads to a more accurate way to estimate D_H from the unstable cycles at the period-doubling critical point.

At the period-doubling critical point the orbit is fractal and marginally stable, not chaotic. Unimodal maps that peak above unity and generate a complete binary tree of ever-contracting intervals in backward iteration have trajectories that are fractal, chaotic, and repelling. The bouncing ball model, just beyond the period-doubling limit, for some initial conditions, has a region of phase space that can be called a nonfractal strange attractor. We turn next to invariant sets that may be fractal, chaotic, or both, and especially to the partial explanation of the popular phrase 'strange attractor'.

14.3 Strange attractors

A path from regular to chaotic motion different from period doubling was proposed by Ruelle, Takens, and Newhouse (RTN). Historically, before period doubling universality was discovered, the RTN path to chaos was proposed as an alternative to Landau's idea of the development of turbulence in the Navier–Stokes equations of hydrodynamics. According to Landau, who only argued qualitatively and intuitively and did not use the Navier–Stokes equations explicitly to back up his argument, turbulence should appear as the consequence of the successive instability of conditionally periodic motions (quasiperiodicity) with infinitely many different frequency ratios. In other words, Landau argued in favor of the need for infinitely many unstable frequencies as a prelude to turbulence. Another way to think of it is that turbulence would require the successive instabilities of tori of dimension n, for $n = 1, 2, 3, \ldots, \infty$.

Ruelle, Takens, and Newhouse replaced the direct study of the Navier–Stokes nonlinear partial differential equations by the study of an infinite system of coupled ordinary differential equations that would follow from the Navier–Stokes equations systematically by a mode-expansion (by an expansion in orthogonal functions of the spatial coordinates, but with time-dependent expansion coefficients). One effect of finite viscosity is to truncate this into a finite but very large number. The Lorenz model follows from the arbitrary truncation to three modes of such an expansion. RTN argued (as Lorenz had shown) that one does not need infinitely many modes, but only three (which includes the possibility of three-frequency problems), for deterministic chaos. It has long been known from observation that fluid turbulence in open flows is a problem of successive instabilities of many vortices of many different sizes, likely with chaotic vorticity in both space and time. A precise and widely accepted definition of fluid turbulence has not yet been found, but turbulence is certainly not described by low ($n \approx 3$) degree of freedom chaos.

We return to the perturbation of an integrable conservative flow by combined damping and driving that can make the flow nonintegrable for large enough control parameter values. In the case of weak enough driving and dissipation foliations are destroyed and replaced by a trapping region, but one foliation may survive in distorted form for a limited range of the perturbation parameter. Inside the trapping region we expect to find one or more attractors of the flow. Motion far from any attracting region is overdamped: dissipation dominates to reduce the amplitudes of large deviations from the attractor. For the RTN scenario to work, we also must have at least two separate frequencies in the dynamics problem. If the system with $n \geq 3$ is autonomous, then at low enough driving the motion is overdamped everywhere in phase space, corresponding to a zero dimensional sink. The motion spirals into the sink due to the presence of at least one frequency in the unperturbed problem. At higher rates of driving the sink becomes a source and one frequency is excited yielding a one dimensional attractor, where motions with amplitudes much greater than the size of the stable limit cycle are overdamped but amplitudes that are small compared with the size of the limit cycle are unstable and grow as a result of negative damping. At even higher driving rates the limit cycle becomes unstable via a Hopf bifurcation and two frequencies are excited, meaning that an attracting 2-torus appears. On the 2-torus the motion is either stable periodic or stable quasiperiodic depending upon the winding number. Both the one and two dimensional tori are the consequence of average energy balance in some region of phase space. RTN showed that the next bifurcation for quasiperiodic orbits actually leads to a three dimensional torus that is, under small perturbations, unstable with respect to decay onto a 'strange attractor'. Motion on a strange attractor is chaotic but not necessarily fractal, and not all fractal invariant sets are chaotic. The bouncing ball model at criticality has a nonchaotic fractal invariant set, the 2^∞ orbit, inside the trapping region (inside the rectangular slice). Some chaotic sets are attracting, others are repelling, and some are neither attracting nor repelling. The logistic map just beyond criticality has a strange nonattracting set that includes a class of orbits that are nonfractal, but fractal trajectories also occur for other choices of initial conditions for the very same control parameter values (we remind the reader of the $D(\lambda)$ spectrum). Unimodal maps that peak above unity and generate a complete binary tree of ever-contracting intervals in backward iteration generate 'strange' trajectories that are fractal, repelling, and chaotic ('strange repeller'). Unimodal maps that peak above unity generate a complete binary tree of intervals that define the geometric or 'physical' pictures of a strange attractor. For that work we refer the reader to the literature (e.g. McCauley, 1993, or Cvitanovic et al., 1988).

We turn next to a model that is motivated by physics, and where the RTN prediction of the transition of quasiperiodic orbits from torus-confinement to chaos can be studied. Given the stable periodic and quasiperiodic orbits on a torus in a two-frequency system, the RTN scenario leaves open the possibility that persisting periodic attractors (periodic orbits on the torus with negative Liapunov exponents) might short-circuit the onset of chaos via unstable quasiperiodicity through the persistence of attracting periodic orbits (stable periodic orbits might persist alongside the unstable quasiperiodic ones). We have to pay serious attention to this possibility because 'mode locking' has been observed in nature. Mode locking is the phenomenon whereby two very weakly coupled nonlinear oscillators (like two pendulum clocks on opposite sides of a wall) tend to synchronize, indicating a rational rather than an irrational frequency ratio. In the section that follows we shall discuss both the quasiperiodic and periodic motions and their respective transitions to chaotic behavior in a model where the RTN prediction of the transition to chaos via quasiperiodicity occurs. We shall also show that, in the same model, attracting periodic orbits persist long after the quasiperiodic ones have become unstable, so that one would have to ask Mother Nature how she chooses her initial conditions to be able to tell whether or not it is possible to see the RTN transition to chaos (instead of stable and unstable periodicity) in observations of two-frequency systems in nature. Measure theorists argue that mathematics is capable of determining which class of initial conditions is likely to be chosen, but we reject that claim as there is no evidence for it from observation (and a computer calculation is at best a simulation that does not qualify either as a real experiment or as an observation of nature).

14.4 The two-frequency problem

The damped, periodically driven pendulum is defined by the Newtonian equation

$$m\ddot{\theta} + \beta\dot{\theta} + \omega_\circ \sin\theta = A + K\cos\omega t \tag{14.60}$$

in which there are two 'bare' frequencies: $\omega_\circ$ and ω. This is also the equation of the Josephson junction in the theory of superconductivity. Instead of studying the system as a flow in a three dimensional phase space, we can derive a two dimensional iterated map called the stroboscopic map

$$\left.\begin{array}{l} \theta_{n+1} = G_1(\theta_n, \dot{\theta}_n) \\ \dot{\theta}_{n+1} = G_2(\theta_n, \dot{\theta}_n), \end{array}\right\} \tag{14.61}$$

where $n = 0,1,2,\ldots$ correspond to times $0,2T,3T,\ldots$ and $T = 2\pi/\omega$ is the period of the external driving force. To see this, simply begin with the solution written in the standard form

$$\left.\begin{array}{l} \theta(t) = \psi(\theta_\circ, r_\circ, t - t_\circ) \\ r(t) = \phi(\theta_\circ, r_\circ, t - t_\circ) \end{array}\right\} \tag{14.62}$$

where $r = d\theta/dt$. Next, let $T = t - t_\circ$ be the period and denote $\theta(nT) = \theta_n$ and $r(nT) = r_n$. Then

$$\left.\begin{array}{l} \theta_n = \psi(\theta_{n-1}, r_{n-1}, T) \\ r_n = \phi(\theta_{n-1}, r_{n-1}, T). \end{array}\right\} \tag{14.63}$$

Because the functions ψ and ϕ are independent of the index n, we have an iterated map of the form (14.61).

It is the choice of strobing time equal to the driving period T that guarantees that the functions G_1 and G_2 do not change with time (with n). The Jacobian J of the stroboscopic map (14.61) is simply

$$J(t + T,t) = \frac{\partial(\theta(t + T),\dot\theta(t + T))}{\partial(\theta(t),\dot\theta(t))} = e^{-\beta T/m}. \tag{14.64}$$

This result follows from

$$\nabla \cdot V = \frac{\partial\dot\theta}{\partial\theta} + \frac{\partial\ddot\theta}{\partial\dot\theta} = -\beta/m \tag{14.65}$$

so that areas in the two dimensional $(\theta_n,\dot\theta_n)$-phase space contract at the constant rate

$$b = e^{-2\pi\beta/m\omega} \tag{14.66}$$

where $\beta > 0$ in the case of ordinary friction.

In order to model the damped-driven pendulum by a two dimensional map that has some qualitatively similar properties, note that $G_1(\theta + 2\pi,r) = 2\pi + G_1(\theta,r)$ while $G_2(\theta + 2\pi,r) = G_2(\theta,r)$. A model satisfying these conditions is provided by the much-studied dissipative standard map,

$$\left.\begin{array}{l} \theta_{n+1} = \theta_n + 2\pi\Omega + Kf(\theta_n) + br_n \\ r_{n+1} = br_n + Kf(\theta_n) \end{array}\right\} \tag{14.67}$$

where $f(\theta + 2\pi) = f(\theta)$ and $J = b$, so that $0 < b < 1$ represents dissipation and $b = 1$ yields the limit of an undamped system. Very small b therefore represents the limit of a very strong dissipation.

If, in the dynamical system, r_n should relax rapidly relative to θ_n so that

$$r_n \cong G_2(\theta_n,r_n) \tag{14.68}$$

for large enough n, so that we could solve to obtain $r_n \approx g(\theta_n)$, then we could instead study the dynamics approximately as a one dimensional circle map

$$\theta_{n+1} = G_1(\theta_n,g(\theta_n)) = F(\theta_n) \tag{14.69}$$

where $F(\theta_n + 2\pi) = F(\theta_n) + 2\pi$. So far, no one has managed to prove rigorously that this approximation is valid for the damped-driven pendulum. However, we can still invent a simple model where the approximation holds: if $f(\theta) = 0$ then we have the linear twist map

$$\left.\begin{array}{l} \theta_{n+1} = \theta_n + 2\pi\Omega + br_n \\ r_{n+1} = br_n \end{array}\right\} \tag{14.70}$$

which has the solution

$$\left.\begin{array}{l} \theta_n = \theta_\circ + 2\pi n\Omega + b^n r_\circ \\ r_n = b^n r_\circ. \end{array}\right\} \tag{14.71}$$

At long times, for $0 < b < 1$, we obtain the linear circle map

$$\theta_n \approx \theta_\circ + 2\pi n\Omega, \tag{14.72}$$

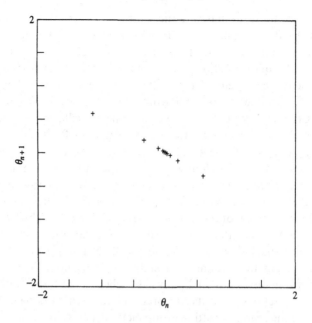

Fig. 14.11 Numerically computed return map for stable motions of the
damped-driven pendulum ($K = 0.3$).

which represents motion on a two dimensional torus if we think of the map as
representing a flow in a higher dimensional phase space. The system (14.67) with finite
r_n therefore describes motion toward an attracting two dimensional torus ($r_n = 0$
means that the motion is on the torus). The computed return map for the damped-
driven pendulum equations in the nonchaotic regime is shown in figure 14.11 and is
linear, as expected.

More generally, the system (14.71) represents attraction by a torus for the case where
$0 < b < 1$. In what follows, we shall therefore analyze a one dimensional circle map

$$\theta_{n+1} = F(\theta_n) = \theta_n + 2\pi\Omega + Kf(\theta_n) \tag{14.73}$$

in order to get an analytic idea of the dynamics of the two-frequency problem on the
torus. Qualitatively, this replacement of a distorted torus by a perfectly circular one
may be a reasonable approximation only so long as we avoid the regime where the one
dimensional map has a multivalued inverse, which means that the parameter K should
obey $K \leq K_c$, where for $K > K_c$ the map F is not uniquely invertible. A nonuniquely
invertible one dimensional map may exhibit deterministic chaos, but a one dimensional
map with a multivalued inverse cannot follow exactly from a set of differential
equations because (with finite Jacobi determinant) we can always integrate the
differential equations backward in time uniquely. In other words, one cannot expect to
investigate the chaotic regime of the damped-driven pendulum by using a one
dimensional map, even if we could get away with it when the motion is nonchaotic.

As a definite model, consider the sine circle map

$$\theta_{n+1} = \theta_n + 2\pi\Omega - K\sin\theta_n \tag{14.74}$$

which reduces at zero driving strength (at $K = 0$) to the linear circle map

$$\theta_{n+1} = \theta_n + 2\pi\Omega. \tag{14.75}$$

If Ω is rational, then the motion is periodic; irrational Ω corresponds to quasiperiodic motion that is ergodic but nonmixing. In what follows, we take $K \geq 0$. Here, as in section 4.3, Ω represents the ratio of two unrenormalized frequencies.

In experiments, 'frequency locking' is typically observed in two-frequency systems, meaning that there is a tendency for the 'effective' or 'renormalized' frequency ratio, which we shall discover below to be the winding number W, to 'lock in' at a rational value, reflecting the tendency of two coupled nonlinear oscillators to synchronize with each other through a weak coupling. This warns us that the RTN prediction, although mathematically very general, may not necessarily be the route to chaos that two-frequency systems in nature will follow: near the end of the section, we shall explain mathematically why stable quasiperiodic orbits may be hard to observe in nature and why, instead, the periodic orbits (orbits with rational winding numbers W) may dominate the dynamics in observations of nature. First, we shall show that, for irrational winding numbers, the sine circle map has a transition to chaos that agrees qualitatively with expectations based upon the RTN prediction (the circle map is supposed to represent the dynamics of stable or marginally stable orbits of the damped-driven swing on the torus). In any case, whether the transition to chaos is by quasiperiodic or by periodic orbits, the study of the dynamics near the transition requires the use of the renormalization group method, and so we shall eventually refer the reader to the literature for the study of critical point properties, because the study of the renormalization group method would take us well beyond the scope of this text.

A generally nonlinear circle map $\theta_n = F(\theta_{n-1})$ is defined by

$$F(\theta_n + 2\pi) = 2\pi + F(\theta_n) \tag{14.76}$$

so that

$$\theta_{n+1} = \theta_n + 2\pi\Omega + Kf(\theta_n) \tag{14.77}$$

where

$$f(\theta_n + 2\pi) = f(\theta_n) \bmod 2\pi. \tag{14.78}$$

In order to analyze an arbitrary circle map, it is useful to introduce the renormalized winding number, or just the winding number W. In general, W and Ω do not coincide because Ω is the exact winding number only when $K = 0$, which describes the linear case. It follows easily that it is not Ω, but rather the winding number

$$W = \lim_{n \to \infty} \frac{F(\theta_n) - \theta_\circ}{2\pi n} \tag{14.79}$$

that reflects periodicity or lack of same. As a simple check, if $\theta_{n+1} = \theta_n + 2\pi\Omega$, then $\theta_n = \theta_\circ + 2\pi n\Omega$ and we retrieve $W = \Omega$, as expected. More generally, for any linear or nonlinear circle map, W tells us the average number of rotations through 2π per iteration simply because $(f(\theta_n) - \theta_\circ)$ is the total distance travelled around the circle during n iterations of the map.

Arnol'd (1965) has shown that, if W is irrational and if F is invertible, then an arbitrary nonlinear circle map $\theta_{n+1} = F(\theta_n)$ is conjugate, by a continuous transformation, to a rotation through some angle ϕ_n, that is, to the linear circle map! In order to formulate this conjugacy, we write

$$\phi_{n+1} = T(\phi_n) = \phi_n + 2\pi W, \tag{14.80}$$

which is a pure rotation with the same winding number W as given by (14.79) for the original map $f(\theta)$. Next, we ask, when is it possible to find a transformation from θ to ϕ such that the conjugacy (14.80) holds for an arbitrary circle map F? That is, we want to construct analytically a transformation

$$\phi = H(\theta) \tag{14.81}$$

so that the mapping R defined by

$$R = HFH^{-1}, \tag{14.82}$$

where

$$\theta_{n+1} = R(\phi_n), \tag{14.83}$$

obeys

$$R(\phi) = \phi + 2\pi W. \tag{14.84}$$

According to Arnol'd (1965, 1983), the condition for this transformation to exist is that F is sufficiently smooth and invertible, and W is irrational. Take care to note that Arnol'd's result is universal: F can be a sine circle map or anything else that obeys (14.76) so long as it is smooth and invertible, and W is irrational. If the transformation is in addition differentiable (which we assume to be true in what follows), then every stable quasiperiodic orbit of the sine circle map has a vanishing Liapunov exponent: Liapunov exponents are invariant under differentiable coordinate transformations, and the linear circle map has a vanishing exponent.

As we already know, noninvertibility of a one dimensional map is necessary for deterministic chaos, so it is only possible for the map

$$\left.\begin{array}{l} F(\theta) = \theta + 2\pi\Omega + Kf(\theta), \\ f(\theta + 2\pi) = f(\theta), \end{array}\right\} \tag{14.85}$$

with $f(\theta) = -\sin\theta$ to be chaotic if $K > 1$. This follows from

$$f'(\theta) = 1 - K\cos\theta, \tag{14.86}$$

which shows that $f(\theta)$ is monotonically increasing for $K < 1$: here, since $f(1) = f(0)$ mod 1, $f(\theta)$ is single-valued, hence invertible. At $K = 1$, $f'(\theta) = 0$ at $\theta = 0$ ($= 2\pi$ mod 1). When $K > 1$ $f(\theta)$ has a maximum and minimum (mod 2π) at

$$\theta = 1 - \cos^{-1}\frac{1}{K} \tag{14.87}$$

and

$$\theta = \cos^{-1}\frac{1}{K}, \tag{14.88}$$

respectively. Therefore, chaos requires $K > 1$ where there is no chance of conjugacy to a map that is manifestly nonchaotic, namely, the linear circle map.

As we shall see below, periodic orbits have no transition to chaos at $K = 1$, only the quasiperiodic ones do. Before discussing the critical point where the transition for quasiperiodicity occurs, we first deduce some essential properties of the periodic and quasiperiodic orbits in the regular regime where $K < 1$.

For one thing, we eventually want to understand why, in nature, oscillations always tend to 'lock in' at some rational frequency $W = P/Q$ for a finite range of Ω, for a fixed value of $K > 0$. Frequency locking means: we shall find that there is a continuum of the pair of numbers (Ω, θ_o) that, within definite limits, yield exactly the *same* winding rational number W for the same fixed value of K! The winding number $W(K, \Omega)$ turns out to be continuous and is monotonically increasing with Ω, but is itself determined by the pair of numbers (K, θ_o). Geometrically, this frequency locking corresponds to wedges of parametric resonance in the (Ω, K)-plane that are called Arnol'd tongues. In what follows, we show how to derive a few Arnol'd tongues in the small K limit. With the system of resonant wedges in mind, we then discuss semi-quantitatively what happens as K approaches unity from below (as the sine circle map approaches noninvertibility). Finally, after all is said and done, we shall return to the question why frequency locking tends to be observed in nature, and will state the reason why periodic rather than quasiperiodic behavior can dominate mathematically in the dynamics.

First, we want to illustrate by an example how periodic motion occurs for a fixed rational winding number W and a fixed value of K along a finite width $\Delta\Omega$ of the Ω-axis, and thereby forms the resonant wedges called Arnol'd tongues in the (Ω, K)-plane. For the sine circle map

$$\theta_{n+1} = \theta_n + 2\pi\Omega - K \sin \theta_n, \tag{14.89}$$

if we set the winding number

$$W = \frac{\theta_n - \theta_o}{\pi n} \tag{14.90}$$

equal to a definite rational number P/Q, then we can in principle solve algebraically for the range of values of Ω over which this latter condition holds for a fixed value of K in the unit interval. Since $W = \Omega$ yields the correct average rotation through 2π when $K = 0$, the idea is to solve for Ω as an expansion in powers of K whenever K is small.

We fix both W and K and then ask for the range of Ω (corresponding to different choices of θ_o) over which the corresponding periodic orbits exist. If $W = 0$, we have a fixed point

$$\theta_o = \theta_o + 2\pi\Omega - K \sin \theta_o \tag{14.91}$$

or

$$2\pi\Omega - K \sin \theta_o = 0, \tag{14.92}$$

which yields

$$\Omega = \frac{K}{2\pi} \sin \theta_o. \tag{14.93}$$

For all such values of Ω, the orbit is periodic with winding number $W = 0$. Note that, for fixed K, this periodicity occurs for all values of Ω that lie within a wedge of width K/π that is centered on $\Omega = 0$ in the (Ω, K)-plane (figure 14.12). All points within this wedge (corresponding, at fixed K, to varying the initial condition θ_o from $-\pi$ to π) have the very same winding number: $W = 0$.

To construct periodic orbits with $W = 1/1$, we must solve

$$\theta_1 = \theta_o + 2\pi\Omega - K \sin \theta_o \tag{14.94}$$

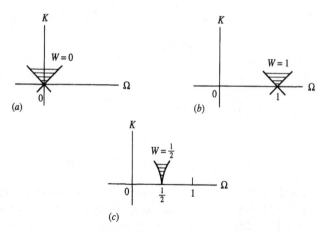

Fig. 14.12 Wedges of parametric resonance (Arnol'd tongues).

with

$$\theta_1 = \theta_o + 2\pi. \tag{14.95}$$

We find that

$$\Omega = 1 + \frac{K}{2\pi}\sin\theta_o \tag{14.96}$$

so that $\theta_o = 0$ corresponds to $\Omega = 1/1$. For a fixed value of K, $\theta_o = 0$ determines the center of a 'resonant wedge' of width K/π, and all points within this wedge correspond to periodic orbits with values of θ_o and Ω different from 0 and 1, but all of these orbits have exactly the same winding number $W = 1/1$ (figure 14.12b).

Consider next the simplest orbit with winding number $W = 1/2$, so that

$$\theta_2 = \theta_o + 1\pi\Omega - K\sin\theta_o + 2\pi\Omega + K\sin(\theta_o + 2\pi\Omega - K\sin\theta_o). \tag{14.97}$$

We set $\Omega = 1/2 + \Delta\Omega$, where $\theta_o = 0$ yields $\Omega = 1/2$, and also obtain

$$4\pi\Delta\Omega - K\sin\theta_o - K\sin(\theta_o + \pi + 2\pi\Delta\Omega - K\sin\theta_o) = 0. \tag{14.98}$$

Further simplification yields, for small K, that

$$\Delta\Omega \approx K^2(\sin 2\theta_o)/8\pi, \tag{14.99}$$

or

$$\Omega = 1/2 + \Delta\Omega \approx 1/2 + K^2(\sin 2\theta_o)/8\pi. \tag{14.100}$$

The intersection of this Arnol'd tongue with a parallel drawn through the value K on the vertical axis in the (Ω,K)-plane has the approximate width $K^2/4\pi$ whenever K is small enough (figure 14.12(c)).

We are therefore led to expect the following general behavior: if

$$\left.\begin{array}{l} F(\theta) = \theta + 2\pi\Omega + Kf(\theta) \\ f(\theta) = f(\theta)\,\mathrm{mod}\,2\pi, \end{array}\right\} \tag{14.101}$$

then under suitable differentiability conditions on f, and for rational winding numbers,

$$W = P/Q, \tag{14.102}$$

we should solve for cycles by using

$$\theta_Q = \theta_\circ + 2\pi Q W = \theta_\circ + 2\pi P \tag{14.103}$$

to obtain

$$\theta_\circ + 2\pi Q\Omega + f^{(Q)}(\theta_\circ,\Omega,K) = \theta_\circ + 2\pi P \tag{14.104}$$

where $f^{(Q)}$ is obtained from functional composition. We can write $\Omega = P/Q + \Delta\Omega$ in order to obtain the shape of the resonant wedge for a cycle with rational winding number $W = P/Q$, where $\Delta\Omega_{max} = \Omega_{max} - \Omega_{min}$ is the width of the wedge. By this method, it is possible to understand that wedges of parametric resonance grow from every rational winding number on the Ω-axis.

When $K = 0$, the quasiperiodic orbits occupy all of the space on the Ω-axis and the periodic orbits are confined to the countable point set where Ω is rational. In order to understand qualitatively how, for finite K, the winding numbers of quasiperiodic orbits can define a Cantor set of finite size, and where the complementary set is a devil's staircase of resonant wedges, we make the following observations. Every rational number gives rise to a resonant wedge of finite width for finite K, and we can choose to organize the rational winding numbers $W = P/Q$ according to numerators Q of increasing size: 1, 1/2, 1/3, 2/3, 1/4, 3/4, 1/5, 2/5, 3/5, 4/5,...,0. This is a nonintuitive property of the continuum that is hard to imagine geometrically, and the closest that we can come to a visualization is to plot the corresponding devil's staircase: if, for fixed $K < 1$, we remove the width of each resonant wedge from the Ω-axis, then what is left over is a Cantor set of bare winding numbers that give rise to quasiperiodic orbits: the removal of infinitely many closed intervals (the resonant wedges) from the unit interval on the Ω-axis leaves behind a Cantor set of open sets. The nonintuitive part is that it is hard to see how wedges of finite width growing from every rational value of Ω do not immediately overlap for any finite value of $K > 0$. The plot of Ω vs K would represent a devil's staircase of resonant wedges, that is geometrically similar to figure 14.10. To show that the wedges do not overlap for $0 < K < 1$ we must compute the sum of the widths of all wedges in $0 < \Omega < 1$ and show that the sum is less than unity. We next estimate the sum of all tongue widths for the case where K is very small.

Notice that for each value of Q there are at most $Q - 2$ different winding numbers (this is an overestimate for Q), excepting the case $W = 1/2$. Assume for simplicity that all cycles with the same denominator Q have tongue widths no greater than on the order of K^Q. We can now show that the devil's staircase takes up an amount of space on the order of K whenever K is small. A simple series summation shows that (to within a constant factor)

$$l_{st} \sim K + K^2 + \sum_{Q=3}^{\infty} (Q - 2)K^Q \sim 3K \tag{14.105}$$

where l_{st} is the amount of space in [0,1] on the Ω-axis that is occupied by the devil's staircase. The result is consistent with the facts that (i) the $K = 0$ set occupies no space, and (ii) the resonant wedges grow continuously and smoothly from the $K = 0$ set. The proof that the wedges take up very little space for very small K is only a variation on the usual proof that the rational numbers take up no space at all (as we showed in the last section). For small K, therefore, the Ω values that give rise to irrational values of the winding number W occupy a Cantor set with finite length $l_c \approx 1 - l_{st} \approx 1 - 3K$, which is nearly the entire unit interval for K small (as indeed it must be). Both the staircase and the complementary Cantor set have fractal dimension equal to unity for small K because both sets have finite length. But what happens to these dimensions as K approaches unity?

Bohr, Bak, and Jensen (1984, 1985) have used a numerical analysis to show that, as $K \to 1^-$, $l_{st} \to 1^-$ as well[1], and have obtained numerical evidence for the scaling law

$$l_c \approx 1 - l_{st} \sim (1 - K)^\beta \text{ as } K \to 1^- \qquad (14.106)$$

for the size l_c of the Cantor set near the point where the circle map becomes noninvertible. In other words, the quasiperiodic orbits at $K = 1$ have winding numbers that originate in bare winding numbers Ω that occupy a fractal set with zero size. Bohr et al. have numerically estimated a fractal dimension of $D \approx 0.87$ for that zero-length Cantor set. For details, we refer the reader to the literature (Cvitanovic et al., 1985, Schuster, 1988, or McCauley, 1993). We return now to the reason why periodicity seems to win over quasiperiodicity in two-frequency systems in nature, at least within the nonchaotic regime.

Frequency locking was first observed and reported in the seventeenth century by the Hollander physicist Christian Huyghens, who observed that two separate clocks on a wall tended to synchronize with each other after a while. This suggests the qualitative idea of an attractor, and indeed the tendency to frequency locking in nature can be understood by using the idea of attracting periodic orbits: because the invertible circle map with an irrational winding number is conjugate to the linear circle map by a coordinate transformation, every stable quasiperiodic orbit necessarily has a vanishing Liapunov exponent (this means that the Liapunov exponent at a wedge boundary is zero, which can be easily checked for the $W = 0$ and 1 orbits discussed above). In other words, stable quasiperiodic orbits of the nonlinear circle map do not attract nearby orbits. However, the stable periodic orbits of the nonlinear circle map have negative Liapunov exponents, which means that they do attract nearby orbits, and this provides a practical mechanism for frequency locking. For example, the Liapunov exponent of all orbits of the sine circle map with $W = 1/1$ is given by

$$\lambda = \ln |1 - [K^2 + (2\pi)^2(\Omega - 1)^2]^{1/2}|, \qquad (14.107)$$

so that the stability range of the orbit with $W = 1/1$ and $\theta_o = 0$, for example, is $0 < K < 2$ because that is the range where simultaneously (i) the 1-cycle is stable because $f'(\theta)$ has magnitude less than unity, consequently (ii) $\lambda < 0$. Note that the stability range for this particular periodic orbit extends to $K = 2$, whereas all quasiperiodic orbits have become unstable (and so $\lambda > 0$ may occur) whenever $K > 1$. Note also that $\lambda = 0$ at the wedge boundaries where

$$\Omega = 1 \pm K/2\pi, \qquad (14.108)$$

as expected.

---

[1] $k \to 1^-$ means 'as k approaches unity from below ($k < 1$ and $k \to 1$).

We want to show next that the orbits with rational winding numbers become unstable through period-doubling bifurcations. Each unstable orbit has a positive Liapunov exponent, but we expect that chaos does not set in until all periodic orbits have become unstable: so long as there is at least one attracting periodic orbit, orbits that are repelled by any of the unstable periodic and unstable quasiperiodic orbits will still have a relatively calm 'resting place' unless one chooses initial conditions that lie precisely on an unstable quasiperiodic or unstable periodic orbit. Choosing initial conditions that lie on an unstable quasiperiodic orbit is not so easy, however: the winding number W is not a parameter that is easily accessible to control, in practice, in the laboratory.

The circumstances with the logistic map are entirely similar: for the fixed point $x = 1 - 1/D$, the Liapunov exponent is $\lambda = \ln|(D - 2)|$ and is negative for $1 < D < 3$, which is the stability range of that fixed point. When $D > 3$, the fixed point has a positive Liapunov exponent but this does not cause any chaos so long as there are other stable cycles for the trajectories to settle onto. The idea of chaos requires bounded motion (self-confined motion, for example) with mutually repelling orbits: for any choice of initial conditions in the relevant region in phase space, all nearby orbits must repel each other because each orbit has its own positive Liapunov exponent. According to Feigenbaum (1980) period doubling may be ubiquitous in nature, as the mechanism in nonlinear dynamics whereby unstable periodicity can develop out of stable periodicity via pitchfork bifurcations. Therefore, let us look next for evidence of period doubling in the sine circle map.

In order to illustrate period doubling for an analytically calculable example, we look at the 1-cycle with $W = \Omega = 1/1$ so that $\theta_o = 0$. This 1-cycle is stable so long as $0 < K < 2$, as we have shown above, and is unstable for $K > 2$. If period doubling occurs, this 1-cycle cannot give birth to a 2-cycle with $W = 1/2$. This would be geometrically impossible, as the two points on the 2-cycle have to coalesce into the single point $\theta_o = 0$ as K approaches 2 from above. The only possibility that is allowed by the geometry of the circle map is that $W = \Omega = 2/2$. This 2-cycle is geometrically possible: the equations of the 2-cycle are (with $\theta_2 = \theta_o + 4\pi$)

$$\left.\begin{array}{l} \theta_1 = \theta_o + 2\pi - K\sin\theta_o \\ \theta_o = \theta_1 - 2\pi - K\sin\theta_1. \end{array}\right\} \tag{14.109}$$

In this case, with $W = 2/2$, θ_1 undershoots θ_o by the amount $K\sin\theta_o$ in a single 2π-revolution, and in the next 2π-revolution $\theta_1 + \pi$ falls short of θ_o by the amount $K\sin\theta_o$, which tells us that we must have $\sin\theta_o = -\sin\theta_1$ for periodicity. With $\theta_o > 0$, this means that $\theta_1 = 2\pi - \theta_o$. This combination permits the coalescence of both θ_o and θ_1 into a single value at the bifurcation point $K = 2$. Here, we quote Feigenbaum (1980), who said it very well:

... for some range of parameter values, the system exhibits an orderly periodic behavior; that is, the system's behavior reproduces itself every period to time T. Beyond this range, the behavior fails to reproduce itself after T seconds; it almost does so, but in fact it requires two intervals of T to repeat itself. That is, the period has doubled to $2T$. This new periodicity remains over some range of parameter values until another critical parameter value is reached after which the behavior almost reproduces itself after $2T$, but in fact, it now requires $4T$ for reproduction ...

Going on with the mathematics, equations (14.109) simplify to read

$$K\sin\theta_o = 2\theta_o, \tag{14.110}$$

which has solutions if $K > 2$ but not if $K < 2$. Next, we look at stability of this 2-cycle that is born of a pitchfork bifurcation at $K = 2$.

For stability, we need $-1 < f'(\theta_o)f'(\theta_1) < 1$, which reduces to the single equation $(1 - K\cos\theta_o)^2 < 1$. Instability occurs whenever $(1 - K\cos\theta_o)^2 = 1$. There are three possibilities, but one is eliminated by the second of the pair of conditions that follow from the numeric solution of

$$1 - K\cos\theta_o = -1. \tag{14.111}$$

Using this condition along with equation (14.110), it follows that $\theta_o = (K^2 - 1)^{1/2}$, and instability occurs whenever the equation

$$\sin\theta_o = \frac{\theta_o^2}{(2 + \theta_o^2)^{1/2}} \tag{14.112}$$

has a solution. The first solution is at $K_1 = 2 (\theta_o = 0)$, the second falls at a parameter value K_2 corresponding to an angle θ_o that falls short of $\pi/2$ by a finite distance. For $K_1 < K < K_2$ the 2-cycle is stable, as can be seen from returning to the stability condition $(1 - K\cos\theta_o)^2 < 1$, which is simultaneously the condition that the 2-cycle with $W = 2/2$ has a negative Liapunov exponent. At $K > K_2$ the 2-cycle persists in unstable form and a stable 4-cycle occurs. We refer the reader to the paper by Perez and Glass (1982) for further details of the rich 'phase diagram' that emerges from period doublings within mode lockings as well as the application of circle map dynamics to heartbeat rhythms.

Exercises

Period doubling (bouncing ball)

1. Why, physically, should the 1-cycle become unstable merely because ϕ_o deviates too far from 0 as γ approaches γ_{max}? What is beneath the mathematics of this instability?

2. Plot qualitatively the motions of the ball and table vs time for the case where $n = 3$, $k = 6$.

3. Determine the upper stability limit γ' for the ball's 2-cycles. If the algebra is too difficult, then determine γ' numerically to within two decimals.

4. Perform numerical integrations (using $n = 1$) and show that the ball's orbit changes systematically from a 1-cycle to a 2-cycle to a 4-cycle as the parameter γ is increased steadily from $\gamma_{min} = 2\pi(1 - \alpha)$. (For example take $\alpha = 1/2$.)

5. Show numerically that, for $\gamma > \gamma_{max}$, the map has two separate solutions that are both different from $\cos\phi = 2\pi n(1 - \alpha)/\gamma$.

6. Obtain numerical evidence for chaotic motion when γ is large enough. Are there any windows of stability (e.g. cycles of odd order) imbedded within the chaotic regime $\gamma > \gamma_\infty$?

Period doubling (logistic map)

7. For the 1-cycle at $x = 1 - 1/D$, show that $\lambda = \ln|2 - D|$. Show, therefore, that the orbit is stable for $1 < D < 3$. Observe that $\lambda = \ln 2$ at $D = 4$.

8. Solve for the location of the 2-cycle and show that the stability range is $D_1 < D < D_2$ with $D_1 = 3$ and $D_2 = 1 + \sqrt{6}$. In particular, compute the Liapunov exponent for the 2-cycle. Show that $\lambda = 1/2 \ln |2D + 4 - D^2|$ and that (i) $\lambda < 0$ for $3 < D < 1 + \sqrt{6}$, (ii) $\lambda = 0$ at $D = 1 + \sqrt{6}$, and (iii) $\lambda = \ln 2$ when $D = 4$.

9. Locate the 'window of stability' of the 3-cycle of the logistic map $f(x) = Dx(1 - x)$.

Circle map

10. For the damped-driven pendulum, choose a set of parameter values $(\beta, \omega_o, \omega)$, integrate the differential equation numerically, and plot θ_{n+1} vs θ_n for different values of K. For K not too large, can you represent your results in the form of a circle map $\theta_{n+1} = \theta_n + 2\pi\Omega - K'f(\theta_n)$? If so, what are Ω and K'? (Hint: transform to dimensionless variables before integrating the differential equation.)

11. For K small and for both $W = 1/3$ and $W = 2/3$, determine $\Omega(K, \theta_o)$ and also the two tongue widths.

12. Show numerically that for $K > K_2$ the 2-cycle with $\Omega = 2/2$ and $W = 2/2$ bifurcates into a 4-cycle. What is the winding number of the 4-cycle?

13. For the orbits with $W = 2/2$ and $K_1 < K < K_2$, prove that the Liapunov exponent vanishes along the wedge boundaries.

15

Hamiltonian dynamics and transformations in phase space

15.1 Generating functions for canonical transformations

If we start with a Lagrangian L, then the formulation of dynamics is covariant with respect to general coordinate transformations in configuration space. The Lagrangian transforms like a scalar under those transformations. Given any Lagrangian in generalized coordinates, one can construct the corresponding Hamiltonian as a function of the coordinates q_i and the canonically conjugate momenta p_i and the phase flow description of dynamics follows. Because the generalized velocities generally do not transform like the components of a vector in configuration space, Hamilton's equations are themselves not covariant with respect to general coordinate transformations in configuration space. Also, if we start with a Hamiltonian system then an arbitrary point transformation of the $2f$ coordinates $(q_1, \ldots, p_f)$ in phase space generally will *not* yield another Hamiltonian system. Whether the transformed equations are or are not Hamiltonian does not take away from their correctness or usefulness: every transformed set of equations is correct if the starting point is correct and if the transformation is carried out without making an error. If we restrict the class by asking for the collection of transformations in phase space whereby any Hamiltonian system is transformed into another Hamiltonian system, then we arrive at the subject of this chapter: the transformations that are called *canonical*.

It is possible to define canonical transformations in the following way. Phase space is a flat differentiable manifold with Cartesian coordinate axes labeled by the canonically conjugate variables $(q_1, \ldots, p_f)$, *whether or not the generalized coordinates q_i are Cartesian or curvilinear in a corresponding Lagrangian configuration space.* Canonical transformations can be defined as transformations that leave invariant the Cartesian phase volume element $d\Omega_f = dq_1 \ldots dq_f dp_1 \ldots dp_f$, and consequently the corresponding incompressibility condition for the velocity in phase space,

$$\nabla \cdot V = \sum_{i=1}^{f} \left(\frac{\partial \dot{q}_i}{\partial q_i} + \frac{\partial \dot{p}_i}{\partial p_i} \right) = 0. \tag{15.1}$$

Canonical transformations are therefore just transformations among Cartesian coordinate systems in the phase space of canonical f degree of freedom systems. Lagrangian configuration space is a metric space and is Riemannian in the presence of constraints. The q-submanifold of phase space is Cartesian without a metric and coincides with Lagrangian configuration space only if the latter is Euclidean *and* only if Cartesian coordinates are used there. Otherwise, the q-subspace of phase space and configuration space are entirely different manifolds. In other words, the reader must recognize that, even though the same variables q_i are used to label axes in both cases, the two spaces are *geometrically* different from each other: one manifold is always flat with Cartesian axes

q_i while the other manifold may be curved and the q_i generally label curvilinear, not Cartesian, axes.

We will not follow the path outlined above but will follow an equivalent path. The invariance of the Cartesian phase volume element can then be derived as a consequence of our starting point. Consider a Hamiltonian system

$$\dot{q}_i = \frac{\partial H}{\partial p_i}, \; \dot{p}_i = -\frac{\partial H}{\partial q_i} \tag{15.2}$$

with f degrees of freedom and consider also a transformation of the $2f$ coordinates in phase space[1]

$$Q_i = \phi_i(q,p,t), \; P_i = \psi_i(q,p,t) \tag{15.3}$$

where (q,p) denotes a point in phase space and therefore is shorthand notation for the $2f$ generalized coordinates and canonical momenta $(q_1,\ldots,q_f,p_1,\ldots,p_f)$. Whether this Hamiltonian description follows from a Lagrangian or simply from the incompressibility condition for some other kind of flow is irrelevant in all that follows. The main idea is the condition that the transformation yields another set of canonically conjugate phase space variables $(Q_1,\ldots,P_f)$, with Hamiltonian $K(Q,P,t)$ so that

$$\dot{Q}_i = \frac{\partial K}{\partial P_i}, \; \dot{P}_i = -\frac{\partial K}{\partial Q_i}. \tag{15.4}$$

Whenever this is true, then we call the transformation (15.3) canonical.

Our starting point for deriving the condition that the transformation (15.3) is canonical is the action principle. We can derive the first set of canonical equations from the action principle with all end points $q_i(t_1)$ and $q_i(t_2)$ fixed,

$$\delta A_1 = \delta \int_{t_1}^{t_2} \left(\sum_{i=1}^{f} p_i \dot{q}_i - H \right) dt = \int_{t_1}^{t_2} \left[\sum_{i=1}^{f} \left(-\dot{p}_i - \frac{\partial H}{\partial q_i} \right) \delta q_i \right.$$
$$\left. + \left(\dot{q}_i - \frac{\partial H}{\partial p_i} \right) \delta p_i \right] dt \tag{15.5}$$

and the second set follows from

$$\delta A_2 = \delta \int_{t_1}^{t_2} \left(\sum_{i=1}^{f} P_i \dot{Q}_i - K \right) dt = \int_{t_1}^{t_2} \left[\sum_{i=1}^{f} \left(-\dot{P}_i - \frac{\partial K}{\partial Q_i} \right) \delta Q_i \right.$$
$$\left. + \left(\dot{Q}_i - \frac{\partial K}{\partial P_i} \right) \delta P_i \right] dt, \tag{15.6}$$

also with all end points $Q_i(t_1)$ and $Q_i(t_2)$ fixed. If both sets of dynamical equations are to follow from varying only the first action A_1, then it follows that the two separate differential forms in the integrands for A_1 and A_2 must differ by a closed differential:

$$dS = \left(\sum_{i=1}^{f} p_i dq_i - H dt \right) - \left(\sum_{i=1}^{f} P_i dQ_i - K dt \right). \tag{15.7}$$

This yields $A_1 = A_2 + S(t_2) - S(t_1)$, so that $\delta A_1 = \delta A_2$. Formally, this is an extension

---

[1] Our transformations are related to those known in the older literature on geometry as contact transformations. The terminology was introduced by Lie and is explained in the texts by Sommerfeld (1964) and Whittaker (1965).

of our earlier treatment of gauge transformations, but whereas the form of S in that case was such as to yield exactly the same equations of motion from both A_1 and A_2, this case is different: K and H are generally not the same function, so that the two different Hamiltonian systems, related by a coordinate transformation, generally provide different mathematical descriptions of the same mechanical system (the exception to the rule occurs in the case of a symmetry transformation, in which case H is invariant).

The function S is called the generating function of the canonical transformation (15.3). It follows that $S(q,Q,t)$ satisfies the conditions

$$p_i = \frac{\partial S}{\partial q_i}, \; P_i = -\frac{\partial S}{\partial Q_i}, \text{ and } K = H + \frac{\partial S}{\partial t}. \tag{15.8}$$

The condition that $dS = L_1 dt - L_2 dt$ is a closed differential is nontrivial: each separate contribution $L_i dt$ to dS is generally *not* closed[2] and generally yields a path-dependent integral A_i. Path-dependence formed the basis for deriving Lagrange's equations of motion from the action principle in chapter 2.

Therefore, the condition that the transformation is canonical is satisfied if the generating function S can be constructed. The generating function provides a description of the transformation in the form

$$p_i = \frac{\partial S}{\partial q_i} = g_i(q,Q,t), \; P_i = -\frac{\partial S}{\partial Q_i} = h_i(q,Q,t). \tag{15.9}$$

These equations must, in principle, be solvable for the original form of the transformation: $Q_i = \phi_i(q,p,t)$ and $P_i = \psi_i(q,p,t)$, as well as for the inverse transformation, where (q,p) are expressed as functions of (Q,P,t). Therefore, the generating function $S(q,Q,t)$ provides a potentially complete description of the canonical transformation, and determination of the generating function determines the new Hamiltonian as well:

$$K = H + \frac{\partial S}{\partial t}. \tag{15.10}$$

Note that the Hamiltonian transforms like a scalar function, $K(Q,P,t) = K(\phi(q,p,t),\psi(q,p,t),t) = H(q,p,t)$, only if the canonical transformation is time-independent. If the reader has followed the discussion of Noether's theorem in chapter 2, then it will come as no surprise that canonical transformations (15.3) that leave the Hamiltonian *invariant*, those yielding $H(q,p,t) = H(Q,P,t)$, which means that K and H are the *same* function, will play a very special role in the search for solutions of Hamilton's equations (chapter 16).

Notice also that, on the right hand side of the expression for dS, there are $4f$ phase space variables (q,p,Q,P), but only $2f$ of these are independent. Above, we have chosen those $2f$ to be (q,Q) in our discussion of the generating function, but this is not the only choice: there are three other possibilities, (q,P), (Q,p), and (p,P), all of which can be reached by Legendre transformations. For example, with the notation $F_1 = S$, then

$$F_2 = F_1 + \sum_{i=1}^{f} Q_i P_i \tag{15.11}$$

also satisfies the condition that the two original differential forms differ by closed differential form, and we have a description of the same canonical transformation in

---

[2] We show in chapter 16 that Ldt is closed for *integrable* canonical systems.

terms of the generating function $F_2(q,P,t)$:

$$p_i = \frac{\partial F_2}{\partial q_i}, \quad Q_i = \frac{\partial F_2}{\partial P_i}, \text{ and } K = H + \frac{\partial F_2}{\partial t}. \qquad (15.12)$$

Notice that this means that when both F_1 and F_2 are expressed in terms of the variables (Q,P,t), then

$$\frac{\partial F_1}{\partial t} = \frac{\partial F_2}{\partial t} \qquad (15.13)$$

because we have the same transformation and the same Hamiltonian K in either case. With

$$F_3 = F_1 - \sum_i q_i P_i \qquad (15.14)$$

we obtain

$$-q_i = \frac{\partial F_3}{\partial p_i}, \quad -P_i = \frac{\partial F_3}{\partial Q_i}, \text{ and } K = H + \frac{\partial F_3}{\partial t} \qquad (15.15)$$

while

$$F_4 = F_3 + \sum_i P_i Q_i \qquad (15.16)$$

yields

$$-q_i = \frac{\partial F_4}{\partial p_i}, \quad Q_i = \frac{\partial F_4}{\partial P_i}, \text{ and } K = H + \frac{\partial F_4}{\partial t}. \qquad (15.17)$$

It is possible to write down the generating functions for several relatively simple canonical transformations. The first and most important of these is the identity transformation which is given, for example, by

$$F_2 = \sum_{i=1}^{f} q_i P_i. \qquad (15.18)$$

Direct calculation yields $Q_i = q_i$ and $P_i = p_i$.

Second, general coordinate transformations in configuration space, the subject of much of chapter 2, are canonical:

$$F_2 = \sum_{k=1}^{f} \phi_k(q_1,\ldots,q_f,t)P_k \qquad (15.19)$$

yields

$$Q_i = \phi_i(q_1,\ldots,q_f,t) \text{ and } p_i = \frac{\partial Q_k}{\partial q_i} P_k, \qquad (15.20)$$

where we have used summation convention in order to illustrate the result in the language of chapter 2 (the canonical momentum transforms like the covariant components of a configuration space vector). This is interesting: Hamilton's equations are not covariant under these transformations, but that does not matter because the general coordinate transformations of Lagrangian theory are only a subset of the much larger collection of canonical transformations. 'Manifest covariance' of dynamical equations is sometimes an important tool, but it is not necessary in order to formulate physics correctly. For the latter, one needs empirically correct invariance principles and equations of motion that reflect them, covariant or not.

Third, orthogonal transformations (including time-dependent transformation to rotating frames) are a subset of general configuration space transformations and therefore also are canonical.

Fourth, we can write a generating function that interchanges the labeling of the phase space variables: $F_1 = \Sigma q_i Q_i$ yields $p_i = Q_i$ and $q_i = -P_i$. In the original formulation of Lagrangian dynamics starting with $L = T - U$, the coordinates q_i label particle positions whereas the canonical momenta p_i are derivatives of the kinetic energy with respect to the generalized velocities dq_i/dt. Therefore, in the Lagrangian description, there is no equivalence of q and p. However, there are Hamiltonian systems where this interchange transformation leaves the Hamiltonian invariant: the simple harmonic oscillator written in the rescaled form $H = (p^2 + q^2)/2$ is only one example. Here, in the phase flow picture, there is no difference whatsoever between the coordinates q and p, and there are other systems with this kind of invariance (point vortices in the plane are another example).

Finally, we can see that the collection of all canonical transformations forms a group. First, starting with

$$\mathrm{d}S = \sum_{i=1}^{f} p_i \mathrm{d}q_i - \sum_{i=1}^{f} P_i \mathrm{d}Q_i + (K - H)\mathrm{d}t, \tag{15.21}$$

we see that the transformations $\mathrm{d}S_{12}(q,Q)$ and $\mathrm{d}S_{23}(Q,q')$ can be carried out sequentially, but also can be combined to yield the direct transformation described by $\mathrm{d}S_{13}(q,q') = \mathrm{d}S_{12}(q,Q) + \mathrm{d}S_{23}(Q,q')$, so that we have a group combination law for successive transformations. The reader can check to verify that this combination law is associative. Above, we have shown that the identity can always be constructed, and it is clear that the inverse of $\mathrm{d}S_{12}(q,Q)$ is $\mathrm{d}S_{21}(Q,q) = -\mathrm{d}S_{12}(q,Q)$. Therefore, all of the conditions for a group are satisfied.

15.2 Poisson brackets and infinitesimal transformations: symmetry, invariance, and conservation laws

We can introduce the relevant Poisson brackets in a simple way. Consider any point function $F(q,p,t)$ of the $2f$ phase space variables and the time t. Then

$$\frac{\mathrm{d}F}{\mathrm{d}t} = \sum_{i=1}^{f} \left(\frac{\partial F}{\partial q_i} \dot{q}_i + \frac{\partial F}{\partial p_i} \dot{p}_i \right) + \frac{\partial F}{\partial t} = \sum_{i=1}^{f} \left(\frac{\partial F \partial H}{\partial q_i \partial p_i} - \frac{\partial F \partial H}{\partial p_i \partial q_i} \right) + \frac{\partial F}{\partial t}, \tag{15.22}$$

which we can write as

$$\frac{\mathrm{d}F}{\mathrm{d}t} = \{F,H\}_{q,p} + \frac{\partial F}{\partial t} \tag{15.23}$$

where

$$\{F,G\}_{q,p} = \sum_{i=1}^{f} \left(\frac{\partial F}{\partial q_i} \frac{\partial G}{\partial p_i} - \frac{\partial F}{\partial p_i} \frac{G}{\partial q_i} \right) \tag{15.24}$$

is called the Poisson bracket of F with G, evaluated in the (q,p)-coordinate system[3]. In particular, Hamilton's equations can be written as

$$\dot{q}_i = \{q_i,H\} \text{ and } \dot{p}_i = \{p_i,H\}. \tag{15.25}$$

In fact, all derivatives in phase space can be expressed in terms of Poisson brackets. The fundamental Poisson bracket relations are

$$\{q_i,p_k\} = \varepsilon_{ik}, \ \{q_i,q_k\} = 0, \text{ and } \{p_i,p_k\} = 0, \tag{15.26}$$

where $\varepsilon_{ik} = \varepsilon(q_i, p_k) = - \varepsilon(p_k, q_i)$ and p_i are called canonically conjugate variables, and we could identify ε_{ik} as the Kronecker delta δ_{ik} were it not for the antisymmetry condition $\{p_k,q_k\} = - \{q_k,p_k\} = - 1$.

The important point to be proven now is that *all* Poisson bracket relations for scalar point functions F and G are invariant under canonical transformations. To show this, we can begin with the generating functions and use them to derive a direct way to determine whether a given set of variables is canonically conjugate. This will lead us automatically to the invariance of Poisson bracket relations.

The idea is to find conditions that guarantee that the transformation from the system

$$\dot{q}_i = \frac{\partial H}{\partial p_i}, \ \dot{p}_i = - \frac{\partial H}{\partial q_i} \tag{15.2}$$

actually yields

$$\dot{Q}_i = \frac{\partial K}{\partial P_i}, \ \dot{P}_i = - \frac{\partial K}{\partial Q_i} \tag{15.4}$$

with

$$K = H + \frac{\partial S}{\partial t}. \tag{15.27a}$$

We illustrate the proof using notation for $f = 1$ for simplicity, but the proof goes through in the same way for $f > 1$. Starting with $Q = \phi(q,p,t)$ and $P = \psi(q,p,t)$, we have

$$\dot{Q} = \frac{\partial \phi}{\partial q} \dot{q} + \frac{\partial \phi}{\partial p} \dot{p} + \frac{\partial \phi}{\partial t} = \frac{\partial Q}{\partial q} \frac{\partial H}{\partial p} - \frac{\partial Q}{\partial q} \frac{H}{\partial q} + \frac{\partial \phi}{\partial t}. \tag{15.28}$$

Using

$$H = K - \frac{\partial S}{\partial t} \tag{15.27b}$$

then yields

$$\dot{Q} = \frac{\partial Q}{\partial q} \frac{\partial K}{\partial p} - \frac{\partial Q}{\partial q} \frac{\partial^2 S}{\partial t \partial p} - \frac{\partial Q}{\partial p} \frac{\partial K}{\partial q} + \frac{\partial Q}{\partial p} \frac{\partial^2 S}{\partial q \partial t} + \frac{\partial \phi}{\partial t}, \tag{15.29a}$$

[3] This differs from the definition of Poisson brackets introduced in chapter 7. We show the connection between the two definitions below.

which we can rewrite as

$$\dot{Q} = \frac{\partial Q}{\partial q}\left(\frac{\partial K}{\partial Q}\frac{\partial Q}{\partial p} + \frac{\partial K}{\partial P}\frac{\partial P}{\partial p}\right) - \frac{\partial Q}{\partial q}\frac{\partial^2 S}{\partial t\partial p} - \frac{\partial Q}{\partial p}\left(\frac{\partial K}{\partial Q}\frac{\partial Q}{\partial q} + \frac{\partial K\partial P}{\partial P\partial q}\right)$$
$$+ \frac{\partial Q}{\partial p}\frac{\partial^2 S}{\partial q\partial t} + \frac{\partial \phi}{\partial t}. \tag{15.29b}$$

We can rearrange this to read

$$\dot{Q} = \frac{\partial K}{\partial Q}\left(\frac{\partial Q}{\partial q}\frac{\partial Q}{\partial p} - \frac{\partial Q}{\partial p}\frac{\partial Q}{\partial q}\right) + \frac{\partial K}{\partial P}\left(\frac{\partial Q}{\partial q}\frac{\partial P}{\partial p} - \frac{\partial Q\partial P}{\partial p\partial q}\right)$$
$$- \frac{\partial Q}{\partial q}\frac{\partial^2 S}{\partial t\partial p} + \frac{\partial Q}{\partial p}\frac{\partial^2 S}{\partial q\partial t} + \frac{\partial \phi}{\partial t}, \tag{15.29c}$$

and also

$$\dot{Q} = \frac{\partial K}{\partial P}\{Q,P\}_{q,p} - \frac{\partial Q}{\partial q}\frac{\partial^2 S}{\partial t\partial p} + \frac{\partial Q}{\partial p}\frac{\partial^2 S}{\partial q\partial t} + \frac{\partial \phi}{\partial t}. \tag{15.29d}$$

From the condition that

$$\dot{Q} = \frac{\partial K}{\partial P}, \tag{15.29e}$$

we obtain two results. First, it is necessary that

$$\{Q,P\}_{q,p} = 1. \tag{15.30}$$

Second, using

$$p = \frac{\partial F_2}{\partial q} \text{ and } Q = \frac{\partial F_2}{\partial P} \tag{15.31}$$

we see that we need

$$\left\{Q,\frac{\partial F_2}{\partial t}\right\}_{q,p} = \frac{\partial^2 S}{\partial P\partial t} \tag{15.32}$$

in order that

$$-\left\{Q,\frac{\partial S}{\partial t}\right\}_{q,p} + \frac{\partial \phi}{\partial t} = -\frac{\partial^2 S}{\partial P\partial t} + \frac{\partial^2 S}{\partial t\partial P} = 0. \tag{15.33}$$

However, it follows from a short calculation that for *arbitrary scalar phase space* functions $F'(Q,P) = F(q,p)$ and $G'(Q,P) = G(q,p)$,

$$\{F',G'\}_{Q,P} = \{F,G\}_{q,p}\{q,p\}_{Q,P}, \tag{15.34}$$

so that *all* Poisson bracket relations are preserved under canonical transformations if the *fundamental* bracket relations (15.26) are preserved. Under these circumstances, it follows that

$$\{Q,F\}_{q,p} = \frac{\partial F}{\partial P}, \tag{15.35}$$

of which the condition (15.32) is only one special case. In other words, the invariance of the fundamental bracket relations under canonical transformations guarantees that all other Poisson bracket relations are preserved as well:

$$\{F',G'\}_{Q,P} = \{F,G\}_{q,p}\{q,p\}_{Q,P} = \{F,G\}_{q,p}. \tag{15.36}$$

Likewise, with $p = \tilde{\psi}(Q,P,t)$, we obtain

$$\dot{p} = \frac{\partial \tilde{\psi}}{\partial Q}\dot{Q} + \frac{\partial \tilde{\psi}}{\partial P}\dot{P} + \frac{\partial \tilde{\psi}}{\partial t} = \frac{\partial p \partial K}{\partial Q \partial P} - \frac{\partial p \partial K}{\partial P \partial Q} + \frac{\partial \tilde{\psi}}{\partial t}. \tag{15.37}$$

Using

$$K = H + \frac{\partial S}{\partial t} \tag{15.27a}$$

yields

$$\dot{p} = \frac{\partial p}{\partial Q}\left(\frac{\partial H}{\partial q}\frac{\partial q}{\partial P} + \frac{\partial H}{\partial p}\frac{\partial p}{\partial P}\right) + \frac{\partial p}{\partial Q}\frac{\partial^2 S}{\partial t \partial P} - \frac{\partial p}{\partial P}\left(\frac{\partial H}{\partial q}\frac{\partial q}{\partial Q} + \frac{\partial H \partial p}{\partial p \partial Q}\right)$$
$$- \frac{\partial p}{\partial P}\frac{\partial^2 S}{\partial Q \partial t} + \frac{\partial \tilde{\psi}}{\partial t}. \tag{15.38a}$$

This can be rearranged to read

$$\dot{p} = -\frac{\partial H}{\partial q}\{q,p\}_{Q,P} + \left\{p,\frac{\partial S}{\partial t}\right\}_{Q,P} + \frac{\partial p}{\partial t}. \tag{15.38b}$$

Using

$$p = \frac{\partial F_2}{\partial q} \text{ and } Q = \frac{\partial F_2}{\partial P}, \tag{15.39}$$

it is necessary that

$$\left\{p,\frac{\partial S}{\partial t}\right\}_{Q,P} + \frac{\partial p}{\partial t} = -\frac{\partial^2 S}{\partial q \partial t} + \frac{\partial^2 S}{\partial t \partial q} = 0 \tag{15.40}$$

and

$$\{q,p\}_{Q,P} = 1 \tag{15.41}$$

in order that

$$\dot{p} = -\frac{\partial H}{\partial q} \tag{15.42}$$

is satisfied.

Continuing similarly, one can prove quite generally that all of the fundamental bracket relations $\{q_i,p_j\} = \delta_{ij}$, $\{q_i,q_j\} = 0$, and $\{p_i,p_j\} = 0$, are preserved under general time-dependent canonical transformations. It follows that all Poisson bracket relations of scalar functions are preserved as well, so that it makes no difference which set of canonically conjugate coordinates and momenta one uses to evaluate them. In other words, we can drop the subscripts $\{,\}_{q,p}$ or $\{,\}_{Q,P}$ on Poisson brackets and use *any* complete canonical set of variables to evaluate a set of Poisson bracket relations.

The invariance of the Euclidean scalar product under the symmetry operations of Euclidean space defines the objects that are left invariant (are left similar to themselves) under those transformations. According to F. Klein, and more generally, the geometric properties of any space are defined by the objects that are left invariant under a group of transformations, so that it is always a transformation group that defines the metric geometry of a space. Straight lines transform into straight lines, circles into circles, and rigid bodies into other rigid configurations of themselves under rigid translations and rigid rotations, the basic symmetry operations of Euclidean space. As we have noted above, the phase space of Hamiltonian systems has many properties similar to Cartesian coordinates in Euclidean space but there is no idea of a distance between two points (no metric) in phase space. However, Klein's idea still applies: the invariance property that defines the geometry of phase space, which was called symplectic geometry by H. Weyl, is simply a certain tensor (section 15.6) that guarantees the preservation of scalar Poisson brackets relations under canonical transformations. In the modern mathematics literature canonical transformations are called symplectic, and canonically conjugate variables are called symplectic coordinates. Since there is no metric, the invariances of symplectic geometry are qualitative, like the idea of the invariance of the nature of a Cartesian torus under differentiable coordinate transformations. Tori turn out to be important for bounded motion in Hamiltonian dynamics (chapter 16) because, with symplectic coordinates $(q_1,\dots,p_f)$ we are always working on a flat manifold, and the torus is the only closed, orientable Cartesian manifold. In phase space in general, we are not interested in the idea of metric invariance but concern ourselves with the more useful idea of topologic invariance, the idea (for example) that simple closed curves are equivalent to a circle, and that integrable systems are equivalent to a set of translations at constant speed. Constant speed translations can occur on any surface, curved or flat, and the closed manifolds of bounded motion of integrable noncanonical flows are typically curved rather than flat.

Consider next a one-parameter continuous group of canonical transformations

$$Q_i = q_i(\alpha) = \phi_i(q,p,\alpha), \ P_i = p_i(\alpha) = \psi_i(q,p,\alpha) \tag{15.43}$$

where each transformation in the group is generated by varying the group parameter α, and where $\alpha = 0$ yields the identity: $Q_i = q_i(0) = q_i$ and $P_i = p_i(0) = p_i$. Assume also that α is a canonical parameter for the transformation group (see chapters 7 and 12 for the definition of a canonical parameter). For an infinitesimal transformation, we write

$$Q_i = q_i + \delta q_i \text{ and } P_i = p_i + \delta p_i \tag{15.44}$$

where

$$\delta q_i = \frac{\partial \phi_i}{\partial \alpha} \delta\alpha \text{ and } \delta p_i = \frac{\partial \psi_i}{\partial \alpha} \delta\alpha. \tag{15.45}$$

Since the fundamental Poisson bracket relations must be satisfied,

$$\{Q_i,P_k\}_{q,p} = \{q_i,p_k\}_{q,p} = \varepsilon_{ik}, \tag{15.46}$$

it follows that

$$\{\delta q_i,\delta p_k\}_{q,p} = 0 \tag{15.47a}$$

to first order in $\delta\alpha$. Written out explicitly, this states, for $i = k$, that

$$\frac{\partial \delta q_i}{\partial q_i} + \frac{\partial \delta p_i}{\partial p_i} = 0 \tag{15.47b}$$

for each $i = 1, \ldots, f$. But this is just the requirement for a divergence-free velocity field $V_i(\alpha) = (dq_i(\alpha)/d\alpha, dp_i(\alpha)/d\alpha)$ in phase space, one where the canonical group parameter α plays the role of the 'time'. That is, the velocity field must be given by a stream function G:

$$\frac{dq_i(\alpha)}{d\alpha} = \frac{\partial G}{\partial p_i} \text{ and } \frac{dp_i(\alpha)}{d\alpha} = -\frac{\partial G}{\partial q_i}. \tag{15.48}$$

The Hamiltonian form of these equations is not accidental: every one-parameter continuous group defines a flow in phase space (see chapter 7), and whenever that flow is phase-volume-preserving (is divergence free[4]), then the flow can be put into canonical form by finding a stream function, or Hamiltonian, G. The Hamiltonian G is called the infinitesimal generator of the group of transformations, and α, the canonical group parameter, is analogous to the time (α obeys an additive group parameter combination law). It follows without fanfare that the Hamiltonian H is the infinitesimal generator of a dynamical system where the canonical group parameter is the time t (this is half of the meaning of 'canonical' in the context of Hamiltonian mechanics – the other half of the meaning is stated below). We now see what the Hamiltonian formulation accomplishes: it places all volume-preserving flow in phase space on an equal footing, because every divergence-free flow defines a set of stream functions G_k that generate the group. For a one dimensional group, whether the group parameter is the time t or some other canonical parameter is unimportant for the development of the mathematical theory.

An example of an infinitesimal generator is the angular momentum about some axis, say, the x_3-axis where (x_1, x_2, x_3) are Cartesian coordinates: with $G = M_3 = \varepsilon_{3ij} x_i p_j = (x_1 p_2 - x_2 p_1)$, where $p_j = m dx_j/dt$, we obtain

$$\frac{dx_1}{d\theta} = \frac{\partial G}{\partial p_1} = -x_2, \quad \frac{dx_2}{d\theta} = \frac{\partial G}{\partial p_2} = x_1, \quad \frac{dp_1}{d\theta} = -\frac{\partial G}{\partial x_1} = -p_2, \quad \frac{dp_2}{d\theta} = -\frac{\partial G}{\partial x_2} = p_1, \tag{15.49}$$

which are the differential equations of an infinitesimal rotation about the x_3-axis in phase space. We leave it as an easy exercise for the reader to integrate these equations and obtain the form of the finite rotations.

An even easier example is that of translations along the Cartesian $(q_1, \ldots, p_f)$-coordinate axes in phase space. If we take $G = -q_k$, then we get a pure translation along the p_k-axis, whereas $G = p_k$ yields a pure translation along the q_k-axis.

Now, we can connect the infinitesimal generator to the generating function. It is easy to verify by direct calculation that the generating function F_2 is given, to first order in $\delta\alpha$, by

$$F_2(q, P) \approx \sum_{i=1}^{f} q_i P_i + \delta\alpha G. \tag{15.50}$$

That is, this expression defines the generating function near the identity (S. Lie introduced and developed the method of studying groups of transformations near the identity). To first order in $\delta\alpha$, we obtain from this the equations of the infinitesimal transformation

---

[4] Liouville's theorem is satisfied by every flow that is generated by a stream function G; that is, the $2f$ dimensional phase space volume element is preserved under the canonical transformations of the group. It is possible to prove even more generally that the phase space volume element is preserved under all canonical transformations.

$$\delta q_i = \delta \alpha \frac{\partial G}{\partial p_i} \text{ and } \delta p_i = -\delta \alpha \frac{\partial G}{\partial q_i}. \tag{15.51}$$

Under the infinitesimal transformation generated by G, a scalar phase space function $F(q,p)$ changes by the amount

$$\mathrm{d}F = \delta \alpha X F = \delta a\{F,G\} \tag{15.52}$$

where the first order differential operator X is defined by

$$X = \{\ ,G\} = \sum_{i=1}^{f} \left(\frac{\partial G}{\partial p_i} \frac{\partial}{\partial q_i} - \frac{\partial G}{\partial q_i} \frac{\partial}{\partial p_i} \right). \tag{15.53}$$

We can construct the finite transformation by iterating the infinitesimal transformation. Since

$$F + \mathrm{d}F = (1 + \delta\alpha X)F, \tag{15.54}$$

iteration with $\delta\alpha = \alpha/N$ yields the finite transformation in the form

$$F(q(\alpha),p(\alpha)) = e^{\alpha X} F(q,p) \tag{15.55}$$

where

$$e^{\alpha X} \approx \left(1 + \frac{\alpha}{N} X \right)^N. \tag{15.56}$$

Equally directly, we can write

$$F(q(\alpha),p(\alpha)) = F(q,p) + \delta\alpha\{F,G\} + \delta\alpha^2\{\{F,G\},G\} + \cdots$$
$$= F(q,p) + \delta\alpha X F + \delta\alpha^2 X^2 F + \cdots = e^{\alpha X} F(q,p) \tag{15.57}$$

where $U(\alpha) = e^{\alpha X}$ is the group operator, and is analogous to the time-evolution operator. Clearly, we assume here that the symmetry generated by G is globally analytic except for isolated singularities (see section 7.6 for a brief discussion of the convergence of series representations of group operators).

In particular, the finite transformations of the group are given by

$$q_i(\alpha) = \phi_i(q,p,\alpha) = e^{\alpha X} q_i(0) \text{ and } p_i(\alpha) = \psi_i(q,p,\alpha) = e^{\alpha X} p_i(0). \tag{15.58}$$

At this stage, it is a good idea to have an example. With $G = M_3 = (x_1 p_2 - x_2 p_1)$ and $p_i = m\mathrm{d}x_i/\mathrm{d}t$, we have the generator of rotations in the (x_1,x_2)-plane, and $\alpha = \theta$ is the rotation angle. Let's calculate $x_1(\theta) = U(\theta)x_1 = \phi_1(x_1,x_2,p_1,p_2,\theta)$ as an exercise. By using the fact (easy to prove) that

$$\{A,BC\} = \{A,B\}C + B\{A,C\}, \tag{15.59}$$

we obtain from the fundamental bracket relations that $Xx_1 = \{x_1,x_1 p_2 - x_2 p_1\} = -x_2$, $X^2 x_1 = -Xx_2 = -\{x_2,x_1 p_2 - x_2 p_1\} = -x_1$, $X^3 x_1 = -Xx_1 = x_2$, and $X^4 x_1 = Xx_2 = x_1$, so that the circle closes. This yields, by resummation of the power series in θ, that

$$x_1(\theta) = \phi_1(x,p,\theta) = x_1 \cos\theta - x_2 \sin\theta, \tag{15.60}$$

so that we have calculated the transformation function ϕ_1 (it is independent of the p_i because rigid rotations do not mix coordinates and momenta). It is equally easy to calculate the other three transformation functions.

There is a deep connection between invariance principles, symmetries, and conservation laws that we can now express in a very beautiful and useful form, a form that follows from treating all group generators on an equal footing with the Hamiltonian, which is the generator of translations in time.

As we have pointed out above, under an infinitesimal transformation generated by G, a scalar phase space function $F(q,p)$ changes by the amount

$$dF = \delta a X F = \delta a \{F,G\} \tag{15.52}$$

where the operator X is defined by

$$X = \{\ ,G\} = \sum_{i=1}^{f} \left(\frac{\partial G}{\partial p_i} \frac{\partial}{\partial q_i} - \frac{\partial G}{\partial q_i} \frac{\partial}{\partial p_i} \right). \tag{15.53}$$

Therefore, functions that are invariant under the transformations generated by G satisfy the first order partial differential equation

$$XF = \{F,G\} = 0. \tag{15.61}$$

It follows from this, combined with the time-evolution law

$$dF/dt = \{F,H\}, \tag{15.62}$$

that *a conserved quantity leaves the Hamiltonian invariant and therefore generates a continuous symmetry operation on the dynamical system.* The point is simple and follows from a two-pronged interpretation: regarded as

$$\frac{dG}{dt} = \{G,H\}, \tag{15.63}$$

$\{G,H\} = 0$ means that G is conserved. Regarded as

$$\frac{dH}{da} = \{H,G\}, \tag{15.64}$$

$\{G,H\} = 0$ means that the Hamiltonian is left invariant by the one-parameter group of transformations generated by G. For example, a rotationally symmetric Hamiltonian is one where $\{M_i,H\} = 0$ for all three components M_i of the angular momentum (because the M_i are the generators of infinitesimal transformations in the three mutually perpendicular Euclidean planes), and this means that all three components of angular momentum are conserved by the motion of the dynamical system. In general, conservation laws generate symmetries of the dynamical system and vice versa. This illustrates the deep connection between symmetry, invariance, and conservation laws that holds both in classical and in quantum mechanics (recall our discussion of Noether's theorem in chapter 2).

More explicitly, let us consider the case where the Hamiltonian H is invariant under rotations in the (x,y)-plane. Here, the angular momentum $M_z = xp_y - yp_x$ is the infinitesimal generator of rotations about the z-axis, and rotational invariance yields the first order partial differential equation

$$\{H,M_z\} = x \frac{\partial H}{\partial y} - y \frac{\partial H}{\partial x} + p_x \frac{\partial H}{\partial p_y} - p_y \frac{\partial H}{\partial p_x} = 0. \tag{15.65}$$

This yields (if we integrate (15.65) using the method of characteristics, as in Duff, 1962, for example) that if we use four Cartesian variables (x,y,p_x,p_y), then H can only have the form $H(p_x^2 + p_y^2, x^2 + y^2, xp_x + yp_y, xp_y - yp_x)$. That is, H can depend only upon the coordinates and momenta in combinations that are left invariant under a rotation about the z-axis.

Finally, we have observed that the Hamiltonian flow is a Lie group flow where the infinitesimal generator is the Hamiltonian H and the group parameter is the time t. That is, the motion of the dynamical system is a canonical transformation in phase space:

$$q_i(t) = \phi_i(q,p,t) \text{ and } p_i(t) = \psi_i(q,p,t) \tag{15.66}$$

where (q,p) denote the $2f$ initial conditions $(q_1(0),\dots,p_f(0))$, and

$$\{q_i(t),p_j(t)\}_{q,p} = \delta_{ij}, \ \{q_i(t),q_j(t)\}_{q,p} = 0, \text{ and } \{p_i(t),p_j(t)\}_{q,p} = 0. \tag{15.67}$$

In section 6.4, in order to prove that the return map

$$\left.\begin{array}{l} q_n = G_1(q_{n-1},p_{n-1}) \\ p_n = G_2(q_{n-1},p_{n-1}), \end{array}\right\} \tag{6.9}$$

for a two degree of freedom Hamiltonian system is area-preserving, that $dq_{n-1}dp_{n-1} = dq_n dp_n$, we showed that this property follows if the fundamental Lagrange brackets

$$[q_j(t),q_k(t)] = 0, \ [p_j(t),p_k(t)] = 0, \ [q_j(t),p_k(t)] = \delta_{jk} \tag{6.24a}$$

are time-translational invariant, which would mean that

$$[q_j(t'),q_k(t')] = 0, \ [p_j(t'),p_k(t')] = 0, \ [q_j(t'),p_k(t')] = \delta_{jk} \tag{6.24b}$$

but where the Lagrange brackets in (6.24a) and (6.24b) are defined with respect to the variables $(q_1(t),\dots,p_f(t))$:

$$[F,G] = \sum_{i=1}^{f} \frac{\partial(q_i(t),p_i(t))}{\partial(F,G)}. \tag{6.22}$$

We now prove that the fundamental Lagrange bracket relations are invariant under arbitrary canonical transformations. Let $u_i = f_i(q_1,\dots,p_f)$ denote any $2f$ variables constructed from the $(q_1,\dots,p_f)$ by transformations f_i. By $2f$ independent variables we mean that

$$\frac{\partial u_j}{\partial u_k} = \delta_{jk}. \tag{15.68}$$

Consider next the sum

$$\sum_{l=1}^{2f} [u_l,u_i]\{u_l,u_j\} = \sum_{l,k,m} \frac{\partial(q_k,p_k)}{\partial(u_l,u_i)} \frac{\partial(u_l,u_j)}{\partial(q_m,p_m)} \tag{15.69a}$$

where we temporarily define Lagrange brackets with respect to canonical variables denoted as $(q_1,\dots,p_f)$:

$$[F,G] = \sum_{i=1}^{f} \frac{\partial(q_i,p_i)}{\partial(F,G)}. \tag{15.70}$$

By direct expansion of (15.69a), requiring only two lines of calculus, we obtain

$$\sum_{l=1}^{2f} [u_l,u_i]\{u_l,u_j\} = \sum_{k,m=1}^{f} \left(\frac{\partial u_j}{\partial p_k}\frac{\partial p_k}{\partial u_i} + \frac{\partial u_j}{\partial q_k}\frac{\partial q_k}{\partial u_i} \right) = \frac{\partial u_j}{\partial u_i} = \delta_{ji}. \tag{15.69b}$$

Next, let $(Q_1,\ldots,P_f)$ be any other set of canonically conjugate variables obtained from $(q_1,\ldots,p_f)$ by a canonical transformation. Since the variables $(Q_1,\ldots,P_f)$ satisfy the fundamental Poisson bracket relations

$$\{Q_i,P_k\} = \varepsilon_{ik}, \ \{Q_i,Q_k\} = 0, \text{ and } \{P_i,P_k\} = 0 \tag{15.69c}$$

with respect to the original variables $(q_1,\ldots,p_f)$, it follows from inserting (15.69b) into (15.69c) that the Lagrange bracket relations

$$[Q_j,Q_k] = 0, \ [P_j,P_k] = 0, \ [Q_j,P_k] = \varepsilon_{jk} \tag{15.71}$$

also hold with respect to the variables $(q_1,\ldots,p_f)$. In other words, like the fundamental Poisson bracket relations, the fundamental Lagrange bracket relations are canonical invariants. Consequently, the iterated map (6.24a) is area-preserving because the integral

$$I_2 = \sum_{i=1}^{f} \oint\!\!\int dq_i dp_i \tag{6.19a}$$

is invariant under canonical transformations.

Consider next the closed line integral

$$\sum_{i=1}^{n} \oint M_i dx_i \tag{15.72}$$

in an n dimensional phase space where M is a vector. By Stokes' theorem

$$\sum_{i=1}^{n} \oint M_i dx_i = \sum_{i,k=1}^{n} \int\!\!\int \left(\frac{\partial M_i}{\partial x_k} - \frac{\partial M_k}{\partial x_i} \right) dx_i dx_k. \tag{15.73}$$

With $x_i = q_i$ and $x_{i+f} = p_i$, $i = 1,\ldots,f$, and also $M_i = p_i$, $i = 1,\ldots,f$ but $M_i = 0$ if $i > f$, then (6.19a) becomes

$$\sum_{i=1}^{f} \oint p_i dq_i = \sum_{i=1}^{f} \int\!\!\int dq_i dp_i, \tag{15.74}$$

which shows that the closed loop integral on the left hand side of (6.19a) is also invariant under canonical transformations. This result is essential for the discussion of integrability in the next chapter.

15.3 Lie algebras of Hamiltonian flows

The generalization to a multiparameter group of transformations is immediate: with m group parameters $\alpha = (\alpha_1,\ldots,\alpha_m)$, then $Q_i = \phi_i(q,p,\alpha)$ and $P_i = \psi_i(q,p,\alpha)$ yield

$$F_2(q,P) \approx \sum_{i=1}^{f} q_i P_i + \sum_{\sigma=1}^{m} \delta\alpha_\sigma G_\sigma \tag{15.75}$$

and the infinitesimal transformations are given by

$$\left.\begin{aligned}
\delta q_i &= Q_i - q_i \approx \sum_{\sigma=1}^{m} \delta\alpha_\sigma \frac{\partial G_k}{\partial q_i} \\
\delta p_i &= P_i - p_i \approx - \sum_{\sigma=1}^{m} \delta\alpha_\sigma \frac{\partial G_\sigma}{\partial p_i},
\end{aligned}\right\} \tag{15.76}$$

which is a generalization of the equations of Hamilton to the case of several different infinitesimal generators G_k in a multiparameter flow. In this case, a scalar phase space function $F(q,p)$ transforms according to $F \to F + dF$ where

$$dF = \sum_{i=1}^{f} \left(\delta q_i \frac{\partial F}{\partial q_i} + \delta p_i \frac{\partial F}{\partial p_i} \right) = \sum_{\sigma=1}^{m} \delta\alpha_\sigma X_\sigma F \tag{15.77}$$

and the operators X_σ are defined by

$$X_\sigma = V \cdot \nabla = \frac{\partial G_\sigma \partial}{\partial p_i \partial q_i} - \frac{\partial G_\sigma \partial}{\partial q_i \partial p_i} = \{ \, , G_\sigma \} \tag{15.78}$$

in terms of the m different velocity fields V_σ^i of the m-parameter group flow

$$V_\sigma^i = \left(\frac{\partial G_\sigma}{\partial p_i}, \, -\frac{\partial G_\sigma}{\partial q_i} \right) \tag{15.79}$$

and the phase space gradient ∇.

Recall that in chapter 7, where we did not restrict to Hamiltonian or even to conservative systems, we arrived quite naturally at the following definition of a Poisson bracket of the velocity fields for two different flows:

$$[V_1, V_2]_k = \sum_{i=1}^{m} \left(V_2^i \frac{\partial V_1^k}{\partial x_i} - V_1^i \frac{\partial V_2^k}{\partial x_i} \right), \tag{15.80}$$

for phase space variables $(x_1, \dots, x_n)$, which we can now take to be $(q_1, \dots, p_f)$ with $n = 2f$. We also showed that Lie's integrability condition, the condition that the point transformations

$$Q_i = \phi_i(q, p, \alpha) \text{ and } P_i = \psi_i(q, p, \alpha) \tag{15.81}$$

exist globally, is satisfied whenever the m different vector fields close under the Poisson bracket operation,

$$[V_\kappa^i(x), V_\gamma^i(x)] = c_{\kappa\gamma}^\tau V_\tau^i(x), \tag{15.82}$$

and thereby form a complete basis set of vectors in the tangent space near the group identity. The coefficients $c_{\kappa\gamma}^\tau$ are constants, the 'structure constants' of the group of transformations.

This definition of the Poisson bracket is more general than the one that we have adopted earlier in this chapter. To demonstrate this, we now derive the bracket $\{G_\alpha, G_\beta\}$ for two infinitesimal generators from the brackets of their corresponding velocity fields:

$$[V_1, V_2]_k = \frac{\partial}{\partial q_k} \{G_1, G_2\} \tag{15.83}$$

for $n = 1, \ldots, f$, and

$$[V_1, V_2]_k = \frac{\partial}{\partial p_k} \{G_1, G_2\} \tag{15.84}$$

for $n = f + 1, \ldots, 2f$, which completes the proof. Furthermore, because the vector fields generated by the G_σ satisfy the Jacobi identity (see chapter 7), so do the infinitesimal generators themselves:

$$\{G_\mu, \{G_\nu, G_\sigma\}\} + \{G_\sigma, \{G_\mu, G_\nu\}\} + \{G_\nu, \{G_\sigma, G_\mu\}\} = 0. \tag{15.85}$$

The integrability condition for the m-parameter group of transformations can also be written (see chapter 7) as the closure condition for infinitesimal operators of the Lie algebra under commutation,

$$(X_\alpha, X_\beta) = c_{\alpha\beta}^\gamma X_\gamma, \tag{15.86}$$

where the constants $c_{\alpha\beta}^\gamma$ are the same structure constants as appear in the Poisson bracket closure condition for the velocity fields above. By using Jacobi's identity, we obtain from the commutator closure conditions the closure conditions

$$\frac{\partial}{\partial x_i} \{G_\alpha, G_\beta\} = c_{\alpha\beta}^\gamma \frac{\partial G_\gamma}{\partial x_i} \tag{15.87}$$

in terms of the 'new' Poisson brackets, which we shall refer to from now on simply as the Poisson brackets. Integration then yields the closure condition in terms of the infinitesimal generators G_σ in the form

$$\{G_\alpha, G_\beta\} = c_{\alpha\beta}^\gamma G_\gamma + \beta_{\alpha\beta}, \tag{15.88}$$

where the $\beta_{\alpha\beta} = -\beta_{\beta\alpha}$ are antisymmetric integration constants.

The constants $\beta_{\alpha\beta}$ are extremely important: they allow us to interpret the fundamental Poisson bracket relations

$$\{q_i, q_k\} = 0, \quad \{p_i, p_k\} = 0, \quad \{q_i, p_k\} = \varepsilon_{ik} \tag{15.89}$$

group theoretically: the fundamental bracket relations, which can be taken as the definition of canonically conjugate variables, mean that any set of $2f$ canonically conjugate variables (q, p) are infinitesimal generators for a complete set of *commuting* translations in the $2f$ dimensional phase space, and that q_i is the canonical parameter for the infinitesimal generator p_i and vice versa. So, we finally can see why the phase space coordinates (q, p) are called canonically conjugate: the meaning is group theoretical. In other words, the quantity $G_i = -q_i$ is the infinitesimal generator of translations along the p_i-axis, and p_i is the generator of translations along the q_i-axis.

An equivalent way to say this is that the fundamental bracket relations merely represent the fact that the coordinates $(q_1, \ldots, p_f)$ define Cartesian axes on the phase space manifold. This is the same as saying that phase space is a 'flat' manifold: the commutation of translations is equivalent to the statement that the parallel displacement of an arbitrary vector along the infinitesimal path from (q, p) to $(q + \delta q, p)$ to $(q + \delta q, p + \delta p)$ yields precisely the same result as does parallel displacement of the same vector along the path (q, p) to $(q, p + \delta p)$ to $(q + \delta q, p + \delta p)$. In contrast, translations along orthogonal axes on a curved manifold do not commute with each other: on the surface of a sphere (or on any curved manifold) two infinitesimal translations along two locally perpendicular axes like the spherical coordinate axes

(θ,ϕ) do not commute with each other (this is the famous 'deficit' obtained in parallel transport of tangent vectors along alternative infinitesimal paths that form a closed loop). Hence, the commutation rules $(M_i,M_j) = \varepsilon_{ijk}M_k$ for rotations in a three dimensional flat manifold can be interpreted as the lack of commutation of translations on a curved two dimensional manifold, the sphere.

Carrying this further, we can identify the generator of rotations in phase space by $M_i = \varepsilon_{ijk}q_jp_k$. It is easy to show that

$$\left.\begin{aligned}
\{M_i,M_j\} &= \varepsilon_{ijk}M_k\\
\{q_i,M_j\} &= \varepsilon_{ijk}q_k\\
\{p_i,M_j\} &= \varepsilon_{ijk}p_k.
\end{aligned}\right\} \tag{15.90}$$

These bracket relations, combined with the fundamental Poisson bracket relations (15.89) show that we have a Lie algebra where the q's, p's, and M's are the generators of a complete set of translations and rotations for the Cartesian axes of phase space. Hamiltonian mechanics allows one to perform translations and rotations on a symplectic manifold because the manifold is flat and the coordinates $(q_1,\dots,p_f)$ are Cartesian. The fact that the q's and p's in phase space are always of Cartesian character has implications for quantization rules: one cannot directly apply the rules of 'canonical quantization' described in the next section except when the quantum system is described in Cartesian coordinates. Transformations to some, but not all, other canonical coordinate systems can be carried after quantization has been performed.

If, as in our study of central potentials in chapter 4, we consider rigid rotations with Cartesian coordinates (x_1,x_2,x_3) in Euclidean space, then the components of angular momentum have the form $M_i = \varepsilon_{ijk}x_jp_k$ where $p_k = mdx_k/dt$ and $\{x_j,p_k\} = \varepsilon_{jk}$. It follows by direct calculation that

$$\{M_i,M_j\} = \varepsilon_{ijk}M_k, \tag{15.91}$$

so that the structure constants of the group of proper rotations in three dimensional Euclidean space (this group is denoted by $SO^+(3)$) has structure constants equal to ε_{ijk}, a fact that we already know from chapter 8. Whenever the Hamiltonian is spherically symmetric, then $\{M_i,H\} = 0$ for $i = 1,2,3$, but we can only use one of these conserved quantities, not all three, to reduce the problem by eliminating a degree of freedom.

The point here is that only compatible conserved quantities, quantities that satisfy both $\{G_\sigma,H\} = 0$ and $\{G_\mu,G_\nu\} = 0$ can be used to eliminate (isolate) degrees of freedom. Compatible conserved quantities are said to be 'in involution' in the older literature on mechanics and Lie algebras, but physicists also call them 'commuting constants of the motion' because the operators defined by any two of them commute whenever the respective Poisson bracket vanishes. Why is compatibility necessary? In other words, why can't we use two or more noncommuting conserved quantities simultaneously to reduce the number of degrees of freedom and thereby simplify the problem to be solved? This was explained in chapter 12 and can also be stated as follows. Let θ_i be the generalized coordinate conjugate to the generator M_i. That M_i is conserved means that θ_i does not appear in H. Hence, $p_i = M_i$ is a conserved canonical momentum. Canonical momenta must commute, $\{p_i,p_j\} = 0$, so that two or more noncommuting generators, whether or not they are conserved, cannot simultaneously serve as canonical momenta. Noncommuting conserved quantities are said to be incompatible, since only one of them can be chosen to be a conserved canonical momentum in the attempt to reduce the dimension of the problem. In the language of chapter 12, either of the θ_i taken alone qualifies as a generalized coordinate, but taken together the pair (θ_1,θ_2) is nonholonomic.

15.4 The Dirac quantization rules

The final empirical discovery that allowed the construction of quantum mechanics was made by Heisenberg, who (as did many others) tried to understand the known facts of atomic spectra in terms of the old Bohr–Sommerfeld theory, a collection of ad hoc rules that no one could derive or understand. Heisenberg tried to apply the theory to transitions between energy levels and deduced a formula that Max Born, his thesis advisor, recognized as the rule for matrix multiplication. Born and Jordan, and independently Dirac, succeeded in interpreting and generalizing Heisenberg's matrix multiplication rule. In his fundamental 1926 paper, Dirac inferred that the correct generalization of Heisenberg's observation is the rule $\mathbf{AB} - \mathbf{BA} = (ih/2\pi)\mathbf{C}_{\{A,B\}}$, where $\mathbf{A}$ and $\mathbf{B}$ are 'q-numbers' ('quantum variables'), as Dirac called them, or self-adjoint operators as we now call them, h is Planck's constant, $\{A,B\}$ is the Poisson bracket of any pair of classical variables A and B and $\mathbf{C}$ is the quantum operator representing that Poisson bracket. This rule, when applied to the fundamental Poisson bracket relations (15.26), yields the fundamental canonical quantization rules,

$$(\mathbf{q}_k,\mathbf{p}_l) = ih\varepsilon_{kl}\mathbf{I}/2\pi, \ (\mathbf{q}_k,\mathbf{q}_l) = 0, \ (\mathbf{p}_k,\mathbf{p}_l) = 0, \qquad (15.92)$$

where the $\mathbf{q}$'s and $\mathbf{p}$'s are taken to be self-adjoint operators and $\mathbf{I}$ is the identity operator[5]. These commutation rules, as stated, are only valid in Cartesian coordinates. In quantum mechanics, the equivalence of Cartesian with other systems of coordinates in phase space is lost (there is either configuration space or momentum space but no phase space in quantum mechanics) and is replaced by the idea of unitary transformations among different coordinate systems.

Dirac (1926) stated that 'The correspondence between the quantum and classical theories lies not so much in the limiting operation $h \to 0$ as in the fact that the mathematical operations on the two theories obey in many cases the same laws.' Here, Dirac refers to the similarity in the algebraic structure of the Poisson bracket rules and the commutation rules for quantum operators. The classical fundamental Poisson bracket rules define the Lie algebra of translations in phase space. Which Lie algebra do the canonical commutation rules represent? For the generators of rigid rotations, there is no essential difference between the Lie algebra of the classical theory and that of the quantum theory, aside from factors of h and the state space that each operates on: if $\mathbf{X}_i$ is our usual anti-self-adjoint generator of rotations, then $\mathbf{M}_i = 2\pi i\mathbf{X}_i/h$ is self-adjoint, so that $(\mathbf{X}_k,\mathbf{X}_l) = \varepsilon_{klm}\mathbf{X}_m$ becomes $(\mathbf{M}_k,\mathbf{M}_l) = (ih/2\pi)\varepsilon_{klm}\mathbf{M}_m$, which is well known in quantum theory. For the generators of translations, the Lie algebra that includes the quantum conditions $\mathbf{q}_k\mathbf{p}_l - \mathbf{p}_l\mathbf{q}_k = ih\mathbf{I}\varepsilon_{kl}/2\pi$ is *not* identical with the one described by the classical conditions $\{q_k,p_l\} = \varepsilon_{kl}$, but the former can be understood as a certain generalization of the latter. We now deduce the Lie algebra that the quantum commutation rules reflect.

Every Hamiltonian system uses group parameters as phase space coordinates: the canonically conjugate variables $(q_1,\ldots,q_f,p_1,\ldots,p_f)$ are themselves the canonical group parameters for a complete set of commuting translations in phase space, the space of the $2f$ q_i and p_i, of which q space is an f dimensional subspace. That the q_i and p_i are simultaneously infinitesimal generators and canonical group parameters of commuting translations is expressed by the fundamental Poisson bracket relations

---

[5] In this chapter we use I rather than E for the identity matrix.

$$\{q_k,p_l\} = \varepsilon_{kl}, \ \{q_k,q_l\} = 0, \ \{p_k,p_l\} = 0. \tag{15.93}$$

From these Poisson bracket rules it follows by direct calculation that the $2f$ infinitesimal operators defined by $X_i = \{ ,q_i\}$ and $Y_k = \{ ,p_k\}$ all commute with each other: $(X_i,X_j) = 0$, $(Y_i,Y_j) = 0$, and $(X_i,Y_j) = 0$. Consider next the action of the operators $U_i(q_i) = \exp[(q_i - q_{i\circ})Y_i]$ and $V_i(p_i) = \exp[(p_i - p_{i\circ})X_i]$ on any analytic phase space function $F(q_{1\circ},...,q_{f\circ},p_{1\circ},...,p_{f\circ})$, where $(q_{1\circ},...,q_{f\circ},p_{1\circ},...,p_{f\circ})$ is any other point in phase space. The action of either operator on F is simply a Taylor expansion in $(q_i - q_{i\circ})$ or $(p_i - p_{i\circ})$, which is equivalent to 'translating' that function from the point $(q_\circ,p_\circ)$ to the point (q,p), and we have used an obvious shorthand notation: we can therefore write $F(q,p) = U(q,p)F(q_\circ,p_\circ)$, where

$$U(q,p) = U_1 \dots U_f V_1 \dots V_f = \exp[\Sigma(q_i - q_{i\circ})Y_i + \Sigma(p_i - p_{i\circ})X_i], \tag{15.94}$$

because all $2f$ of the operators U_l and V_k commute with each other. Now, let us compare this set of circumstances with the analogous one in quantum mechanics.

In quantum mechanics, $\mathbf{q}_k$ and $\mathbf{p}_l$ are self-adjoint operators obeying the fundamental commutation rules

$$(\mathbf{q}_k,\mathbf{p}_l) = ih\varepsilon_{kl}\mathbf{I}/2\pi, \ (\mathbf{q}_k,\mathbf{q}_l) = 0, \ (\mathbf{p}_k,\mathbf{p}_l) = 0. \tag{15.92}$$

Here, we must abandon the $2f$ dimensional phase space of classical Hamiltonian mechanics: quantum mechanics requires an *infinite* dimensional abstract space of *complex* vectors Ψ. Otherwise, the fundamental commutation rule $(\mathbf{q}_k,\mathbf{p}_k) = ih\mathbf{I}/2\pi$ cannot be satisfied. The group theoretic interpretation of the fundamental commutation rules (15.92) follows from the commutation rules of the unitary operators $U(q) = e^{iqq}$ and $U(p) = e^{ipp}$, because the quantum analog of a canonical transformation in phase space is a unitary transformation in the infinite dimensional space of state vectors Ψ, as was first emphasized by Dirac. If $(\mathbf{q},\mathbf{p}) = ic$ is a 'c-number', which means that it is the identity operator, then the operators of finite translations $U(q) = e^{iqq}$ and $U(p) = e^{ipp}$ obey $U(q)U(p) = e^{iqq + ipp - icqp/2}$ *and therefore fail to commute only to within a phase factor*. This commutation to within a phase factor corresponds to the fact that we cannot define a single physical state in quantum mechanics by a single vector $\Psi = \psi e^{i\theta}$ with real phase θ. Instead, we must think of a physical state in quantum mechanics as the collection of all 'rays' $e^{i\theta}\Psi$ with Ψ fixed, but permitting all possible different real phases θ. The reason for this is that $\Psi' = U(q)U(p)\Psi$ belongs to the same *physical* state as does $\Psi'' = U(p)U(q)\Psi$, because Ψ' and Ψ'' differ only be a phase factor, and *the absolute phase of a state in quantum mechanics is undeterminate*. In other words, commutation of the operators $U(q)$ and $U(p)$ to within a phase factor corresponds to the indeterminacy of the absolute phase θ of any vector $\Psi = \psi e^{i\theta}$, that is, to a certain kind of gauge invariance. This was first pointed out by Weyl (1960).

According to Weyl, the gauge invariance described above falls into the general case of 'ray representations' of Lie groups, where a single state $\{e^{i\theta}\Psi\}$ is defined by the collection of 'rays' $e^{i\theta}\Psi$ with all different possible phases (see Hammermesh, 1962). Lie's integrability condition is, in this case, replaced by the more general closure condition $(X_\mu,X_\nu) = c^\lambda_{\mu\nu}X_\lambda + \beta_{\mu\nu}$, where the $\beta_{\mu\nu}$'s are antisymmetric constants[6]. The group

---

[6] Note that even this commutation rule has its analog in the classical Poisson bracket formalism (see equation (15.88)), and that in both cases the constants β_{ik} represent commuting translations.

theoretic interpretation of the fundamental commutation rules (15.92) of quantum mechanics is therefore as follows: with $c_{\mu\nu}^{\lambda} = 0$, we have a 'complete set of commuting translations in ray space', where the constant $\beta_{\mu\nu} \propto h$ on the right hand side of the commutation rule (15.92) for $\mathbf{q}_k$ and $\mathbf{p}_k$ corresponds to the arbitrariness in the phase of any quantum state vector Ψ. In other words, the operators $\mathbf{q}$ and $\mathbf{p}$ are the infinitesimal generators of the finite 'commuting' translations $U(q)$ and $U(p)$ in the infinite dimensional ray-space. It follows by inference from the analysis of chapter 12 that the corresponding canonical group parameters q and p constitute a sort of generalization to Hilbert space of the idea of a pair of *nonholonomic* coordinates in phase space or configuration space. Either member of the pair can be used alone as an holonomic coordinate to form *either* representations $\Psi(q) = \langle q, \Psi \rangle$ in q-space, *or* representations $\Phi(p) = \langle p, \Psi \rangle$ in p-space. The consequence of trying to use both 'quantally non-holonomic coordinates' simultaneously is the uncertainty principle $\Delta q \Delta p \geq h/2\pi$. That one cannot use two or more of the rotation angles α_i simultaneously as generalized coordinates in the classical theory of rigid rotations, and that one cannot construct representations of state vectors as a simultaneous function of q and p in quantum mechanics, have the same source: both pairs of 'coordinates' are nonholonomic in the sense that they are group parameters corresponding to pairs of noncommuting operators. In one case, the operators act in configuration space or phase space (the classical case) whereas in the other case they act in an infinite dimensional abstract linear vector space.

The gauge invariance represented by arbitrariness in phase θ of the state vector, or 'wave function' $\Psi(q) = \langle q, \Psi \rangle$, is of fundamental importance. For example, the superfluidity of ^{4}He at temperatures less than $T_\lambda \approx 2.1$ K corresponds to 'broken gauge invariance', and the velocity of the superflow, to a first approximation, is given by the Bohr–Sommerfeld result

$$\vec{v}_s = h\nabla\theta/2\pi, \tag{15.95}$$

where θ is the phase of a certain one-body probability amplitude (in the next chapter, we will see that this represents a classical integrability condition). Superfluidity requires a certain 'stiffness' against dissipation. The dissipation occurs through a mechanism called 'phase slip', which is the quantum analog of the mechanism of vorticity creation in the boundary layers of a normal viscous fluid. The points where the magnitude of $\Psi(q)$ vanishes are, as Dirac showed, points where the phase θ is undefined. These are 'topologic singularities' that normally can be identified as singular vortex lines with circulation quantized according to the old Bohr–Sommerfeld rule (the quantization rules makes $\Psi(q)$ single-valued). Topologic singularities will always appear at large enough times, which means that every superflow is only metastable and that all 'persistent currents' will eventually decay if you wait long enough (the currents can, under the right conditions, persist for years with no apparent decay). The point is that the decay time for the superflow is long whenever the system is very 'stiff' against the creation of topologic singularities that can move transverse to the flow.

One may ask: why do electrons and other particles obey different laws of motion than macroscopic billiard balls? This viewpoint is qualitatively Aristotelian. The viewpoint of physics is that these 'particles' are submicroscopic *singularities of fields*, not microscopic billiard balls, and to know their laws of motion one must ask Mother Nature via experiment.

15.5 Hidden symmetry: superintegrability of the Kepler and isotropic harmonic oscillator problems

When there are not only f, but $f + 1$ global, independent, and commuting isolating integrals, then the system is 'overintegrable' *if* one of those isolating integrals eliminates the quasiperiodic orbits. We have shown how this works for the Kepler problem in chapter 4. We now discuss in more detail the two central force systems with this property, the Kepler problem and the isotropic harmonic oscillator. The aim here is to exhibit the origin of the nontrivial $(f + 1)$th conservation law in a 'hidden symmetry'. Both Hamiltonians are rotationally invariant, but each Hamiltonian is, in addition, invariant under a larger group of transformations that is not obvious upon superficial inspection.

Beginning with the Kepler Hamiltonian

$$H = \tilde{p}p/2m + k/|x|,\tag{15.96}$$

rotational invariance in three dimensional Euclidean space (configuration space) is expressed by

$$\{M_k,H\} = 0.\tag{15.97}$$

If we introduce the Laplace–Runge–Lenz vector

$$A = -\Omega_p L - kmx/|x|\tag{15.98}$$

where

$$\Omega_p = \begin{pmatrix} 0 & p_3 & -p_2 \\ -p_3 & 0 & p_1 \\ p_2 & -p_1 & 0 \end{pmatrix},\tag{15.99}$$

then we showed by direct calculation in chapter 4 that

$$\frac{\mathrm{d}A}{\mathrm{d}t} = 0,\tag{15.100}$$

that the vector A is conserved, which means also that

$$\{A_1,H\} = 0,\tag{15.101}$$

expressing invariance of the Hamiltonian under infinitesimal transformations generated by the components A_i of A.

Furthermore,

$$\{A_k,M_l\} = \varepsilon_{kli}A_i,\tag{15.102}$$

and if we introduce the vector D

$$D = A/(2m|H|)^{1/2}\tag{15.103}$$

in place of A, then we obtain also

$$\{\tilde{D}D,M_i\} = 0\tag{15.104}$$

and

$$\{\tilde{M}M,D_i\} = 0.\tag{15.105}$$

In other words, H and the lengths squared, D^2 and M^2, of D and M all have vanishing Poisson brackets with each other and therefore can be used to reduce the problem to a one dimensional curve, as was done in chapter 4. Of course, we need that D^2 is analytic (in particular, single-valued) in order to get single-valuedness of the orbit equation.

Now, we are in a position to ask: what is the symmetry group of the Kepler problem? With only the angular momentum M_i, H is invariant under $SO^+(3)$, the group of single-valued real rotations in three dimensions. However, we can show by a lengthy calculation that

$$\left.\begin{array}{l} \{M_i,M_j\} = \varepsilon_{ijk}M_k \\ \{D_i,M_j\} = \varepsilon_{ijk}D_k \\ \{D_i,D_j\} = \varepsilon_{ijk}M_k \end{array}\right\} \tag{15.106}$$

that the six generators D_l and M_k leave the Kepler Hamiltonian invariant close under commutation and therefore define a Lie algebra! What is the Lie group generator by exponentiating these six generators? It is $O^+(4)$, the group of proper rotations in a four dimensional Euclidean space, because each commutation rule describes rotations, each generator generates a rotation in a plane, and with six generators there are six independent planes. But that is just what's needed for rotations in four dimensions. However, this 'hidden symmetry' is nonintuitive because the abstract fourth dimension is not a direction in configuration space or phase space. By writing

$$\left.\begin{array}{l} J_k = \dfrac{M_k + iD_k}{2} \\[3mm] K_k = \dfrac{M_k - iD_k}{2}, \end{array}\right\} \tag{15.107}$$

we can see that $SO^+(4)$ splits into two commuting sub-algebras:

$$\left.\begin{array}{l} \{J_l,K_k\} = 0 \\ \{J_l,J_k\} = \varepsilon_{lkm}J_m \\ \{K_l,K_k\} = \varepsilon_{lkm}J_m, \end{array}\right\} \tag{15.108}$$

which exhibits explicitly that there are two independent analytic conserved quantities that are compatible with each other and with the Hamiltonian. This explains why the Kepler problem has only periodic orbits, a fact that is predicted by Bertrand–Königs theorem (see chapter 4).

Next, we treat the other problem that is very special from the standpoint of Bertrand's theorem: the isotropic harmonic oscillator.

We begin with the two dimensional anisotropic oscillator where $H = H_1 + H_2$ with

$$H_i = p_i^2/2 + \omega_i^2 x_i^2/2. \tag{15.109}$$

With the generator

$$M_3 = \varepsilon_{3jk}x_j p_k \tag{15.110}$$

of rotations about the x_3-axis perpendicular to the (x_1,x_2)-plane,

$$\{M_3,H\} = x_1 x_2(\omega_1^2 - \omega_2^2) \tag{15.111}$$

for the isotropic oscillator, $\omega_1 = \omega_2$, and we have rotational invariance:

$$\{M_3,H\} = 0. \tag{15.112}$$

Hence, M_3 is conserved as we already know from chapter 4. Introducing the second rank tensor

$$A_{ij} = (p_i p_j + m^2 \omega^2 x_i x_j)/2m, \tag{15.113}$$

we see also that the components A_{ij} of A are generators of transformations that leave H invariant because

$$\{A_{ij}, H\} = 0. \tag{15.114}$$

Now, it is easy to get the orbit equation algebraically without ever performing an integral: it follows by an easy calculation that

$$\tilde{x}(HI - A)x = |M|^2/2 \tag{15.115}$$

and also that

$$A_{jk}M_k = 0. \tag{15.116}$$

The latter result means that A is confined to the plane of the orbit, the (x_1, x_2)-plane.

To go further, we diagonalize $HI - A$ where I is the identity tensor. Denoting by y_i the two eigenvectors, we have

$$\tilde{y}(HI - A)y = y_1^2(H - \lambda_1) + y_2^2(H - \lambda_2) = M^2/2. \tag{15.117}$$

Note that this is the equation of an ellipse if $H - \lambda_i > 0$. In order for this to be true,

$$H - \lambda_{1,2} = \{H \pm [H^2 + 4(2A_{12}^2 - H_1 H_2)]^{1/2}\}/2 \tag{15.118}$$

should be positive, which means that

$$2A_{12}^2 - H_1 H_2 < 0 \tag{15.119}$$

is required. This reduces to the condition

$$\omega^2(x_1 p_2 - x_2 p_1^2) > 0, \tag{15.120}$$

which is true. Therefore, we have for the orbit the equation of an ellipse,

$$\frac{y_1^2}{a^2} + \frac{y_2^2}{b^2} = 1. \tag{15.121}$$

What is the symmetry group of the two dimensional isotropic oscillator? Noting that $A_{15} = A_{21}$, there are only three independent components of A, A_{11}, A_{22}, and A_{15}. Writing

$$\left. \begin{array}{l} K_1 = A_{12}/\omega \\ K_2 = (A_{22} - A_{11})2\omega \\ K_3 = M_3/2, \end{array} \right\} \tag{15.122}$$

we find that

$$H^2 = 4\omega^2(K_1 + K_2 + K_3). \tag{15.123}$$

Furthermore, the generators K_i close under commutation,

$$\{K_i, K_j\} = \varepsilon_{ijk}K_k, \tag{15.124}$$

yielding the Lie algebra of either $SO^+(3)$ or $SU(2)$. $SU(2)$ is the group of unitary transformations in a complex two dimensional space. Without further analysis, one cannot decide between these two groups.

Proceeding to the three dimensional isotropic oscillator, where $H = H_1 + H_2 + H_3$, we note that the tensor A is still symmetric, $A_{ij} = A_{ji}$, so that A has at most six independent components. Furthermore, an easy calculation yields

$$\{A_{ij}, M_k\} = \varepsilon_{jkl}A_{il} + \varepsilon_{ikl}A_{jl} \tag{15.125}$$

and

$$\{A_{ij}, A_{lk}\} = \frac{\omega^2}{4}(\delta_{jl}\varepsilon_{ikm}M_m + \delta_{jl}\varepsilon_{jkm}M_m + \delta_{il}\varepsilon_{ikm}M_m + \delta_{jk}\varepsilon_{ilm}M_m + \delta_{ik}\varepsilon_{jkm}M_m) \tag{15.126}$$

where $M_i = \varepsilon_{ijk}x_j p_k$, so that combined with the bracket relations

$$\{M_i, M_j\} = \varepsilon_{ijk}M_k, \tag{15.127}$$

the three angular momenta M_i and the six independent components A_{ij} of the tensor A close under the Poisson bracket operation and therefore form a Lie algebra. Which algebra is it?

In fact, there are only eight independent quantities because, for example, A_{33} can be rewritten as $A_{33} = H - A_{11} - A_{22}$. Defining eight new phase space functions as

$$
\left.
\begin{aligned}
A_\circ &= (2A_{33} - A_{11} - A_{22})/\omega \\
A_{\pm 1} &= \pm(-1)(A_{13} \pm iA_{23})/\omega \\
A_{\pm 2} &= (A_{11} - A_{22} \pm 2iA_{12})/\omega \\
M_\circ &= M_3 \\
M_{\pm 1} &= M_1 \pm iM_2,
\end{aligned}
\right\} \tag{15.128}
$$

it is possible to show by a longer but still easy calculation that these six generators also close under the Poisson bracket operation, yielding

$$
\left.
\begin{aligned}
\{M_\circ, M_{\pm 1}\} &= -(\pm i)M_{\pm 1} \\
\{A_\circ, M_{\pm 1}\} &= -(\pm i)A_{\pm 1} \\
\{M_\circ, A_\circ\} &= 0 \\
\{M_{\pm 1}, M_{-(\pm 1)}\} &= -(\pm 2i)M_\circ \\
\{A_{\pm 1}, A_{-(\pm 1)}\} &= (i/2)M_\circ \\
\{A_\circ, A_{\pm 2}\} &= 0 \\
\{A_{\pm 1}, A_{\pm 2}\} &= (i/2)M_{\pm 1} \\
\{A_{\pm 1}, A_{-(\pm 2)}\} &= -(1/2)M_{\pm 1} \\
\{A_{\pm 2}, A_{-(\pm 2)}\} &= 0
\end{aligned}
\right\} \tag{15.129}
$$

and that the resulting commutation rules represent the Lie algebra of the $SU(3)$ group, the group of unitary transformations (without reflection) in a three dimensional complex vector space. Rewritten in Cartan's canonical form, the generators are given as

$$
\left.
\begin{aligned}
H_1 &= \tfrac{1}{6}(\sqrt{3}M_3\cos\alpha + A_\circ\sin\alpha) \\
H_2 &= \tfrac{1}{6}(\sqrt{3}M_3\sin\alpha - A_\circ\cos\alpha) \\
E^\lambda_{\pm 1} &= \frac{1}{4\sqrt{3}}(M_{\pm 1} \pm 2\lambda A_{\pm 1}) \\
E_{\pm 2} &= \frac{1}{2\sqrt{6}}A_{\pm 2}
\end{aligned}
\right\} \tag{15.130}
$$

where α may be chosen so that one of the H_i and either the E's or E^{λ}'s (for one value of λ) may be chosen to yield a triplet of operators that look like $SU(2)$ generators, while the other H_i commutes with the triplet. In quantum mechanics, this formulation of the isotropic oscillator has found application in the classification of nuclear spectra.

In general, the Hamiltonian of the n dimensional isotropic oscillator (with $n \geq 2$) is left invariant by the transformations of the group $SU(n)$.

Summarizing, Bertrand's theorem holds because the Kepler problem and the isotropic three dimensional simple harmonic oscillator each have a nontrivial $(f + 1)$th conserved quantity that is analytic in the orbit parameters, which forces the orbit equation to be single-valued (finite multivaluedness would also yield periodicity).

Hidden symmetries were of much interest in the 1960s because of the prediction of the existence of an unstable particle ('resonance'), the Ω^- particle, by the use of the Lie algebra of $SU(3)$ in quantum field theory. An experiment performed at a later date verified the existence of the Ω^-, in agreement with the prediction. The particle was previously 'missing' in the sense that the $SU(3)$ theory predicted a certain decimet of particles, but only nine of the particles were known from experiment at the time of the theoretical prediction.

15.6 Noncanonical Hamiltonian flows

We have shown how the theory of canonically conjugate Hamiltonian systems can be formulated naturally from the standpoint of Lie algebras of Poisson brackets. More generally, we showed in chapter 7 that Poisson brackets and Lie algebras provide a natural algebraic framework for discussing vector fields that describe both conservative and dissipative flows, without any need for a Hamiltonian system with canonically conjugate variables. We show next, surprisingly enough, that there are noncanonical Poisson bracket formulations with a corresponding Hamiltonian for an interesting class of noncanonical flows. This formulation applies to certain flows whether dissipative or conservative. These are the so-called noncanonical Hamiltonian flows on Poisson manifolds.

We can choose words to introduce noncanonical Hamiltonian flows in either of two ways: we can speak of generalizing the symplectic case, or we can speak of restricting the more general nonsymplectic case in order to reach the symplectic one. We regard the latter as the better choice because the Poisson manifolds include the symplectic manifolds as a special case. The restriction of arbitrary phase flows to the class of noncanonical[7] Hamiltonian flows is based upon the construction of a particular Lie algebra of vector fields that is an invariant sub-algebra of all the Lie algebra of all possible phase flows.

It is conceptually beneficial to begin by exhibiting the analogy between a Euclidean space and a Lie algebra. The transformations that define Euclidean (or Cartesian) space, global rigid translations and global rigid rotations of global Cartesian coordinate systems, leave the Euclidean scalar product invariant. The rules are given simply in terms of a vector inner product rule $(\mathbf{a},\mathbf{b})$ that is symmetric for any two vectors $\mathbf{a}$ and $\mathbf{b}$ and that maps two vectors into a number:

---

[7] It would be more accurate to say 'not-necessarily canonical' rather than noncanonical, but this phraseology would be too cumbersome.

(1) commutivity: $(\mathbf{a},\mathbf{b}) = (\mathbf{b},\mathbf{a})$
(2) associativity: $(\lambda\mathbf{a},\mathbf{b}) = \lambda(\mathbf{a},\mathbf{b})$
(3) distributivity: $(\mathbf{a}_1 + \mathbf{a}_2,\mathbf{b}) = (\mathbf{a}_1,\mathbf{b}) + (\mathbf{a}_2,\mathbf{b})$.

No inner product rule is needed for the abstract definition of a Lie algebra of vectors. The idea of an inner product is replaced by an outer product rule $[\mathbf{a},\mathbf{b}]$ that is antisymmetric for any two vectors $\mathbf{a}$ and $\mathbf{b}$, and where that rule maps the two vectors into a third vector $\mathbf{c} = [\mathbf{a},\mathbf{b}]$. The rules for Lie algebra of vectors are simply:

(1) anticommutivity: $[\mathbf{a},\mathbf{b}] = - [\mathbf{b},\mathbf{a}]$
(2) associativity: $[\lambda\mathbf{a},\mathbf{b}] = \lambda[\mathbf{a},\mathbf{b}]$
(3) distributivity: $[\mathbf{a}_1 + \mathbf{a}_2,\mathbf{b}] = [\mathbf{a}_1,\mathbf{b}] + [\mathbf{a}_2,\mathbf{b}]$
(4) Jacobi identity: $[[\mathbf{a},\mathbf{b}],\mathbf{c}] + [[\mathbf{b},\mathbf{c}],\mathbf{a}] + [[\mathbf{c},\mathbf{a}],\mathbf{b}] = 0$.

Note that the cross product rule of elementary vector algebra is a totally antisymmetric outer product rule, so that for orthonormal Euclidean unit vectors $\hat{\mathbf{e}}_i$ in three dimensions $[\hat{\mathbf{e}}_i,\hat{\mathbf{e}}_j] = \hat{\mathbf{e}}_i \times \hat{\mathbf{e}}_j = \varepsilon_{ijk}\hat{\mathbf{e}}_k$ satisfies the rules for a Lie algebra, the algebra of infinitesimal rotations, since $c_{ij}^k = \varepsilon_{ijk}$. The 'angular momentum commutation rules' in quantum mechanics, the commutation rules for the generators of infinitesimal rotations, are sometimes written in this form (see Landau and Lifshitz, 1977).

A Poisson manifold can be thought of as a phase space along with a certain Poisson bracket algebra of functions on that manifold and is defined as follows: given the flat manifold (phase space) of a dynamical system, a Poisson manifold is described by defining an antisymmetric bracket operation $\{g,h\}$ on differentiable functions g and h defined on the manifold, so that $\{g,h\} = f$ is also a function on the same manifold. The bracket operation must satisfy the following rules

(1) bilinearity: $\{g,h\}$ is bilinear in g and h
(2) antisymmetry (anticommutivity): $\{g,h\} = - \{h,g\}$
(3) Jacobi's identity: $\{\{f,g\},h\} + \{\{h,f\},g\} + \{\{g,h\},f\} = 0$
(4) Leibnitz's rule: $\{fg,h\} = f\{g,h\} + g\{f,h\}$.

Rules 1–3 define a Lie algebra. Leibnitz's rule is also satisfied if we define the Poisson bracket by a gradient operation,

$$\{g,h\} = - \widetilde{\nabla g}B\nabla h, \tag{15.131a}$$

where the gradient is defined with respect to whatever phase space coordinates are in use. Note that rule (2) requires that B must be a completely antisymmetric tensor, $\tilde{B} = - B$. Transformations that leave Poisson bracket operations invariant are called Poisson transformations. One may call the bracket operation (15.131a) a Poisson bracket, a Lie bracket, or a Lie–Poisson bracket. Jacobi's identity is not automatically satisfied by an arbitrary antisymmetric tensor B, but we shall come to that point in good time.

Beginning with the differential equations of any flow written in the vector form

$$\frac{\mathrm{d}x}{\mathrm{d}t} = V(x_1,\dots,x_n) \tag{15.132}$$

where V is an n-component vector in phase space, which is a flat vector space with a Cartesian inner product rule, we can arrive systematically at the definition (15.131a) of

Poisson brackets of functions in phase space by considering the Lie algebra of the class of flows with the restricted form $V_i = B\nabla H_i$, where B is a totally antisymmetric matrix that is the same for every flow in the class, and where H_i is a scalar function that defines the particular flow under consideration in that class. We can motivate the study of this class of flows algebraically by the following observation: the condition that a particular flow $V = B\nabla H$ leaves a function G invariant is

$$\frac{dG}{dt} = \tilde{V}\nabla G + \frac{\partial G}{\partial t} = \widetilde{\nabla H}\tilde{B}\,\nabla G + \frac{\partial G}{\partial t} = 0. \tag{15.133}$$

If H is time-independent, then it follows that H is a conserved quantity because $\widetilde{\nabla H}B\nabla H = 0$ due to the antisymmetry of B. From a restricted point of view, this corresponds to the class of dynamical systems that have at least one global time-dependent conservation law. From a larger perspective, H may depend upon the time t and is then not a conserved quantity. We shall adhere in what follows mainly to the restricted case because we can always generalize to a phase space of $n + 1$ dimensions if need be.

Next, we show how to arrive at the Lie bracket definition (15.131a) by generalizing the reasoning of section 15.3 that led from the general Poisson brackets of vector fields in an arbitrary phase space (chapter 7) to the canonical Poisson brackets of functions of canonically conjugate variables $(q_1,\ldots,p_f)$ on a symplectic manifold. The Poisson bracket rule (7.44a) of chapter 7 that describes the Lie algebra of all possible different flows $V_1, V_2, V_3,\ldots$ in phase space can be written as

$$V_3 = [V_1, V_2] = \tilde{V}_2\nabla V_1 - \tilde{V}_1\nabla V_2, \tag{7.44b}$$

where, as is usual in phase space, we assume a Cartesian inner product (but with no metric) $\tilde{V}_1 V_2 = V_{1k}V_{2k}$ for two vectors V_1 and V_2. Now, we restrict ourselves to the class of phase flows in our n dimensional space that is defined by vector fields of the form $V_i = B\nabla H_i$, where B is a totally antisymmetric tensor that is common to every flow in the collection and where the H_i are scalar functions in phase space that distinguish the different flows from each other within the class. The Poisson bracket rule for two such flows now reads

$$[V_1, V_2] = \widetilde{\nabla G_2}B\nabla B\nabla G_1 - \widetilde{\nabla G_1}B\nabla B\nabla G_2 \tag{15.134}$$

and defines a third flow $[V_1, V_2] = V_3$, but V_3 generally does not have the form of the tensor B operating on the gradient of a third scalar function H_3. If we restrict the class by requiring closure under the Poisson bracket operation, if we enforce that

$$[B\nabla H_1, B\nabla_2] = \widetilde{\nabla H_2}B\nabla B\nabla H_1 - \widetilde{\nabla H_1}B\nabla B\nabla H_2 = B\nabla H_3 \tag{15.135}$$

so that flows of the form $V_i = B\nabla H_i$ form a Lie algebra that is an invariant sub-algebra of the algebra of all possible phase space flows, then a short calculation shows that (15.135) follows from (15.134) with $H_3 = -\widetilde{\nabla H_1}B\nabla H_2$, but *only* if the antisymmetric tensor B satisfies the nontrivial condition

$$B_{jm}\frac{\partial B_{li}}{\partial x_j} + B_{ij}\frac{\partial B_{lm}}{\partial x_i} + B_{lj}\frac{\partial B_{mi}}{\partial x_j} = 0, \tag{15.136}$$

which reminds us of a Jacobi identity. Indeed, if we require that the Lie–Poisson bracket rule (15.131a) satisfies the Jacobi identity (3) above, then we arrive precisely at the condition (15.136), which is algebraically very interesting. Furthermore, with (15.136) in force,

$$[B\nabla H_1, B\nabla H_2] = \widetilde{\nabla H}_2 B\nabla B\nabla H_1 - \widetilde{\nabla H}_1 B\nabla B\nabla H_2 = B\nabla(-\widetilde{\nabla H}_1 B\nabla H_2)$$
$$= B\nabla\{H_1, H_2\}, \tag{15.137}$$

so that the Poisson brackets (15.131a) of functions g and h defined on the Poisson manifold represent the same Lie algebra as do vector fields of the form $V_i = B\nabla H_i$. Actually, we should be a little more careful: Poisson bracket relations of the form

$$\{g,h\} = -\widetilde{\nabla g} B\nabla h + C_{gh} \tag{15.131b}$$

follow from our reasoning, where $C_{gh} = -C_{hg}$ is an antisymmetric constant. We know that these antisymmetric constants are of the utmost importance in the canonical case.

We satisfy the Jacobi identity (15.136) for the case where $n = 3$ in order to gain some insight into the nontriviality of the condition on the tensor B. Writing the antisymmetry condition on the 3×3 matrix B in terms of an axial vector b, $B_{ij} = \varepsilon_{ijk} b_k$, allows us to rewrite (15.136) as

$$\widetilde{b}\Omega_b = 0 \tag{15.138}$$

where

$$\Omega_b = \begin{pmatrix} 0 & b_3 & -b_2 \\ -b_3 & 0 & b_1 \\ x_2 & -b_1 & 0 \end{pmatrix}, \tag{15.139a}$$

which is recognizable in Gibbs' notation as the condition

$$\vec{b} \cdot \nabla \times \vec{b} = 0 \tag{15.139b}$$

for the differential form $b_i dx_i$ to be integrable via an integrating factor M, the condition that $b = \nabla G/M$ so that $M(b_i dx_i) = dG$. In fact, M is Jacobi's first multiplier because it follows immediately that $V = B\nabla H = -\Omega_b \nabla H$ or, in Gibbs' notation, $\vec{V} = \nabla G \times \nabla H/M$. If both H and G are time-independent then the flow defined by the vector field V is integrable because both H and G are conserved. This is not surprising: in three dimensions, the existence of one conservation law implies the existence of a second one and integrability follows if each conservation law isolates one coordinate in terms of the remaining variables. For example, torque-free tops with different moments of inertia and therefore different kinetic energies can all have the same angular momentum, so that we can take $H = M^2$ and $G = T$, although we can take $H = T$ and $G = M^2$ just as well, showing that the choice of conservation law that one calls the Hamiltonian is generally not unique for a noncanonical flow.

Although for $n > 3$ (15.136) is a nontrivial condition on the tensor B and generally does not lead to an integrability condition on an axial vector field, it is a condition that is satisfied by a large enough class of flows to be of interest. All such flows, whether canonical or noncanonical, whether dissipative or conservative, are Hamiltonian where H_i is the Hamiltonian: for any function f of the phase space variables $(x_1, \ldots, x_n)$ we can write

$$\frac{df}{dt} = \{f,H_i\} + \frac{\partial f}{\partial t} = \widetilde{\nabla f} B \nabla H_i + \frac{\partial f}{\partial t} = \widetilde{V} \nabla f + \frac{\partial f}{\partial t}. \tag{15.140}$$

In particular, the equations of motion of the flow have the Hamiltonian form

$$\frac{dx_k}{dt} = \{x_k,H_i\} = \widetilde{\nabla x_k} B \nabla H_i = (B \nabla H_i)_k = V_k \tag{15.141}$$

and the Poisson bracket relations for the n position variables are simply

$$\{x_i,x_j\} = B_{ij}. \tag{15.142}$$

Lie's approach to this problem was the inverse of ours: his idea was to specify the commutation rules $\{x_i,x_j\}$ and then to study the consequences, whereas we have treated the commutation rules (15.142) as the consequence of the particular dynamics problem of interest. From Lie's perspective, one would start by studying commutation rules of the form $\{x_i,x_j\} = c_{ij}^k x_k$, so that $B_{ij} = c_{ij}^k x_k$ is linear in the phase space variables x_i. We leave it as an exercise for the reader to determine whether closure under commutation of the x_i with constant structure coefficients c_{ij}^k is consistent with dissipative flows or if it always requires a conservative flow. Lie was the first to study Poisson manifolds under the heading of 'function groups' (see Eisenhart, 1961).

From the perspective of a single dynamical system, the Hamiltonian may not be unique: any conserved quantity H_i that depends on all n of the variables x_i will do the job. However, if one starts by defining the form of B, which means that one defines the entire class of flows under consideration, then of course there may be only one Hamiltonian for a given system. Note also that every symplectic manifold is a Poisson manifold but that the reverse does not hold: we can easily construct noncanonical Hamiltonian systems with odd or even dimension n that have no possible representation in terms of canonically conjugate variables $(x_1,\dots,y_f) = (x_1,\dots,x_{2f})$ and Hamilton's canonical equations

$$\frac{dx_i}{dt} = \frac{\partial H}{\partial y_i}, \frac{dy_i}{dt} = -\frac{\partial H}{\partial x_i}. \tag{15.143}$$

In order to arrive at the canonical formulation (15.143), the phase space must be of even dimension *and* the tensor B must be a constant matrix with a very special form to be exhibited below. For a noncanonical system, the tensor B is usually not constant.

For an example of a noncanonical Hamiltonian flow, consider Euler's equations for the motion of a rigid body in a constant gravitational field (see section 9.5):

$$\frac{dM'}{dt} - \Omega'M' = -\Omega_x \cdot F' \tag{9.47a}$$

along with

$$\frac{dF'}{dt} = \Omega'F' \tag{9.48a}$$

where

$$\Omega = \begin{pmatrix} 0 & \omega_3' & -\omega_2' \\ -\omega_3' & 0 & \omega_1' \\ \omega_2' & -\omega_1' & 0 \end{pmatrix} \tag{9.49}$$

and where

$$\Omega_{x'} = \begin{pmatrix} 0 & x'_3 & -x'_2 \\ -x'_3 & 0 & x'_1 \\ x'_2 & -x'_1 & 0 \end{pmatrix}, \tag{9.50}$$

so that the phase space is the manifold of the six unknowns $(M'_1,\ldots,F'_3)$ or $(\omega'_1,\ldots,F'_3)$ because x' is a constant vector in the principal axis frame. The total energy is given by

$$E = \frac{\tilde{\omega}'I'\omega'}{2} + \tilde{x}'F' \tag{9.51}$$

and is always conserved if gravity is the only force acting on the body. We can therefore choose $H = E$ as the noncanonical Hamiltonian, but only for a definite choice of the tensor B in (15.136). In order to determine the form of B we require that $df/dt = \{f,H\}$ yields the right equation of motion for any phase space function f. In particular, we can construct the tensor B by forcing the Poisson bracket formulation to yield the right equations of motion for L'_i (or ω'_i) and F'_i. With a bit of foresight as to what we're up against, we start by assuming that

$$\{g,h\} = \widetilde{\nabla_{M'}g}B_\circ\nabla_{M'}h + \widetilde{\nabla_{M'}g}B_1\nabla_{F'}h + \widetilde{\nabla_{F'}g}B_1\nabla_{M'}h + \widetilde{\nabla_{F'}g}B_2\nabla_{F'}h \tag{15.144}$$

where $B_\circ$, B_1, and B_2 are three completely antisymmetric 3×3 matrices: this expression satisfies rules (1)–(4) above for the six dimensional $(M'_1,\ldots,F'_3)$-phase space. By restriction to the case where $x' = 0$, the torque-free case, we reproduce the correct result (9.47a) from $dM_i/dt = \{M_i,H\}$ only if $B_\circ = -\Omega'$. Next, with $x' \neq 0$ we can reproduce equation (9.48) from $dF_i/dt = \{F_i,H\}$ only if $B_1 = -\Omega'_{F'}$ (meaning that $B_{1ij} = -\varepsilon_{ijk}F'_k$) along with $B_2 = 0$. The desired noncanonical Poisson bracket operation is therefore given by

$$\{g,h\} = -\widetilde{\nabla_{M'}g}\Omega'\nabla_{M'}h - \widetilde{\nabla_{M'}g}\Omega'_{F'}\nabla_{F'}h - \widetilde{\nabla_{F'}g}\Omega'_{F'}\nabla_{M'}h. \tag{15.145}$$

The fundamental Lie–Poisson brackets are given by

$$\left.\begin{aligned} \{M'_i,M'_j\} &= -\varepsilon_{ijk}M'_k \\ \{M'_i,F'_j\} &= -\varepsilon_{ijk}F'_k \\ \{F'_i,F'_j\} &= 0 \end{aligned}\right\} \tag{15.146}$$

and represent the Lie algebra of a complete set of infinitesimal translations and rotations in the six dimensional Cartesian phase space of the rigid body (compare with equations (15.89) and (15.90)). This is just the Lie algebra of the Euclidean group, a subset of the Galilean group. The noncanonical Hamiltonian formulation holds whether or not the rigid body equations correspond to an integrable case. Even though there are at least three conservation laws even in the nonintegrable case we have used only one conservation law in order to formulate a Hamiltonian description: we choose the energy because it depends upon all six of the variables, whereas the conserved quantity $\tilde{F}'F'$ depends upon only three of the variables. For the reader who worries that we have treated a six dimensional problem by using 3×3 matrices, note that we can write the Poisson brackets correctly by using the 6×6 matrix

$$B = \begin{pmatrix} \Omega' & \Omega'_{F'} \\ \Omega'_{F'} & 0 \end{pmatrix}.$$ (15.147)

An integrable noncanonical flow, whether conservative or dissipative, can be described quite generally as a noncanonical Hamiltonian flow on a Poisson manifold if there is at least one conserved quantity H that depends upon all of the n coordinates. Any integrable flow described infinitesimally by the differential equations

$$\frac{dx}{dt} = V(x_1,\ldots,x_n),$$ (15.148)

has $n-1$ isolating integrals, and we can use $n-3$ of them to reduce the problem to the study of a flow in three dimensions, where the velocity field of the reduced problem then has the form $V_i = \varepsilon_{ijk}\partial h/\partial x_j\partial g/\partial x_k/m$, and where g and h are the remaining two invariants of the n dimensional flow. Let us take $g(x_1,x_2,x_3) = C_1$ as the confining surface so that h is the streamline equation of the surface $g = C_1$; h generally depends only upon two variables, say x_1 and x_2 (one can as well interchange the roles of h and g). The point is that g serves as a noncanonical Hamiltonian (as with a canonical system, the flow takes place on a constant Hamiltonian surface): defining the noncanonical Poisson bracket operation by

$$\{\alpha,\beta\} = -\widetilde{\nabla\alpha}\Omega_b\nabla\beta$$ (15.149)

where

$$\Omega_b = \frac{1}{m}\begin{pmatrix} 0 & \partial h/\partial x_3 & -\partial h/\partial x_2 \\ -\partial h/\partial x_3 & 0 & \partial h/\partial x_1 \\ \partial h/\partial x_2 & -\partial h/\partial x_1 & 0 \end{pmatrix}$$ (15.150)

(because $b = \nabla h/m$), we see quite generally that

$$\frac{d\alpha}{dt} = \{\alpha,g\} = \widetilde{V}\nabla\alpha,$$ (15.151)

which is the right time-evolution equation if the function α has no explicit time-dependence. In particular, the flow itself is described by the Poisson bracket equations

$$\frac{dx_i}{dt} = \{x_i,g\} = \widetilde{V}\nabla x_i = V_i.$$ (15.152)

Note that $\{h,g\} = 0$ follows and is the compatibility relation between the confining surface and the streamline equation on that surface, in perfect analogy with commuting conserved quantities in canonical Hamiltonian dynamics. Note also that the basic Lie–Poisson bracket relations are given by

$$\left.\begin{array}{l} \{x_i,x_j\} = -\varepsilon_{ijk}b_k \\[2mm] \{b_i,b_j\} = -\varepsilon_{lmn}b_n\dfrac{\partial b_i}{\partial x_l}\dfrac{\partial b_j}{\partial x_m} \\[2mm] \{x_i,b_j\} = -\varepsilon_{imn}b_n\dfrac{\partial b_j}{\partial x_m}, \end{array}\right\}$$ (15.153)

where $b = \Delta h/M$, so that the generators of translations on a Poisson manifold generally do not commute with each other and generally do not close under commutation. Therefore, the six generators $(x_1,\dots,b_3)$ generally do not give rise to an integrable six-parameter Lie group.

We emphasize that there is no restriction to conservative systems: the system under consideration may be dissipative but must be integrable with time-independent conservation laws. We can easily exhibit the Poisson manifold/noncanonical Hamiltonian description for an integrable dissipative system by considering a linear flow that spirals out of a source at the origin due to the velocity field $V = (x_1 - x_2, x_1 + x_2)$. As we stated in chapter 3, $H(x_1,x_2) = \ln r - \theta = C$ is conserved, where $x_1 = r\cos\theta$ and $x_2 = r\sin\theta$. The idea is to use H as a noncanonical Hamiltonian by defining the Poisson bracket relation $\{\alpha,\beta\}$ to reproduce the correct equations of motion:

$$\frac{d\alpha}{dt} = \{\alpha,H\} = \tilde{V}\nabla\alpha. \tag{15.154}$$

In particular, $dx_i/dt = \{x_i,H\} = V_i$ must hold. With

$$\{\alpha,\beta\} = -\vec{b}\cdot\nabla\alpha \times \nabla\beta \tag{15.155}$$

we get the right answer if the vector $\vec{b}$ is perpendicular to the phase plane and has the magnitude $b = 1/M$ where M is the integrating factor for the differential form $dx_1 V_2 - dx_2 V_1 (= dG/M)$. The fundamental bracket relations in this case are given by $\{x_1,x_2\} = 1/M = -r^2$, or by $\{r,\theta\} = -r^3$. It is clear that the Hamiltonian for a source/sink problem will generally be a multivalued function that is undefined at the location of a source or a sink. Closed sets of Poisson brackets generally do not follow from flows defined by dissipative systems. For the Dirac-quantization of a noncanonical dynamical system, one needs a closed set of Lie brackets.

Dirac's quantization rule uses the Lie algebra structure of a closed set of Poisson bracket relations in phase space to define a new Lie algebra by an algebraic correspondence rule and can be applied to noncanonical as well as to canonical classical systems: according to Dirac, one must construct self-adjoint operators **A** and **B** that satisfy the commutation relation $[\mathbf{A},\mathbf{B}] = (ih/2\pi)\mathbf{C}$, where the operator **C** is constructed by replacing the quantities in the classical Poisson bracket $C = \{A,B\}$ by self-adjoint operators. As is well known, this replacement is not unique except in trivial cases because there is no unique prescription for deciding the ordering of the operators $\mathbf{X}_i$ in a given scalar function $C(x_1,\dots,x_n)$ in order to construct an operator $\mathbf{C}(\mathbf{X}_1,\dots,\mathbf{X}_n)$ that is symmetric in the $\mathbf{X}_i$. For example, $C(A,B) = AB$ is unambiguously replaced by $\mathbf{C} = (\mathbf{AB} + \mathbf{BA})/2$, but there is no unique replacement by operators of the expression $C(A,B) = \cos(AB)$. Even in the case where the problem of nonuniqueness does not arise (nonuniqueness occurs in both canonical and noncanonical cases) one cannot expect to be able to apply Dirac's quantization rule without doing some hard work in the noncanonical case. To illustrate this, consider the integrable conservative flow defined by the torque-free rigid body with energy $E = M_1/I_1 + M_2^2/I_2 + M_3^2/I_3$. The three Poisson variables M_i do not commute with each other ($\{M_i,M_j\} = -\varepsilon_{ijk}M_k \neq 0$ if $i \neq j$) and therefore the three corresponding quantum operators cannot be diagonalized simultaneously. In the symmetric case where $I_1 = I_2$, we can rewrite the energy as $E = M^2/2I_1 + M_3^2(1/2I_3 - 1/2I_1)$. Here, since M^2 commutes with M_3, the application of Dirac's quantization rule leads directly and easily to a *simple formula* for the energy levels in terms of the eigenvalues of the self-adjoint operators $\mathbf{M}_2$ and $\mathbf{M}_3$ by quantizing the system according to the noncanonical Poisson bracket relations $\{M_i, M_j\} = -\varepsilon_{ijk} M_k$. The noncanonical approach is used by Landau and Lifshitz (1977), who go on to discuss the quantization of the completely asymmetric torque-free top via the same noncanonical Poisson bracket relations (the minus sign in $\{M_i, M_j\} = -\varepsilon_{ijk} M_k$ occurs because we work in a rotating frame). Landau's treatment (actually O. Klein's treatment, as Landau points out) of the completely asymmetric torque-free rigid body does *not* lead to a simple formula for the energy levels of the asymmetric top: because the three M_i do not commute with each other, the best that one can do is to compute the energy levels numerically one at a time. Landau's treatment shows, however, that one does not need canonically conjugate coordinates and momenta in order to quantize a system via Dirac's rules: all that one needs is a Lie algebra, Cartesian variables, and closure under commutation of the vector fields that appear in any set of Lie–Poisson bracket relations (closure under commutation, preferably with constant structure coefficients c_{ij}^k so that the state vectors and probability amplitudes (the 'wave functions') are determined by an integrable Lie group). What $\{M_i, M_j\} = -\varepsilon_{ijk} M_k$ and the canonical Poisson bracket rules $\{q_i, p_j\} = 0$, $\{q_i, q_j\} = 0$, and $\{p_i, p_j\} = 0$ have in common is that both sets of commutation rules describe Lie algebras: the former describe the complete set of three noncommuting translations on the surface of a two-sphere, a Poisson manifold that is Riemannian, whereas the canonical bracket rules describe a complete set of $2f$ commuting translations on a $2f$ dimensional symplectic manifold that is a flat differentiable manifold.

The problem of quantizing the completely asymmetric top in the presence of gravity is a very interesting unsolved problem that can be formulated by applying Dirac's rule to the Lie algebra (15.146), the algebra of the Euclidean group in three dimensions, to construct the corresponding quantum Lie algebra. Because this system is nonintegrable, it cannot be solved by using any method that relies upon constructing state vectors and energy eigenvalues by the simultaneous diagonalization of a complete set of commuting self-adjoint operators that commute with the Hamiltonian E.

We end this section with a discussion of the symplectic case, where we have a Hamiltonian system with f generalized coordinates x_i and f canonical momenta y_i ('symplectic coordinates') and where the Poisson brackets have the structure

$$\{g,h\} = \sum_{i=1}^{f} \left(\frac{\partial g}{\partial x_i} \frac{\partial h}{\partial y_i} - \frac{\partial g}{\partial y_i} \frac{\partial h}{\partial x_i} \right). \tag{15.156a}$$

Writing $(x_1, \ldots, x_{2f}) = (x_1, \ldots, y_f)$, these brackets have the Poisson manifold structure

$$\{g,h\} = -\widetilde{\nabla g} B \nabla h \tag{15.156b}$$

with the totally antisymmetric tensor B given by $B_{ij} = 0$ if both indices fall into the range $[1,f]$ or if both fall into the range $[f+1, 2f]$, $B_{ij} = -1$ if $i = 1$ to f with $j = f + 1$ to $2f$, and $B_{ij} = 1$ if $i = f + 1$ to $2f$ with $j = 1$ to f. For $f = 1$

$$B = \begin{pmatrix} 0 & -1 \\ 1 & 0 \end{pmatrix} \tag{15.157}$$

and for $f = 2$ we get

$$B = \begin{pmatrix} 0 & 0 & 1 & 0 \\ 0 & 0 & 0 & 1 \\ -1 & 0 & 0 & 0 \\ 0 & -1 & 0 & 0 \end{pmatrix} = \begin{pmatrix} 0 & I \\ -I & 0 \end{pmatrix}. \tag{15.158}$$

For $f > 2$ the structure of the symplectic tensor B should be fairly clear from this example. The main point is: for rotations and translations in a Euclidean space, the Euclidean inner product of two vectors is invariant (the Euclidean metric is invariant). In a symplectic space (the phase space of a canonical Hamiltonian system), the totally antisymmetric tensor B that defines the Lie–Poisson bracket $\{g,h\} = -\widetilde{\nabla g} B \nabla h$ is left invariant by canonical transformations: B is the same $f \times f$ matrix for every set of canonically conjugate coordinates and momenta (for every set of 'local symplectic coordinates'). This summarizes briefly the old and interesting subject of canonical transformations in up to date terminology.

As Weinstein (1983) has pointed out, many of the modern ideas about Poisson manifolds and Poisson algebras were implicit in the work of Sophus Lie. Perhaps the reader has by now begun to appreciate the depth of Lie's (according to Hamel, 1904, then seemingly unsubstantiated) claim: 'Die Prinzipien der Mechanik haben einen gruppentheoretischen Ursprung.' Quantum mechanics is also included, but today we would take the liberty to replace Lie's choice of adjective 'gruppentheoretischen' by the more accurate one 'liealgebraischen'.

Exercises

1. (a) Show that the transformation

    $$Q = p + iaq, \ P = \frac{p - iaq}{2ia}$$

    is canonical and find the generating function F_2.
    (b) Use the transformation to formulate the linear harmonic oscillator problem in the variables (Q,P).

2. For which constants α and β is the transformation

    $$Q = \frac{\alpha p}{x}, \ P = \beta x^2,$$

    canonical? Find the generating function F_1.

3. Let $A = -\frac{1}{2}\Omega_B x$ denote the vector potential for the constant magnetic field B.

    (a) If v_j are the Cartesian components of the velocity of a particle in the magnetic field, then evaluate the Poisson brackets

    $$[v_i, v_j], \ i \neq j = 1,2,3.$$

(b) If p_i is the canonical momentum conjugate to x_i, evaluate the Poisson brackets

$$[x_i,v_j], \quad [p_i,v_j],$$
$$[x_i,\dot{p}_j], \quad [p_i,\dot{p}_j].$$

(Goldstein, 1980)

4. (a) Prove that the Poisson bracket of two constants of the motion is itself a constant of the motion even when the constants depend on time explicitly.
   (b) Show that if the Hamiltonian and a quantity F are constants of the motion, then the nth partial derivative of F with respect to t must also be a constant of the motion.
   (c) As an illustration of this result, consider the uniform motion of a free particle of mass m. The Hamiltonian is certainly conserved, and there exists a constant of the motion

$$F = x - \frac{pt}{m}.$$

   Show by direct computation that the partial derivative of F with t, which is a constant of the motion, agrees with $[H,F]$.

(Goldstein, 1980)

5. (a) For a single particle show directly, i.e., by direct evaluation of the Poisson brackets, that if u is a scalar function only of r^2, p^2, and $\mathbf{r} \cdot \mathbf{p}$, then

$$[u,M] = 0.$$

   (b) Similarly show directly that if F is a vector function,

$$F = ur + vp - w\Omega_x p,$$

   where u, v, and w are scalar functions of the same type as in part (a), then

$$[F_i,M_j] = \varepsilon_{ijk}M_k.$$

(Goldstein, 1980)

6. Continuing from equation (15.42), complete the proof that the fundamental Poisson bracket relations $\{q_i,p_j\} = \varepsilon_{ij}$, $\{q_i,q_j\} = 0$, and $\{p_i,p_j\} = 0$ are preserved in form under canonical transformations.

7. Verify equations (15.82) and (15.83), and therefore also equation (15.87).

8. Verify the Poisson bracket relations for the components of the Laplace–Runge–Lenz vector

$$A = -\Omega_p M - mk\hat{x}.$$

9. Verify that the components of the two dimensional matrix A defined by (15.113) for the two dimensional isotropic harmonic oscillator are constants of the motion.

10. Verify all of the Poisson bracket relations stated for the $\{f,g\} = \nabla f B \nabla g$ in section 15.6. In particular, show that (15.136) must be true in order that the Jacobi identify $\{\{f,g\},h\} + \{\{h,f\},g\} + \{\{g,h\},f\} = 0$ is satisfied.

11. For a Lie algebra defined by $\{J_\mu, J_v\} = C_{\mu v}^\lambda J_\lambda$, show that $\{C, J_\mu\} = 0$ if $C = g^{\mu v} J_\mu J_v$

where $g^{\mu v}$ is the inverse of the matrix $g_{\mu v} = C_{\mu \beta}^\alpha C_{v \alpha}^\beta$. Show also that for a 'rigid body' defined by the energy

$$T = \tilde{J} I^{-1} J / 2,$$

where I is diagonal and constant, that $\{C, T\} = 0$ so that C is conserved by torque-free motion.

12. Consider flows where the tensor $B = -\tilde{B}$ is defined by $B_{ik} = \{x_i, x_k\} = C_{ik}^j x_j$. Are dissipative flows allowed?

16

Integrable canonical flows

If you can't define it precisely, then you don't know what you're talking about.

Julian Palmore[1]

16.1 Solution by canonical transformations

The solution of Hamilton's equations via a canonical transformation, for example by the Hamilton–Jacobi method, requires the construction of some formally prescribed coordinate system $(Q_1, \ldots, P_f)$, whose existence *always* requires that a (generally unstated) integrability requirement must be satisfied by the flow in question. This, we expect on quite general grounds from section 13.2.

In the traditional discussion of the Hamilton–Jacobi method and action-angle variables found in most mechanics texts, integrability requirements are always ignored in favor of purely formal manipulations that may be misleading and inapplicable. Here, we follow Jacobi, Liouville (1809–1882), Lie, and Einstein and discuss the integrability requirements for solution via canonical transformations explicitly. Whittaker (1965) makes the integrability requirement clear in his presentation of Liouville's theorem on involution systems, but also makes incorrect statements about integrability requirements earlier in his text. Goldstein (1980) does not define integrability vs nonintegrability at all and makes several statements that are completely wrong. Landau and Lifshitz (1976) recognize a demand for integrability when they see it, but then make the mistake of assuming implicitly that all canonical systems are integrable. A completely integrable canonical flow is one where some of the desired methods of generating function construction work because the required generating functions exist globally.

In Goldstein's popular text (1980), generating functions that produce canonical transformations to three different canonical coordinate systems are discussed in an effort to formulate alternative ways to express the complete solution of a canonical Hamiltonian system. In two cases, the solution in the new coordinate system is trivially integrable, and this presumably should be the motivation for trying to construct the corresponding generating functions. In particular, Goldstein prefaces his formulation of the motivation for the Hamilton–Jacobi method by stating that we can search for the canonical transformation from the canonical variables $(q(t),p(t))$ at time t back to the initial conditions (q_o,p_o) at time $t_o < t$, and states that this is the most general procedure. That for a canonical phase flow the transformation defined by $q_{i_o} = U(-t)q_i(t)$ and $p_{i_o} = U(-t)p_i(t)$ exists (if a Lipshitz condition is satisfied) and is canonical is beyond the realm of disputation. The generating function $F_2(q(t),p_{i_o}) = F_1(q(t),q_{i_o}) + \Sigma p_{i_o}q_{i_o}$ has the infinitesimal generator H, is given locally by

$$F_2 \approx \Sigma q_{i_o}p_i(t) + \delta t H + \ldots, \tag{16.1}$$

[1] George & Harry's in New Haven, ca. 1966.

and generates

$$
\left.
\begin{aligned}
q_{i_o} &= \frac{\partial F_2(q(t),p_o)}{\partial p_{i_o}}, \\
p_i(t) &= \frac{\partial F_2(q(t),p_o)}{\partial q_i(t)}.
\end{aligned}
\right\}
\tag{16.2}
$$

If the generating function F_2 could be constructed globally, then the idea would then be to obtain the f equations $q_{i_o} = U(-t)q_i(t)$ directly from (16.2) and then invert them to find the solution in the form $q_i(t) = \phi_i(q_o,p_o,t)$ and $p_i(t) = \psi_i(q_o,p_o,t)$. That some of this is, as Goldstein writes, both true and most general is beyond dispute. However, Goldstein's advice amounts merely to advocating solution by integrating Hamilton's equations backward in time, a procedure that is completely equivalent in complexity to integrating forward in time! Therefore, what Goldstein presumably wants us to assume is that the $2f$ initial conditions can play the role of $2f$ nontrivial conserved quantities. This, we know from section 13.2, would constitute a false even if historic hope. The reason for the confusion in Goldstein can be traced to the complete lack of any distinction between integrable and nonintegrable flows. We can characterize his approach as one where (i) all Hamiltonian systems are assumed implicitly to be globally integrable, and (ii) no distinction is made between local and global constants of the motion. Even worse, the generating function F_2 defined locally by (16.1) does not always exist globally.

The idea behind one particular Hamilton–Jacobi method of solution is the construction of a function $F_2(q,P)$ that treats a complete set of $2f$ canonical variables (Q,P) as $2f$ constants: for any Hamiltonian $H(q,p,t)$, the proposal is made to transform from an arbitrary set of $2f$ canonical variables (q,p) to a new set of $2f$ canonical variables defined by $Q_i = \beta_i = $ constant and $P_i = \alpha_i = $ constant, defining Hamilton's 'equilibrium problem'. Here, if Goldstein were right, the $2f$ β's and α's would be allowed to be the $2f$ initial conditions (q_o,p_o). The solution would then be expressible in the form

$$
\left.
\begin{aligned}
p_i &= \frac{\partial F_2(q,\alpha,t)}{\partial q_i} \\
\beta_i &= \frac{\partial F_2(q,\alpha,t)}{\partial \alpha_i}
\end{aligned}
\right\}
\tag{16.3}
$$

where

$$
\left.
\begin{aligned}
\dot{\beta}_i &= \frac{\partial K}{\partial \alpha_i} = 0 \\
\dot{\alpha}_i &= -\frac{\partial K}{\partial \beta_i} = 0,
\end{aligned}
\right\}
\tag{16.4}
$$

and where the choice $K = 0$ can be imposed for consistency. Although the equilibrium problem turns out to be well defined for globally integrable Hamiltonian systems, we *cannot* identify the constant variables (β,α) as the initial conditions (q_o,p_o). We will show below that, in any global solution via canonical transformation, at most f of the $2f$ conserved quantities may be trivial local constants of the motion like the initial conditions (q_o,p_o). In particular, at least f (but not $2f - 1$) of the first integrals must be compatible globally conserved quantities.

Now for the integrability requirement: if we set $K = 0$ in

$$K = H(q,p,t) + \frac{\partial F_2}{\partial t} = h\left(q, \frac{\partial F_2}{\partial q}, t\right) + \frac{\partial F_2}{\partial t}, \qquad (16.5)$$

then we obtain the time-dependent Hamilton–Jacobi partial differential equation

$$h\left(q, \frac{\partial F_2}{\partial q}, t\right) + \frac{\partial F_2}{\partial t} = 0 \qquad (16.6)$$

for the global generating function F_2 that solves the equilibrium problem (16.4). Because we have chosen $K = 0$ and $dQ_i/dt = d\beta_i/dt = 0$, it is necessary for the existence of $F_2(q,P)$ that

$$dF_2 = \sum_{i=1}^{f} p_i dq_i - H dt = L dt \qquad (16.7)$$

be a *closed* differential, but the integrand $L dt$ of the action of an arbitrary Hamiltonian system is *not* a closed differential!

That $L dt$ must be closed is an integrability condition that must be satisfied if the Hamilton–Jacobi method equation is to be solvable globally, meaning solvable for all finite times. Whenever the action is path-dependent, then the generating function F_1 defined by (16.7) does not exist globally as a function of the q's, α's and t and a solution of the Hamilton–Jacobi equation, if it exists at all, will be limited to times that are much smaller than a period of the motion. In the nonintegrable case, the action may be a path-dependent *functional* analogous to the 'quasi coordinates' of chapter 12. We show in the next section how the integrability condition demanded by (16.7), the equality of a complete set of second cross-partial derivatives, is satisfied whenever the Hamiltonian dynamical system has at least f global, commuting isolating integrals that hold for all finite times.

Most mechanics textbook writers follow the outdated literature by not making it clear that the f commuting conservation laws needed for complete integrability cannot be replaced arbitrarily by f trivial local constants provided by any f of the $2f$ initial conditions[2]. Einstein, who was not a traditional theorist, explained the missing integrability condition in 1917 in his criticism of Bohr–Sommerfeld quantization, but Liouville had already provided a more complete explanation in the nineteenth century. In the era before Poincaré discovered a geometric example of nonintegrability that is chaotic, there seems to have been no clear idea that solutions of differential equations could be anything but completely integrable, meaning solution by 'reduction to quadratures' (in section 13.2, we pointed out this assumption in the texts by Whittaker and Eisenhart). On this basis of historic confusion, wrong statements about the use of initial conditions as constants of the motion were made by many physicists and mathematicians who wrote about classical mechanics (the list excludes Arnol'd, Einstein, Kolmogorov, Koopman, Moser, von Neumann, Poincaré, and Wintner). The

---

[2] Along with assuming implicitly that all Hamiltonian systems are integrable, the popular and stimulating text by Landau and Lifshitz leads the reader to confuse the two entirely different generating functions defined by (16.1) and (16.7): the first does not exist globally while the second one exists globally only for an integrable system: see equations (43.6 and 43.7) on p. 139 of *Mechanics*, third edition, by Landau and Lifshitz (1976).

modern approach to integrability can be found in the monograph by Fomenko (1988), which unfortunately presumes a very advanced knowledge of differential geometry, topology, and Lie algebras.

16.2 The Hamilton–Jacobi equation

We consider next the global solution of Hamilton's equilibrium problem. The analysis holds for time-dependent Hamiltonians, time-dependent conservation laws, and bounded or unbounded motion in phase space.

We begin with a Hamiltonian system where, for reasons to become clear as we proceed, there are at least f first integrals of the motion $G_\sigma(q_1,\ldots,p_f,t) = C_\sigma$ satisfying the canonical commutation rules $\{G_\mu,G_\nu\} = 0$. The system $\{G_\sigma\}$ is called an 'involution system' in the older literature. We need not assume that the first integrals G_σ and the Hamiltonian are time-independent. The f quantities G_σ must commute with each other but are permitted to be time-dependent first integrals of the motion,

$$\frac{\mathrm{d}G_\sigma}{\mathrm{d}t} = \{G_\sigma,H\} + \frac{\partial G_\sigma}{\partial t} = 0, \tag{16.8}$$

so that $\{G_\sigma,H\}$ is not assumed to vanish. The G_σ generate f canonical transformations, but these transformations generally do not leave the Hamiltonian invariant.

We assume also that the f first integrals $G_\sigma = C_\sigma$ are isolating integrals so that they can be rewritten in the form $p_i = f_i(q_1,\ldots,q_f,C_1,\ldots,C_f,t)$, where the C_i are constants independent of the time, and the functions f_i are generally two-valued. The f isolating integrals

$$p_i - f_i(q_1,\ldots,q_f,C_1,\ldots,C_f,t) = 0 \tag{16.9}$$

must all commute with each other, $\{p_i - f_i, p_k - f_k\} = 0$, and from this it follows that

$$\{p_k, f_i\} + \{f_k, p_i\} = 0, \tag{16.10}$$

from which follows

$$\frac{\partial f_i}{\partial q_k} = \frac{\partial f_k}{\partial q_i}. \tag{16.11}$$

The latter result is part of the integrability condition required to make $\mathrm{d}F_1 = L\mathrm{d}t$ exact: it guarantees that the canonical momentum is the q-subspace gradient of a multivalued potential $\Phi(q,C)$, so that $\Sigma p_i \mathrm{d}q_i = \Sigma f_i \mathrm{d}q_i$ is a closed differential in the q's. However, the entire differential form $L\mathrm{d}t = p_i \mathrm{d}q_i - H\mathrm{d}t$ must be closed in the f q's *and* the time t if the time-dependent Hamilton–Jacobi differential equation is to be completely integrable. In addition, therefore, observe that

$$\dot{p}_i = \sum_1^f \frac{\partial f_i}{\partial q_k} \cdot \frac{\partial H}{\partial p_k} + \frac{\partial f_i}{\partial t} = \sum_1^f \frac{\partial f_k}{\partial q_i} \cdot \frac{\partial H}{\partial p_k} + \frac{\partial f_i}{\partial t}, \tag{16.12}$$

so that

$$-\frac{\partial H}{\partial q_i} - \sum_1^f \frac{\partial f_k}{\partial q_i} \cdot \frac{\partial H}{\partial p_k} = \frac{\partial f_i}{\partial t}. \tag{16.13}$$

Notice that the left hand side of equation (16.13) is just the total derivative of h with respect to q_i, where $H(q,p,t) = h(q,\partial f/\partial q,t)$:

$$-\frac{dh}{dq_i} = \frac{\partial f_i}{\partial t}. \tag{16.14}$$

This means that when we replace p_i by f_i everywhere in the integrand of the classical action, Ldt, we obtain

$$Ldt = \sum_1^f p_i dq_i - Hdt = \sum_1^f f_i dq_i - hdt, \tag{16.15}$$

and by (16.11) and (16.14) this particular differential form, considered as a function of the $f + 1$ variables $(q_1,\ldots,q_f,t)$ is *closed*,

$$Ldt = \sum_1^f f_i dq_i - hdt = d\Phi(q_1,\ldots,q_f,C_1,\ldots,C_f,t), \tag{16.16}$$

which is the content of Liouville's involution theorem. Under these conditions, the classical Hamiltonian system is completely integrable in the sense that the Hamilton–Jacobi equation has a global solution (the action is path-independent), and the motion in phase space is confined to an $f + 1$ dimensional surface that is the intersection of the f compatible surfaces $G_\sigma(q_1,\ldots,p_f,t) = C_\sigma$ along with the time axis, in the $2f + 1$ dimensional $(q_1,\ldots,p_f,t)$-phase space.

As Whittaker (1965) shows by an example from rigid body theory (see p. 325 of his text), one can use Liouville's result to solve a certain class of integrable problems as follows. Considered as a function of all $2f + 1$ of its variables we have

$$d\Phi = \sum_{i=1}^f \left(\frac{\partial\Phi}{\partial q_i}dq_i + \frac{\partial\Phi}{\partial C_i}dC_i\right) + \frac{\partial\Phi}{\partial t}dt. \tag{16.17}$$

Comparison with the expression $Ldt = d\Phi$ above suggests that we ask whether the f quantities

$$b_\sigma = \frac{\partial\Phi(q,C,t)}{\partial C_\sigma} \tag{16.18}$$

also are constants of the motion. Note that

$$\frac{db_\sigma}{dt} = \sum_{i=1}^f \left(\frac{\partial^2\Phi}{\partial q_i\partial C_\sigma}\frac{dq_i}{dt} + \frac{\partial^2\Phi}{\partial C_i\partial C_\sigma}\frac{dC_i}{dt}\right) + \frac{\partial^2\Phi}{\partial t\partial C_\sigma} = \sum_{i=1}^f \frac{\partial^2\Phi}{\partial q_i\partial C_\sigma}\frac{dq_i}{dt} + \frac{\partial^2\Phi}{\partial t\partial C_\sigma} \tag{16.19}$$

along streamlines because each C_σ is conserved. Since Φ is a point function of its $2f + 1$ variables, we can freely interchange orders of integration to obtain

$$\frac{db_\sigma}{dt} = \frac{\partial}{\partial C_\sigma}\left(\sum_{i=1}^f \frac{\partial\Phi}{\partial q_i}\frac{dq_i}{dt} + \frac{\partial\Phi}{\partial t}dt\right) = \frac{\partial}{dC_\sigma}\left(\sum_{i=1}^f p_i\dot q_i - H\right), \tag{16.20}$$

which we would expect to vanish. Can we prove this?

That the fb_σ are also constants of the motion follows from an invariance property of Pfaffian differential forms (Whittaker, 1965, pp. 307, 308, and 325) but we can obtain that result by a more familiar method as follows. By writing $Ldt = d\Phi - \Sigma b_\sigma dG_\sigma/dt$ the action principle takes on the form

$$\delta A = \delta \int_{t_1}^{t_2} \left(\sum_{i=1}^{f} p_i \dot{q}_i - H \right) dt = -\delta \int_{t_1}^{t_2} \sum_{\sigma=1}^{f} b_\sigma \dot{G}_\sigma dt. \tag{16.21}$$

Choosing as coordinates the noncanonical variables $(q_1, \ldots, q_f, C_1, \ldots, C_f)$ and choosing variations with fixed end points, we obtain from $\delta A = 0$ the $2f$ differential equations

$$\frac{dG_\sigma}{dt} = 0, \quad \frac{db_\sigma}{dt} = 0, \tag{16.22}$$

which provides us with a noncanonical description of Hamilton's equilibrium problem in terms of f nontrivial constants G_σ and f constants b_σ, most or all of which may be trivial integration constants (compare (16.22) with the canonical description (16.4)).

An alternative way to see that the b's are constants is to use $\Phi(q,C,t) = F_2(q,P,t)$ where, in the original statement of the equilibrium problem, the f Q's and f P's denoted by (β,α) are all constants corresponding to $K = 0$. Here, we have

$$b_\sigma = \frac{\partial \Phi}{\partial C_\sigma} = \sum_{k=1}^{f} \frac{\partial F_2}{\partial P_k} \frac{\partial P_k}{\partial C_\sigma} = \sum_{k=1}^{f} Q_k \frac{\partial P_k}{\partial C_\sigma}. \tag{16.23}$$

The b_σ are allowed to be trivial local constants that, together with the f global constants C_σ, give us the $2f$ constants necessary to pin down a single solution of Hamilton's equations, and have the form $b_\sigma = g_\sigma(q_1, \ldots, q_f, C_1, \ldots, C_f, t)$. The idea is to invert these f equations and solve for the f generalized coordinates in the form $q_i(t) = g_i(t, C_1, \ldots, C_f, b_1, \ldots, b_f)$. Then, one substitutes these f equations into $p_i = f_i(q_1, \ldots, q_f, C_1, \ldots, C_f, t)$ in order to obtain f equations of the form $p_i(t) = h_i(t, C_1, \ldots, C_f, b_1, \ldots, b_f)$ that, together with the f $q_i(t)$, give us the complete solution, in the desired form, of the Hamiltonian dynamics problem for a definite choice of $2f$ initial conditions (q_o, p_o) expressed via the constants (C, b).

We illustrate the method for two trivially integrable examples: first, for a particle with mass m in a potential $U(q)$, where $H(q,p) = p^2/2m + U(q) = E$ is constant. With one constant of the motion $H = E$, we obtain from integrating $L dt = p dq - H dt$, which is here exact, that

$$\phi(q,E,t) = \int dq [2m(E - U(q))]^{1/2} - Et. \tag{16.24a}$$

The second constant is trivial, is given by

$$b = \frac{\partial \phi}{\partial E} = \int \frac{\sqrt{m} dq}{[2(E - U(q))]^{1/2}} - t, \tag{16.24b}$$

and the result should be compared with the elementary result from section 3.6 where

$$t - t_o = \int_{q_o}^{q} \frac{\sqrt{m} dq}{[2(E - U(q))]^{1/2}}. \tag{16.24c}$$

Next, we illustrate the method of separation of variables. Consider a one dimensional particle of unit mass in a potential $U(q) = -\mu/q$, so that $H = p^2/2 - \mu/q$. The Hamilton–Jacobi equation is

$$\frac{1}{2} \left(\frac{\partial \Phi}{\partial q} \right)^2 - \frac{\mu}{q} + \frac{\partial \Phi}{\partial t} = 0. \tag{16.25}$$

Because the Hamilton H is time-independent we can separate variables in the form $\Phi = S(q) + f(t)$ where $f(t) = - Et = \mu t/q_o$. A simple integration then yields

$$S(q) = (2\mu q_o)^{1/2} \sin^{-1}(q/q_o)^{1/2} + [2\mu q(q_o - q)/q_o]^{1/2}. \tag{16.26}$$

There are two trivial integration constants q_o and $b = \partial\Phi/\partial q_o$, and we can also compute $p = \partial\Phi/\partial q$.

In general, for a time-independent Hamiltonian, complete separation of variables in the time-independent Hamilton–Jacobi equation

$$h(q,\partial\Phi/\partial q) = E, \tag{16.27}$$

where $F_2 = \Phi - Et$ would amount to assuming that

$$\Phi(q_1,\ldots,q_f,C_1,\ldots,C_f) = \sum_{i=1}^{f} \Phi_i(q_i,C_1,\ldots,C_f), \tag{16.28}$$

but this is not at all required by the integrability conditions (16.11) that make the reduced action integrand $d\Phi = \Sigma p_i dq_i$ a closed differential form. In other words, there may be integrable systems that are not separable in any canonical coordinate system (see Landau and Lifshitz, 1976, and Gutzwiller, 1990, for examples of separable problems).

16.3 Integrability and symmetry

Die Prinzipien der Mechanik haben einen gruppentheoretischen Ursprung.

Marius Sophus Lie

We show next how f commuting isolating integrals put completely integrable motion on a flat f-dimensional manifold in phase space. The idea is to consider a time-independent Hamiltonian $H(q,p)$ and also a time-independent generating function $F_2(q,P) = \Phi(q,C)$, which solves the time-independent Hamiltonian–Jacobi equation

$$H(q,p) = h\left(q,\frac{\partial\Phi}{\partial q}\right) = K(P) \tag{16.29}$$

where

$$\dot{P}_k = \frac{\partial K}{\partial Q_k} = 0 \text{ and } \dot{Q}_k = \frac{\partial K}{\partial P_k} = \omega_k = \text{constant}. \tag{16.30a}$$

The solution of (16.29) is then given by the reduced action

$$\Phi(q,C) = \sum_{i=1}^{f} \int p_i dq_i \tag{16.31a}$$

if the integrability condition (16.11) holds, so that the integral in (16.31a) is path-independent, if it does not encircle a singularity of an isolating integral.

Historically, the canonical coordinates (Q,P) in (16.30a) have been labeled 'action-angle variables' although the so-called angle variables are really translation variables on a flat manifold. Our method of proof that the Lie transformation to the f translation variables $Q_k = \omega_k t + Q_{k_o}$ exists for an integrable canonical system is based upon the use of the operators that generate f commuting symmetries and is a marriage of the methods of sections 12.1 and 15.2: we shall parameterize integrable motion by using f

group parameters, simultaneously using the Poisson brackets of the f corresponding conserved quantities to construct the operators that define the required coordinate transformations and yield the 'angle variables'. In the next section we show how to construct the canonically conjugate action variables.

In order to exhibit the generality of the result, which does not need an underlying Lagrangian mechanics at all (see also Duff, 1962, and Sneddon, 1957), we begin from the very general standpoint of a canonical Hamiltonian system, a conservative dynamical system with f degrees of freedom that is given by the generalization to $2f$ dimensions of the idea of a stream function in two dimensional ideal fluid hydrodynamics:

$$\dot{x}_i = \frac{\partial H}{\partial y_i}, \ \dot{y}_i = -\frac{\partial H}{\partial x_i}, \tag{16.32a}$$

where $H = H(x_1,\ldots,x_f,y_1,\ldots,y_f,t)$, so that the Cartesian forms of the divergence-free condition

$$\nabla \cdot V = \sum_{i=1}^{f} \left(\frac{\partial \dot{x}_i}{\partial x_i} + \frac{\partial \dot{y}_i}{\partial y_i} \right) = \sum_{i=1}^{f} \left(\frac{\partial^2 H}{\partial x_i \partial y_i} - \frac{\partial^2 H}{\partial y_i \partial x_i} \right) = 0 \tag{16.33}$$

and phase volume element $d\Omega = dx_1 \ldots dx_f dy_1 \ldots dy_f$ are preserved both by the flow and by all canonical transformations. The Hamiltonian equations (16.32a) are called canonical, and the variables (x_i,y_i) representing one degree of freedom are called canonically conjugate. The time is the canonical group parameter for the Hamiltonian H that generates the flow infinitesimally, and the phase space coordinates $(x_1,\ldots,x_f,y_1,\ldots,y_f)$ are simultaneously infinitesimal generators *and* canonical group parameters for a complete set of commuting translations along the $2f$ Cartesian (x,y)-axes in phase space, as we showed in chapter 15 by interpreting the fundamental Poisson bracket relations. Here, we use the notation $(x_1,\ldots,x_f,y_1,\ldots,y_f)$ to emphasize that the canonically conjugate variables, whatever their curvilinear origin in Lagrangian dynamics, are always *defined* to be Cartesian coordinates in phase space, which is a flat differentiable manifold of $2f$ dimensions. It is unnecessary to assume that there is a corresponding Lagrangian, or that $H = T + U$ or even that our Hamiltonian represents a Newtonian mechanical system. We shall see in what follows that there will be no mathematical distinction at all between the Hamiltonian H and the f first integrals G_σ that commute with it and lead to complete integrability.

Suppose that there are f time-independent, functionally independent and global constants of the motion $G_\sigma(x,y) = C_\sigma$ holding for all times t, that commute with each other and with a time-independent Hamiltonian $H(q,p)$, so that $\{G_\mu,G_\nu\} = 0$, where $\{,\}$ denotes the Poisson bracket operation of chapter 15. Then

$$X_\sigma = \{,G_\sigma\} = \sum_{i=1}^{f} \left(\frac{\partial G_\sigma \partial}{\partial y_i \partial x_i} - \frac{\partial G_\sigma \partial}{\partial x_i \partial y_i} \right) \tag{16.34}$$

is the infinitesimal operator of the one-parameter symmetry group generated by the conserved quantity G_σ. Consider next an f-parameter transformation where the kth pair is denoted by $(x_k(\theta_1,\ldots,\theta_f),y_k(\theta_1,\ldots,\theta_f))$ and the θ_σ are the *canonical* parameters of the f different one-parameter transformation groups generated by the f constants of the motion G_σ. The transformed pair $(x_k(\theta_1,\ldots,\theta_f),y_k(\theta_1,\ldots,\theta_f))$ is canonically conjugate by construction: the infinitesimal transformation that generates the finite transformation is simply

$$\delta x_k \approx \sum_{\sigma=1}^{f} \delta\theta_\sigma \frac{\partial G_\sigma}{\partial y_k}, \ \delta y_k \approx -\sum_{\sigma=1}^{f} \delta\theta_\sigma \frac{\partial G_\sigma}{\partial x_k} \quad (16.35a)$$

and the generating function $F_2(x_1,\ldots,x_f,y_1(\theta_1,\ldots,\theta_f),\ldots,y_f(\theta_1,\ldots,\theta_f))$, to lowest order in $\delta\theta_\sigma$, is given by

$$F_2 \approx \sum_{i=1}^{f} x_i y_i(\theta_1,\ldots,\theta_f) + \sum_{\sigma=1}^{f} \delta\theta_\sigma G_\sigma \quad (16.35b)$$

and differs from the generating function F_2 that is implicit in equations (16.29) and (16.31). If the f independent conservation laws $G_\sigma = C_\sigma =$ constant generate f different flows, then each separate infinitesimal transformation can be integrated to yield the finite one-parameter transformation in the form $U_\sigma(\theta_\sigma) = e^{\theta_\sigma X_\sigma}$ (no summation convention here), so that $x_k(\theta_1,\ldots,\theta_f) = U_1(\theta_1)\ldots U_f(\theta_f)x_k$. Since $[X_\sigma,X_\tau] = 0$ where $[X_\sigma,X_\tau] = X_\sigma X_\tau - X_\tau X_\sigma$ denotes the commutator of the two operators, we have $[U_\sigma,U_\tau] = 0$, so that $U_1(\theta_1)\ldots U_f(\theta_f) = U(\theta_1,\ldots,\theta_f) = e^{\theta_1 X_1 + \cdots + \theta_f X_f}$ also follows. Therefore, we have f commuting finite *translations* U_σ. Whenever the σth flow has one or more singularities on the imaginary θ_σ-axis, as is typical in the theory of differential equations, then the radii of convergence of the exponential formulae $U_\sigma(\theta_\sigma) = e^{\theta_\sigma X_\sigma}$ applied to the f canonically conjugate pairs (x_i,y_i) will be determined by the nearest singularity on the imaginary θ_σ-axis, but analytic continuation may be used to extend the formulae to larger (real) values of θ_σ.

Seen purely mathematically, each conserved quantity G_σ is also a Hamiltonian with a (real) canonical 'time-like' parameter θ_σ, where the flow is generated by the Hamiltonian system

$$\frac{dx_i}{d\theta_\sigma} = \frac{\partial G_\sigma}{\partial y_i}, \ \frac{dy_i}{d\theta_\sigma} = -\frac{\partial G_\sigma}{\partial x_i}. \quad (16.32b)$$

Furthermore, each G_σ must be an *integrable* Hamiltonian because it is in involution with f other Hamiltonians, the $f - 1$ other G_λ and also H. It is clear that a nonintegrable Hamiltonian cannot generate a symmetry of an integrable Hamiltonian system, nor can we assume any special properties of the flow generated by G_σ without attributing the very same properties automatically and implicitly to the flow generated by the Hamiltonian H whose flow we want to analyze. From the phase space standpoint, there is no distinction at all between H and the f constants of the motion G_σ that commute with it. Now for the marriage preparations.

In the theory of rigid body rotations (chapters 9 and 12), the transformation $x' = Rx$ from inertial to body-fixed Cartesian axes is given by an orthogonal three by three matrix $R(\theta_1,\theta_2,\theta_3)$ where the θ_i are the canonical group parameters corresponding to three noncommuting subgroups generated by matrices M_i that obey the commutation rule $[M_i,M_j] = \varepsilon_{ijk}M_k$ (one can also use other parameterizations of the $SO^+(3)$ group, but this is unimportant). The equation $dR/dt = R\Omega$ defines the rotation generator $\Omega = \varepsilon_{ijk}\omega_k$, and in the inertial frame the Cartesian position vector x of the pth particle in the rigid body then obeys $dx/dt = -\Omega x$. It also follows that $dR^{-1}/dt = -\Omega R^{-1}$, so that $x(t) = R^{-1}(\theta_1,\theta_2,\theta_3)x(0)$. Therefore, $R^{-1}(\theta_1,\theta_2,\theta_3)$ *is the time-evolution operator* for the dynamical system, and the motion has been parameterized quite naturally through the time dependence of the three group parameters θ_i.

The useful analogy of Hamiltonian dynamics with rigid body dynamics is as follows: in rigid body theory, the transformations $R(\theta_1, \theta_2, \theta_3)$ connect all possible states of a rotating body in configuration space. Therefore, R^{-1} qualifies to serve as the time-evolution operator. In the present case, the transformations $U(\theta_1, \dots, \theta_f)$ leave invariant the f dimensional surface on which the motion occurs (integrable Hamiltonian motion is confined geometrically to the intersection of the f compatible surfaces G_σ = constant) and also connect all states of the Hamiltonian system on that surface. Therefore, we can use $U(\theta_1, \dots, \theta_f)$ as the time-evolution operator for our Hamiltonian system just by letting the group parameters θ_σ depend upon the time t, exactly as we did in the theory of rigid body motions.

Our method of treatment of integrable systems is completely analogous to the parameterization of rigid body motion by a set of parameters of $SO^+(3)$. In the latter case, the rotation operators do not commute but the starting point is in both cases the same: in an f degree of freedom system, f independent group parameters θ_σ can be used to parameterize the dynamics. In our case, the f operators commute so that we can write

$$
\begin{aligned}
x_k(\theta_1 + \Delta\theta_1, \dots, \theta_f + \Delta\theta_f) &= U_1(\Delta\theta_1), \dots, U_f(\Delta\theta_f) x_k(\theta_1, \dots, \theta_f) \\
&= U(\Delta\theta_1), \dots, \Delta\theta_f) x_k(\theta_1, \dots, \theta_f),
\end{aligned}
\tag{16.36}
$$

where $U(\Delta\theta_1, \dots, \Delta\theta_f) = \exp(\Delta\theta_1 X_1 + \dots + \Delta\theta_f X_f)$. Because the operators commute, the variables θ_σ and $\Delta\theta_\sigma$ on the right hand side of (16.36) combine additively, $U(\Delta\theta_1, \dots, \Delta\theta_f) U(\theta_1, \dots, \theta_f) = U(\theta_1 + \Delta\theta_1, \dots, \theta_f + \Delta\theta_f)$, just as does the time t in the time-evolution operator. Therefore, if we use the group operator $U(\Delta\theta_1, \dots, \Delta\theta_f)$ as the time-evolution operator by letting the increments $\Delta\theta_\sigma$ depend directly upon the time t, then the only possibility is that $\Delta\theta_\sigma = \omega_\sigma \Delta t$ where $\Delta t = t - t_o$ and the frequencies ω_σ are constants[3], independent of the time and can depend upon the initial data only through the constants of the motion G_σ.

So far, we have not had to assume that the f conserved quantities G_σ are isolating integrals (we have had no need to express the canonical momenta y_i in terms of the f generalized coordinates x_i and the f constants C_i).

Because the f operators X_σ commute with each other, the parameters θ_σ are holonomic (this follows from the analysis of section 12.1). Therefore, we can choose the f variables $\theta_\sigma(t) = \omega_\sigma t + \theta_\sigma$ as *generalized coordinates* for our f degree of freedom Hamiltonian system. The $2f$ variables $G_\sigma(x,y) = C_\sigma$ and $\theta_\sigma(t) = \omega_\sigma t + \theta_\sigma$ provide a complete solution of the dynamics problem, albeit not a canonically conjugate one.

To get a canonical description of integrable motion as f commuting translations, we must also construct the f canonical momenta P_σ conjugate to the f constant velocity translations $Q_\sigma = \theta_\sigma$. We denote the corresponding canonical momenta P_σ by I_σ. Since the transformation from the (θ, I) coordinates to the (x, y) coordinates has no explicit time-dependence, the scalar relation $K(\theta, I) = H(x, y)$ defines the new Hamiltonian K. *It follows immediately that K can depend only upon the momenta I_σ because the generalized velocities $\omega_\sigma = d\theta_\sigma/dt$ are constants.* By Hamilton's equations for the canonical variables θ_σ and I_σ,

$$
\dot{I}_\sigma = -\frac{\partial K}{\partial \theta_\sigma} = 0 \quad \text{and} \quad \dot{\theta}_\sigma = \frac{\partial K}{\partial I_\sigma} = \omega_\sigma = \text{constant},
\tag{16.30b}
$$

---

[3] If $\theta = f(t)$ and both θ and t are additive, then it follows easily that $f'(t) = $ constant, so that $f(t) = at + b$, where a and b are constants.

it follows that the momenta I_σ are also constants. *The immediate result is that every integrable canonical flow is differentially equivalent to a set of f independent translations along the f Cartesian axes labeled by the new coordinates θ_σ.*

Consider next the geometry of the flat f dimensional manifold in phase space to which integrable motion is confined. Because the f commuting translations θ_σ, $\{\theta_\sigma, \theta_\tau\} = 0$, require a *flat* f dimensional manifold for their existence, the surface defined by the intersection of the f surfaces $G_\sigma(q,p) = C_\sigma$ is therefore flat: the coordinates θ_σ are simply Cartesian coordinates on that manifold (see chapter 11).

Stated in other words, there is only one possibility for bounded motion that is consistent with a *constant* velocity field $V = (\omega_1, \ldots, \omega_f, 0, \ldots, 0)$. Because no points of the f dimensional manifold are deleted, the f axes $(\theta_1, \ldots, \theta_f)$ are necessarily Cartesian and the underlying manifold is therefore flat: integrable motion defines a globally parallel flow on a flat manifold where the motion is a constant speed translation. Flatness rules out surfaces of finite curvature like spheres or spheres with handles. For bounded motion, the f dimensional manifold must be closed because a body's position in phase space cannot go to infinity even though the Cartesian coordinates $\theta_\sigma(t)$ go linearly to infinity as t goes to infinity. The only remaining possibilities for bounded integrable motion are therefore flat manifolds that close on themselves, f dimensional Cartesian tori.

The representation of bounded integrable motion requires that there be a time τ_σ such that $\theta_\sigma(n\tau_\sigma) = \omega_\sigma n\tau_\sigma + \theta_\sigma = \theta_\sigma$ modulo the distance $\omega_\sigma\tau$, for $n = 1, 2, \ldots$. Geometrically, this is the same as saying that the Cartesian axes θ_σ must wrap around in cylindrical fashion and close on themselves. That Cartesian axes that close on themselves describe a torus for $f = 1$ and $f = 2$ was explained qualitatively, which is to say topologically, in chapter 6. That this is true for $f \geq 3$ is a matter of definition of higher dimensional tori with zero curvature, which we cannot picture except via intersections with two or three dimensional subspaces.

Because bounded motion occurs along the f Cartesian axes θ_σ on a closed manifold, we are free to express the motion on the confining f-surface in terms of an appropriate set of eigenfunctions generated by our f commuting translation operators[4]. If we write $L_\sigma = -iX_\sigma = -i\, d/d\theta_\sigma$ (the last term $d/d\theta_\sigma$ follows because θ_σ is the canonical group parameter for the translations generated by X_σ) and choose periodic boundary conditions for the angular variable θ_σ, then the operator L_σ is self-adjoint (see Stakgold, 1979, for self-adjoint first order linear differential operators). The same boundary conditions make the transformations U_σ orthogonal. The unnormalized eigenfunctions of L_σ are therefore $e^{im_\sigma\theta_\sigma}$ where m_σ runs from negative to plus infinity through the integers, including zero. The unnormalized simultaneous eigenfunctions Ψ of the f commuting operators L_σ are given by

$$\Psi = \exp(im_1\theta_1 + \ldots + im_f\theta_f). \tag{16.37}$$

---

[4] Koopman (1931) proved the correspondence of the generally *nonlinear* phase-volume-preserving flows of Hamiltonian systems with one-parameter groups of *linear* unitary transformations in the Hilbert space of functions of the phase space coordinates $(x_1, \ldots, y_f)$. He went on to prove that 'angle variables' cannot be constructed in the completely nonintegrable case where there are no conserved quantities other than H, and that in the case where the unitary operator U_t representing the system's time-evolution in Hilbert space has a pure point spectrum (the integrable case), the spectral decomposition of U_t provides a means of representing the canonically conjugate pairs by trigonometric series. In a later paper by Koopman and von Neumann (1932), it is noted that the case of a 'mixing' system, corresponds to nonintegrability (no angular variables) where there is 'everywhere dense chaos', and that for this case one does not need a *large* value of f, that $f = 2$ suffices.

Since the eigenfunctions (16.37) form a complete set, we can use them to expand the initial conditions in (16.36) in the form

$$x_k(\theta_1,\ldots,\theta_f) = \sum_{\{m_\sigma\}=-\infty}^{\infty} x_{k,m_1,\ldots,m_f} e^{(im_1\theta_1+\ldots+im_f\theta_f)} \tag{16.38}$$

where $\{m_\sigma\}$ denotes the eigenvalues $(m_1,\ldots,m_f)$.

The reader should note that we have *not* assumed that the flow generated by G_σ is periodic in the translation coordinate θ_σ: to study that particular flow via equations (16.32b), one cannot arbitrarily impose periodic boundary conditions on the trajectory defined by the f canonically conjugate pairs $(x_\sigma(\theta_\sigma),y_\sigma(\theta_\sigma))$ as a function of the 'time' θ_σ. Instead, the expansion (16.38) means that we have parameterized the f-torus to which the Hamiltonian flow of equation (16.32a) is confined by f Cartesian coordinates θ_σ. This agrees qualitatively with the fact that there are no sources or sinks in a Hamiltonian flow. The absence of sources and sinks (the absence of repelling and attracting fixed points of (16.32a)) is required by virtue of the fact that every Hamiltonian flow preserves the phase volume element $d\Omega = dx_1\ldots dy_f$ (Liouville's theorem), which directly implies Poincaré's recurrence theorem for bounded flows: every phase trajectory returns arbitrarily often to the neighborhood of any point that the system has visited in the past whenever the phase space motion is bounded. The final result, *obtained from group theory and geometry*, is the generalized Fourier series expansion

$$x_k(t) = \sum_{\{m_\sigma\}=-\infty}^{\infty} x_{k,m_1,\ldots,m_f} e^{[im_1(\omega_1\Delta t+\theta_1)+\ldots+im_f(\omega_f\Delta t+\theta_f)]} \tag{16.39}$$

The expansion provides the universal quantitative realization of the qualitative content of Poincaré's recurrence theorem for integrable systems. The failure to prove convergence of series of the form (16.39) for nonintegrable perturbations of an integrable system was a famous problem that led eventually to the Kolmogorov–Arnol'd–Moser theorem (see chapter 17).

The operators U_σ are orthogonal (strictly speaking, the corresponding linear operators in Hilbert space are orthogonal), are commuting translation operators on the torus, so they also can be diagonalized simultaneously. In order to obtain explicit expression for the frequencies ω_σ, one must first derive the Hamiltonian K in terms of the action variables. It is not surprising that the expansion (16.39) can be obtained by group theory since the special functions of mathematical physics follow quite generally from group theory (see Talman, 1968, for other examples).

From equations (16.37) and (16.38) it follows that torus motion is separable in the same sense that we understand separability in the theory of boundary-value problems: the simultaneous eigenfunctions of several operators are products of eigenfunctions of the separate operators. There is, so far as we know, no corresponding requirement that the Hamilton–Jacobi equation should separate in a special coordinate system.

If we can construct the Hamiltonian K in terms of the action variables I_σ, then we can find the frequencies by differentiation. As we have noted above, the action variables, the canonical momenta conjugate to the generalized coordinates θ_σ, are not given directly by the f conserved quantities G_σ but are provided instead by the f infinitesimal generators I_σ that yield the operators X_σ directly in diagonal form. Each canonically conjugate pair θ_σ and I_σ must satisfy the Poisson bracket relations

$$\{\theta_\sigma, \} = \frac{\partial}{\partial I_\sigma} \text{ and } \{ ,I_\sigma\} = \frac{\partial}{\partial \theta_\sigma} \tag{16.40}$$

in order that $\{\theta_\sigma, I_\kappa\} = \varepsilon_{\sigma\kappa}$.

Finally, note that we cannot choose H to be one of the f conserved quantities G_σ in the generating function F_2 defined by (16.35b): with $\delta\theta_\sigma = \omega_\sigma \delta t$, the infinitesimal transformations (16.35a) become Hamilton's equations

$$\dot{x}_k = \sum_{\sigma=1}^{f} \omega_\sigma \frac{\partial G_\sigma}{\partial y_k}, \quad \dot{y}_k = - \sum_{\sigma=1}^{f} \omega_\sigma \frac{\partial G_\sigma}{\partial x_k} \tag{16.35c}$$

so that $H = \Sigma \omega_\sigma G_\sigma$, where the frequencies ω_σ are dependent upon the f G_σ's (in action-angle variables with $f \geq 2$, one cannot assume that the transformed Hamiltonian $K(I_1,\ldots,I_f)$ is itself an action variable). Therefore, H is itself functionally dependent upon the f independent generators of its symmetries. Since the anti-self-adjoint operators X_σ generate f orthogonal translations on the torus, the Hamiltonian operator $H = \{,H\} = \Sigma \omega_\sigma X_\sigma$, the generator of time-translations of the Hamiltonian system (16.32a), is also anti-self-adjoint and generates constant velocity translations on the torus (or, for unbounded motion, on a flat manifold that is infinite in extent).

In what follows, we revert to the usual physics notation (q,p) for the canonical variables (x,y).

16.4 Action-angle variables

We begin with an involution system, with f commuting isolating integrals in the form $p_i - f_i(q_1,\ldots,q_f,C_1,\ldots,C_f) = 0$. The commutation rules $\{p_i - f_i, p_k - f_k\} = 0$ yield

$$\{p_k, f_i\} + \{f_k, p_i\} = 0, \tag{16.10}$$

from which it follows that

$$\frac{\partial f_i}{\partial q_k} = \frac{\partial f_k}{\partial q_i}. \tag{16.11}$$

This last condition means that the Poincaré–Cartan integral invariant

$$\Delta\Phi = \oint \sum_{i=1}^{f} p_i dq_i \tag{16.31b}$$

of section 15.2 is path-independent for different closed paths that loop around the same singularity without cutting across any new ones, because (16.11) guarantees that the integrand $\Sigma p_i dq_i = d\Phi$ is a closed differential. Clearly, for a nonvanishing integral (16.31b) the potential Φ cannot be analytic and must be multivalued. Furthermore, the canonical momentum must be the q-space gradient of the multivalued potential:

$$p = \nabla\Phi. \tag{16.41a}$$

We shall see below that we can use either (16.11) or (16.41a) as the condition that defines an integrable canonical system: for nonintegrable systems neither of these conditions holds, as Einstein reminded the Bohr–Sommerfeld school of quantization (who apparently ignored Liouville's involution theorem) in 1917.

A completely integrable canonical system is one where the canonical momentum in the q-subspace of phase space defines the streamlines of a flow defined by a vortex potential Φ (the multivaluedness of Φ may reflect vortices in the flow pattern). The integrability condition (16.11) is essential for the *global* reduction of the Schrödinger equation of quantum mechanics to the Hamilton–Jacobi equation in any semi-classical limit (see ter Haar, 1971, for the usual argument that connects Schrödinger with Hamilton–Jacobi equations). In other words, one can define a completely integrable canonical Hamiltonian quantum system as one that has a Hamilton–Jacobi semi-classical limit that holds for all finite times. The helium atom, a three-body problem, is nonintegrable (as was proven by Poincaré) and therefore has no such Hamilton–Jacobi limit, even though local Hamilton–Jacobi descriptions may be possible over times short compared with a period of the motion. Therefore, the classical helium atom cannot be described using action-angle variables and cannot be quantized semi-classically by the Bohr–Sommerfeld method.

We now concentrate on the generating function F_2 that yields the transformation to action-angle variables defined by

$$\dot{I}_k = -\frac{\partial K}{\partial \theta_k} = 0 \text{ and } \dot{\theta}_k = \frac{\partial K}{\partial I_k} = \omega_k = \text{constant.} \tag{16.42}$$

If we consider the path-independent invariant $\Delta\Phi$ defined by the closed loop integral

$$\Delta\Phi = \sum_{i=1}^{f} \oint_{C_k} p_i \mathrm{d}q_i = \sum_{i=1}^{f} \oint_{C_k} I_i \mathrm{d}\theta_i, \tag{16.43}$$

and then integrate around any simply connected closed path C_k where $\Delta\theta_k = 2\pi$ and $\Delta\theta_j = 0$ for $j \neq k$, then

$$I_k = \frac{1}{2\pi} \sum_{i=1}^{f} \oint_{C_k} p_i \mathrm{d}q_i. \tag{16.44}$$

The condition that this result defines f canonical momenta I_k is that (16.44) is independent of the choice of integration path C_k, which requires that the integrand $\Sigma p_i \mathrm{d}q_i$ must be a closed differential $\mathrm{d}\Phi$ of a multivalued function Φ. The required integrability condition (16.41a) is guaranteed by (16.11). The necessary and sufficient condition for (16.41a) to hold is that there exist f global, functionally independent, commuting isolating integrals G_σ. Each canonical momentum I_σ is equal to the change $\Delta\Phi$ in the multi-valued potential Φ for any path where $\nabla\theta_k = 0$ if $k \neq \sigma$ but where $\Delta\theta_\sigma = 2\pi$. The transformation to action-angle variables is just the Lie transformation for the special case of an integrable canonical system. The price of getting the Lie transformation for only f conservation laws is that the f conservation laws must be compatible, must all commute with each other.

The integral on the right hand side of (16.44) is nonvanishing because the f canonical momenta p_i are two-valued. For integrable motion and for finite times, the phase space trajectory intersects the (q_i, p_i)-plane finitely many times and the points fall on a simply closed curve, as we explained for the case of $f = 2$ in section 6.1 (see figure 6.1). For example, for a free particle enclosed within a one dimensional box of width L with perfectly reflecting walls, $p = mv$ or $p = -mv$, depending upon the direction of travel. In this case, $I = mvL/\pi$, $K(I) = H(p) = p^2/2m = \pi^2 I^2/2mL^2$ and so $\omega = \pi^2 I/m^2 = 4\pi v/L$.

In principle, one should be able to use the generating function $F_2(q,P)$ to transform back to the original variables: from equation (16.41a) we have

$$p_k = \frac{\partial \Phi}{\partial q_k} \qquad (16.41b)$$

so that we can identify the two functions $F_2(q,I) = \Phi(q,C)$. Again, it follows from (16.41b) that the solution of the time-independent Hamilton–Jacobi equation (16.29) is given by the 'reduced action'

$$\Phi(q,C) = \int^f \sum_{i=1}^f p_i dq_i, \qquad (16.45)$$

where the p_i are expressed by (16.41b) as functions of (q,C) rather than in terms of (q,I). The idea of carrying out the global coordinate transformation would be to use

$$\theta_k = \omega_k t + \theta_{k_o} = \frac{\partial F_2}{\partial I_k} = g_k(q,I) \qquad (16.46)$$

to solve for all f of the generalized coordinates q_k as functions of t and the f action variables I_j, and also to substitute that result into

$$p_k = \frac{\partial F_2}{\partial q_k} = h_k(q,I), \qquad (16.47a)$$

which would then yield the canonical momenta p_k as functions of t and the f conserved momenta I_k. Clearly, we have to assume that (16.46) determines q_k as a function of t and the $2f$ constants θ_{k_o} and I_k. If this determination holds, then all of the p's as well (through (16.47a)) are determined completely as functions of the time and the initial data through the $2f$ constants $(\theta_{1_o}, \ldots, \theta_{f_o}, I_1, \ldots, I_f)$, f of which are generally trivial while the remaining f are nontrivial and global.

Forging fearlessly ahead with the example of a particle in a box, $F_2(x,I) = L(\omega t + \theta_o)/\pi$ so that $\theta = \omega t + \theta_o = \pi x/L$. This yields $x = vt + x_o$ as it must. With

$$p = \begin{cases} mv, 0 < t < \tau/2 \\ -mv, \tau/2 < t < \tau \end{cases} \qquad (16.48)$$

where $\tau = 2L/v$ and $I = mvL/\pi = \pm pL/\pi$, we find $F_2(x,I) = \pm \pi I x/L$. This yields in turn that $\theta = \omega t + \theta_o = \pm \pi x/L$, so that

$$x = \begin{cases} vt, 0 < t < \tau/2 \\ -vt + 2L, \tau/2 < t < \tau \end{cases}, \qquad (16.49)$$

and also that

$$\theta = \begin{cases} \omega t, 0 < t < \tau/2 \\ -\omega t - 2\pi, \tau/2 < t < \tau \end{cases}, \qquad (16.50)$$

where $\omega = 2\pi/\tau = 2\pi v/L$. Equation (16.39) takes the form

$$x(t) = \sum_{n=-\infty}^{\infty} x_n e^{in\omega t} = \sum_{n=0}^{\infty} [a_n \cos(n\omega t) + b_n \sin(n\omega t)] \qquad (16.51)$$

since x is real ($x_n^* = x_{-n}$), and because $x(t)$ is symmetric about $t = \tau/2$ it follows that $b_n = 0$ for all n. From the orthogonality in function space of $\cos(n\omega t)$ and $\sin(n\omega t)$ it follows, if we compute the Fourier coefficients for the dimensionless quantity $x(t)/L$ rather than $x(t)$, that

$$b_n = \frac{4}{\tau} \int_0^\pi \frac{vt}{L} \cos(n\omega t) dt = \begin{cases} -4/n^2\pi^2, & n \text{ odd} \\ 0, & n \text{ even} \end{cases} \qquad (16.52)$$

(see Stakgold, 1979, Margenau and Murphy, 1956, or Sommerfeld, 1964, for orthogonal functions). The periodicity of the motion of the particle in the box is therefore expressed analytically by the final result

$$x(t)/L = 1 - \frac{4}{\pi^2} \sum_{\substack{n=0 \\ (n \text{ odd})}}^\infty \frac{1}{n^2} \cos(n\omega t), \qquad (16.53)$$

which obeys the periodicity condition $x(t + N\tau) = x(t)$ for $N = 0,1,2,\dots$. The velocity is then given by

$$v(t)/L = \frac{4}{\pi^2} \sum_{\substack{n=0 \\ (n \text{ odd})}}^\infty \frac{1}{n} \sin(n\omega t) \qquad (16.54)$$

and represents a periodic function.

Success with the free-particle model gives us the necessary confidence to go further, but is there a catch?

For an f degree of freedom system and an arbitrary choice of symplectic coordinates $(q_1,\dots,p_f)$, the f conservation laws in the form $G_\sigma(q_1,\dots,p_f) = C_\sigma = $ constant must determine all f of the f canonical momenta as functions of the f generalized coordinates q_i, yielding f two-valued isolating integrals of the form $p_i = f_i(q_1,\dots,q_f,C_1,\dots,C_f)$. This alone, however, does not permit us to use

$$p_k = f_k(q_1,\dots,q_f,C_1,\dots,C_f) = \frac{\partial F_2(q_1,\dots,q_f,I_1,\dots,I_f)}{\partial q_k} \qquad (16.47\text{b})$$

to identify the generating function $F_2(q_1,\dots,q_f,I_1,\dots,I_f)$ merely by integration, because $C_\sigma = G_\sigma \neq I_\sigma$. The only way to go further by this method is to be able to use the results of the conservation laws in the form $p_i = f_i(q_1,\dots,q_f,C_1,\dots,C_f)$ *directly* to perform the integrations (16.43) and (16.45) in order to identify the f action variables I_σ and the generating function F_2. We emphasize that the condition (16.11) guarantees that the differential form $\Sigma p_i dq_i = \Sigma f_i(q_1,\dots,q_f) dq_i$ is closed for any *arbitrary* choice of generalized coordinates $(q_1,\dots,q_f)$ taken as the starting point.

Returning now to the question whether there might be a 'catch' in practice, even if you have the luck to discover them, one or more of the f conservation laws $G_\sigma(q,p) = C_\sigma$ may be such complicated and interwoven functions of the $2f$ q's and p's that the solution for the p's as functions of the q's and C's may be extremely difficult or even impossible, except numerically. The closed form solution of algebraic equations of degree higher than three is nontrivial, not to mention transcendental equations. What exists mathematically is not necessarily conveniently solvable, but it is mathematical existence, not solvability, that casts integrable motion into the form of f independent constant velocity translations. In spite of any practical difficulty in determining the frequencies as functions of the f action variables, equation (16.39) guarantees that the motion on the torus is stable periodic or stable quasiperiodic. The value of qualitative analysis is that it helps us to understand the geometric nature of the motion even in the absence of analytic success.

Solvable, integrable, and separable are three different ideas. All of the systems considered in this text are solvable, whether or not integrable. We have seen above that integrable Hamiltonian systems are always separable in the sense that the simultaneous eigenfunctions of f commuting operators factor into a product of eigenfunctions, but there is also a different idea of separability that is prevalent in the literature on integrability. There, one tries to use the symmetry implied by the f conservation laws to find a coordinate system $(q_1, \ldots, p_f)$ in phase space where the f degrees of freedom become decoupled in the sense that

$$p_k = \frac{\partial F_2(q_k, I_1, \ldots, I_f)}{\partial q_k} = \sum_{i=1}^{f} \frac{\partial F_{2k}(q_k, I_1, \ldots, I_f)}{\partial q_k} = \sum_{i=1}^{f} f_i(q_k, I_1, \ldots, I_f), \quad (16.55)$$

which means that the function F_{2k} depends only upon q_k and not at all upon the other q_j *in at least one special coordinate system*. In this case, in the integrals (16.43) and (16.45), *each term* in the integrand's sum, $p_i dq_i = f_i(q_i, C_1, \ldots, C_f) dq_i$, is exact. This restriction would require that the generating function F_2 that satisfies the generally nonlinear but time-independent Hamilton–Jacobi partial differential equation

$$h\left(q, \frac{\partial F_2}{\partial q}\right) = K(I) \quad (16.56)$$

must then satisfy separability in the form

$$F_2(q_1, \ldots, q_f, I_1, \ldots, I_f) = \sum_{i=1}^{f} F_{2i}(q_i, I_1, \ldots, I_f). \quad (16.57)$$

Our experience with second order linear partial differential equations conditions us to expect that symmetry implies separability in at least one special coordinate system. Here, however, we face a nonlinear first order partial differential equation. A separable Hamilton–Jacobi system is always integrable but there is no known implication from the integrability condition (16.11) above that the Hamilton–Jacobi equation must be separable in some special coordinate system. We refer the reader to Gutzwiller's book (1990) for a discussion of the difficulties met in trying to separate variables for the Toda lattice, a known integrable problem in Hamiltonian dynamics (see also Toda, 1981).

Having developed the necessary transformation theory for autonomous Hamiltonian systems, we next study the details of some well-known integrable systems in action-angle variables. We also discuss perturbations of integrable systems that yield globally nonintegrable flows that are still integrable in fragmented parts of phase space in chapter 17. The integrable systems that we study in the next section are atypical, as they have separable Hamilton–Jacobi equations in at least one coordinate system. It was in part because both the Kepler and isotropic harmonic oscillator problems are separable in two *different* coordinate systems that Einstein was led to his now-famous discovery of equation (16.41a)[5].

Euler's solid body is integrable in two entirely different formulations. Whittaker (1965) exhibits the integrability of the canonical description of the flow in Euler angles by finding three commuting conservation laws. With three compatible conservation

---

[5] The older statement of the Bohr–Sommerfeld quantization rule is that each closed line integral

$$\oint p_i dq_i$$

should be quantized. To eliminate the ambiguity in quantizing the Kepler and oscillator problems, Einstein pointed out that this is wrong and that it is instead the integral invariant (16.31b) that should be quantized, and that (16.41a) rules out the consideration of both helium atoms and statistical mechanics!

laws, one can construct the canonical variables (I,θ) that describe the motion as a constant velocity flow on a three dimensional torus in the six dimensional $(\phi,\theta,\psi,p_\phi,p_\theta,p_\psi)$-phase space. In section 9.2, in contrast, we used the three Cartesian angular velocities ω_i as dynamical variables and showed that the motion is confined to a two dimensional sphere in a three dimensional $(\omega_1,\omega_2,\omega_3)$-phase space. There is no continuous deformation of a two-sphere into a 3-torus. This is consistent with the fact that the three velocities ω_i are nonintegrable. Their time integrals are three non-holonomic variables α_i, whereas the Euler angles are generalized coordinates (are holonomic coordinates). The transformation $\omega_i dt = \lambda_{i1}d\phi + \lambda_{i2}d\theta + \lambda_{i3}d\psi$ is nonintegrable and therefore does not give rise to three functions $\alpha_i = f_i(\phi,\theta,\psi)$ that would permit the construction of a topology-violating global coordinate transformation. The two different topologic descriptions of this integrable system are not equivalent, which shows that purely topologic descriptions of canonical flows are not necessarily unique.

16.5 Central potentials

For the simple harmonic oscillator in one dimension

$$H = (p^2 + \omega^2 q^2)/2 = K(I) \tag{16.58}$$

with

$$p = \frac{\partial S}{\partial q}, \ \theta = \frac{\partial S}{\partial I} \tag{16.59}$$

we find that

$$I = \frac{1}{2\pi}\oint p dq = \int_{q_{min}}^{q_{max}} [2(E - \omega^2 q^2)]^{1/2}dq = E/\omega \tag{16.60}$$

so that $K = E = I\omega$. Hamilton's equations of motion have the form

$$\dot{I} = -\frac{\partial K}{\partial \theta} = 0, \ \dot{\theta} = \frac{\partial K}{\partial I} = \omega \tag{16.61}$$

with

$$\theta = \omega t + \theta_\circ = \frac{\partial S}{\partial I}. \tag{16.62}$$

Consider next a central potential problem. Since the motion is confined to a plane in Euclidean space we have

$$H(r,\phi,p_r,p_\phi) = K(I_1,I_2). \tag{16.63}$$

Here, we have a two dimensional torus with translation variables θ_1 and θ_2 in the four dimensional (r,θ,p_r,p_θ)-phase space. In this case

$$I_1 = \frac{1}{2\pi}\oint p_r dr = \frac{1}{\pi}\int_{q_{min}}^{q_{max}} [2(E - V(r)]^{1/2}dr \tag{16.64}$$

with $V(r) = M_z^2/2\mu r^2 + U(r)$. It follows that

$$I_2 = \frac{1}{2\pi}\oint p_\phi d\phi = M_z, \tag{16.65}$$

and to go further one must specify the central potential $U(r)$. For the Kepler problem

$$U(r) = - Gm_1m_2/r \tag{16.66}$$

one can show (see ter Haar, 1971) that

$$K(I_1, I_2) = - 2\pi^2 G^2 \mu m_1^2 m_2^2/(I_1 + I_2)^2. \tag{16.67}$$

This problem exhibits 'accidental degeneracy': note that $\omega_1 = \omega_2 = \omega$ where

$$\omega = \frac{\partial K}{\partial I_i} = \frac{4\pi^2 G^2 \mu m_1^2 m_2^2}{(I_1 + I_2)^3}. \tag{16.68}$$

This means there are no quasiperiodic orbits because $\omega_1/\omega_2 = 1$ for all energies and angular momenta. Accidental degeneracy corresponds to a hidden symmetry that yields an $(f + 1)$th global conservation law that is analytic in the orbit parameters (see chapter 4). Also, there are two different coordinate systems where the Hamiltonian is separable (see Whittaker, 1965).

The isotropic harmonic oscillator also separates in two different coordinate systems, Cartesian and spherical, and exhibits accidental degeneracy due to an analytic conservation law that reflects a hidden symmetry.

16.6 The action integral in quantum theory

Semi-classical approximations to Schrödinger's equation via the Hamilton–Jacobi equation are well known, but Dirac also showed how the action functional arises in quantum mechanics without making any approximations (p. 127 of *Quantum Mechanics*). Feynman developed Dirac's idea in q-space and arrived at the path-integral formulation of quantum theory. There, the action functional is not restricted to the path in the q-subspace of phase space that makes the action an extremum: trajectories other than the classical one that satisfies Hamilton's equations are allowed in quantum mechanics.

In Feynman's path integral approach to quantum mechanics, one writes

$$K(q, q_o, t - t_o) \sim \sum_{\text{paths}} e^{2\pi i A/h} \tag{16.69}$$

where q stands for the f coordinates $(q_1, \ldots, q_f)$ in the q-subspace of phase space (the full (q,p)-phase space does not exist in quantum mechanics), K is the 'propagator', or Green function, and A denotes the classical action functional. In general, each path in the sum in (16.69) generates a different number for the action A between the fixed end points q and q_o. The absolute square of the propagator K yields the probability that a particle that is known to be at the point q_o in q-space at time t_o may be found at the point q at a later time t (see Feynman and Hibbs, 1965).

With the end points in q-space fixed, the 'sum' in (16.69) must be understood mathematically as a functional integral over all possible continuous paths that connect the two end points (functional integrals were first introduced by N. Wiener in the related context of Brownian motion and diffusion equations (see Kac, 1959)). The smooth Newtonian path that solves Hamilton's equations is only one possible path out of a continuum of generally nondifferentiable paths in the sum. The tangent vector dq/dt, hence also the canonical momentum $p = \partial L/\partial \dot{q}$, does not exist at a definite position q along most trajectories in the functional integral, in agreement with the uncertainty principle.

In order to get the prefactor right in (16.69), one must evaluate the function space integral. We concentrate here only on the complex exponential $e^{2\pi iA/h}$ in the integrand. If the classical system is integrable, then Liouville's integrability theorem tells us, for paths that are consistent with the classical conservation laws (and hence are differentiable), that the action $A = \Phi(q,C,t) - \Phi(q_o,C,t_o) = \Delta\Phi$ is path-independent where C denotes the required f conserved quantities $(C_1,\ldots,C_f)$. If one approximates the path integral of an integrable system by summing only over these paths, then aside from a prefactor one obtains

$$K(q,q_o,t - t_o) \sim \sum_{\text{paths}} e^{2\pi iA/h} \sim e^{2\pi i\Delta\Phi/h} \qquad (16.70)$$

with a *path-independent* exponent $\Delta\Phi$. For an integrable system, in this approximation, the propagator is the complex exponential of a classically determined quantum phase difference $\Delta\Phi$ that itself depends only upon the end points. For a nonintegrable system, no such simplicity is possible, except perhaps locally over times that are short compared with a period of the classical motion. There, the details of the paths enter into the evaluation of the path integral because the action integrand Ldt is globally nonintegrable.

There is at least one case where the approximation to the phase indicated in (16.70) is valid: if the Lagrangian is Gaussian (free particle or harmonic oscillator) then the result is exact. Also, if the system is integrable and the Lagrangian is not quadratic, but one evaluates the path integral by using a stationary phase approximation, then again the classical phase difference is independent of path.

For example, integrability is clearly seen in the free-particle Feynman propagator in one dimension,

$$K(q,q_o,t - t_o) = \left[\frac{m}{2\pi ih(t - t_o)}\right]^{1/2} e^{\pi im(q - q_o)^2/h(t - t_o)}, \qquad (16.71)$$

where $A = mv^2(t - t_o)/2 = m(x - x_o)^2/2(t - t_o)$ with $q = x$ for one dimensional motion. The same is true for the simple harmonic oscillator and other quadratic examples studied in Feynman and Hibbs (1965). With a quadratic Lagrangian (Gaussian path integral), whether you evaluate the Lagrangian first and then do the path integral or vice versa is irrelevant, because the same classical phase difference $\Delta\Phi$ falls out in either case.

Recent studies have used the path integral in order to study nonintegrable quantum dynamics (any quantum system that has no global Hamilton–Jacobi description) from the standpoint of the classification and computation of the unstable periodic orbits of the corresponding nonintegrable classical system. There, the periods and certain other symbolic dynamics properties are topologic invariants and provide a coordinate-free description of the chaotic classical dynamics (see Friedrich, 1992, and references therein). A fascinating approximate result has been obtained from this method: the semi-classical reproduction of the energy spectrum of the quantum helium atom. The classical problem is a nonintegrable three-body system of two electrons and a doubly charged positive nucleus. Action-angle variables and Bohr–Sommerfeld quantization do not exist in this case. In the calculation, the classical problem is approximated by a one dimensional problem where the two electrons move only on opposite sides of the nucleus. This problem is chaotic, and the quantum energy spectrum has been reproduced by computing the low order unstable periodic orbits in order to evaluate Feynman's path integral approximately. This problem and other nonintegrable ones in atomic physics represent an extremely interesting frontier of research in nonlinear dynamics and quantum mechanics.

Exercises

1. Solve the simple harmonic oscillator

$$H(q_1 p) = p^2/2m + m\omega^2 q^2/2$$

 by the Hamilton–Jacobi method. Show that

$$\Phi = \tfrac{1}{2}m\omega \cosec[\omega(t - t_o)]\{(a^2 + q^2)\cos[\omega(t - t_o)] - 2aq\}.$$

 Verify that $p = \dfrac{\partial\Phi}{\partial q}, \; H = \dfrac{\partial\Phi}{\partial t}$

 and also compute

$$b = -\frac{\partial\Phi}{\partial a}.$$

 (Whittaker, 1965)

2. Use separation of variables $S(q_1, q_2, P_1, P_2) = S_1(q_1) + S_2(q_2) - Et$ to solve the Hamilton–Jacobi equation for the two dimensional isotropic harmonic oscillator.

 (ter Haar, 1971)

3. Use the Hamilton–Jacobi equation to obtain the orbit equation for a particle moving in a two dimensional potential $U = k/r$. (Hint: use coordinates $(u,v) = (r + x, r - x)$.)

 (ter Haar, 1971).

4. Compute the action variables and the Hamiltonian $K(I_1, I_2, I_3)$ for the three dimensional harmonic oscillator. Show that all three frequencies are equal to each other.

5. $L = \tfrac{1}{2}(\dot\theta^2 + \dot\phi^2 \sin^2\theta)$ defines geodesic motion on the unit sphere. In Lagrangian configuration space, ϕ and θ are coordinates on a surface of constant positive curvature, the two-sphere. Show that this does not conflict with the fact that $(\phi, \theta, p_\phi, p_\theta)$ can be treated as Cartesian coordinates in a four dimensional phase space:

 (a) Show that $L\, dt$ is exact and find $\Phi(\phi, \theta, p_\phi, E)$.
 (b) Find the orbit equation $\phi = f(\theta, p_\phi, E)$. This defines a third conserved quantity that eliminates quasiperiodic orbits on the 2-torus in phase space.
 (c) Identify the translation variables (θ_1, θ_2) where $\dot\theta_i = \omega_i(I_1, I_2) = $ constant, and then construct the action variables I_k (and also $K(I_1, I_2)$).

17

Nonintegrable canonical flows

The Irregularity of the Moon's Motion hath been all along the just Complaint of Astronomers; and indeed I have always look'd upon it as a great Misfortune that a Planet so near to us as the Moon is, and which be so wonderfully useful to us by her Motion, as well as by her Light and Attraction (by which our Tides are chiefly occasioned) should have her Orbit so unaccountably various, that it is in any manner vain to depend on any Calculation of an Ellipse, a Transit, or an Appulse of her, tho never so accurately made.

Isaac Newton (1702)

17.1 The Henón–Heiles model

Every cookbook can be encoded as a finite-length computer program. Even if one can prove that a system is a flow in phase space as we did for a class of driven-dissipative dynamical systems in chapter 13, we have at our disposal no cookbook recipe for deciding whether the system is completely integrable. In practice, one has recourse to two imperfect tests and neither can play the role of an infallible final authority.

One test is numeric and negative. There, one searches phase space for evidence of deterministic chaos by computing Poincaré sections numerically for a lot of different initial conditions (or looks for divergent time series with nearby initial conditions). No numerical search can be complete, and the absence of chaos does not necessarily imply complete integrability.

The second method is analytic but local. Here, one studies the poles of solutions in the complex time plane. This method was introduced and used with success by S. Kowalevskaya in rigid body theory. In the papers by Tabor (1989), integrable cases of both the Lorenz and Henón–Heiles models are discovered by the use of that method. For a comprehensive discussion of integrability and methods of locating analytic first integrals, we refer the reader to the very advanced book by Fomenko (1988).

We follow the numerical path by reviewing the results that were obtained for a model introduced by Henón and Heiles in astronomy (published in the same era as Lorenz's original work):

$$H = (p_x^2 + p_y^2)/2 + U = E = \text{constant} \tag{17.1}$$

with

$$U = (Ax^2 + By^2 + 2Dx^2y - 2Cy^3/3)/2. \tag{17.2}$$

The known completely integrable cases occur for (i) $D/C = 0$, and all values of A, B, (ii) $D/C = -1$, $A/B = 1$, (iii) $D/C = -1/6$, and all values of A,B, and (iv) $D/C = -1/16$, $A/B = 1/16$ (see Tabor, 1989). We therefore consider next a case that is not known to be completely integrable.

For the parameter values $A = B = C = D = 1$,

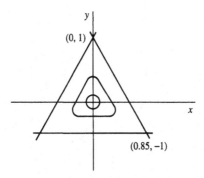

Fig. 17.1 Equipotentials of the Henón–Heiles problem.

$$U = (x^2 + y^2 + 2x^2y - 2y^3/3)/2 \qquad (17.3)$$

and there are four equilibria at $(0,0)$, $(0,1)$, $(-\sqrt{3/4}, -1/2)$, and $(\sqrt{3/4}, -1/2)$. The origin $(0,0)$ is an elliptic point in the linear approximation whereas $(0,1)$ is a saddle-point of U so that the motion is unbounded for $y > 1$. We consider the region of bounded motion $0 < y < 1$, corresponding to low enough energies E. The equipotentials in figure 17.1 show that uniformly bounded motion occurs if $E < 1/6$ (the second derivatives of the potential U show that escape to infinity is possible only if $E > 1/6$). We know that the model is integrable in the small oscillation approximation where the potential U is approximated as quadratic, and so we expect regular motions for low enough energies. At higher energies still obeying $E < 1/6$ the door is left open for more interesting behavior.

We study a Poincaré map numerically as the energy E is increased. The numerically generated map indicates a sequence of bifurcations that lead eventually to the appearance of sensitivity with respect to small changes in initial conditions. Figure 17.2 shows the results of numerical computations for $E = 1/12$, $1/8$, and $1/6$ for several different initial conditions. The results represent a return map where a point (y, p_y) is plotted whenever x passes through 0 with $p_x > 0$. For $E = 1/24$ (see Berry, 1978) the computed motion falls on smooth curves that can be identified as intersections of tori with the (y, p_y)-plane. The computation shows four elliptic and three hyperbolic points. Only elliptic and hyperbolic points occur as equilibria in an area-preserving iterated map (see chapter 6). For $E = 1/12$, nothing qualitatively new occurs but for $E = 1/8$ we see evidence of bifurcations: there are at least four new elliptic points, but in addition some of the iterations appear to be distributed irregularly, suggesting chaotic orbits. The maximum Liapunov exponent of an orbit can be estimated numerically and crudely from the time required for two nearby time series, e.g. for $x(t)$ vs t, for two nearby initial conditions to diverge from each other $\lambda \approx t_d^{-1}$ where t_d is the time needed for the difference between the two series to become on the order of unity. For $E = 1/6$, the disorder in the phase plane generated by pseudo-orbits has increased due to period-doubling bifurcations where elliptic points have become hyperbolic. The scatter will not disappear if we limit the calculations to integration times short enough that one can recover one or more digits of the initial data in a backward integration of Hamilton's equations: uniformly bounded pseudo-orbits with positive Liapunov exponents mean that the trajectories are chaotic. Correspondingly, the exponential instability of nearby orbits cannot be eliminated by a correct computation that avoids

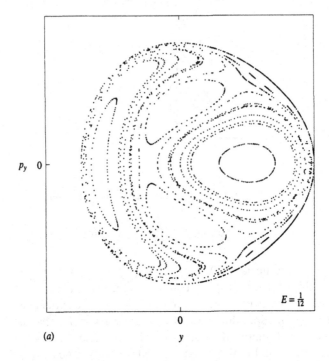

$$p_y \quad 0$$

$$E = \tfrac{1}{12}$$

$$0$$

(a)   y

Fig. 17.2 Poincaré sections of the Henón–Heiles model for energies
E = 1/12. 1/8, and 1/6.

the uncontrollable errors committed by using floating-point arithmetic. The map is area-preserving and chaotic, so there must be a positive and a negative Liapunov exponent of equal magnitudes. These exponents are not constants but instead vary with initial conditions from one class of exact trajectories to another.

How much of the computed regular behavior indicated in figure 17.2 can be accounted for by a perturbation treatment that assumes integrability? This question is motivated by the expectation that not all tori are destroyed immediately as the perturbation strength C is increased from zero. In an integrable canonical system, f dimensional invariant tori foliate the $2f$ dimensional phase space, making chaotic motion impossible, but under the weakest possible perturbation that makes the system nonintegrable some tori disappear suddenly while others persist for stronger perturbations. The proof of existence and method of computing the persisting tori from perturbation theory were worked out theoretically by Kolmogorov, Arnol'd, and Moser before the Henón–Heiles model was introduced. Next, we discuss canonical perturbation theory and then survey the KAM theorem, which is a theorem about perturbation expansions.

17.2 Perturbation theory and averaging

In chapter 16 we introduced two generating functions that are useful if the model is integrable. We now consider a generating function, one that locally solves a time-independent Hamilton–Jacobi partial differential equation, in the search for 'nearly integrable' systems. A nearly integrable canonical system is one where there are a lot of persisting tori that can be described by perturbation theory, and where the chaotic

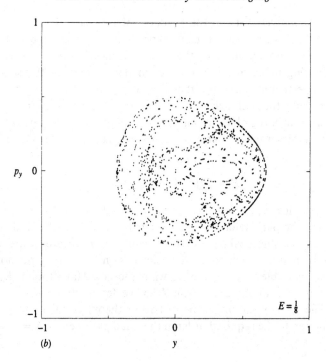

(b)

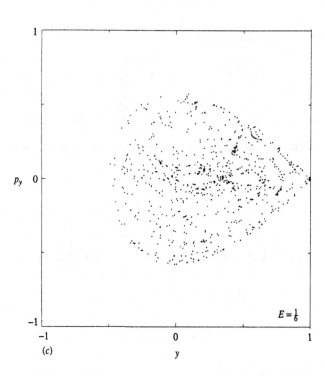

(c)

motions occupy relatively little of the available phase space. Those tori that persist under perturbation can be described by perturbing the integrable system (the low energy quadratic approximation to the Henón–Heiles model is an example). The resulting generating function is 'localalized' in phase space in the sense that it is confined to undestroyed tori and therefore does not hold for arbitrary initial conditions, especially those that yield chaotic motions in small portions of phase space.

Beginning with an integrable canonical system with Hamiltonian $H_o(J_1,\ldots,J_f)$ in action-angle variables, we consider a perturbation that makes the system nonintegrable,

$$H = H_o(J) + \varepsilon H_1(J,\theta) + \ldots \tag{17.4}$$

so that the variables $(\theta_1,\ldots,J_f)$ are not action-angle variables for the perturbed Hamiltonian H. The parameter ε is to be treated as small in what follows.

We now construct canonical perturbation theory, the Poincaré–von Zeipel method of studying perturbations via canonical transformations. We try to transform to a new set of canonical variables $(\bar\theta,\bar J)$ where we want the new Hamiltonian K, to $O(\varepsilon)$, to depend on $\bar J$ alone, for an obvious reason. Also, we denote the generating function by $F_2 = S(\theta,\bar J)$. The idea of canonical perturbation theory is to try to construct the generating function to the required order in the small parameter ε, $S = S_o + \varepsilon S_1 + \ldots$, where

$$S_o = \sum_{i=1}^{f} \theta_i \bar J_i \tag{17.5}$$

generates the identity transformation and

$$\left. \begin{aligned} J_i &= \frac{\partial S(\theta,\bar J)}{\partial \theta_i} = \bar J_i + \varepsilon \frac{\partial S_1(\theta,\bar J)}{\partial \theta_i} + \ldots, \\[2mm] \theta_i &= \frac{\partial S(\theta,\bar J)}{\partial \bar J_i} = \bar\theta_i + \varepsilon \frac{\partial S_1(\theta,\bar J)}{\partial \bar J_i} + \ldots \end{aligned} \right\} \tag{17.6a}$$

By direct substitution, these equations can be rewritten approximately as

$$\left. \begin{aligned} J_i &= \bar J_i + \varepsilon \frac{\partial S_1(\bar\theta,\bar J)}{\partial \theta_i} + \ldots, \\[2mm] \bar\theta_i &= \frac{\partial S(\theta,\bar J)}{\partial \bar J_i} = \theta_i - \varepsilon \frac{\partial S_1(\bar\theta,\bar J)}{\partial \bar J_i} + \ldots \end{aligned} \right\} \tag{17.6b}$$

where the unperturbed Hamiltonian can be rewritten formally as

$$H_o(J) = H_o\left(\bar J + \varepsilon \frac{\partial S_1}{\partial \theta} + \ldots \right)$$

$$\approx H_o(\bar J) + \varepsilon \frac{\partial H_o}{\partial \bar J_i} \frac{\partial S_1}{\partial \theta_i} + \ldots \tag{17.7a}$$

and where we begin now to use the summation convention. The result is that

$$H(\bar\theta,\bar J) \approx H_o(\bar J) + \varepsilon\left(\omega_i(\bar J)\frac{\partial S_1}{\partial \theta_i} + H_1(\bar\theta,\bar J) \right) + \ldots = K_o(\bar J) + K_1(\theta,J) + \ldots \tag{17.7b}$$

and the aim is to find a way to identify $H(\bar{\theta},\bar{J}) \approx K(\bar{J})$. To do this, the method of averaging, an uncontrolled approximation, must be used.

We define

$$\langle H_1 \rangle = \frac{1}{2\pi} \int_0^{2\pi} H_1 d\theta \tag{17.8}$$

where $H_1 = \langle H_1 \rangle + \{H_1\}$ so that $\{H_1\}$ is the oscillatory part of the perturbation H_1. Note also that

$$K_1 = \omega_i(\bar{J}) \frac{\partial S_1}{\partial \theta_i} + H_1 \tag{17.9}$$

and that

$$\int_0^{2\pi} \frac{\partial S_1}{\partial \theta} d\theta = 0. \tag{17.10}$$

The idea is to try to choose S so that K_1 is independent of $\bar{\theta}$. Therefore, with

$$K_1 = \omega_i(\bar{J}) \frac{\partial S_1}{\partial \theta_i} + \langle H_1 \rangle + \{H_1\} \tag{17.11}$$

we choose

$$\omega_i(\bar{J}) \frac{\partial S_1}{\partial \theta_i} + \langle H_1 \rangle = 0. \tag{17.12}$$

If this approximation holds, then K is approximately dependent only on the f variables $\bar{J}_i$. Expanding in Fourier series

$$\left. \begin{aligned} H_1 &= \sum_{m_k=-\infty}^{\infty} H_{1k}(\bar{J}) e^{i\vec{m}\cdot\vec{\theta}}, \\[2mm] S_1 &= \sum_{m_k=-\infty}^{\infty} S_{1k}(\bar{J}) e^{i\vec{m}\cdot\vec{\theta}}, \end{aligned} \right\} \tag{17.13}$$

we find that

$$S_{1\vec{m}} = i \frac{H_{1\vec{m}}}{\vec{m}\cdot\vec{\omega}(\bar{J})} \tag{17.14}$$

which leads to the famous problem of small denominators in perturbation theory. Clearly, for rational frequency ratios perturbation theory fails completely but divergence of the perturbation series may also occur for irrational frequency ratios.

17.3 The standard map

The Henón–Heiles system shows evidence of a sequence of bifurcations. As the energy E is increased, the computed motion for some initial conditions is not confined to tori but appears to be disorderly: slightly different initial conditions yield entirely different completely deterministic trajectories after only a short time. For low enough energies the motion is regular (integrable), but as E is increased the size of the regular regions in phase space decreases while regions consisting of a scatter of points (the disorder) grow.

In order to attempt to describe the change from integrable to nonintegrable and even chaotic motion, we start with a model that has an integrable limit, the perturbed circle map. Circle maps were introduced in chapter 4.

If we begin with a two degree of freedom integrable system in action-angle variables, then $H_o(J_1, J_2) = E$, $\theta_i = \omega_i t + \theta_{io}$ and the motion is either periodic or quasiperiodic on a two dimensional torus in the four dimensional phase space. The circle map follows from the following Poincaré section: consider the intersection of the orbit with the plane transverse to the torus of radius J_1, whenever θ_1 advances through 2π. If we denote J_2 by J_n and θ_2 by θ_n in the nth intersection of the orbit with the plane, then the Poincaré map of any integrable canonical system is given by the linear circle map

$$\left.\begin{array}{l} J_{n+1} = J_n \\ \theta_{n+1} = \theta_n + 2\pi\Omega \end{array}\right\} \tag{17.15}$$

where $\Omega = \omega_1/\omega_2$. If the bare winding number Ω is rational then the θ_n-motion is periodic and the orbit consists of a finite set of points. If Ω is irrational then the θ_n motion is quasiperiodic, the orbit generates a dense (but not necessarily uniformly dense) set of points whose closure is the entire circle as $n \to \infty$. This is the universal nature of the return map for any integrable two degree of freedom canonical system with bounded motion.

We know that the return map of a canonical system is area-preserving (chapters 6 and 15). In the case of a two dimensional return map Liouville's theorem can be used directly to exhibit the area-preserving property:

$$\frac{\partial(J_{n+1}, \theta_{n+1})}{\partial(J_n, \theta_n)} = 1, \tag{17.16}$$

which is just the Poisson bracket relation for the variables (θ_n, J_n).

Next, we perturb the integrable system by generalizing to a circle map of the form

$$\left.\begin{array}{l} J_{n+1} = J_n + \varepsilon f(J_{n+1}, \theta_n) \\ \theta_{n+1} = \theta_n + 2\pi\Omega(J_{n+1}) + \varepsilon g(J_{n+1}, \theta_n). \end{array}\right\} \tag{17.17}$$

The extra terms represent a perturbation that makes the Hamiltonian flow non-integrable, like the nonquadratic potential energy term in the Hénon–Heiles model. The perturbed twist map must be area-preserving, and so (17.16) must be satisfied at least to O(ε). Using the Poisson bracket notation for (17.16) we find that

$$\begin{aligned} J &= \{J_{n+1}, \theta_n + 2\pi\Omega + \varepsilon g\} \\ &= \{J_{n+1}, \theta_n\} + \varepsilon\{J_{n+1}, g\} \\ &= \{J_n + \varepsilon f, \theta_n\} + \varepsilon\{J_{n+1}, g\} \\ &= 1 + \varepsilon\{f, \theta_n\} + \varepsilon\{J_{n+1}, g\}. \end{aligned} \tag{17.18}$$

Therefore, we impose the condition

$$\{\theta_n, f\} + \{g, J_{n+1}\} = 0 \tag{17.19}$$

to O(ε^2) = 0.

Since

$$\{f, \theta_n\} = \frac{\partial f}{\partial J_n} = \frac{\partial f}{\partial J_{n+1}} \frac{\partial J_{n+1}}{\partial J_n} \approx \frac{\partial f}{\partial J_{n+1}} + O(\varepsilon) \tag{17.20}$$

and since

$$\{g, J_{n+1}\} = \{g, J_n\} + \varepsilon\{g, f\} = \{g, J_n\} + O(\varepsilon) = \frac{\partial g}{\partial \theta_n} + O(\varepsilon), \qquad (17.21)$$

the area-preserving condition requires that

$$\frac{\partial f}{\partial J_{n+1}} + \frac{\partial g}{\partial \theta_n} = 0. \qquad (17.22a)$$

At this point we are still free to model perturbed twist maps by different choices of f and g, so long as we satisfy the area-preserving constraint (17.22a).

A map that is discussed in the literature (see Lichtenberg and Lieberman, 1983, and references therein) follows from assuming that

$$\frac{\partial f}{\partial J_{n+1}} = 0 \text{ and } g = 0, \qquad (17.22b)$$

yielding

$$\left. \begin{array}{l} J_{n+1} = J_n + \varepsilon f(\theta_n) \\ \theta_{n+1} = \theta_n + 2\pi\Omega(J_{n+1}). \end{array} \right\} \qquad (17.23)$$

We can now develop the perturbation theory as follows. We linearize about a period one fixed point of the map $J_{n+1} = J_n = J_\circ$, so that $\Omega(J_\circ) = m$ is an integer (see chapter 14 for fixed points of maps). Then, we set $J_n = J_\circ + \Delta J_n$ to obtain

$$\left. \begin{array}{l} \Delta J_{n+1} \cong \Delta J_n + \varepsilon f(\theta_n) \\ \theta_{n+1} = \theta_n + 2\pi m + 2\pi\Omega'(J_\circ)\Delta J_{n+1}. \end{array} \right\} \qquad (17.24)$$

If then we set $I_n = 2\pi\Omega'\Delta J_n$ then we find that

$$\left. \begin{array}{l} I_{n+1} = I_n + 2\pi\Omega'\varepsilon f(\theta_n), \\ \theta_{n+1} = \theta_n + 2\pi m + I_{n+1} \\ = \theta_n + I_{n+1}, \quad \text{mod } 2\pi. \end{array} \right\} \qquad (17.25)$$

With $k = 2\pi\varepsilon f_{max}$ and $f^* = f/f_{max}$ we obtain the map

$$\left. \begin{array}{l} I_{n+1} = I_n + kf^*(\theta_n) \\ \theta_{n+1} = \theta_n + I_{n+1}, \quad \text{mod } 2\pi. \end{array} \right\} \qquad (17.26)$$

The Chirikov–Taylor model, also known as the standard map, follows from the choice $f^*(\theta) = \sin\theta$. This model is discussed in the texts by Lichtenberg and Lieberman (1983) and Zaslavskii et al. (1991). For different choices of the force constant K and initial conditions, motions qualitatively reminiscent of those shown in figure 17.2 can be reproduced by the standard map, as is illustrated in figure 17.3.

17.4 The Kolmogorov–Arnol'd–Moser (KAM) theorem

Whenever $\varepsilon = 0$ the system is integrable. The base winding number Ω defines an invariant torus. Which tori persist under weak perturbation, and which of them suddenly disappear? In other words, when does stable periodic or stable quasiperiodic motion with winding number Ω persist under a perturbation that introduces some degree of nonintegrability into phase space? The KAM theorem states a sufficient condition for the persistence of an invariant torus with winding number Ω under sufficiently weak perturbation. The KAM condition is not intuitive because it is one of

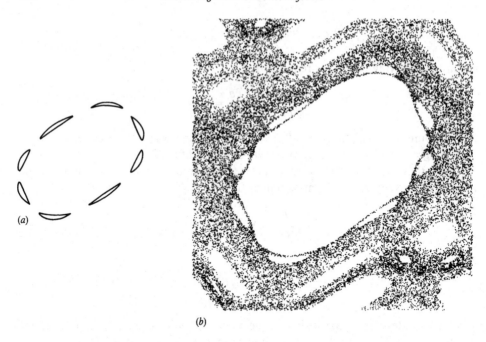

(a)

(b)

Fig. 17.3(a) Plot of the standard map for $k = 0.95$, $J_\circ = 0.3$, and $\theta_\circ = 0.5$.
(b) Plot of the standard map for $k = 1.2$, $J_\circ = 0.1$ and $\theta_\circ = 0.1$.

'sufficient incommensurability' of the winding number with respect to approximations by rational winding numbers. Winding numbers and their continued fraction expansions are explained in chapter 4.

For the nonintegrable perturbed system, consider the systematic approximation of an irrational winding number Ω by a sequence of rational approximations $p_1/q_1, p_2/q_2, \ldots, p_n/q_n, \ldots$. If, as $n \to \infty$, we can show that

$$|\Omega - p_n/q_n| > K(\varepsilon)/q_n^{5/2} \qquad (17.27)$$

as $\varepsilon \to 0$, then the orbit with irrational winding number Ω survives, according to KAM, as a stable quasiperiodic orbit under the perturbation and so, therefore, does the corresponding torus. The quantity $K(\varepsilon)$ is a positive constant dependent on the perturbation strength ε. Rapidly convergent sequences of rational approximations are generated by continued fractions (Newton's method produces faster convergence by skipping some of the rational approximates). The closely related problem of the dissipative circle map was discussed in chapter 14. Dissipative circle maps can be derived from standard maps by making the latter dissipative. The theory of both systems has been discussed extensively by Arnol'd (1963, 1978). According to the condition (17.27), almost all irrational tori should survive under weak perturbation: for a weak enough perturbation, almost all of phase space should remain foliated by tori with quasiperiodic orbits. This can be understood by calculating the relative size of the intervals where the KAM condition fails.

We can ask: what is the fraction of the unit interval where the condition

$$|\Omega - p_n/q_n| \leq K(\varepsilon)/q_n^{5/2} \qquad (17.28)$$

holds as $n \to \infty$ when $\varepsilon \ll 1$? If we denote the length of that point set by $L(\varepsilon)$, then we find that

$$L(\varepsilon) < K(\varepsilon) \sum_{q=1}^{\infty} q^{-5/2} \approx K(\varepsilon) \qquad (17.29)$$

since the $q^{-5/2}$ series converges. Hence, the size of the gaps where the KAM condition (17.27) is violated vanishes as $\varepsilon \to 0$.

What happens to the periodic and quasiperiodic orbits that do not survive? Of the periodic orbits, some remain stable while others become unstable periodic. Discussions can be found in the literature under the heading of the Poincaré–Birkhoff theorem. We know from the numerical work on the Henón–Heiles problem that new pairs of stable periodic orbits emerge as older stable orbits are destroyed via pitchfork bifurcations (period-doubling bifurcations). Pitchfork bifurcations were discussed in chapter 3. Period doubling for area-preserving maps has been discussed by MacKay in Cvitanovic (1984).

An example of an irrational winding number that is destroyed under perturbation is given by π, which has a rapidly convergent continued fraction expansion. On the other hand, the famous Golden Mean, or any other winding number whose continued fraction expansion is populated asymptotically by unity,

$$\Omega = \cfrac{1}{\dots + \cfrac{1}{1 + \cfrac{1}{1 + \cfrac{1}{1 + \dots}}}}, \qquad (17.30)$$

will survive for small enough ε because rational approximations to these winding numbers are the most slowly convergent of all (computable) winding numbers. This class of winding numbers is sometimes mislabeled as 'the most irrational'. This is misleading because most irrational numbers that can be defined to exist cannot be computed by any algorithm, including continued fractions (see McCauley, 1993, chapter 11).

17.5 Is clockwork possible in the solar system?

We now return to the theme of chapters 1 and 4. The basis for the Platonic belief in a stable, clockwork universe was advanced superficially by Newton's explanation of Kepler's two-body problem: Keplerian elliptic orbits are the epitome of clockwork. The Newtonian law of gravity was derived from the assumption that the planetary orbits *are* perfect ellipses. The underpinning for imagining the Newtonian solar system as either perfect clockwork or even quasiperiodic clockwork was destroyed two centuries later by Poincaré's treatment of the three-body problem: mathematicians have not, so far, been able to prove that even a highly idealized model of our solar system (using point masses) is absolutely stable. Because of chaotic motions in the phase space of the mathematical problem of three point masses interacting gravitationally, we cannot even prove that the corresponding idealized (and oversimplified) model of the earth–moon–sun system is stable. Results of numerical integrations of solar system models are of doubtful value unless the computer program can recover initial conditions via backward integration. The same applies to all cosmological models of the universe.

Newton was aware of the irregularities that arose in the comparison of known planetary orbits with the expectation of Keplerian ellipses and circles. He formulated the three-body problem and used (the equivalent of the first term in) a perturbation expansion to try to explain the known anomalies of the moon's orbit, but failed to explain all of them. Newton believed God must intervene from time to time to reset 'initial conditions' in order to maintain the stability of the solar system.

The Newtonian model of the solar system is nonintegrable. Some initial conditions yield stable orbits, others yield chaotic ones. To date, we still do not understand how the observed orbits of planets in our solar system, most of which are approximately circular and (perhaps only) apparently stable as expected on the basis of the Keplerian two-body problem, should fall out of an idealized mathematical description of the solar system based upon point masses. Modern workers have tried to understand the observed planetary orbits from the standpoint of the KAM theorem (see Sussman and Wisdom, 1992).

Exercises

1. Reproduce the Poincaré sections (Fig. 17.2) and compute the time series $x(t)$ vs t for the Henón–Heiles model. From the time series, estimate the length of the time interval over which backward integration reproduces your initial conditions to within at least one-digit accuracy.

2. For the simple pendulum in a uniform gravitational field,

$$H = \frac{p^2}{2m} - mgl(\cos\phi - 1) = \frac{p^2}{2m} + \frac{mgl}{2}\phi^2 - \frac{mgl}{4!}\phi^4 + \frac{mgl}{6!}\phi^6\ldots,$$

solve the problem to lowest order in canonical perturbation theory, taking

$$H_1 = -\frac{mgl}{4!}\phi^4 + \ldots.$$

Start by showing in the action-angle variables of the simple harmonic oscillator that

$$H = \omega_o I - \frac{1}{6}\frac{I^2}{m}\sin^4\theta + \ldots$$

where $\omega_o = \sqrt{gl}$, and therefore that we can write

$$H_1 = -\frac{I^2}{48m}(3 - 4\cos 2\theta + \cos 4\theta).$$

Show finally that the new frequency is given approximately by

$$\bar{\omega} = \frac{\partial K}{\partial \bar{I}} = \omega_o - \frac{\bar{I}}{8m}$$

and that the O(ε) correction to the generating function is given by

$$S_1 = -\frac{I^2}{192\omega_o m}(8\sin 2\theta - \sin 4\theta).$$

(Lichtenberg and Lieberman, 1983)

3. Reproduce the Poincaré sections shown in the text for the standard map.

18

Simulations, complexity, and laws of nature

18.1 Integrability, chaos, and complexity

Phase space flows divide into two basic classes: completely integrable and 'nonintegrable' (incompletely integrable). In the first case the motion is equivalent to a single time-translation for all finite times (chapter 3). In the second case singularities of the conservation laws prevent this degree of simplicity (section 13.2).

Nonintegrable systems can be divided further into two classes: those that are in principle predictable for all times, and those that are not. The first class includes deterministic dynamics where there is a generating partition[1]. An example, the asymmetric tent map, was given in chapter 14. A chaotic dynamical system generates infinitely many different statistical distributions through the variation of classes of initial conditions. Only the differentiable distribution (corresponding to the invariant density) is generated by initial data that occur with measure one. The fragmented statistical distributions, including multifractal ones, occur for the measure zero set of initial conditions.

Whenever a dynamical system is computable (McCauley, 1993), which is the only case that can be used computationally and empirically, then the measure zero initial conditions are of dominant importance. There is, as yet, no known case in nature where the differentiable probability distribution describes the experiments or observations performed on a driven-dissipative chaotic dynamical system. The generating partition forms the support of every probability distribution that occurs mathematically, and can be discovered (when it exists) through backward iteration of the Poincaré map.

When there is a generating partition then dynamical systems divide further into topologic universality classes where the long-time behavior and statistical distribution can be discovered via symbolic dynamics. For example, the logistic map $f(x) = Dx(1 - x)$ with $D_c < D < 4$ belongs to the same universality class as the Henón map (Cvitanovic et al., 1988) whereas the logistic map with $D \geq 4$ belongs to the universality class of the binary tent map (McCauley, 1993). A single symbol sequence corresponds to a single initial condition, and all of these maps are in principle perfectly predictable for all times merely by reading a symbol sequence letter by letter (McCauley, 1993, Moore, 1990).

According to von Neumann (1966), a dynamical system is complex whenever it is easier to build than to describe mathematically. Under this definition chaotic dynamical systems with a generating partition (the logistic map and the Henón map) are not complex, whereas an embryo or a living cell should be.

As in parts of chapters 6 and 14, we must think of the dynamical system as a computer (automaton) and the initial condition as the program. The system's long-time behavior is encoded in the far-away digits of the initial condition in some base of

[1] In McCauley (1993) a generating partition is called the 'natural partition'. See also chapter 14 in this book.

arithmetic. For example, the automaton defined by the binary tent map performs a trivial computation: in binary arithmetic it either reads a bit or else flips all bits and then reads one (chapter 6). The bakers' transformation, an area-preserving map of the unit square, is computationally equally trivial: it merely reads each bit in the initial condition written as a two-sided binary sequence (McCauley, 1993).

Complex dynamical systems are not necessarily chaotic and may have a high degree of computational capability. Moore (1990) has shown that a simple generalization of the bakers' transformation is equivalent to a Turing machine and therefore has universal computational capability: through its dynamics, it can compute any digital program that can be written algorithmically. The program is encoded digitally as the initial condition. For example, the map can compute an exact replica of its own behavior: there, both the map and initial condition are encoded as the initial condition for the map. Such systems are not merely chaotic and are not necessarily chaotic. They have no generating partition (no solvable symbolic dynamics) and the long-time dynamical behavior is undecidable in Turing's sense: given an algorithm for an initial condition as an infinite digit string, there is *in principle* no way to predict or even understand the long-time behavior of the map, statistically or otherwise, in the absence of a digit by digit calculation. It is impossible to classify and understand the different possible statistics that may arise from different programs and there are certainly no universal scaling laws for arbitrary sets of initial conditions. *Given perfect knowledge of the initial condition (in the form of an algorithm), it is impossible to say anything at all about the long-time behavior of the map in advance of a detailed calculation for a specific initial condition.* Moore states that such a system can be realized physically by billiard balls reflected from mirrors. So far, there exists no explicit construction of a model defined by a particle in a potential well with three or more degrees of freedom exhibiting such behavior.

We have just described the case of *maximum* computational complexity. There must be lesser degrees of complexity as well. Aside from the Turing machine, and related ideas like NP-completeness (see Garcy and Johnson, 1979), there is as yet no precise definition of complexity or degrees of complexity. In particular, no one has yet constructed a dynamics model that simulates *time-evolution of complexity* that might mimic organic evolution from simple cells and bacteria to plants, fish, and animals. More fundamentally, the complexity exhibited by a living cell is still far beyond the limit of all known dynamical models. The degree of computational complexity needed to describe a living cell is unknown.

18.2 Simulations and modelling without invariance

Consider an economic system ... we do grant that the probability distribution governing these doings definitely exists – i.e., there is a true objective law that governs the way the economic system behaves.

J. L. Casti (1990)

Irregularities are common in nature. Stable periodicity and quasiperiodicity are extremely rare. Mathematically, very simple-looking nonlinear differential equations with only three variables, and also noninvertible one dimensional maps, can generate chaotic motions deterministically. Certain simple two dimensional maps and Newtonian dynamical systems are capable of maximum computational complexity. Fluid turbulence provides numerous examples of irregularities in both space and time in a precisely defined Newtonian dynamical system: we know how to formulate fluid mechanical time-evolution according to Newton's laws of motion, but infinitely many degrees of freedom represented by nonlinear partial differential equations are the stumbling block in our attempt to understand turbulence mathematically. We do not yet understand nonlinear partial differential equations of the second order, or even the behavior of nonlinear dynamical systems with finitely many degrees of freedom (like the double pendulum or the three-body problem) well enough to be able to derive most of the important features of fluid turbulence from Newtonian physics in a systematic way that starts from energy and momentum conservation and makes controlled approximations.

We do understand how the simplest abstract one dimensional chaotic dynamical system (the binary tent map) can generate all possible histograms that can be constructed simply by varying the initial conditions. All possible statistical distributions are included in the distributions of blocks of digits in decimal expansions of initial conditions. The corollary is that statistics, taken alone, that are generated by an unknown dynamical system are inadequate to infer the dynamical law that regulates the behavior (McCauley, 1993, chapters 7 and 9).

In principle, it is possible to extract the universality class, and even an iterated map, from a chaotic time series generated by some Newtonian systems. In practice, the data must be extremely precise in order to eliminate all possibilities other than the correct one, or even to narrow down the choices to a single topologic universality class. Given the most accurate existing data on a fluid dynamical system, the unique extraction of the universality class from a chaotic time series has yet to be accomplished, showing how hard the problem is in reality (see Chhabra et al., 1989, and references therein, and also Gunaratne, 1990).

Beyond the realm of physics, irregularities also abound in the behavior of individuals and groups of individuals. Given what we know about laws of inanimate nature, is it reasonable to expect that we can extract universal laws of human behavior from time series or statistics representing socio-economic behavior? Is it possible to reduce some aspects of human behavior, collective or individual, to a set of universal formulae, or even to a finite set of invariable rules?

System theorists and economists often speak of the economy, which depends on human behavior and man-made rules, as if it were a mechanical system like the weather, the abstract mathematical equivalent of N interacting inanimate bodies that cannot choose to obey or not to obey laws of motion. The weather is regulated deterministically by the thermo-hydrodynamic equations with boundary and initial conditions for the earth's atmosphere, but there are as yet no known, empirically established universal economic or other behavioral laws of nature. By disregarding Galileo's historic and fruitful severing of the abstract study of inanimate motion from imprecise notions like youths alearning, acorns asprouting, and markets emerging, some mathematical economists try to describe the irregularities of individual and collective human nature as if the price movements of a commodity, determined by human decisions and man-made political and economic rules, would define mathematical variables and abstract universal equations of motion analogous to ballistics and astronomy, or analogous to a cooking recipe (which also can be written as a finite number of rules, and therefore as a finite-length computer program). In contrast with mathematically lawful physical phenomena, where there are conservation laws and invariance principles, everything of socio-economic interest changes uncontrollably. An exception might be Pavlovian–Skinnerian stimulus-response like advertising stimulated consumerism that implicitly rules out creatively intelligent behavior like unbounded language/idea production, and the discovery of new mathematical ideas (see Hadamard, 1954, for the latter).

The mere existence of statistics, and even the existence of an underlying mechanism, does not necessarily imply that there is an underlying mathematical *law* that can be discovered. It has not been proven that human thought and behavior *are* completely mechanical, but even if that *were* the case there are still mechanical systems with behavior too complex to be described by differential equations, maps, or finite sets of rules (including finite-length computer programs) with simple or arbitrary (usually misnamed 'random') initial conditions. Social behavior, even if it were mechanical, is certainly not predictable and may not even be effectively computable in the following sense: the observed statistics vary with time, and any mechanism that governs their behavior need not be so simple that it can be compressed into a 'formula' that is shorter than merely writing down the statistics as the individual or collective behavior is observed daily. This kind of behavior is generated by both simple and complicated deterministic automata that are capable of universal computation, and their long-time behavior cannot be understood in advance of the occurrence of that behavior. Given an arbitrary program (initial condition), the future behavior is not predictable in advance, nor is it qualitatively understandable in advance of the calculation that generates the statistics iteration by iteration (which also cannot be known before they occur, because they vary from initial condition to initial condition). In contrast, the differential equations and coupled iterated maps studied in dynamical systems theory usually can be reduced to automata that are too simple to be programmed to compute anything other than their own orbits, automata that are too limited in 'behavior' to simulate even the simplest creative decision-making by the human brain.

Our discussion centers on what Wigner (1967) has called 'the unreasonable effectiveness of mathematics' in describing the motion of inanimate objects in nature, and that fields far beyond physics have tried unsuccessfully to imitate by *postulating* laws that are not derivable from the universal laws of physics and are not empirically correct over long time intervals. Examples of the latter are 'invisible hands', 'efficient markets', 'rational markets with infinite foresight', 'behavior modification', 'Marxist determinism' and so on. Rather than merely assuming that the socio-economic sciences are susceptible to abstract and precise mathematization in the style of physics, chemistry, and genetics, it is interesting to ask why *any* aspect of nature, animate or inanimate, should obey precise and relatively simple mathematical rules like Newton's laws of motion or Schrödinger's equation. In other words, what is required of some aspects of nature in order that we can discover that there is systematic 'behavior' ('motion' in Aristotle's sense) that can be identified via controlled, repetitive experiments or repeated observations, and may be described as if it evolved in time in an abstract and relatively *simple* machine-like fashion (according to Descartes' dream), analogous to Keplerian orbits or Galilean trajectories of apples? After all, many system theorists and economists commonly assume that the economy operates like a well-defined dynamical system (deterministic or stochastic), the equivalent of an automaton that is too simple to simulate any kind of creative behavior, including the violation of politically legislated laws.

This is a strange assumption. Without human brains and human agreements economic behavior could not exist, whereas Newton's laws of motion would still hold universally. Like the rules of arithmetic, we cannot change laws of nature merely by making decisions or by thinking wishfully. There is an unbridgeable gap between natural irregular behavior like chaos in the gravitational three-body problem or the double pendulum, whose motions are dictated by universal local laws that can be written as a finite number of simple mathematical formulae, and statistics that are collected from the observation of unpredictability that is caused by human convention, invention, and intervention like the buying and selling of commodities and real estate, the education of a student, or discovering a new mathematical theorem. Universal laws that are generalizations of simple regularities of nature like parabolic trajectories and Keplerian orbits differ markedly from human-created systems of merely conventional behavior, meaning learned or agreed-on behavior, which can always be violated. As Wigner has pointed out, there is a deep and unavoidable connection between the strict reproducibility of experiments (as in kinematics), or reproducibility of observations (as in astronomy), invariance principles like translational and rational invariance, and mathematical laws of motion.

Wigner has pointed out that without translational and rotational invariance in space and translational invariance in time, the fundamental results from which the *universal laws of nature* follow (like Keplerian orbits and Galilean trajectories) could not have been discovered. In particular, no new or old laws of nature have been discovered by analyzing chaotic or complex time series.

Experiments involving gravity are reproducible precisely because absolute position and absolute time are irrelevant as initial conditions, which is the same as saying that Euclidean space is homogeneous and isotropic and that the flow of time is homogeneous. The translational invariance of the law of inertia $d\mathbf{p}/dt = 0$ is just a way of saying that the law of inertia can be verified regardless of where, in absolute space and absolute time, you perform the experiment. *The law of inertia is the foundation of physics*: from it follows Newton's second law of motion, and not the reverse. If absolute time and absolute position were relevant initial conditions, then the law of inertia would not hold and there would have been no Galilean or Keplerian regularities, and consequently no universal laws of mechanics could have been discovered by Newton.

Mathematical models that are merely postulated without the support of empirically discovered invariance principles have led neither to qualitative understanding nor to long-term predictability. Without empirically valid invariance principles, there are no fundamental constants in an empirical model. We do not yet know how to derive the fundamental constants of physics from a more fundamental theory, but the argument of sections 1.2 and 1.4 indicates that both rotational and translational invariance are necessary in order that the gravitational constant G should be the same for all gravitating bodies in the solar system. In models of the economy, nothing that is nontrivial is left invariant by the time-evolution of the system (there are no known nontrivial conservation laws of human behavior). Without the relativity principle discovered by Galileo, which is just a restatement of the law of inertia, Newton would not have known how to write down a correct, universal differential equation of motion describing forced motion.

In fields far beyond physics where there is no analog of the first law of motion but mathematical models are postulated anyway, there is still always the assumption that there are underlying machine-like regularities, 'laws', that are accurately reflected by the model even though no lawful regularities have ever been discovered. Imagine a mathematical model of the economy (or of any other sort of social behavior) that is based upon differential equations or some other dynamical systems approach (iterated maps, cellular automata[2], and stochastic models provide three other popular methods). The model will depend upon unknown parameters and also upon initial conditions. Over short enough time intervals, any limited set of empirical data can be reproduced by a model with enough unknown parameters. When the parameters are determined by fitting the model to a specific set of initial conditions, then over long enough time intervals the parameters that should then be treated as constants in the model must be allowed to vary in a completely unforeseeable way in both time and space in order to try to account for the newer statistics that arise daily. For example, the parameter values in a model that were chosen to describe last year's stock market behavior or last year's economy will not allow the model to be used to predict stock market behavior or the economy correctly during the present year even if the initial conditions can be correctly chosen. Even if, by accident, next week's stock market prediction should turn out to be correct (as is occasionally and accidentally the case), the chosen model will eventually and systematically predict the market incorrectly, and also will provide no real insight into how the stock market or the economy really work. The consequence of the lack of invariance is that the 'constants' in these models must be changed in unforeseeable ways in order to use the 'two-weeks-ago model' to try to account for the consequences of last week's observed behavior. That there are fundamental constants like the universal gravitational constant G in physics is not accidental and could not be true if there were no underlying invariance principles like the isotropy and homogeneity of space.

To illustrate the danger inherent in modelling and simulations that are not guided by empirically established laws of nature, let us imagine that Newton knew Kepler's laws of planetary motion but that Galileo and Descartes had never lived. Consequently, we shall calculate as if the law of inertia, the action–reaction principle, and the principle of equivalence were unknown. In order to represent the only physical idea available, that absolute velocity is a natural state and that motion must be maintained by a cause, Newton could have written

$$\beta \vec{v} = \vec{F}, \tag{1.21}$$

where β is the resistance to planetary motion provided by the fictitious ether. He would have had neither the empirical nor philosophic motivation to write down anything else (we can presume that Newton learned his analytic geometry from Fermat). We now state a simple fact: all three of Kepler's laws and the observed planetary positions can be accounted for *in detail* by this Aristolian law of motion, although it obeys no relativity principle and therefore provides a completely false 'explanation' of planetary motions and gravity. The proof follows easily from the method of analysis of the Kepler problem used in chapter 1. Merely assume the formula for a conic section,

$$\frac{1}{r} = C(1 + \varepsilon \cos \theta), \tag{1.6}$$

which describes Kepler's first law if $\varepsilon < 1$, and then derive the force law that produces this orbit in accordance with Aristotle's law of motion (1.21). The calculation is elementary and is assigned as exercise 3. The result, in cylindrical coordinates, is the noncentral force law

---

[2] The universal computational capability of certain cellular or other automata must not be confused with the entirely different uses of the word universality in deterministic chaos and critical phenomena.

$$\vec{F} = (M_z C\varepsilon \sin\theta, M_z/r, 0) \qquad (1.21)$$

where $M_z = \beta r d\theta/dt$ is constant in the simplest case, and would be the angular momentum if the linear friction constant β of the ether could be replaced by the mass m of the planet orbiting the fixed sun (there is enough free choice in the angular component F_θ of $\vec{F}$ that we can satisfy Kepler's second law by choosing M_z to be constant). Plato's ideal circular orbits appear if $\varepsilon = 0$, and we also obtain the correct hyperbolic orbits of scattering problems whenever $\varepsilon > 0$. The proof that Kepler's third law also follows is identical to the proof in the Newtonian case, as neither conservation of energy (which does not follow from (1.21)) nor the correct force law for gravity is needed to derive either of those two laws from the Newtonian description of Keplerian motion.

The Aristotelian model of the solar system gets the positions right while getting the velocities wrong, and cannot describe vertical free-fall on earth (or the accompanying horizontal motion at constant velocity) without changing the force law, but it still does a better job of prediction than does the best known model of the economy: no arbitrary time-variation of the three parameters in the model is necessary, even at long times. By starting with an orbit equation that includes precession of the perihelion of Mercury, we could derive another force law that would agree perfectly with the known data. Clearly, Newton would have rejected the result out of hand although such curve-fitting is welcome in less demanding quarters.

Perhaps Newton, who also performed experiments, would eventually have discovered the law of inertia. Without Descartes and the spread of his mechanistic philosophy, however, the witch- and heretic-hunters might eventually have adequately hindered or even burned a Deist like Newton, completely eliminating the possibility of knowledge of universal laws of nature in our time. The oft-heard assertion that 'someone else eventually would have discovered all of the important ideas anyway' assumes either that history does not matter or else that history plays an organic role in creating ideas. It automatically presumes a society that is open enough to the free exchange and discussion of new ideas to provide the intellectual resources like well-stocked libraries that are necessary for informing and stimulating creative thought. It also ignores the fact that Newton's genius has not yet been matched, including in our own time. As Whitehead said, we are still living off the intellectual capital of the seventeenth century. It seems that this capital of physics cannot be so easily borrowed by the soft sciences, which really should be renamed the 'hard' sciences because no mathematical laws are available there that would permit any universal understanding of socio-economic phenomena.

The Aristotelian force law demonstrates that mere agreement with limited empirical data provides no evidence at all that a forced 'fit' by an arbitrarily chosen mathematical model explains the underlying phenomena that have produced the data. Given a limited set of data, it is often possible to invent an arbitrary model with enough parameters to reproduce the data. It may even be possible to achieve limited short-term predictability with an arbitrary model, but nothing but misunderstanding of the underlying phenomena can result from a model that is not grounded in empirically verifiable invariance principles.

18.3 History in the context of mechanics

In the index to the six hundred odd pages of Arnold Toynbee's *A Study of History*, abridged version, the names of Copernicus, Galileo, Descartes, and Newton do not occur.

Arthur Koestler (1959)[3]

We return briefly to the theme of chapter 1, where we discussed the beginning of theoretical and experimental physics ('reductionism') in the context of western history.

Newton, along with Faraday and Maxwell, lie buried in Westminster Abbey in London. Tycho Brahe's bones were not transported back to Denmark. He mercifully died before he could know of Kepler's greater success as an astronomer, and was buried in the Tyn Church in Prague. Descartes, who loved to lie in bed each morning and think, died shortly after tutoring the demanding, nail-hard, and unphilosophic Christina regularly at five a.m. in the freezing palace library. He was buried briefly in Stockholm but eventually landed (after a visit to Sant-Geneviéve) in the Church of Sainte-Germain des Prés. Late in life, perhaps with a feeling of nothing more of significance to do, Kepler fell prey to the wandering ways of the degenerated men in his family, died in the beautiful old Bavarian city of Regensburg and was buried in the Lutheran churchyard there. The church and graveyard were destroyed during the Thirty Years War, the attempt by the Swedish army under Christina's father, Gustav Adolph, to make Germany and Switzerland safe for Protestantism.

Christina eventually had enough of Puritanism, abdicated the throne and converted to Catholicism in Innsbruck on the way to live in Rome. Under the influence of Descartes, and other French Catholic libertines in Stockholm, she had come to believe that Catholicism offered more freedom than Lutheranism. In Rome Christina discovered the competing neoPuritanism of the Counter Reformation.

Earlier in Friuli, the Inquisitor attempted to lead Domenico Scandella, or Menocchio, into revealing the real source of his ideas as the devil, the presumed ruler of all things earthly:

Inquisitor: *Do you believe in an earthly paradise?*
Menocchio: *I believe that the earthly paradise is where there are gentlemen who have many possessions and live without working.*

Carlo Ginzburg (1992).

Menocchio, the miller whose half-baked ideas about the origin of the universe seem more interesting than the purely rhetorical pedantic pablum that was propagated by his learned Inquisitors, was subjected to two trials that lasted over parts of several years. He spent three years in prison as the consequence of explaining his personal views on cosmology to an ecclesiastic judge from the lovely old city of Vittorio Veneto, and ended life as a broken old man abandoned by his family, demonstrating that worse luck sometimes follows bad luck.

---

[3] In the unabridged ten volumes of Toynbee's grand scheme of history (1947), there are three brief references to Copernicus, two to Galileo, three to Newton, and none at all to Kepler! Spengler's *Decline of the West*, whose second chapter is entitled 'The meaning of numbers', pays much more attention to the role of science in history: in the abridged version there are five references to Euclid, two to Archimedes, three references to Copernicus, nine to Desargues and Descartes, twenty-one to Newton and Leibnitz, seven to Gauss, five to Hertz, Kirchhoff, and Kelvin, four to Cauchy, six to Lagrange, Riemann, and Klein, one to Planck, and twenty-seven all together to God, Jesus, Mohammed, Charlemagne, Luther, and Calvin. That might represent a somewhat reasonable balance were it not for the fact that Einstein, who revolutionized scientific thought in Spengler's lifetime, is not mentioned at all, in spite of the fact that Minkowski and Lorentz are.

Galileo, who also spoke his mind confidently but had a high position and was well respected, was also arrested and incarcerated: first in the Villa Medici, then in an apartment overlooking St Peter's Square, then formally in the Grand Duke's villa at Trinita del Monte, later in an apartment in a palace in Siena. Finally, he was returned to his farm at Arcetri and his house in Firenze where he took up dynamics again having abandoned it for fifteen years during his campaign in favor of the Copernican system. He rests in the Church of Santa Croce in Firenze in the good intellectual company of Machiavelli and Michaelangelo.

Paradise . . . was . . . the invention of the relatively leisured class. . . . Work is the condition for equality. . . . bourgeois and Marxist ideals of equality presume a world of plenty, they demand equal rights before a cornucopia . . . to be constructed by science and the advancement of knowledge. . . . The peasant ideal of equality recognizes a world of scarcity . . . mutual fraternal aid in struggling against this scarcity and a just sharing of what the world produces. Closely connected with the peasant's recognition, as a survivor, of scarcity is his recognition of man's relative ignorance. He may admire knowledge and the fruits of knowledge but he never supposes that the advance of knowledge reduces the extent of the unknown. . . . Nothing in his experience encourages him to believe in final causes. . . . The unknown can only be eliminated within the limits of a laboratory experiment. Those limits seem to him to be naive.

John Berger, *Pig Earth*

Thomas Jefferson, a colonial product of The Enlightenment, was a self-taught scholar of the Old English language and English history in the time of *Ælfræd cyning*, Alfred the Great. Jefferson saw the settlement of America in the historic context of the *Völkerwanderung*[4]. Unlike the flamboyant Descartes, who was adept with both sword and pen, Jefferson was not a soldier. He was an ineffective war governor during the American War of Independence who enjoyed playing music and discussing philosophy with captured 'Hessian' officers and their visiting families held under house arrest in Virginia. He later served as envoy to France and adapted so well to European life that he did not want to return home. He did return, and became the first (and perhaps last) American President to be shaped more by Newtonian ideas that by the Reformation and Counter Reformation[5].

Sophus Lie, a solitary researcher who established no school but whose ideas still attract many followers, became physically weakened following a mental breakdown that fell on him due to his belief that his dominating colleague and strong supporter, Felix Klein, had taken the credit for his own implicit idea of defining geometry by invariance under coordinate transformations. He had always missed the Norwegian *heimat* and he finally abandoned cosmopolitan Leipzig for Oslo and died there soon afterward.

The bones of Karl der Grosse supposedly lie in a golden casket high above the floor of the octagonal cathedral that he ordered to be built in Aachen. His Frankish linguistic researches did not survive: the son who best succeeded him politically spoke only Latin and despised the tribal folklore assembled by his father.

The German farmers from the three Forest Cantons who presumably met at the *Rütli* around the end of the thirteenth century and swore, among other things, to give

---

[4] The *Völkerwanderung* did not end within the British Isles until the thirteenth-century defeat of the Norwegians by the Scots, and was superficially brought to a conclusion within Switzerland by the *Walser* migrations of the thirteenth century (later Walser settlements can be found in many other parts of the world).

[5] He said of Bacon, Locke, and Newton: 'I consider them as the three greatest men that ever lived, without any exception.'

up the ancient tribal tradition of feuding among themselves, also inspired the revival and transmission into the future of another old tribal tradition: in 1314 the herders of Schwyz ransacked the monastery at Einsiedeln and kidnapped the monks, who held legal title to the nearby Alpine pastures. In 1315 a small band of cow and sheep herders defeated a much larger Austrian Army at Morgarten, an act that was repeated systematically and often enough over the next two hundred years to give birth to the Swiss Confederation. Communal grazing rights and direct democracy persist in some Alpine villages.

John Berger, in *About Looking*, tries to make us aware of the extent to which abstraction has replaced many deep-rooted practices in daily life since the beginning of the industrial revolution:

The nineteenth century, in western Europe and North America, saw the beginning of a process, today being completed by corporate capitalism, by which every tradition which has previously mediated between man and nature was broken.

A similar viewpoint was expressed over seventy years ago by the poetic historian Spengler[6], who in *Decline of the West* characterized western European/North American civilization as one where the entire countryside is dominated by a few enormous cities called megalopolises. The abstract driving force of a megalopolis is the spirit of money-making, an abstraction that has become so common that it is regarded as 'practical'. In a single uncontrolled approximation, traditions and other ideas of the countryside that interfere with progress defined as economic development are rejected as unrealistic or irrelevant in the face of a single, quantitative position whose units are dollars, yen, and marks. Spengler managed to characterize our era of western civilization reasonably well without any quantitative mathematics, although the Lie–Klein idea of invariance of geometry under coordinate transformations may have inspired his speculative comparison of entirely different cultures and their later civilizations (he apparently studied mathematics and physics in universities, as well as art, literature and music).

The following (end of a) dialogue can be found in the book by Cowan et al. (1994) *Complexity; Metaphors, Models, and Reality* (p. 16) about complex adaptable systems in biology, economics, and other fields:

Lloyd: *Why do we have measures of complexity? Certainly, if you have a measure of complexity that's just a number that says, 'This object is so much complex,' that doesn't tell you anything about the system. On the other hand, assume that you have a particular state that you want to attain – say, a slightly better state of the economy than we have now – and you want to know how complicated that problem is to solve, and you're able to measure complexity. One of the Murray Gell-Mann's proposals for how complicated a problem is, is 'what's the minimum amount of money you'd need to solve it?'*

Anderson: *Well, that's proportional to computer time.*

Lloyd: *Perhaps the unit of complexity should be 'money'. And if you are able, in some sense, to formalize the difficulty of solving this problem, of getting the economy better, and you find you can*

---

[6] Following Goethe, an anti-Newtonian who followed Plato and Aristotle, Spengler regarded human societies as organisms moving toward a 'destiny', a vague idea of organic determinism that Goethe believed to be in conflict with mechanistic time-evolution that proceeds via local cause and effect. In trying to distinguish between global 'destiny' and local 'cause and effect', Spengler was not aware of the idea of attractors in mathematics, which depend precisely upon local cause and effect. In 'system theory', which permeates modern sociological thinking, there is also confusion generated by the mythology that treating 'the whole' is possible without knowing the local interactions that determine the behavior of the whole machine.

measure its complexity in terms of dollars, or maybe yen, then that kind of measure could be extremely useful.

Anderson: *Yes, I agree. I was actually being purposely provocative. I hoped to be, anyway.*

The prediction of a computable chaotic trajectory is limited, decimal by decimal, by computation time, but there are also integrable many-body problems that are not complex and also require large amounts of dollars, yen or marks. It is also stated in the same book (p. 11) that (low dimensional chaos) 'is basically not complex in a true sense: it has not settled down to a simple fixed point, indeed, but the number of bits necessary for specification of where you are is highly limited.' This claim is partly false: the binary specification of a single state x_n in the logistic map $f(x) = 4x(1 - x)$ requires precisely $N(n) = 2^n(N_o - 2) + 2$ bits, where N_o is the number of bits in any simple initial condition $x_o = 0.\varepsilon_1 \cdots \varepsilon_{N_o} 000 \cdots$. If the string representing x_n is arbitrarily truncated to $m \leq N(n)$ bits, then after on the order of m iterations the first bit (and all other bits) in $x_{n'}$, where $n' \approx n + m$, is completely wrong. Multiplication of two finite binary strings of arbitrary length cannot be carried out on any fixed-state machine, and if multiplication is done incorrectly at any stage then the bits in x_n cannot be known *even to one-bit accuracy*. The complexity of a dynamical system, like fractal dimensions and Liapunov exponents, probably cannot be satisfactorily described by a single number. As we said above, von Neumann (1966), who stimulated the discussion of complex machines over forty years ago[7], stated that machines of low complexity are easier to talk about (mathematically) than to build (the logistic and Hénon maps), whereas the opposite is true of systems with a high degree of complexity (living cells, the growth of embryos).

In the transition from village to modern and now postmodern life, direct contact with raw nature has largely been replaced by artificial simulations of life: television, zoos, Hollywood movies, contrived novels, Disneyland, and virtual reality now provide the common basis for daily experience. Even in the hard sciences, computer simulations are often confused with experiments and observations of nature.

Today, in the west, as the culture of capitalism abandons its claim to be a culture and becomes nothing more than an Instant-Practice, the force of time is pictured as the supreme and unopposed annihilator.

John Berger, *About our Faces*

Postmodernism has cut off the present from all futures. The daily media adds to this by cutting off the past.

John Berger, *Keeping a Rendezvous*

The recent postmodern advertisement of 'the end of history' was anticipated by Spengler, who argued that all traditional cultures evolve irreversibly toward rootless destinies that are developed civilizations dominated by the megalopolis.

History is not arbitrary, random, or irrelevant but it also is not mathematically describable. The universality of mathematical laws of motion, so far as anyone has demonstrated, does not extend beyond the bounds of physics and physics-based sciences like chemistry and genetics. Even Darwin's 'laws' (in contrast with Mendel's)

---

[7] According to Freeman Dyson (1992), the now-common assumption that biological replication can be studied independently of metabolism (e.g. Eigen's school), goes back to (an interpretation of Gödel's theorem by) von Neumann. The idea that it makes sense to try to describe life at the molecular level in terms of information alone, without any consideration of energy, was characterized extremely well in a lecture (1995 Geilo NATO-ASI) on protein-folding: We shall now approximate a mountain landscape by a golf course.

have a socio-economic rather than physico-chemical origin. The best that we can say is that previous conditions ('initial conditions') matter. In the biologic-historic realm 'initial conditions' are of the utmost importance:

The genetic information which assures reproduction works against dissipation. The sexual animal is like a grain of corn – is a conduit of the past into the future. The scale of that span over millennia and the distance covered by that temporal short circuit which is fertilization are such that sexuality opposes the impersonal passing of time and is antithetical to it.

John Berger, *About our Faces*

Exercises

Consider the 'Aristotelian' law of motion

$$\beta \frac{d\vec{x}}{dt} = \vec{F},$$

1. (a) If $\vec{F} = -\nabla U$ where U is the gradient of a scalar, show that whenever U is translationally invariant the analog of momentum conservation is that the position $\vec{R}$ of the center of mass is a constant, representing the idea of the state of rest as the natural state of a force-free body.
    (b) Show that there is no relativity principle, that there are no transformations to moving frames that leave the force-free equation of motion

    $$\frac{d\vec{x}}{dt} = 0$$

    invariant.

2. (a) Show that momentum conservation follows from $d\vec{R}/dt = 0$ (translational invariance) in the form

    $$\sum_l m_l \vec{v}_l = 0.$$

    (b) Show also that the momentum conservation law

    $$\sum_l m_l \vec{v}_l = 0$$

    follows from the equation of motion combined with the action–reaction principle, whether or not $U(\vec{x})$ is translationally invariant.

3. Use the formula for a conic section to show that Kepler's first law follows from the force law $\vec{F} = (M_z C\varepsilon \sin\theta, M_z/r, 0)$ in cylindrical coordinates, and that Kepler's second and third laws follow as well.
    Observe that circular orbits ($\varepsilon = 0$) with constant velocity imply a constant tangential force and vanishing radial component of force. In physics, the opposite is required: a constant radial force that points toward the center of the orbit, along with vanishing tangential component of force are required for uniform circular motion.

4. Observe that the force law of exercise 3 cannot be used to account for projectile trajectories on earth. Show instead that Galileo's parabolic trajectories are reproduced by the force law $\vec{F} = \{\text{constant}, \pm \beta \sqrt{[2(E - y^2)]}\}$ where $E = v_y^2/2 + gy = \text{constant}$ is maintained by a balance of driving against friction.

Bibliography

Chapter 1

Classical mechanics

Arnol'd, V. I., *Mathematical Methods of Classical Mechanics*, transl. from Russian, Springer-Verlag, Berlin (1989).
Goldstein, H., *Classical Mechanics*, Addison-Wesley, Reading, Mass. (1980).
Hamel, G., *Mechanik*, Teubner, Stuttgart (1912).
Knudsen, J. M. and Hjorth, P. G., *Elements of Newtonian Mechanics*, Springer-Verlag, Heidelberg (1995).
Landau, L. D. and Lifshitz, E. M., *Mechanics*, Pergamon, Oxford (1976).
Symon, K., *Mechanics*, Addison-Wesley, Reading, Mass. (1960).
Whittaker, E. T., *Analytical Dynamics*, Cambridge University Press (1965).

Symmetry and invariance principles

Saletan, E. and Cromer, A., *Theoretical Mechanics*, Wiley, New York (1971).
Wigner, E. P., *Symmetries and Reflections*, University of Indiana Press, Bloomington (1967).

Coordinate transformations and geometry

Arnol'd, V. I., *Geometrical Methods in the Theory of Ordinary Differential Equations*, Springer-Verlag, New York (1983).
Buck, C., *Advanced Calculus*, McGraw-Hill, New York (1956).
Dubrovin, B. A., Fomenko, A. T., and Novikov, S. P., *Modern Geometry – Methods and Applications. Part I. The Geometry of Surfaces, Transformation Groups, and Fields*, transl. by R. G. Burns, chapter 1, Springer-Verlag, New York (1984).
Margenau, H. and Murphy, G. M., *The Mathematics of Chemistry and Physics*, van Nostrand, Princeton (1964).

From ancient to medieval mathematics

Brun, V., *Alt er Tall, Matematikkens historie fra oldtid til renessanse*, Universitetsforlaget, Lommedalen (1981).
Danzig, T., *Number, The language of science*, Doubleday Anchor Books, Garden City, NJ (1956).
Ekeland, I., *Zufall, Glück, und Chaos*, transl. from French, chapter 1, Carl Hanser Verlag, München (1991).

The mechanics–mathematics revolution in western Europe

Arnol'd, V. I., *Huygens and Barrow, Newton, and Hooke*, transl. from Russian, Birkhäuser-Verlag, Basel (1990).

Arnol'd, V. I., *Ordinary Differential Equations*, Springer-Verlag, Berlin (1992).
Barbour, J., *Absolute or Relative Motion*, vol. I, Cambridge University Press (1989).
Bell, E. T., *The Development of Mathematics*, McGraw-Hill, New York (1945).
Bell, E. T., *Men of Mathematics*, Simon and Schuster, New York (1965).
Cajori, F., editor, *Sir Issac Newton's Mathematical Principles of Natural Philosophy and His System of the World*, transl. from Latin, Dover, New York (1969).
Caspar, M., *Johannes Kepler*, transl. from German by C. D. Hellman, Dover, New York (1993).
Dunham, W., *Journey through Genius*, Penguin, New York (1991).
Galileo, G., *Dialogue Concerning the Two Chief World Systems*, transl. from Italian by S. Drake, 2nd revised edn, University of California Press, Berkeley (1967).
Gindikin, S. G., *Tales of Physicists and Mathematicians*, transl. from Russian by A. Shuchat, Birkhäuser, Boston (1988).
Holton, G., *Science and Anti-Science*, chapter 4 'The Jefferson Research Program', Harvard University Press, Cambridge, Mass. (1993).
Hoyle, F., *Nicolaus Copernicus*, Heinemann, London (1973).
Koestler, A., *The Sleepwalkers*, Macmillan, New York (1959).
Kuhn, T. S., *The Copernican Revolution*, Harvard University Press, Cambridge, Mass. (1957).
Mach, E., *The Science of Mechanics*, transl. from German, Open Court, LaSalle, Ill. (1942).
Weinstock, R., *American Journal of Physics* **50**, 610 (1982).
Westfall, R. S., *Never at Rest, A Biography of Isaac Newton*, Cambridge University Press (1980).

The Cartesian revolution

Davis, P. J. and Hersh, R., *Descartes' Dream: The World According to Mathematics*, Harcourt Brace Janovich, San Diego (1986).
Descartes, R., 'Principles of Philosophy', in *Descartes* by J. Cottingham, R. Stoothoff, and D. Murdoch, Cambridge University Press (1988).
Stolpe, S., *Christina of Sweden*, transl. from Swedish, Macmillan, New York (1966).

Creativity in problem-solving

Hadamard, J., *The Psychology of Invention in the Mathematical Field*, Dover, New York (1954).

Plato, Puritanism, and the devil

Calvin, J., in *Witchcraft in Europe*, ed. by Kors, A. C. and Peters, E., University of Philadelphia Press (1972).
Collingwood, R. G., *The Idea of Nature*, chapter 2, Oxford University Press, London (1945).
Dijksterhuis, E. J., *The Mechanization of the World Picture, Pythagoras to Newton*, transl. from Hollandisch, chapters 1–4, Princeton University Press (1986).
Einhard, *The Life of Charlemagne*, transl. from Latin, Penguin, New York (1984).
Luther, M., in *Witchcraft in Europe*, ed. by Kors, A. C. and Peters, E., University of Philadelphia Press (1972).
Oberman, H. A., *Luther: Man Between God and the Devil*, transl. from German, Doubleday, New York (1992).
Plato, *Timaeus*, transl. by F. M. Cornfeld, The Liberal Arts Press, New York (1959).

Aristotle, Scholasticism, magic, and witches

Aquinas, T. in Magic and the World of Nature in *Witchcraft in Europe*, ed. by Kors, A. C. and Peters, E., University of Philadelphia Press (1972).

Collingwood, R. G., *The Idea of Nature*, chapter 3, Oxford University Press, London (1945).

Dijksterhuis, E. J., *The Mechanization of the World Picture, Pythagoras to Newton*, transl. from Hollandisch, chapters 2–4, Princeton University Press (1986).

Eco, U., *Travels in Hyperreality*, chapter 6, Harcourt Brace, Orlando (1986).

Ginzburg, C., *The Cheese and the Worms, The Cosmos of a Sixteenth Century Miller*, transl. from Italian, Johns Hopkins University Press, Baltimore (1992).

Kramer, H. and Sprenger, J., *Malleus Malificarum*, transl. from Latin, Pushkin Press, London (1948).

Trevor-Roper, H., *The Crisis of the Seventeenth Century*, Harper and Row, New York (1968).

Weizsäcker, C. F. von, Foreword to *Information and the Origin of Life*, by Küppers, B.-O., transl. from German, MIT Press, Cambridge, Mass. (1990).

Tribal freedom and direct democracy

Jones, G., *A History of the Vikings*, Oxford University Press (1991).

Thürer, G., *Free and Swiss*, transl. from German, University of Miami, Coral Gables (1971).

Chapter 2

Variational calculus

Bliss, G. A., *Calculus of Variations*, Math. Assoc. of America, La Salle, Ill. (1962).

Dubrovin, B. A., Fomenko, A. T., and Novikov, S. P., *Modern Geometry – Methods and Applications. Part I. The Geometry of Surfaces, Transformation Groups, and Fields*, transl. by R. G. Burns, Springer-Verlag, New York (1984).

Hamel, G., *Theoretische Mechanik*, 2d Aügabe, Springer-Verlag, Berlin (1967).

Lanczos, C., *The Variational Principles of Mechanics*, University of Toronto Press (1949).

Margenau, H. and Murphy, G. M., *The Mathematics of Chemistry and Physics*, vol. I, van Nostrand, Princeton (1956).

Function space and functionals

Liusternik, L. and Sobolev, V., *Elements of Functional Analysis*, Ungar, New York (1961).

Lagrange's and Hamilton's equations

Arnol'd, V. I., *Mathematical Methods of Classical Mechanics*, transl. from Russian, Springer-Verlag, Berlin (1989).

Corben, H. C. and Stehle, P., *Classical Mechanics*, Wiley, New York (1960).

Goldstein, H., *Classical Mechanics*, Addison-Wesley, Reading, Mass. (1980).

Sommerfeld, A., *Vorlesungen über Theoretische Physik, Bd. 1 Mechanikk*, Dieterich, Wisbaden (1947–52).

Whittaker, E. T., *Analytical Dynamics*, chapters 3, 4 and 9, Cambridge University Press (1965).

Noether's theorem, transformations, and symmetries

Engel, F., *Über die zehn allgemeinen Integrale der klassischen Mechanik*, pp. 270–5, Nachr. Kgl. Ges. Wiss., Göttingen (1916).

Mercier, A., *Analytical and Canonical Formalism in Physics*, Dover, New York (1963).

Saletan, E. J. and Cromer, H. A., *Theoretical Mechanics*, Wiley, New York (1971).

Fluid mechanics and Hamiltonian systems

Lin, C. C., *On the Motion of Vortices in Two Dimensions*, University of Toronto Press (1943).
Milne-Thompson, L. M., *Theoretical Aerodynamics*, Dover, New York (1958).

Chapter 3

Arnol'd, V. I., *Mathematical Methods of Classical Mechanics*, transl. from Russian, Springer-Verlag, Berlin (1989).
Arnol'd, V. I., *Ordinary Differential Equations*, MIT Press, Cambridge, Mass. (1981).
Bell, E. T., *The Development of Mathematics*, McGraw-Hill, New York (1945).
Bell, E. T., *Men of Mathematics*, Simon and Schuster, New York (1965).
Bender, C. M. and Orszag, S. A., *Advanced Mathematical Methods for Scientists and Engineers*, chapter 4, McGraw-Hill, New York (1978).
Buck, C., *Advanced Calculus*, McGraw-Hill, New York (1956).
Burns, S. A. and Palmore, J. I., *Physica* **D37**, 83 (1989).
Davis, H. T., *Nonlinear Integral and Differential Equations*, chapters 4, 10, and 11, Dover, New York (1960).
de Almeida, O., *Hamiltonian Systems: Chaos and Quantization*, Cambridge University Press (1988).
Dubrovin, B. A., Fomenko, A. T., and Novikov, S. P., *Modern Geometry – Methods and Applications. Part I. The Geometry of Surfaces, Transformation Groups, and Fields*, transl. by R. G. Burns, Springer-Verlag, New York (1984).
Duff, G. F. D., *Partial Differential Equations*, University of Toronto Press (1962).
Fomenko, A. T., *Integrability and Nonintegrability in Geometry and Mechanics*, transl. from Russian by M. V. Tsaplina, section 1 of chapter 1, Kluwer, Dordrecht (1988).
Gradstein, I. M. and Ryzhik, I. S., *Tables of Integrals, Series, and Products*, transl. from Russian, Academic, New York (1965).
Ince, E. L., *Ordinary Differential Equations*, chapters I–IV and appendix, Dover, New York (1956).
Jordan, D. W. and Smith, P., *Introduction to Nonlinear Ordinary Differential Equations*, chapters 1 and 2, Clarendon Press, Oxford (1977).
Kilmister, C. W., *Hamiltonian Dynamics*, Wiley, New York (1964).
Klein, F. and Lie, S., *Mathematische Annalen* **4**, 80 (1871).
Lie, S., *Forhand. Vid.-Sels. Christiania* 198 (1874); 1 (1875).
McCauley J. L., *Chaos, Dynamics, and Fractals: an algorithmic approach to deterministic chaos*, sections 1.1–1.3, and 2.1, Cambridge University Press (1993).
Moore, R. A., *Introduction to Differential Equations*, Allyn & Bacon, Boston (1962).
Olver, P. J., *Applications of Lie Groups to Differential Equations*, Springer, New York (1993).
Percival, I. and Richards, D., *Introduction to Dynamics*, Cambridge University Press (1982).
Tabor, M., *Chaos and Integrability in Nonlinear Dynamics*, Wiley, New York (1989).
Whittaker, E. T., *Analytical Dynamics*, chapters 3, 4, and 9, Cambridge University Press (1965).
Whittaker, E. T. and Watson, G. N., *Modern Analysis*, Cambridge University Press (1963).
Wintner, A., *The Analytical Foundations of Celestial Mechanics*, section 194–202, Princeton University Press (1941).

Chapter 4

Arnol'd, V. I., *Mathematical Methods of Classical Mechanics*, transl. from Russian, Springer-Verlag, Berlin (1989).
Ekeland, I., *Zufall, Glück und Chaos*, Carl Hanser Verlag, München (1991).
Goldstein, H., *Classical Mechanics*, Addison-Wesley, Reading, Mass. (1980).

Kac, M., *Statistical Independence in Probability, Analysis, and Number Theory*, The Carus Math. Monogr. No. 12, pub. by The Mathematical Society of America, distr. by John Wiley & Sons (1959).

Martinez-Y-Romero, R. P., Núñez-Yépez, H. N., and Salas-Brito, A. L., *Eur. J. Phys.* **15**, 26 (1992).

McCauley, J. L., *Chaos, Dynamics, and Fractals: an algorithmic approach to deterministic chaos*, Cambridge University Press (1993).

McIntosh, H. V., *American Journal of Physics*, **27**, 620 (1959).

Whittaker, E. T., *Analytical Dynamics*, Cambridge University Press (1965).

Chapter 5

Arnol'd, V. I., *Geometrical Methods in the Theory of Ordinary Differential Equations*, Springer-Verlag, Berlin (1983).

Arnol'd, V. I., *Mathematical Methods of Classical Mechanics*, transl. from Russian, Springer-Verlag, Berlin (1989).

Bender, C. M. and Orszag, S. A., *Advanced Mathematical Methods for Scientists and Engineers*, McGraw-Hill, New York (1978).

Goldstein, H., *Classical Mechanics*, Addison-Wesley, Reading, Mass. (1980).

Ince, E. L., *Ordinary Differential Equations*, Dover, New York (1956).

Margenau, H. and Murphy, G. M., *The Mathematics of Chemistry and Physics*, vol. I, van Nostrand, Princeton (1956).

ter Haar, D., *Elements of Hamiltonian Mechanics*, Pergamon, Oxford (1964).

Whittaker, E. T., *Analytical Dynamics*, chapter 7, Cambridge University Press (1965).

Chapter 6

Armstrong, G., *Basic Topology*, Springer-Verlag, New York (1983).

Arnol'd, V. I., *Mathematical Methods of Classical Mechanics*, transl. from Russian, Springer-Verlag, Berlin (1989).

Arnol'd, V. I., *Geometrical Methods in the Theory of Ordinary Differential Equations*, Springer-Verlag, Berlin (1983).

Borel, E., *Les Nombres Inaccessibles*, Gauthier-Villars, Paris (1952).

Cvitanovic, P., Gunaratne, G., and Procaccia, I., *Phys. Rev.* **A38**, 1503 (1988).

Delachet, A., *Contemporary Geometry*, Dover, New York (1962).

Gnedenko, B. V. and Khinchin, A. Ya., *An Elementary Introduction to the Theory of Probability*, transl. from Russian, Dover, New York (1962).

Jordan, D. W. and Smith, P., *Introduction to Nonlinear Ordinary Differential Equations*, chapter 9, Clarendon Press, Oxford (1977).

Kac, M., *Statistical Independence in Probability, Analysis, and Number Theory*, The Carus Math. Monogr. No. 12, pub. by The Mathematical Society of America, distr. by John Wiley & Sons (1959).

Knuth, D. E., *The Art of Computer Programming II: Semi-Numerical Algorithms*, Addison-Wesley, Reading, Mass. (1981).

MacKay, R. M. and Meiss, J. D., *Hamiltonian Dynamical Systems*, Adam Hilger Ltd, Bristol (1987).

McCauley, J. L., *Chaos, Dynamics, and Fractals: an algorithmic approach to deterministic chaos*, chapters 2, 3, 9, and 11, Cambridge University Press (1993).

Minsky, M. L., *Computation, Finite and Infinite Machines*, Prentice-Hall, Englewood Cliffs, NJ (1967).

Niven, I., *Irrational Numbers*, The Carus Math. Monogr. No. 11, pub. by The Mathematical Society of America, distr. by John Wiley & Sons (1956).

Schuster, H. G., *Deterministic Chaos*, Physik-Verlag, Weinheim (1984).

Whittaker, E. T., *Analytical Dynamics*, chapter 10, Cambridge University Press (1965).

Chapter 7

(* indicates biographical writings on Lie and his works.)

Arnol'd, V. I., *Mathematical Methods of Classical Mechanics*, transl. from Russian, Springer-Verlag, Berlin (1989).

Bender, C. M. and Orszag, S. A., *Advanced Mathematical Methods for Scientists and Engineers*, McGraw-Hill, New York (1978).

Burke, W. L., *Applied Differential Geometry*, Cambridge University Press (1985).

Duff, G. F. D., *Partial Differential Equations*, University of Toronto Press (1962).

Eisenhart, L. P., *Continuous Groups of Transformations*, Dover, New York (1961).

*Engel, F., Sophus Lie *Norsk matematisk tidsskrift* **4**, 97 (1922).

Hammermesh, M., *Group Theory*, pp. 469–77 Addison-Wesley, Reading, Mass. (1962).

*Holst, E., Træk av Sophus Lies ungdomsliv *Ringeren* **9** (1899).

Ince, E. L., *Ordinary Differential Equations*, Dover, New York (1956).

*Jahren, B., Sophus Lie 150 år *Normat* **4**, 145 (1992).

Lie, S. and Engel, F., *Theorie der Transformationsgruppen*, vols. 1 and 2. Teubner, Leipzig (reprinted in 1930).

Lie, S. and Scheffers, G., *Vorlesungen über Differentialgleichungen mit bekannten infinitesimalen Transformationen*, Teubner, Berlin (1891).

McCauley, J. L., *Chaos, Dynamics, and Fractals: an algorithmic approach to deterministic chaos*, Cambridge University Press (1993).

Olver, P. J., *Applications of Lie Groups to Differential Equations*, Springer, New York (1993).

Racah, G., *Group Theory and Spectroscopy*, Princeton lecture notes (1951).

Sneddon, I. N., *Elements of Partial Differential Equations*, McGraw-Hill, New York (1957).

Tabor, M., *Chaos and Integrability in Nonlinear Dynamics*, Wiley, New York (1989).

Wybourne, B. G., *Classical Groups, for Physicists*, Wiley, New York (1974).

Chapter 8

Bell, E. T., *The Development of Mathematics*, McGraw-Hill, New York (1945).

Delachet, A., *Contemporary Geometry*, Dover, New York (1962).

Eddington, A. S., *The Mathematical Theory of Relativity*, Cambridge University Press, London (1963).

Goldstein, H., *Classical Mechanics*, Addison-Wesley, Reading, Mass. (1980).

Kibble, T. W. B., *Classical Mechanics*, Longman, London (1985).

Klein, F., *Vergleichende Betrachtungen über neuere geometrische Forschungen*, Deichert, Erlangen (1872).

Klein, F. and Sommerfeld, A., *Theorie des Kreisels*, Teubner, Leipzig (1897).

Margenau, H. and Murphy, G. M., *The Mathematics of Chemistry and Physics*, vol. I, van Nostrand, Princeton (1956).

McCauley, J. L., *American Journal of Physics* **45**, 94 6, (1977).

Saletan, E. and Cromer, A., *Theoretical Mechanics*, Wiley, New York (1971).

Weyl, H., *Philosophy of Mathematics and Natural Science*, pp. 67–91, Athenaeum, New York (1963).

Wigner, E. P., *Symmetries and Reflections*, University of Indiana Press, Bloomington (1967).

Yaglom, I. M., *Felix Klein and Sophus Lie*, transl. from Russian, Birkhäuser, Boston (1988).

Chapter 9

Arnol'd, V. I., *Mathematical Methods of Classical Mechanics*, transl. from Russian, Springer-Verlag, Berlin (1989).

Arnol'd, V. I. (ed.) *Dynamical Systems III*, Springer-Verlag, Heidelberg (1993).

Bender, C. M. and Orszag, S. A., *Advanced Mathematical Methods for Scientists and Engineers*, McGraw-Hill, New York (1978).

Fomenko, A. T., *Integrability and Nonintegrability in Geometry and Mechanics*, transl. from Russian by M. V. Tsaplina, chapter 1, section 1, Kluwer, Dordrecht (1988).

Goldstein, H., *Classical Mechanics*, Addison-Wesley, Reading, Mass. (1980).

McCauley, J. L., *Chaos, Dynamics, and Fractals: an algorithmic approach to deterministic chaos*, chapter 2, Cambridge University Press (1993).

Saletan, E. J. and Cromer, H. A., *Theoretical Mechanics*, Wiley, New York (1971).

Steeb, W.-H. and Louw, J. A., *Chaos and Quantum Chaos*, chapter 9, World Scientific, Singapore (1986).

Tabor, M., *Nature* **310**, 277 (1984).

Whittaker, E. T., *Analytical Dynamics*, Cambridge University Press (1965).

Whittaker, E. T. and Watson, G. N., *Modern Analysis*, Cambridge University Press (1963).

Chapter 10

Adler, R., Bazin, M., and Schiffer, M., *Introduction to General Relativity*, chapter 1, McGraw-Hill, New York (1965).

Buck, C., *Advanced Calculus*, McGraw-Hill, New York (1956).

Dubrovin, B. A., Fomenko, A. T., and Novikov, S. P., *Modern Geometry – Methods and Applications. Part I. The Geometry of Surfaces, Transformation Groups, and Fields*, transl. by R. G. Burns, chapters 3 and 4, Springer-Verlag, New York (1984).

Eddington, A. S., *The Mathematical Theory of Relativity*, Cambridge University Press, London (1963).

Hauser, W., *Introduction to the Principles of Mechanics*, Addison-Wesley, Reading, Mass. (1965).

Kilmister, C. W., *Hamiltonian Dynamics*, Wiley, New York (1964).

Klein, F., *Vergleichende Betrachtungen über neuere geometrische Forschungen*, Deichert, Erlangen (1872).

Margenau, H. and Murphy, G. M., *The Mathematics of Chemistry and Physics*, vol. I, van Nostrand, Princeton (1956).

McCauley, J. L., *American Journal of Physics* **45**, 94 (1977).

Weyl, H., *Philosophy of Mathematics and Natural Science*, pp. 67–91, Athenaeum, New York (1963).

Wigner, E. P., *Symmetries and Reflections*, University of Indians Press, Bloomington (1967).

Chapter 11

Adler, R., Bazin, M., and Schiffer, M., *Introduction to General Relativity*, McGraw-Hill, New York (1965).

Anderson, J. L., *Principles of Relativity Physics*, Academic, New York (1967).

Becker, R. and Sauter, F., *Electromagnetic Fields and Interactions*, transl. from German, Blaisdell, New York (1964).

Bergmann, P. G., *Introduction to the Theory of Relativity*, Prentice-Hall, New York (1942).

Buck, C., *Advanced Calculus*, McGraw-Hill, New York (1956).

Cartan, E., *On manifolds with an affine connection and the general theory of relativity*, transl. from French, Biblios, Naples/Atlantic Highlands, NJ (1986).

Dubrovin, B. A., Fomenko, A. T., and Novikov, S. P., *Modern Geometry – Methods and Applications. Part I. The Geometry of Surfaces, Transformation Groups, and Fields*, transl. by R. G. Burns, chapters 3 and 4, Springer-Verlag, New York (1984).

Eddington, A. S., *The Mathematical Theory of Relativity*, Cambridge University Press, London (1963).

Einstein, A., Zur elektrodynamik bewegter Körper, *Annalen der Physik* **17**, 891–921 (1905).

Einstein, A., Die Grundlage der allgemeinen Relativitätstheorie, *Annalen der Physik* **49**, 111 (1916).

Einstein, A. and others, *The Principle of Relativity*, transl. from German, Dover, New York (1923).

Einstein, A., *The Meaning of Relativity*, Princeton University Press (1953).

Eisenhart, L. P., *Continuous Groups of Transformations*, Dover, New York (1961).

Fock, V. A., *The Theory of Space, Time and Gravitation*, transl. from Russian, Pergamon, New York (1959).

Kaempffer, W., *Concepts in Quantum Mechanics*, sections 20 and 21, Academic, New York (1965).

Kretschmann, E., *Ann. Phys.* **53**, 575 (1917).

Lindley, D., *The End of Physics*, Basic Books, New York (1993).

Mercier, A., *Analytical and Canonical Formalism in Physics*, Dover, New York (1963).

Misner, C. W., Thorn, K. S., and Wheeler, J. A., *Gravitation*, W. H. Freeman & Co., San Francisco (1973).

Schutz, B. F., *Geometrical Methods of Mathematical Physics*, Cambridge University Press (1980).

Shapere, A. and Wilczek, F., *Geometric Phases in Physics*, World Scientific, Singapore (1989).

Utiyama, R., *Phys. Rev.* **101**, 1597 (1956).

Weyl, H., *Philosophy of Mathematics and Natural Science*, pp. 67–91, Athenaeum, New York (1963).

Wigner, E. P., *Symmetries and Reflections*, University of Indiana Press, Bloomington (1967).

Chapter 12

Arnol'd, V. I. (ed.), *Dynamical Systems III*, Springer-Verlag, Heidelberg (1993).

Arnol'd, V. I., Kozlov, V. V., and Neishtadt, A. I., *Mathematical Aspects of Classical and Celestial Mechanics*, in *Dynamical Systems III*, ed. by V. I. Arnol'd, Springer-Verlag, Heidelberg (1993).

Boltzmann, L., Über die Form der Lagrangeschen Gleichungen für nichtholonome, generalisierte Koordinaten, *Sitzungsberichte der Wiener Akademie* Bd. CXI. Abt. IIa. 'Dez. 1902'.

Campbell, J. E., *Continuous Groups*, Chelsea, New York (1966).

Desloge, E. A., *Classical Mechanics*, vol. 2, pp. 79–732, Wiley, New York (1982).

Eisenhart, L. P., *Continuous Groups of Transformations*, Dover, New York (1961).

Goldstein, H., *Classical Mechanics*, 2nd edn, Addison-Wesley, Reading, Mass. (1980).

Hamel, G., Die Lagrange-Eulerschen Gleichungen der Mechanik, *Zeitschr. für Math. und Physik* **50**, 1–57, (1904).

Hamel, G., *Theoretische Mechanik*, 2nd edn, pp. 473–506, Springer-Verlag, Berlin (1967).

Kilmister, C. W., *Hamiltonian Dynamics*, pp. 40–6, Wiley, New York (1964).

Magnus, W., On the Exponential Solution of Differential Equations of a Linear Operator, *Communications on Pure and Applied Math.*, vol. VII, 660 and 663–73 (1954).

Marsden, J. E., *Lectures on Mechanics*, Cambridge University Press (1992).

McCauley, J. L., *Chaos, Solitons, & Fractals* **4**, 1845 (1994).

Merzbacher, E., *Quantum Mechanics*, pp. 162 and 358, Wiley, New York (1965).

Saletan, E. and Cromer, A., *Theoretical Mechanics*, Wiley, New York (1971).

Whittaker, E. T., *Analytical Dynamics*, Cambridge University Press (1965).

Chapter 13

Armstrong, G., *Basic Topology*, Springer-Verlag, New York (1983).

Arnol'd, V. I., *Ordinary Differential Equations*, transl. from Russian by R. A. Silverman, pp. 48, 49, 76 and 77, MIT, Cambridge (1979–81).

Arnol'd, V. I., *Geometric Methods in the Theory of Ordinary Differential Equations*, transl. from Russian by Joseph Szücs, Springer-Verlag, New York (1983).

Arnol'd, V. I., *Mathematical Methods of Classical Mechanics*, transl. from Russian, Springer-Verlag, Berlin (1989).

Arnol'd, V. I., (ed.) *Dynamical Systems III*, Springer-Verlag, Heidelberg (1993).

Arnol'd, V. I., Kozlov, V. V., and Neishtadt, A. I., *Mathematical Aspects of Classical and Celestial Mechanics*, in *Dynamical Systems III*, ed. by V. I. Arnol'd, Springer-Verlag, Heidelberg (1993).

Buck, C., *Advanced Calculus*, McGraw-Hill, New York (1956).

Burns, S. A. and Palmore, J. I., *Physica* **D37**, 83 (1989).

Dubrovin, B. A., Fomenko, A. T., and Novikov, S. P., *Modern Geometry – Methods and Applications. Part I. The Geometry of Surfaces, Transformation Groups, and Fields*, transl. by R. G. Burns, chapters 2 and 5, Springer-Verlag, New York (1984).

Dubrovin, B. A., Fomenko, A. T., and Novikov, S. P., *Modern Geometry – Methods and Applications. Part II. The Geometry and Topology of Manifolds*, transl. by R. G. Burns, chapters 1–4, Springer-Verlag, New York (1985).

Duff, G. F. D., *Partial Differential Equations*, p. 26, University of Toronto Press (1962).

Eisenhart, L. P., *Continuous Groups of Transformations*, pp. 33–4, Dover, New York (1961).

Fomenko, A. T., *Integrability and Nonintegrability in Geometry and Mechanics*, transl. from Russian by M. V. Tsaplina, Kluwer, Dordrecht (1988).

Hadamard, J., *J. Math. Pure and Applied* **4**, 27 (1898); *Soc. Sci. Bordeaux Proc. Verb.* 147 (1898).

Hammermesh, M., *Group Theory*, Addison-Wesley, Reading, Mass. (1962).

Ince, E. L., *Ordinary Differential Equations*, Dover, New York (1956).

Jordan, D. W. and Smith, P., *Introduction to Nonlinear Ordinary Differential Equations*, chapter 9, Clarendon Press, Oxford (1977).

Kilmister, C. W., *Hamiltonian Dynamics*, Wiley, New York (1964).

Klein, F., *On Riemann's Theory of Algebraic Functions and their Integrals*, transl. from German, Dover, New York (1963).

Lie, S. and Scheffers, G., *Vorlesungen über Differentialgleichungen mit bekannten infinitesimalen Transformationen*, Teubner, Berlin (1891).

Lorenz, E., *J. Atm. Sci.*, **20**, 130 (1963).

McCauley, J. L., *Chaos, Dynamics, and Fractals: an algorithmic approaches to deterministic chaos*, Cambridge University Press (1993).

McCauley, J. L., *Chaos, Solitons & Fractals* **4**, 1969 (1994); **5**, 1493 (1995).

McCauley, J. L., *Physica A*, to be published (1997).

Olver, P. J., *Applications of Lie Groups to Differential Equations*, p. 30, Springer-Verlag, New York (1993).

Palmore, J., private communication (1996).

Racah, G., *Group Theory and Spectroscopy*, Princeton lecture notes (1951).

Schuster, H. G., *Deterministic Chaos*, 2nd revised edn, VCH Verlagsgesellschaft, Weinheim (1988).

Sinai, Ya. G., *Russian Math. Surveys*, **25**, 137 (1970).

Sneddon, I., *Elements of Partial Differential Equations*, p. 66, McGraw-Hill, New York (1957).

Sommerfeld, A., *Mechanics*, Academic, New York (1964).

Stefanni, H. and MacCullum, M., *Differential Equations, their solution using symmetries*, Cambridge University Press (1989).

Tabor, M., *Nature*, **310**, 277 (1984).

Tabor, M., *Chaos and Integrability in Nonlinear Dynamics*, Wiley, New York (1989).

Tabor, M. and Weiss, J., *Phys. Rev.*, **A24**, 2157 (1981).

Whittaker, E. T., *Analytical Dynamics*, pp. 53 and 397, Cambridge University Press (1965).

Zakharov, V. E., *What is Integrability?*, chapters 2 and 3, Springer-Verlag, Berlin (1991).

Chapter 14

Asterisks indicate references on Cantor sets and on Cantor's idea of the continuum.

Arnol'd V. I., *Trans. Am. Math. Soc.* **46**, 213 (1965).

Arnol'd, V. I., *Geometric Methods in the Theory of Ordinary Differential Equations*, Springer-Verlag, Berlin (1983).

*Boas, R. P., *A Primer of Real Functions*, The Carus Math. Monographs No. 13, Wiley, New York (1960).

Bohr, T., Bak, P., and Jensen, M. H., *Physica Scripta* **T9**, 50 (1985); *Phys. Rev.* **A30**, 1960 (1984); *Phys. Rev.* **A30**, 1970 (1984).

Cvitanovic, P. and Søderberg, B., *Physica Scripta* **32**, 263 (1985).

Cvitanovic, P., Jensen, M. H., Kadanoff, L. P., and Procaccia, I., *Phys. Rev. Lett.* **55**, 343 (1985).

Cvitanovic, P., Gunaratre, G., and Procaccia, I., *Phys. Rev.* **A38**, 1503 (1988).

Devaney, R. L., *An Introduction to Chaotic Dynamical Systems*, Benjamin, Menlo Park, USA (1986).

*Dunham, W., *Journey through Genius*, chapter on G. Cantor, Penguin, New York (1991).

Feigenbaum, M. J., Los Alamos Science (1980).

*Gelbaum, B. R. and Olmsted, J. H., *Counterexamples in Analysis*, Holden-Day, San Francisco (1964).

Glass, L. and Perez, R., *Phys. Rev. Lett.* **48**, 1772 (1982).

Green, J. M., *J. Math, Phys.* **20**, 1183 (1980).

Grossmann, S. and Thomae, S., *Z. Naturforschung* **32A**, 1353 (1977).

Guckenheimer, J. and Holmes, P., *Nonlinear Oscillations, Dynamical Systems and Bifurcations of Vector Fields*, Springer-Verlag (1983).

Holmes, P. J., *J. Sound Vibr.* **84**, 173–89 (1982).

Hu, B., *Phys. Reports* **31**, 233 (1982).

Hu, B., in *Proc. of the 1986 Summer School on Statistical Mechanics*, ed. by C. K. Hu, Academia Sinica (1987).

*McCauley, J. L., *Chaos, Dynamics and Fractals*, chapters 4–9, Cambridge University Press (1993).

*Minsky, M. L., *Computation, Finite and Infinite Machines*, chapter 9, Prentice-Hall, Englewood Cliffs, NJ (1967).

Perez, R. and Glass, L., *Phys. Lett.* **90A**, 441 (1982).

Pierań ski, P. and Malecki, J., *Phys. Rev.* **A34**, 582 (1986).

Schuster, H. G., *Deterministic Chaos*, VCH Verlagsgesellschaft, Weinheim (1988).

Chapter 15

Anderson, P. W., *Rev. Mod. Phys.* **38**, 298 (1966).

Arnol'd, V. I., *Mathematical Methods of Classical Mechanics*, transl. from Russian, Springer-Verlag, Berlin (1989).

Arnol'd, V. I., Kozlov, V. V., and Neishtadt, A. I., *Mathematical Aspects of Classical and Celestial Mechanics*, in *Dynamical Systems III*, ed. by V. I. Arnol'd, Springer-Verlag, Heidelberg (1993).

Dirac, P. A. M., *Proc. Roy. Soc.* **A110**, 561–9, (1926); **A109**, 642–53 (1926).

Dubrovin, B. A., Fomenko, A. T., and Novikov, S. P., *Modern Geometry – Methods and Applications. Part I. The Geometry of Surfaces, Transformation Groups, and Fields*, transl. by R. B. Burns, chapters 2 and 5, Springer-Verlag, New York (1984).

Duff, G. F. D., *Partial Differential Equations*, University of Toronto Press (1962).

Eisenhart, L. P., *Continuous Groups of Transformations*, Dover, New York (1961).

Engel, F., Über die zehn allgemeinen Integrale der klassischen mechanik, pp. 270–5, *Nachr. Kgl. Ges. Wiss.*, Göttingen (1916).

Fisher, D. S., Huse, D. A., Fisher, M. P. A., and Fisher, D., Are Superconductors Really Superconducting? *Nature* **358**, 553 (1992).

Fomenko, A. T., *Integrability and Nonintegrability in Geometry and Mechanics*, transl. from Russian by M. V. Tsaplina, chapter 1, Kluwer, Dordrecht (1988).

Goldstein, H., *Classical Mechanics*, Addison-Wesley, Reading, Mass. (1980).

Gutzwiller, M. C., *Chaos in Classical and Quantum Mechanics*, Springer-Verlag, New York (1990).

Hamel, G., *Zeitschr. für Math. und Physik* **50**, 1 (1904).

Hamel, G., *Theoretische Mechanik*, Springer-Verlag, Berlin (1949, 1967).

Hammermesh, M. *Group Theory*, Addison-Wesley, Reading, Mass. (1962).

Kaempffer, W., *Concepts in Quantum Mechanics*, Academic, New York (1965).

Kilmister, C. W., *Hamiltonian Dynamics*, Wiley, New York (1964).

Landau, L. D. and Lifshitz, E. M., *Quantum Mechanics, nonrelativistic theory*, 3rd edn, transl. from Russian by J. B. Sykes and J. S. Bell, Pergamon Press, Oxford (1977).

Marsden, J. E., *Lectures on Mechanics*, Cambridge University Press (1992).

Martinez-Y-Romero, R. P., Núñez-Yépez, H. N., and Salas-Brito, A. L., *Eur. J. Phys.* **15**, 26 (1992).

McIntosh, H. V., *American Journal of Physics*, **27**, 620 (1959).

Pollard, H., *Mathematical Introduction to Celestial Mechanics*, chapter 3, sections 1–3, Prentice-Hall, Engelwood Cliffs, NJ (1966).

Sommerfeld, A., *Mechanics*, Academic, New York (1964).

Tabor, M. and Treve, Y. M., eds., *Mathematical Methods in Hydrodynamics and Integrability in Dynamical Systems*, American Institute of Physics, New York (1982).

Weinstein, A., *J. Diff. Geom*, **18**, 523–57 (1983) and *Exp. Math.* **1**, 95–6 (1983).

Weyl, H., *The Theory of Groups and Quantum Mechanics*, transl. from German, Dover, New York (1960).

Whittaker, E. T., *Analytical Dynamics*, Cambridge University Press (1965).

Wybourne, B. G., *Classical Groups for Physicists*, Wiley, New York (1974).

Chapter 16

Arnol'd, V. I., *Mathematical Methods of Classical Mechanics*, transl. from Russian, Springer-Verlag, Berlin (1989).

Arnol'd, V. I., Kozlov, V. V., and Neishtadt, A. I., *Mathematical Aspects of Classical and Celestial Mechanics*, in *Dynamical Systems III*, ed. by V. I. Arnol'd, Springer-Verlag, Heidelberg (1993).

Berry, M. V. and Ford, J., in *Topics in Nonlinear Dynamics*, ed. by S. Jorna, American Institute of Physics, New York (1987).

Dirac, P. A. M., *Quantum Mechanics*, 4th edn, Oxford University Press, London (1958).

Duff, G. F. D., *Partial Differential Equations*, pp. 58–62, University of Toronto Press (1962).

Einstein, A., *Verh. Deutsche Phys. Ges.* **19**, 82 (1917).

Feynman, R. P. and Hibbs, A. R., *Quantum Mechanics and Path Integrals*, McGraw-Hill, New York (1965).

Fomenko, A. T., *Integrability and Nonintegrability in Geometry and Mechanics*, transl. from Russian by M. V. Tsaplina, Kluwer, Dordrecht (1988).

Friedrich, H., *Physics World* 32–6 (Apr. 1992).

Goldstein, H., *Classical Mechanics*, Addison-Wesley, Reading, Mass. (1980).

Gutzwiller, M. C., *Chaos in Classical and Quantum Mechanics*, Springer-Verlag, New York (1990).

Hamel, G., *Zeitschr. für Math. und Physik* **50**, 1 (1904).

Hamel, G., *Theoretische Mechanik*, Springer-Verlag, Berlin (1949, 1967).

Kac, M. (ed.), *Probability and Related Topics in Physical Sciences*, chapter IV and appendix II, Interscience, New York (1959).

Kaempffer, W., *Concepts in Quantum Mechanics*, Academic, New York (1965).

Kilmister, C. W., *Hamiltonian Dynamics*, Wiley, New York (1964).

Koopman, B. O., *Proc. Nat. Acad. of Sci.* **17**, 316–18 (1931).

Koopman, B. O. and von Neumann, J., *Proc. Nat. Acad. of Sci.* **18**, 255–63 (1932).

Landau, L. D. and Lifshitz, E. M., *Mechanics*, Pergamon, Oxford (1976).

Lichtenberg, A. J. and Lieberman, M. A., *Regular and Stochastic Motion*, chapter 5, Springer-Verlag, New York (1983).

Lie, M. A. and Scheffers, G., *Vorlesungen über Differentialgleichungen mit bekannten infinitesimalen Transformationen*, Teubner, Berlin (1891).

Margenau, H. and Murphy, G. M., *The Mathematics of Chemistry and Physics*, vol. I, van Nostrand, Princeton (1956).

McCauley, J. L., *Chaos, Solitons, & Fractals* **4**, 2133 (1994); **5**, 1493 (1995).

Moser, J. K., *Memoirs Am. Math. Soc.* **81**, 1 (1968).

Moser, J. K., *Math. Intelligencer* **1**, 65 (1978).

Pollard, H., *Mathematical Introduction to Celestial Mechanics*, chapter 3, sections 4 and 5, Prentice-Hall, Engelwood Cliffs, NJ (1966).

Sneddon, I., *Elements of Partial Differential Equations*, pp. 78–80, McGraw-Hill, New York (1957).

Sommerfeld, A., *Mechanics*, Academic, New York (1964).

Stakgold, I., *Green Functions and Boundary Value Problems*, Wiley-Interscience, New York (1979).

Steeb, W.-H. and Louw, J. A., *Chaos and Quantum Chaos*, chapter 9, World Scientific, Singapore (1986).

Tabor, M., *Chaos and Integrability in Nonlinear Dynamics*, Wiley, New York (1989).

Talman, J. D., *Special Functions, A Group Theoretic Approach*, Benjamin, New York (1968).

ter Haar, D., *Elements of Hamiltonian Mechanics*, Pergamon, Oxford (1971).

Toda, M., *Theory of Nonlinear Lattices*, chapter 5, Springer-Verlag, Berlin (1981).

Whittaker, E. T., *Analytical Dynamics*, Cambridge University Press (1965).

Wintner, A., *The Analytical Foundations of Celestial Mechanics*, sections 194–202 and 227–40, Princeton University Press (1941).

Chapter 17

Arnol'd, V. I., *Russ. Math. Surv.* **18**, 85 (1963).

Arnol'd, V. I., *Mathematical Methods of Classical Mechanics*, Springer-Verlag, transl. from Russian, Berlin (1989).

Arnol'd, V. I. and Avez, A., *Ergodic Problems in Classical Mechanics*, chapters 1 and 2, Benjamin, New York (1968).

Berry, M. V., in *Topics in Nonlinear Dynamics*, ed. by S. Jorna, American Institute of Physics, New York (1987).

Cvitanovic, P., (ed.) *Universality in Chaos*, Adam Hilger, London (1984).

Fomenko, A. T., *Integrability and Nonintegrability in Geometry and Mechanics*, transl. from Russian by M. V. Tsaplina, Kluwer, Dordrecht (1988).

Gutzwiller, M. C., *Chaos in Classical and Quantum Mechanics*, Springer-Verlag, New York (1990).

Hénon, M., in *Chaotic Behavior of Deterministic Systems*, ed. by G. Ioos, R. G. Helleman, and R. Stora, North Holland, Amsterdam (1983).

Kolmogorov, A. N., *Dokl. Akad. Nauk. SSSR* **98**, 525 (1954); English version in: *Proc. of the 1954 International Congress of Mathematics*, North-Holland, Amsterdam (1957).

Lichtenberg, A. J. and Lieberman, M. A., *Regular and Stochastic Motion*, chapter 5, Springer-Verlag, New York (1983).

MacKay, R. M. and Meiss, J. D., *Hamiltonian Dynamical Systems*, Adam Hilger Ltd, Bristol (1987).

McCauley, J. L., *Chaos, Dynamics, and Fractals: an algorithmic approach to deterministic chaos*, chapters 2, 3, 9, and 11, Cambridge University Press (1993).

Moser, J., *Göttingen math. Phys.* **K1**, 1 (1962).

Niven, I., *Irrational Numbers*, The Carus Mathematical Monographs No. 11, Math. Assoc. Of America, Rahway (1956).

Peterson, I., *Newton's Clock: Chaos in the Solar System*, W. H. Freeman & Co., New York (1993).

Steeb, W.-H. and Louw, J. A., *Chaos and Quantum Chaos*, chapter 12, World Scientific, Singapore (1986).

Sussman, G. J. and Wisdom, J., *Science* **257**, 56 (1992).

Tabor, M., *Chaos and Integrability in Nonlinear Dynamics*, Wiley, New York (1989).
Whittaker, E. T., *Analytical Dynamics*, chapter 14, Cambridge University Press (1965).
Zaslavskii, G. M., Sagdeev, R. Z., Usikov, D. A., and Chernikov, A. A., *Weak Chaos and Quasi-Regular Patterns*, Cambridge University Press (1991).

Chapter 18

Chaotic dynamics with generating partitions, topologic universality

Cvitanovic, P., Gunaratne, G., and Procaccia, I., *Phys. Rev.* **A38**, 1503 (1988).
McCauley, J. L., *Chaos, Dynamics, and Fractals: an algorithmic approach to deterministic chaos*, Cambridge University Press (1993).

Complex dynamical systems

Bennett, C. H., *Nature* **346**, 606 (1990).
Garcy, M. R. and Johnson, D. S., *Computers and Intractability: A Guide to the Theory of NP-completeness*, Freeman, San Francisco (1979).
Lipton, R. J. *Science* **268**, 542 (1995).
McCauley, J. L., *Physica A*, **237**, 387 (1997).
McCauley, J. L., in *Discrete Dynamics in Nature and Society*, **1**, 17 (1997).
Moore, C., *Phys. Rev. Lett.* **64**, 2354 (1990).
Moore, C., *Nonlinearity* **4**, 199 and 727 (1991).

Invariance principles and laws of nature

Wigner, E. P., *Progress of Theoretical Physics* **11**, 437 (1954).
Wigner, E. P., *Symmetries and Reflections*, University of Indiana Press, Bloomington (1967).

Equations of motion from chaotic time series

Chhabra, A., Jensen, R. V., and Sreenivasan, K. R., *Phys. Rev.* **A40**, 4593 (1989).
Gunaratne, G. H., Universality beyond the onset of chaos, in *Soviet and American Perspectives on Chaos*, ed. by D. Campbell, American Institute of Physics, New York (1990).

Beyond behaviorism

Hadamard, J., *The Psychology of Invention in the Mathematical Field*, Dover, New York (1954).

System theory and socio-economic theory

Ackoff, R. L., *Redesigning the Future*, Wiley-Interscience, New York (1974).
Anderson, P. W., Arrow, K. J., and Pines, D., *The Economy as an Evolving Complex System*, Addison-Wesley, Redwood City (1988).
Arthur, W. B., *Increasing Returns and Path Dependence in the Economy*, University of Michigan Press, Ann Arbor (1994).
Bak, P. and Paczuski, M., *Complexity, Contingency, and Criticality*, Proc. Nat. Acad. Sci. **92**, 6689 (1995).
Bertalanffy, L. von, *General Systems Theory*, G. Braziller, New York (1968).
Casti, J. L., *Searching for Certainty, What Scientists can know about the Future*, Wm. Morrow & Co., New York (1990).

Casti, J. L. and Karlqvist, A., *Beyond Belief: Randomness, Prediction and Explanation in Science*, CRC Press, Boca Raton (1991).

Collins, R., *Theoretical Sociology*, Harcourt Brace Janovich, San Diego (1988).

Cootner, P. (ed.), *The Random Character of Stock Market Prices*, MIT, Cambridge, Mass. (1964).

Cowan, G. A., Pines, D., and Meltzer, D., *Complexity; Metaphors, Models, and Reality*, Addison-Wesley, Reading, Mass. (1994).

Eigen, M. and Winkler, R., *Das Spiel: Naturgesetzte steuern den Zufall*, R. Piper Verlag, München (1965).

Kelly, K., *Out of Control, The New Biology of Machines, Social systems, and the Economic World*, Addison-Wesley, Reading, Mass. (1994).

Peters, E. E., *Chaos and Order in the Capital Markets*, Wiley, New York (1991).

Mechanism too complex for simple formulae with arbitrary initial data

Chomsky, N., *Language and Mind*, Harcourt Brace Janovich, New York (1968).

Eigen, M. and Winkler, R., *Das Spiel: Naturgesetzte steuern den Zufall*, R. Piper Verlag, München (1965).

Hopkin, D. and Moss, B., *Automata*, Elsevier North-Holland (1976).

Turing, A., *Proc. Lon. Math. Soc.* (2) **42**, 230 (1937).

von Neumann, J., *Theory of Self-Reproducing Automata*, University of Illinois Press, Urbana (1966).

A bit of history

Bonjour, E., Offler, H. S., and Potter, G. R., *A Short History of Switzerland*, Oxford University Press, London (1952).

Caspar, M., *Johannes Kepler*, transl. from German by C. D. Hellman, Dover, New York (1993).

Dyson, F., *From Eros to Gaia*, Pantheon, New York (1992).

Ginzburg, C., *The Cheese and the Worms, The Cosmos of a Sixteenth Century Miller*, transl. from Italian, Johns Hopkins University Press, Baltimore (1992).

Holton, G., *Science and Anti-Science*, chapters 5 and 6 on 'the end of science' and 'anti-science', Harvard University Press, Cambridge, Mass. (1993).

Koestler, A., *The Sleepwalkers*, Macmillan, New York (1959).

McCauley, J. L., Science, freedom, simulations and the unbearable triviality of postmodernism, *Gulf Coast VIII*, I. Winter (1995–1996).

Muller, H. J., *The Uses of the Past*, Oxford University Press, New York (1952).

Olby, R., *Origins of Mendelism*, 2nd edn, University of Chicago Press (1985).

Spengler, O., *Decline of the West*, transl. from German, Oxford University Press (1965).

Stolpe, S., *Christina of Sweden*, transl. from Swedish, Macmillan, New York (1966).

Toynbee, A., *A Study of History*, abridgement of vols. I–IV by O. C. Somerwell, Oxford University Press (1947).

Westfall, R. S., *Never at Rest, A Biography of Isaac Newton*, chapter 1, Cambridge University Press (1980).

Young, R. M., *Darwin's Metaphor*, Cambridge University Press (1985).

Books by John Berger

About our Faces, Vintage Press, New York (1984).

About Looking, Vintage Press, New York (1991).

Keeping a Rendezvous, Vintage Press, New York (1992).

Pig Earth, Vintage Press, New York (1992).

Index

accidental degeneracy 427
action-angle variables 421
action functional 50, 75, 427
action at a distance 20, 290
action–reaction principle 15
affine transformations 226
angular momentum 15, 26, 225
'angle variables' (translation
 coordinates) 156, 418
Archimedes 9
area-preserving maps 170–4
areal velocity 19
Aristarchus 5
Aristotelian law of motion 24,
 446
Aristotle 1, 6
Arnol'd, V. I. 136, 313
Arnol'd tongues 366
attractors 93, 321, 329, 359
Augustine 4
automata 181, 350, 441
autonomous system 82
axial vector 225

backward iteration 349, 353
basic theorem of the theory of
 ordinary differential
 equations 313
beats 154
Bernoulli shift 143
Bertrand–Königs theorem
 134-8
bifurcations 118, 133, 338, 370
Bohr–Sommerfeld quantization
 421
bouncing ball model 334
Brachistochrone 49
Brahe 10

canonical group parameter 190
canonical transformations 373
canonically conjugate variables
 73, 109
Cantor function 356
Cantor sets 347
Cartesian spaces (Cartesian
 manifolds) 81, 90, 159, 389,
 405, 416
center of mass 28
central forces 16, 127, 426
chaos
 deterministic 105, 177, 419

semi-classical limit of
 quantum theory 429
chaotic orbits, exact
 computations 175, 350,
 451
charge conservation 264
Chasles's theorem 223
Chirikov–Taylor model 437
Christoffel symbols 253
circle maps
 linear 131, 237
 nonlinear 363
clockwork 134, 439
coarse-graining of phase space
 182
commutators 198, 301, 389
compatibility of conservation
 laws of canonical systems
 158, 389, 412
complexity 441
computable numbers 82
configuration space, Lagrangian
 54
conic sections 16
connections
 affine 280
conservation laws 21, 65, 110,
 111, 190, 309, 317
conservative force 22
constraint forces 51
constraints
 holonomic 60
continued fractions 131, 439
contravariant components of
 vectors 249
coordinate systems
 global 53, 202
 local 277
coordinate tranformations 30,
 37, 53, 85, 162, 171, 183,
 250, 417
Copernicus 10
covariance 40, 57, 217, 248, 268
covariance vs invariance 41, 68,
 247, 293
covariant components of vectors
 249
covector 250, 281
cycles 340

damped and driven-dissipative
 systems 93, 326, 334, 361

degrees of freedom f (f
 canonically conjugate pairs
 of variables) 73
derivative
 covariant 212, 282
 exterior 317
 Lie 212
Descartes 3, 13, 25
determinism 21, 29, 82, 85
devil's staircase 357
differentials
 closed 108, 374, 411, 413
 exact 22, 107
dimensions of
 attractors/repellors 93, 96,
 100, 348
Dirac quantization 390, 404
dynamical system (abstract) 82

economics, mathematical
 modelling 443
effective potential 120, 128, 236
eigenvalues, eigenvectors 150
electrodynamics 262
electromagnetic field tensor 273
elliptic orbits 16, 43 (ex. 13) 136,
 140, 394–5
elliptic point 96, 121
energy 23, 28, 74
entropy, Boltzmann 354
equatorial flattening 219
equilibria 94, 149
Euclidean space 21, 30
Euler angles 232
Euler–Lagrange equations 49,
 304, 326
Euler's equations 227, 401
Euler's theorem 199, 223
Euler's top 228

Feynman's path integral
 (Dirac–Feynman
 quantization) 427
Fibonacci 5
field, vector 20
fixed points 336
flat spaces (flat manifolds) 94
floating-point errors xiv, 82,
 165–9, 179
flows in phase space 86, 331
flows
 conservative 91

Printed in the United States
By Bookmasters